CCIE LAB PRACTICE KIT

CCIE Lab Practice Kit

Michael Satterlee
Stephen Hutnik

Osborne/**McGraw-Hill**
New York Chicago San Francisco
Lisbon London Madrid Mexico City
Milan New Delhi San Juan
Seoul Singapore Sydney Toronto

Osborne/**McGraw-Hill**
2600 Tenth Street
Berkeley, California 94710
U.S.A.

To arrange bulk purchase discounts for sales promotions, premiums, or fund-raisers, please contact Osborne/**McGraw-Hill** at the above address. For information on translations or book distributors outside the U.S.A., please see the International Contact Information page immediately following the index of this book.

CCIE Lab Practice Kit

1 2 3 4 5 6 7 8 9 0 CUS CUS 0 1 9 8 7 6 5 4 3 2 1
Book p/n 0-07-212765-1 and CD p/n 0-07-212764-3
parts of
ISBN 0-07-212766-X

Publisher
Brandon A. Nordin

Vice President & Associate Publisher
Scott Rogers

Acquisitions Editor
Steve Elliot

Acquisitions Coordinator
Alex Corona

Technical Editors
Henry Benjamin, CCIE
Randy Benn, CCIE

Project Manager
Dave Nash

Production
MacAllister Publishing Services, LLC

Cover Design
Will Chan

To my wife Mary Jo, whose love and support were my greatest resources when writing this book. –Mike

To Robin, Bobby, Andrew, and Lauren for all of their love and support. –Steve

ACKNOWLEDGMENTS

Many thanks to our editor Steve Elliot.

We want to thank George Kovachi from Adtran for lending us an Atlas 800 ISDN switch.

CONTENTS

ABOUT THE AUTHORS

Stephen Hutnik, is a Senior Network Engineer at AT&T Global Network Services, where he is responsible for development, testing, and training for the Global backbone of the AT&T Network. He is also an adjunct Professor of Telecommunication at Pace University, and is the co-author of the *All-in-One Cisco CCIE Lab Study Guide*

Michael Satterlee, CCIE, is a Senior Network Architect at AT&T IBM Global Network Services, where he is responsible for the architecture and design of a Global MPLS network. He is a co-author of the *All-in-One Cisco CCIE Lab Study Guide*.

ABOUT THE REVIEWERS

As the leading publisher of technical books for more than 100 years, McGraw-Hill prides itself on bringing you the most authoritative and up-to-date information available. To ensure that our books meet the highest standards of accuracy, we have asked a number of top professionals and technical experts to review the accuracy of the material you are about to read.

We take great pleasure in thanking the following technical reviewers for their insight:

Henry Benjamin, CCIE #4695, is a Cisco Certified Internetwok Engineer, under the Cisco Systems CCIE program. Henry is an IT Engineer for Cisco Systems based in Sydney, Australia. He has more than 9 years experience in Cisco networks including large IP networks and internal Cisco IT design solutions. Henry holds a Bachelor of aeronautical engineering degree from Sydney University. Henry has authored himself and been involved with many other Cisco certification titles. "*This review is dedicated to the loving memory of my mum, Anna Benjamin.*"

Randall Benn, CCIE #1637 and (CCSI) #97128, CCDA is a System Engineer, with Emergent Technologies, a networking and application software reseller providing turnkey solutions, specializing in security consulting & analysis, and data/voice/video convergence. His responsibilities include providing pre- and post-sales support in designing, deploying, troubleshooting, and maintaining customer solutions and delivering end-user training through classroom and on-site delivery.

CHAPTER 1

As Close As You Can Get

After writing the *All-in-One CCIE Lab Study Guide,* we received many comments from networking professionals that were in the process of studying for their CCIE lab exam. The candidates loved the hands-on labs in the first book but wanted more detailed and intricate scenarios.

We decided to write a lab workbook that contains six CCIE sample scenarios. The sample scenarios are as close as we can get to the actual exam without violating the Cisco confidentiality agreement. It is our intent that after going through the material the candidate should feel comfortable enough to attempt the challenging two-day exam.

CCIE Lab Exams

The CCIE lab exam is a challenging, hands-on assessment of your internetworking skills. It costs $1,000 in the United States and stretches over two days. Before you can sign up for the lab exam, you must pass the CCIE qualification exam. Unlike the computer-administered exam, CCIE lab exams are only offered through Cisco locations. CCIE routing and switching candidates have their choice of numerous locations. The exams are standardized among sites, and selecting the location is a matter of geographical preference.

CCIE Lab Locations

Routing and Switching Labs

San Jose, California, USA, (800) 553-NETS or (408) 526-8063, ccie_ucsa@cisco.com

Research Triangle Park, North Carolina, USA, (800) 553-NETS or (408) 527-7177, ccie_ucsa@cisco.com

Halifax, Nova Scotia, Canada, (800) 553-NETS or (902) 492-8811, ccie_ucsa@cisco.com

North Sydney, New South Wales, Australia, +61 2 9935 4128, ccie_apt@cisco.com

Brussels, Belgium, +32 2 778 46 70, ccie_emea@cisco.com

Beijing, China, +86 1 0648 92398, ccie_apt@cisco.com

Tokyo, Japan, +81 3 5219 6409, ccie@cisco.co.jp

Capetown, South Africa, +32 2 778 46 70, ccie_emea@cisco.com

ISP Dial Labs

Halifax, Nova Scotia, Canada, (800) 553-NETS or (902) 492-8811, ccie_ucsa@cisco.com

San Jose, California, USA, (800) 553-NETS or (408) 526-8063, ccie_ucsa@cisco.com

WAN Switching Labs

San Jose, California, USA, (800) 553-NETS or (408) 526-8063, ccie_ucsa@cisco.com

Stockley Park, United Kingdom, +32 2 778 46 70, ccie_emea@cisco.com

According to Jeff Buddemeier, Cisco's technical lead for the CCIE program, the lab setup incorporates five routers and a Catalyst switch. Each candidate has his or her own rack and patch panel. You will also receive a set of Cisco documentation to use throughout the exam. You cannot bring any other notes or documentation into the exam with you.

Your first task will be to create a network to specification. This will take up all of the first day and half of the second. Halfway through the second day, while you are out of the room, the exam proctor will insert faults into your network, and you will have to find and fix them, as well as be able to document the problems and their resolutions.

The exam has a total of 100 possible points. To pass, you must achieve a score of 80 or better. You must achieve a passing score on each section of the exam to be allowed to progress to the next section. For example, a perfect score on the first day would be 45 points. You have to earn at least 30 of them to be allowed to return for the first part of day two. Table 1-1 shows the scoring breakdown.

The lab starting time varies depending upon the location, but it is somewhere between 8:00 and 9:00 a.m. each day, and the lab runs for 7 1/2 hours. You are given a half-hour break for

Table 1-1
CCIE Lab Exam Scoring

Day	Task	Points	Total So Far	Minimum Score to Continue
1	build	45	45	30
2 (part I)	build	30	75	55
2 (part II)	troubleshooting	25	100	80 or better to pass

lunch. A proctor will be in the room to clarify questions and handle any emergencies that may arise, but basically you are on your own.

The failure rate for this exam is high. According to Buddmeier, only about 20 percent of the candidates pass it on the first attempt. On average, CCIE candidates require four to five lab exams before they earn a passing score. Think of your first time through as a learning experience, and if you manage to pass, that is a bonus. There is also no limit on the number of times you can retake the exam.

As with all certification exams, lab exam content and structure are subject to change, so when you are ready to consider taking the lab exam, it's best to get the latest information from Cisco. Cisco's Web site contains specific instructions about how to prepare for each of the CCIE lab and qualification exams.

It cannot be stressed enough that you must get lots of hands-on practice if you hope to pass this exam. If you do not have equipment to practice on at work, you will have to set up a home lab or find another way to gain access to the equipment.

Format of the Book

Each chapter follows a similar format and contains the following topics:

- Technology overview
- IOS requirements
- Detailed list of equipment needed
- Test questions
- Detailed answer guide
- Complete configurations at the end of the chapter and on the CD-ROM

How to Use This Book

Sample tests are provided at the back of the book. You should treat them as actual timed exams and use a Cisco CD-ROM only as a reference. If you get stuck on one topic, move on. Then at the end of the exam, go back and correct your answers using the answer guide. Some topics are covered in each lab, such as subnetting. These are the topics that you need to know cold. It is our intent that if you know the subject material covered in this book, you will be able to pass the lab exam. In case you get stuck, the complete configurations are loaded on the CD-ROM provided.

CD-ROM

The included CD-ROM contains all the configurations that are presented in this book. You can cut and paste the configurations into your routers, thereby saving time. Each chapter has its own directory on the CD-ROM. The file naming convention is router_name.txt.

CHAPTER 2

A Case Study in BGP

Introduction

The *Border Gateway Protocol* (BGP) is an inter-*Autonomous System* (AS) routing protocol and its primary function is to exchange network reachability information with other BGP speakers. The following case study provides complex BGP configuration scenarios to aid the student preparing for the CCIE Lab.

Overview

BGP is a path vector inter-AS routing protocol that is based on distance vector algorithms. (For the purposes of this chapter, *inter* means routing between entities and *intra* means routing within an entity.) An AS is a collection of routers or end stations that are under the same administrative control and viewed as a single entity. The reason BGP is called a path vector protocol is that the BGP routing information carries a sequence of AS numbers, which indicate the path the route has traversed. This information is used to construct a graph of AS connectivity from which routing loops can be pruned.

The interior gateway routing protocols (RIP, OSPF, IGRP, EIGRP) were designed to operate in a single AS that is under a single administrative control. BGP was introduced to facilitate a loop-free exchange of routing information between ASs while controlling the expansion of routing tables through *Classless Inter-Domain Routing* (CIDR) and providing a structured view of the Internet through the use of ASs.

In a sense, the Internet could have been a large *Open Shortest Path First* (OSPF) network. If this were the case, then all organizations that participated in it would have to adhere to the same administrative policies. By segregating the Internet into multiple ASs, one can create one large network that consists of smaller, more manageable networks. Within these smaller networks, or ASs, an organization's unique rules and administrative policies can be applied. A unique number assigned by an Internet registry identifies each AS.

Figure 2-1 displays two *Internet service providers* (ISPs), Xnet and Ynet, each of which consists of multiple networks running multiple *Interior Gateway Protocols* (IGPs). Each service provider is assigned an AS number by an Internet registry that represents its entire network. When Company Xnet and Ynet want to exchange routing information, they do so using BGP.

Company Ynet advertises networks 2.0.0.0 and 3.0.0.0 to company Xnet, and the routes are marked as originating from AS 200. Company Xnet does not need to have a full topological view of company Ynet nor does it need to understand Ynet's internal routing or policies. It simply knows that networks 2.0.0.0 and 3.0.0.0 are in AS 200.

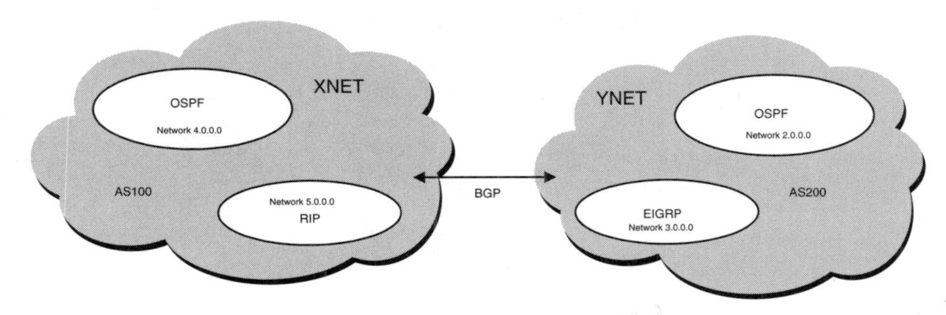

Figure 2-1 Autonomous Systems (ASs)

BGP Terminology

Before diving into the intricate details of BGP, it is important to have a clear understanding of key terms and concepts, some of which are used interchangeably.

External BGP (EBGP) versus Internal BGP (IBGP) Although BGP was designed to be used between ASs (EBGP), it is often used within an AS (IBGP) to carry information between border routers running EBGP to other ASs. This enables all BGP path attributes to be maintained across the AS.

By definition, an EBGP neighbor is a router with an administrative and policy control that is outside of your AS. An IBGP neighbor is a router that is under the same administrative control (see Figure 2-2).

Classless Interdomain Routing (CIDR) This was developed to address the explosive growth of IP addresses present in IP routing tables on Internet routers and the exhaustion of IP address space. CIDR is an address allocation scheme that eliminates the concept of a network class within BGP. In CIDR, an IP network is represented by a prefix, which is the IP address and a number that indicates the leftmost contiguous significant bits in the address. For example, in Figure 2-3, 256 Class C networks are present on service provider A's network. Without CIDR, the service provider must advertise each network individually. Using CIDR, service provider A can advertise all of these networks with one classless advertisement (200.10.0.0 /16), as shown in Figure 2-4.

Supernet A supernet is a network advertisement whose prefix boundary contains fewer bits than the network's natural mask. For example, in Figure 2-4, the natural network mask for the Class C network 200.10.1.0 is 255.255.255.0. However, when you represent it as 200.10.0.0 /16, the mask is 16, which is less than 24. Hence, it is termed a supernet advertisement.

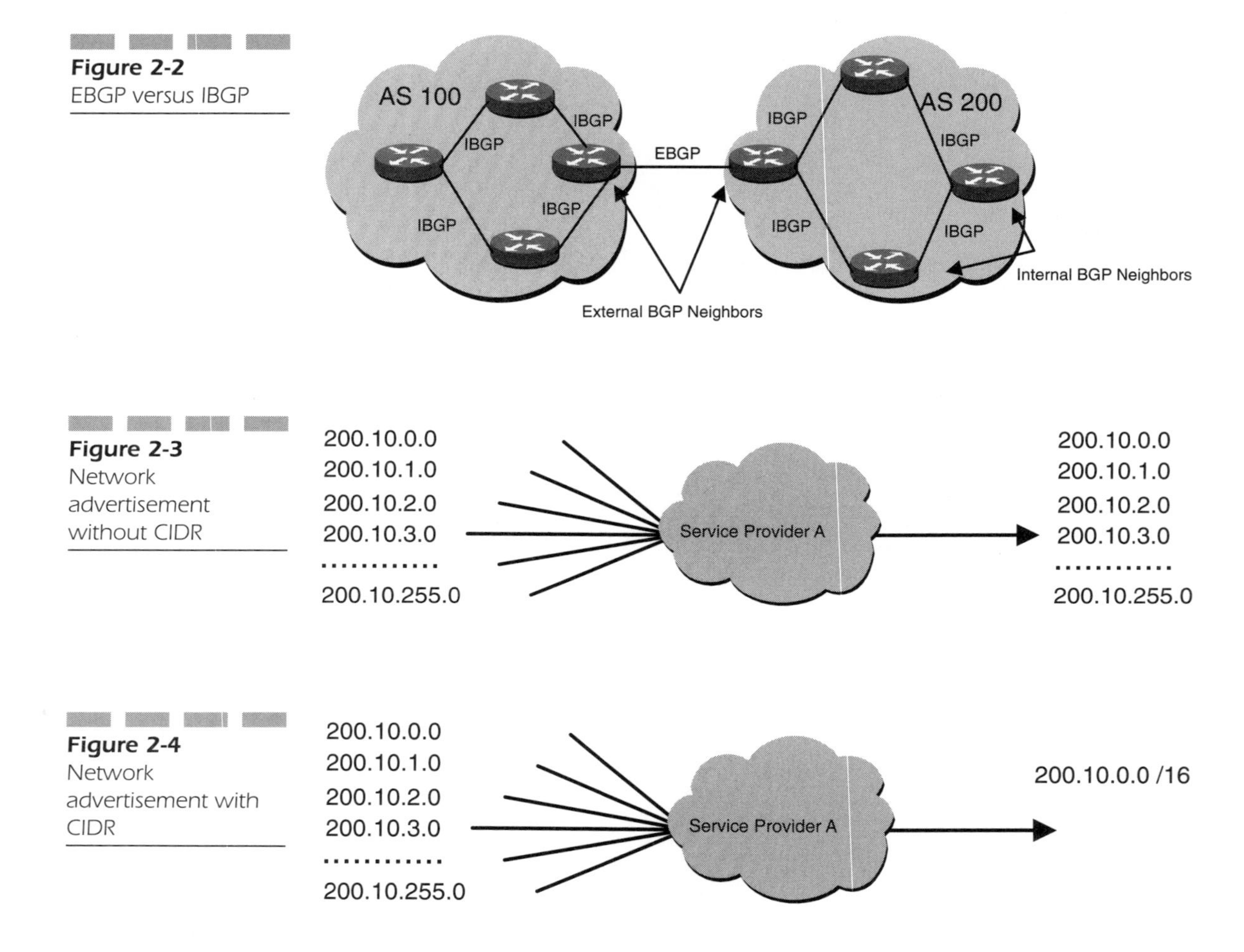

Figure 2-2 EBGP versus IBGP

Figure 2-3 Network advertisement without CIDR

Figure 2-4 Network advertisement with CIDR

IP Prefix An IP prefix is an IP network address along with an indication of the number of bits that makes up the network number. For example, 10.0.0.0 /8 is an IP prefix.

Network Layer Reachability Information (NLRI) This is how BGP supports classless routing (CIDR). The NLRI is part of the BGP update message and is used to list a set of destinations that is reachable. The NLRI field in the BGP update message contains two tuples: <length, prefix>. The length is the number of bits in the mask and the prefix is the IP address. The two combined represent the network number. For example, the network 10.0.0.0 /8 would be advertised in the NLRI field of a BGP update message as<8,10.0.0.0>.

Autonomous System (AS) An AS is a group of routers or hosts that are under the same administrative control and policies. AS numbers are assigned by an Internet registry.

Synchronization Before BGP can announce a route, the route must be present in the IP routing table. In other words, BGP and IGP must be in sync before the networks can be advertised. Cisco enables BGP to override the synchronization requirement with the command **no synchronization**. This allows BGP to announce routes that are known via BGP but are not in the routing table. The reason that this rule exists is that it is important for the AS to be consistent with the routes it advertises.

In Figure 2-5, RouterA and RouterB are the only routers running BGP. If synchronization is disabled on RouterB, it will advertise network 1.0.0.0 /8 to AS 200. When RouterD wants to send traffic to network 1.0.0.0, it sends the packet to RouterB, which does a recursive lookup in its IP routing table and forwards the packet to RouterC. Because RouterC is not running BGP, it has no visibility to network 1.0.0.0 and therefore drops the packet. This is why BGP requires synchronization between BGP and IGP. Care must be taken when disabling synchronization. If an AS is a transit AS, all routes should be running fully meshed IBGP before synchronization is disabled.

Figure 2-5 Synchronization within an AS

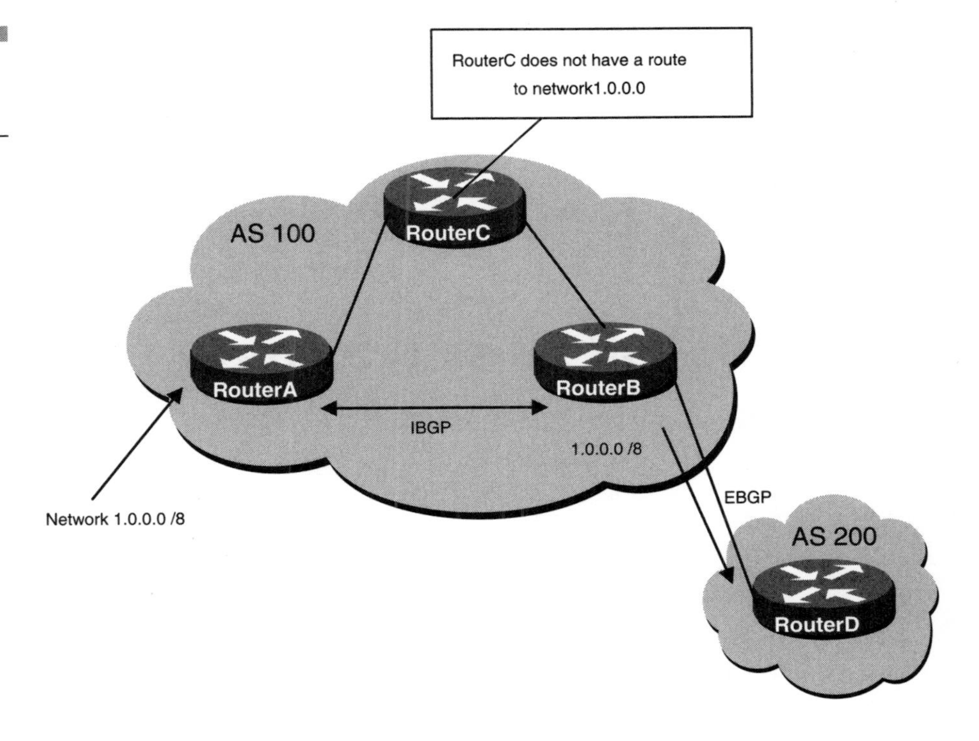

Technology Overview

BGP is an inter-AS routing protocol whose primary function is to exchange network reachability information with other BGP speakers (a BGP speaker is any device that is configured for BGP). BGP uses the *Transmission Control Protocol* (TCP) as its transport protocol (port 179) that provides reliable data transfer between the BGP speakers.

Two BGP routers form a transport protocol connection between one another. The two routers are called neighbors or peers. Once the transport connection is formed, the peer routers exchange messages to open and confirm connection parameters. It is in this stage that the routers exchange information on the BGP version number, the AS number, the hold time, the BGP identifier, and other optional parameters. If the peers disagree on any of the parameters, a notification error is sent and the peer connection does not get established.

If the peer routers agree upon the parameters, then the entire BGP routing table is exchanged using UPDATE messages. The UPDATE messages contain a list of destinations reachable via each system (NLRI) along with path attributes for each route. The path attributes contain information such as the origin of the route and the degree of preference. Path attributes will be covered in great detail later in this chapter.

The BGP table is valid for each peer for the duration of the BGP connection. If any routing information changes, the neighbor router uses incremental updates to convey this information. BGP does not require that routing information be periodically refreshed. If no routing changes occur, BGP peers only exchange keepalive packets, which are sent periodically to ensure that the connection is kept active.

Case Study 1: BGP

IOS Requirements

BGP-4 first became available in IOS release 10.0, but this case study was performed using IOS 11.2.

Equipment Needed

The following equipment is needed to perform this case study:

- One Catalyst 1900 Ethernet switch
- One terminal server with one Ethernet and one serial port

- One Cisco router with one serial port and one Ethernet port
- Three Cisco routers with two serial ports and one Ethernet port
- One Cisco router with one Ethernet port and three serial ports
- One Cisco router with one Ethernet port and four serial ports
- One peripheral router with one Ethernet port and three serial ports. This router (R8) will be used as the Frame Relay switch as well as the BGP neighbor. (The student does not configure this router and should use the configuration provided on the CD-ROM.) Table 2-1 displays each router's interface requirements.
- Cisco IOS 11.2
- A PC running a terminal emulation program for connecting to the console port of the terminal server
- Six Ethernet cables
- Seven Cisco DTE/DCE crossover cables
- One Cisco Rolled cable

Table 2-1

Interface requirements for each router

R1	1 Serial, 2 Ethernet
R2	2 Serial, 1 Ethernet
R3	4 Serial, 1 Ethernet
R4	2 Serial
R5	2 Serial, 1 Ethernet
R6	1 Serial, 1 Ethernet
R7	1 Serial, 1 Ethernet (terminal server)
R8	3 Serial, 1 Ethernet (Frame Relay switch)

Physical Connectivity Diagram

Figure 2-6 shows the physical connectivity for the routers in this case study.

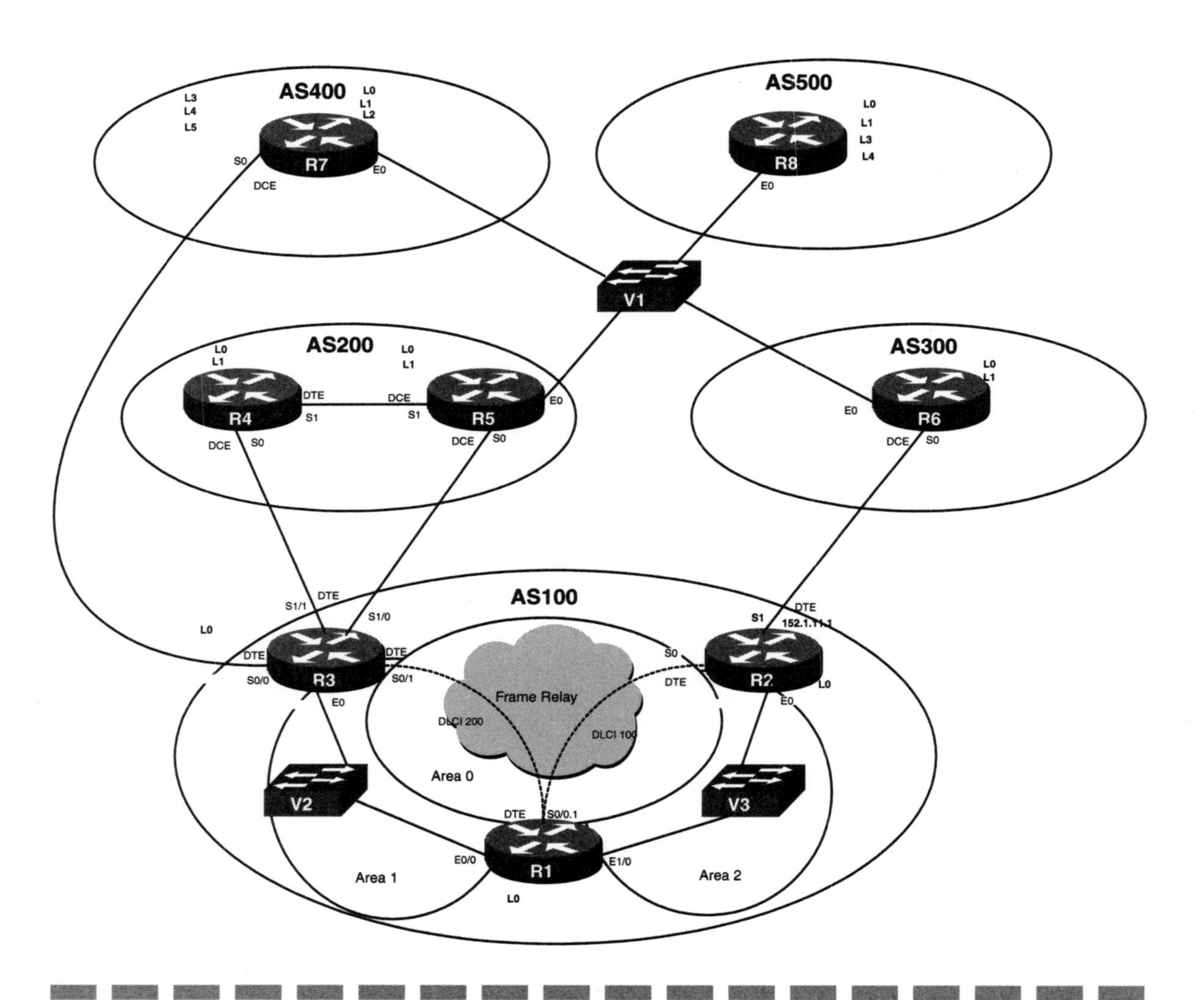

Figure 2-6
Physical connectivity

Notes

- All IP addresses will be assigned from the class B network 152.1.0.0.
- No static or default routes should be used in this case study unless explicitly stated.
- Use Process ID 64 for all OSPF configurations.
- Do not use the network point-to-multipoint command under OSPF.
- No configuration is needed on R8, load the configuration that is provided on the CD.
- Read through the entire case study before beginning.
- Make sure you save your configurations regularly.

Questions

1. Cable the routers according to Figure 2-6 and load the configuration provided on the CD to R8. The configuration provided is from a 4500 with an Ethernet interface and a four-port serial network module. If you plan on using a router other than a 4500 or one that has a different way of numbering its interfaces, such as a 3620, the interfaces provided in the configuration will need to be renamed.
2. Erase any existing configuration from the routers and set each router's hostname. The name will be the corresponding router number. For example, router 1 will be R1.
3. Set the enable password on each router to be **cisco**.
4. Configure the terminal server (R7) so that all routers can be accessed by name through the terminal server. The port number of the terminal server attaches to the corresponding router. That is, port 1 of the terminal server goes to the console port of R1, and the Catalyst switch goes to port 9. (Use the Ethernet IP address as the IP address for reverse telneting.)
5. Place Ethernet interface 0 of R5, R6, R7, and R8 in Virtual Local Area Network (VLAN) 1. The routers should be attached in ascending order, starting with port 1 on the Catalyst switch.
6. Configure the router so that if an incorrect command is entered, the router does not try to translate it to an IP address.
7. The Mac addresses of the Ethernet interfaces attached to the VLAN should have the following format (routernumber. ASnumber .0000). For example, R7 would be set to

```
0007. 0400.0000.
```

8. Configure the Catalyst so that only the designated router can attach to the specific port.
9. Verify that step 8 works by swapping the cables between port 1 and port 2 of the Catalyst switch. The ports should go into blocking mode.
10. Normalize the port.
11. Assign IP addresses to Ethernet 0 of R5, R6, R7, and R8 from network 152.1.1.0. Do not allocate more addresses than needed; use the first available subnet and assign the addresses in ascending order starting with R5.
12. Place Ethernet 0 of R1 and R3 in VLAN 2; place Ethernet 1 of R1 and Ethernet 0 of R2 in VLAN 3.
13. Connect the rest of the routers serially as per Figure 2-6. Make sure that the correct ports are configured for DTE/DCE. The DCE clock will be set to 64K on all routers.
14. R3 and R2 are connected via Frame Relay to R1. Use physical interfaces on R3 and R2 and a logical interface on R1. Assign addresses to each interface using subnet 152.1.10.0. Do not allocate more addresses than needed; use the first subnet available and assign the addresses in ascending order, starting with R1. Configure the interfaces for Frame Relay. Make sure that the only DLCIs that are used are the ones provided. Make sure that all routers can reach each other.

    The following is the physical connectivity of R1, R2, and R3 to the Frame Relay Switch (R8).

Frame Switch (R8) interface S0	R1 Interface S0/0
Frame Switch (R8) interface S0	R2 Interface S0
Frame Switch (R8) interface S0	R3 Interface S0/1

15. Assign IP addresses to Ethernet 0 of R3 and R1 using the next available subnet in 152.1.8.0. The subnet should contain 48 host addresses, R1 should use the first address in the subnet, and R3 should use the last one.
16. Assign IP addresses to Ethernet 1 of R1 and Ethernet 0 of R2 using the first available subnet in 152.1.9.0. The subnet should contain 128 host addresses. R1 should use the first address in the subnet and R2 should use the last one.
17. Assign IP addresses to Serial 0 on R7 and serial S0/0 on R3 using the first available subnet in 152.1.20.0. The address should be assigned out in ascending order, starting with R3.
18. Assign IP address to the serial interfaces connecting R3 and R4, use the first available subnet in 152.1.12.0. The subnet should contain the minimal hosts needed and the IP addresses should be assigned in ascending order, staring with R3.
19. Assign IP addresses to the serial interfaces connecting R3 and R5, use the next available subnet in 152.1.12.0. The subnet should contain the minimal hosts needed and the IP addresses should be assigned in ascending order, staring with R3.

20. Assign an IP address to Serial 1 on R4 and R5 using the next available subnet in 152.1.12.0. The subnet should contain the minimal hosts needed and the IP addresses should be assigned in ascending order, starting with R4.
21. Assign an IP address to Serial 1 of R2 and Serial 0 of R6 using the first available subnet in 152.1.11.0. The subnet should contain the minimal hosts needed and the IP addresses should be assigned in ascending order, starting with R2.
22. Assign the following loopback addresses to the corresponding routers:

    R7 Loopback 0 152.1.20.7 /32

    R7 Loopback 1 152.1.20.17 /28

    R4 Loopback 0 152.1.12.4 /32

    R4 Loopback 0 152.1.13.1 /24

    R5 Loopback 0 152.1.12.5 /32

    R5 Loopback 1 152.1.14.1 /29

    R3 Loopback 0 152.1.10.3 /32

    R2 Loopback 0 152.1.10.2 /32

    R1 Loopback 0 152.1.10.1 /32

    R6 Loopback 0 152.1.11.6 /32

    R6 Loopback 1 152.1.11.129 /26

IGP Protocols

23. Place the serial connections between R3, R2, and R1 in OSPF Area 0.
24. Place VLAN 2 in OSPF area 1 and VLAN 3 in OSPF area 2. R1 should be the DR for Area 1 and R2 should be the DR in Area 2.
25. R1 should prefer the Frame Relay link over the Ethernet connection. All other connections should prefer the interface with the highest bandwidth.
26. Change the Hello interval on R3 to two minutes.
27. Configure EIGRP on the interface connecting R3 to R7. Do not send advertisements out any other interface. Enable EIGRP on all interfaces on R7, but do not send updates out the interface connecting to VLAN1.
28. R7 should be able to see all the networks behind R3. R3 only needs to see network 152.1.20.16/28 via R7.
29. Make sure R7 does not advertise any routes back to R3 that were learned from R3 and vice versa. Do not rely on split horizons!
30. Enable EIGRP using process 56 on R4 and R5. Don't run EIGRP on the interfaces that are connected to other ASs.

31. Make sure that the loopback interfaces on R2, R1, and R3 are being propagated via OSPF. OSPF should not be run on these interfaces, and the routes should not show up as OSPF external routes.

EBGP/IBGP

32. Place R3, R2, and R1 in AS 100. R1 will peer with R2 and R3. R1 is the only router in AS100 that should have two IBGP neighbors.
33. Place R4 and R5 in AS 200. R4 should peer with R5 and R3.
34. Place R6 in AS 300 and R7 in AS 400
35. R7 should BGP peer with R6, R5, and R8
36. R5 should BGP peer with R7, R8, R6, and R3.
37. R6 should BGP peer with R7, R8, R5, and R2.
38. R7, R5, and R6 should all see the following networks via BGP from R8:

    Network 148.1.0.0

    Network 148.2.0.0

    Network 148.3.0.0

    Network 148.4.0.0

39. AS400 should advertise local network 152.1.20.0 via BGP and AS300 should advertise local network 152.1.11.128 via BGP.
40. AS 200 should advertise local network 152.1.13.0 /24 and 152.1.14.0 / 29.
41. AS100 should only accept local routes from AS300. All other routes should be accepted via AS200.
42. AS100 should accept a default route from AS300 in case of a link failure between A200 and AS100. The default should not be advertised outside of AS100.
43. Advertise network 152.1.9.0 via IBGP on R1 and R2. Do not use the redistribute command.
44. AS 100 should advertise all local networks via BGP. To prevent AS 100 from being used as a transit network, AS300 should only accept local networks from AS100.
45. AS100 prefers that network 152.1.9.0 be reachable by the outside world via the link between AS100 and AS300.
46. Under no circumstances can AS100 be used as a transit network for AS200 to reach AS300. To accomplish this, no additional configurations should be added to any router other than R2.

47. AS200 and AS300 should not pass network 152.1.9.0. No additional configuration should be used to achieve this outside of AS100.
48. R3 should use R5 to reach network 152.1.14.0 and R4 to reach network 152.1.13.0. No additional configuration should be used to achieve this outside of AS200.
49. R7 should originate and advertise network 148.0.0.0, 148.5.0.0, 148.6.0.0, and 148.7.0.0. Add four loopback interfaces to R7: loopback 2 IP address 148.0.0.1/16, loopback 3 IP address 148.5.0.1 /16, loopback 4 IP address 148.6.0.1 /16, and loopback 5 IP address 148.7.0.1 /16.
50. R5 should advertise as few prefixes for 148.x.0.0 as needed.
51. R3 is running an IGP (EIGRP) on the private link between it and AS400. Make sure that the IGP route to network 152.1.20.16 is favored over the EBGP route.
52. The configuration on R3 is growing extremely large and approaching the limitation of the router's NVRAM. Reduce the size of the configuration on R3.
53. Configure R3 so that informational messages are logged to the internal buffer. Make sure the router can store up to 16K of information and the information is time stamped.
54. Configure R3 so that SNMP information can only be read by host 192.1.1.1.
55. Network 152.1.13.0 is suspected of being unstable, causing BGP UPDATE and WITHDRAWN messages to be repeatedly propagated on the network. Configure R3 so that if the network continues to FLAP, it will be withdrawn.
56. Enable R3 so that when its neighbors make a BGP change, the change will take effect without resetting the BGP TCP session. R3 should be able to initiate the change locally; no commands should have to be entered on the neighboring routers.

Answer Guide

1. *Cable the routers according to Figure 2-6 and load the configuration provided on the CD to R8. The configuration provided is from a 4500 with an Ethernet interface and a four-port serial network module. If you plan on using a router other than a 4500 or one that has a different way of numbering its interfaces, such as a 3620, the interfaces provided in the configuration will need to be renamed.*
2. *Erase any existing configuration from the routers and set the hostname on each one. The name will be the corresponding router number; for example, router 1 will be R1.*
   The first step is to erase the configuration for all the routers as well as the Catalyst switch. To erase the configuration on the router, perform the following command and then reload the router:

```
R1#write erase or R1#erase startup-config
```

This command erases the startup configuration stored in NVRAM on all platforms except the Cisco 7000 series, the Cisco 7200 series, and the Cisco 7500 series. On those models, the router erases or deletes the configuration pointed to by CONFIG_FILE environment variable. If the CONFIG_FILE variable points to NVRAM, then the router erases NVRAM. If the CONFIG_FILE environment variable specifies a flash memory device and a configuration filename, then the command deletes the named file.

Erasing the configuration on the Catalyst 1900 is a bit different. From the main menu, select **M** for menus.

```
Enterprise Edition Software
Ethernet Address:       00-D0-C0-69-5A-40

PCA Number:             73-3121-01
PCA Serial Number:      FAB03163GAT
Model Number:           WS-C1924-A
System Serial Number:   FAB0319T0KT
Power Supply S/N:       PHI031004WD
PCB Serial Number:      FAB03163GAT,73-3121-01
-------------------------------------------------

1 user(s) now active on Management Console.

        User Interface Menu

     [M] Menus
     [K] Command Line
     [I] IP Configuration

Enter Selection: M
```

From the Menu option select **S** for system configuration.

```
Catalyst 1900 - Main Menu

     [C] Console Settings
     [S] System
     [N] Network Management
     [P] Port Configuration
     [A] Port Addressing
     [D] Port Statistics Detail
     [M] Monitoring
     [V] Virtual LAN
     [R] Multicast Registration
     [F] Firmware
     [I] RS-232 Interface
     [U] Usage Summaries
     [H] Help
     [K] Command Line

     [X] Exit Management Console

Enter Selection:S
```

From the Menu select **F** to reset to factory defaults. This command will reset the switch with factory defaults. All system parameters will revert to their default factory settings. All static and dynamic addresses will be removed.

```
----------------------- Settings ---------------------------------------
     [N] Name of system                              cat
     [C] Contact name
     [L] Location
     [S] Switching mode                              FragmentFree
     [U] Use of store-and-forward for multicast      Disabled
     [A] Action upon address violation               Suspend
     [G] Generate alert on address violation         Enabled
     [I] Address aging time                          300 second(s)
     [P] Network port                                None
     [H] Half duplex back pressure   (10-mbps ports) Disabled
     [E] Enhanced congestion control (10-mbps ports) Disabled

    ----------------------- Actions ----------------------------------------
     [R] Reset system                      [F] Reset to factory defaults
     [V] Reset VTP to factory defaults     [T] Reset to enable Bridge Groups
    ----------------------- Related Menus ----------------------------------
     [B] Broadcast storm control           [X] Exit to Main Menu

Enter Selection: F
```

The next step is to set the hostname on each router. To do this, perform the following global configuration command:

```
router(config)#hostname R5
```

This is the same for the Catalyst. From the main menu, select **K**, Command line:

```
Enterprise Edition Software
Ethernet Address:       00-D0-C0-69-5A-40

PCA Number:             73-3121-01
PCA Serial Number:      FAB03163GAT
Model Number:           WS-C1924-A
System Serial Number:   FAB0319T0KT
Power Supply S/N:       PHI031004WD
PCB Serial Number:      FAB03163GAT,73-3121-01
-------------------------------------------------

1 user(s) now active on Management Console.

        User Interface Menu

     [M] Menus
     [K] Command Line
     [I] IP Configuration

Enter Selection: K
```

Enter the following global configuration command:

```
cat(config)#hostname cat
```

3. *Set the enable password on each router to be cisco.*

When a router is first configured, no default enable password exists. The enable password is used to keep unauthorized users from changing the configuration. To set the enable password, use the following global configuration command:

```
R5(config)#enable password cisco
```

4. *Configure the terminal server (R7) so that all routers can be accessed by name through it. The port number of the terminal server attaches to the corresponding router; that is, port 1 of the terminal server goes to the console port of R1, and the cat switch goes to port 9. Use the Ethernet IP address as the IP address for reverse telneting.*

The terminal server provides access to all your test routers via reverse telnet, which is the process of using telnet to make a connection out an asynchronous port.

All the routers in the case study as well as the Catalyst switch will have their console port connected directly to one of the 16 asynchronous interfaces on the Cisco 2511RJ terminal server using a standard Cisco console rolled cable. The routers will be accessed using a reverse telnet connection. To make a reverse telnet connection, you telnet to any active IP address on the box followed by 200x, where x is the port number that you want to access (Telnet 1.1.1.1 2001).

Use the following line commands to configure the terminal server:

```
R7(config)#line 1 16
R7(config-line)#transport input all
R7(config-line)#no exec
```

The command **transport input all** specifies the input transport protocol. By default on IOS 11.1 and later, the transport input is set to none, while prior to 11.1, the default was all. If the transport input is left to none, you will receive an error stating that the connection is refused by the remote host. The command no exec enables only outgoing connections for the line. This prevents the device that you are attached to from sending out unsolicited data. If the port receives unsolicited data, an EXEC process will start that makes the line unavailable.

The next step is to configure hostnames on R7 so you can attach to any router simply by typing in the router name. The Cisco IOS software maintains a table of host names and their corresponding addresses. Similar to a DNS server, you can statically map host names to an IP address. This is very useful and saves a lot of keystrokes when you have multiple devices connected to the terminal server.

The following global configuration commands are used to set the host names:

```
R7(config)#ip host r1 2001 152.1.1.3
R7(config)#ip host r2 2002 152.1.1.3
R7(config)#ip host r4 2004 152.1.1.3
R7(config)#ip host r5 2005 152.1.1.3
R7(config)#ip host r6 2006 152.1.1.3
R7(config)#ip host r7 2007 152.1.1.3
R7(config)#ip host r8 2008 152.1.1.3
R7(config)#ip host cat 2009 152.1.1.3
```

5. *Place Ethernet interface 0 of R5, R6, R7, and R8 in VLAN 1. The routers should be attached in ascending order, staring with port 1 on the Catalyst switch.*

   This step requires no configuration on the Catalyst because, by default, all ports are in VLAN 1.

6. *Configure the router so that if an incorrect command is entered, the router does not try to translate it to an IP address.*

   By default, the IOS will try to resolve a name to an IP address. If the user mistypes a command, such as typing in "wrete" instead of "write," the IOS will attempt to resolve the name through a domain name server.

```
R7#wrete
Translating "wrete" . . . domain server (255.255.255.255)
% Unknown command or computer name, or unable to find computer address
```

   To disable this function, use the following global configuration command:

```
R7(config)#no ip domain-lookup
```

7. *The* Media Access Control (*MAC) addresses of the Ethernet interfaces attached to the VLAN should have the following format (routernumber. asnumber .0000). For example, R7 would be set to*

```
0007. 0400.0000.
```

   The Cisco IOS enables you to manually set the MAC address on an interface. This is particularly useful for troubleshooting and monitoring. To set the MAC address, use the following interface command:

```
R7(config-if)#mac-address 0007.0300.0000
```

8. *Configure the Catalyst so that only the designated router can attach to the specific port.*

   Catalyst MAC address security enables the Catalyst switch to block input to an Ethernet port when the MAC address of a station attempting to access the port is different from the configured MAC address. When a port receives a frame, the Catalyst compares the source address of that frame to the secure source address learned or configured for the port. By default, when port security is enabled, the Catalyst learns the first device that attaches to the port; this will be the MAC address that the Catalyst will permit. When a source address change occurs, the port is disabled, and the LED for that port turns orange. When the port is re-enabled, the port LED turns green. If the MAC address is not given, the command turns on the learning mode, so that the first MAC address shown on the port becomes the secure MAC address.

To enable port security on the Catalyst 1900, perform the following commands:

From the main menu, select **M** for menus:

```
Enterprise Edition Software
Ethernet Address:       00-D0-C0-69-5A-40

PCA Number:             73-3121-01
PCA Serial Number:      FAB03163GAT
Model Number:           WS-C1924-A
System Serial Number:   FAB0319T0KT
Power Supply S/N:       PHI031004WD
PCB Serial Number:      FAB03163GAT,73-3121-01
-------------------------------------------------

1 user(s) now active on Management Console.

        User Interface Menu

     [M] Menus
     [K] Command Line
     [I] IP Configuration

Enter Selection: M
```

Select **P** for port configuration:

```
   [C] Console Settings
     [S] System
     [N] Network Management
     [P] Port Configuration
     [A] Port Addressing
     [D] Port Statistics Detail
     [M] Monitoring
     [V] Virtual LAN
     [R] Multicast Registration
     [F] Firmware
     [I] RS-232 Interface
     [U] Usage Summaries
     [H] Help
     [K] Command Line

     [X] Exit Management Console

Enter Selection: P
```

Enter the port number:

```
Identify Port:  1 to 24[1-24], [AUI], [A], [B]:
Select [1 - 24, AUI, A, B]: 3
```

Enter **A** for port addressing:

```
Catalyst 1900 - Port 1 Configuration

        Built-in 10Base-T
        802.1d STP State:  Forwarding     Forward Transitions:  2

    ----------------------- Settings ---------------------------------------
     [D] Description/name of port
     [S] Status of port                              Enabled
     [F] Full duplex                                 Disabled
     [I] Port priority (spanning tree)               128 (80 hex)
     [C] Path cost (spanning tree)                   100
     [H] Port fast mode (spanning tree)              Enabled

    ----------------------- Related Menus ----------------------------------
     [A] Port addressing          [V] View port statistics
     [N] Next port                [G] Goto port
     [P] Previous port            [X] Exit to Main Menu

Enter Selection: A
```

Enter **S** for address security:

```
    ----------------------- Settings ---------------------------------------
     [T] Address table size                          Unrestricted
     [S] Addressing security                         Disabled
     [U] Flood unknown unicasts                      Enabled
     [M] Flood unregistered multicasts               Enabled

    ----------------------- Actions ----------------------------------------
     [A] Add a static address
     [D] Define restricted static address
     [L] List addresses
     [E] Erase an address
     [R] Remove all addresses

     [C] Configure port           [V] View port statistics
     [N] Next port                [G] Goto port
     [P] Previous port            [X] Exit to Main Menu

Enter Selection:S

Addressing security serves to restrict the usage of the port to a specified, static set of
addresses.  Network management alerts as well as suspension or disablement of the port may
result from address violations.
```

Enable address security:

```
Addressing security may be [E]nabled or [D]isabled:

Current setting ===> Disabled

    New setting ===> Enabled
```

Now enter **A** to add a static address:

```
----------------------- Settings ---------------------------------------
  [T] Address table size                          1
  [S] Addressing security                         Enabled
  [U] Flood unknown unicasts                      Enabled
  [M] Flood unregistered multicasts               Enabled

----------------------- Actions ----------------------------------------
  [A] Add a static address
  [D] Define restricted static address
  [L] List addresses
  [E] Erase an address
  [R] Remove all addresses

  [C] Configure port           [V] View port statistics
  [N] Next port                [G] Goto port
  [P] Previous port            [X] Exit to Main Menu

Enter Selection:  A
```

Enter the Mac address that you want permitted.

```
This command adds a static address for the port.
```

Enter address:

```
Enter address (6 hex octets:  hh hh hh hh hh hh):  00 07 03 00 00 00
```

Enter **L** to verify the configuration:

```
Catalyst 1900 - Port 1 Addressing

        Dynamic addresses:  0    Static addresses:  2

----------------------- Settings ---------------------------------------
  [T] Address table size                          2
  [S] Addressing security                         Enabled
  [U] Flood unknown unicasts                      Enabled
  [M] Flood unregistered multicasts               Enabled

----------------------- Actions ----------------------------------------
  [A] Add a static address
  [D] Define restricted static address
  [L] List addresses
  [E] Erase an address
  [R] Remove all addresses

  [C] Configure port           [V] View port statistics
  [N] Next port                [G] Goto port
  [P] Previous port            [X] Exit to Main Menu

Enter Selection: L

Type    Address              Accepted source ports
Static   00-07-03-00-00-00     Unrestricted
```

9. *Verify step 8 is working by swapping the cables between port 1 and port 2. The port should go into blocking.*

   In order to verify that the port security function is working, swap the cables between port 1 and port 2. If port security is configured correctly, the port should go into blocking mode, which is indicated by the orange LED above the port.

10. *Normalize the port.*

   In order to get the port out of blocking mode, it must be reset. To do this, log into the Catalyst and go into the Port Configuration menu. Below is the output, note that the status of the port is Suspended-violation.

   To reset the port, select S and then Enable. The LED on the port should return to green, indicating a normal state.

```
    Catalyst 1900 - Port 1 Configuration

    Built-in 10Base-T
    802.1d STP State:  Blocking     Forward Transitions:  2

  ----------------------- Settings --------------------------------------
   [D] Description/name of port
   [S] Status of port                                Suspended-violation
   [F] Full duplex                                   Disabled
   [I] Port priority (spanning tree)                 128 (80 hex)
   [C] Path cost (spanning tree)                     100
   [H] Port fast mode (spanning tree)                Enabled

  ----------------------- Related Menus ---------------------------------
   [A] Port addressing              [V] View port statistics
   [N] Next port                    [G] Goto port
   [P] Previous port                [X] Exit to Main Menu

Enter Selection: S
```

11. *Assign IP addresses to Ethernet 0 of R5, R6, R7, and R8 from network 152.1.1.0. Do not allocate more addresses than needed, use the first available subnet, and assign the addresses in ascending order, starting with R5.*

   The next step is to assign IP addresses to the Ethernet interface of R5, R6, R7, and R8. The question specifies to use the first available subnet from 152.1.1.0 and to not allocate more address space than needed. Because you need four addresses, you must use a 29-bit mask. This gives you a range of addresses from 152.1.1.0 to 152.1.1.7. Although this provides eight addresses, the first and last ones are reserved:

   152.1.1.0←**Network Address reserved**

   152.1.1.1←**R5**

   152.1.1.2←**R6**

152.1.1.3←**R7**

152.1.1.4←**R8**

152.1.1.5←**(unused)**

152.1.1.6←**(unused)**

152.1.1.7←**Broadcast Address reserved**

12. *Place Ethernet 0 of R1 and R3 in VLAN 2. Then place Ethernet 1 of R1 and Ethernet 0 of R2 in VLAN 3.*

In the previous step, you did not need to make any configuration changes to the Catalyst, because by default all ports are in VLAN 1. You must configure Catalyst ports 5 and 6 so that they are in VLAN 2 and also configure ports 7 and 8 in VLAN 3.

The first step is to add two new VLANs, in this case VLAN 2 and VLAN3. To do this select **V** from the menu.

```
        Catalyst 1900 - Main Menu

     [C] Console Settings
     [S] System
     [N] Network Management
     [P] Port Configuration
     [A] Port Addressing
     [D] Port Statistics Detail
     [M] Monitoring
     [V] Virtual LAN
     [R] Multicast Registration
     [F] Firmware
     [I] RS-232 Interface
     [U] Usage Summaries
     [H] Help
     [K] Command Line

     [X] Exit Management Console

Enter Selection:V
```

Enter **A** to add a VLAN:

```
        Catalyst 1900 - Virtual LAN Configuration

    ----------------------- Information ------------------------------------
     VTP version: 1
     Configuration revision: 2
     Maximum VLANs supported locally: 1005
     Number of existing VLANs: 6
     Configuration last modified by: 10.10.4.234 at 00-00-0000 00:00:00

    ----------------------- Settings ---------------------------------------
     [N] Domain name
     [V] VTP mode control              Server
     [F] VTP pruning mode              Disabled
```

```
     [O] VTP traps                       Enabled

    ----------------------- Actions ----------------------------------------
     [L] List VLANs                      [A] Add VLAN
     [M] Modify VLAN                     [D] Delete VLAN
     [E] VLAN Membership                 [S] VLAN Membership Servers
     [T] Trunk Configuration             [W] VTP password
     [P] VTP Statistics                  [X] Exit to Main Menu

Enter Selection:  A
```

Select the VLAN type to be added, in this case it is Ethernet:

```
This command selects the type of VLAN to be added.

The following VLAN types can be added:

[1]Ethernet, [2]FDDI, [3]Token-Ring, [4]FDDI-Net, or [5]Token-Ring-Net

Select a VLAN type [1-5]: Ethernet
```

Select **N** and then enter the VLAN number.

```
        Catalyst 1900 - Add Ethernet VLAN

    ----------------------- Settings ----------------------------------------
     [N] VLAN Number                     3
     [V] VLAN Name                       VLAN0003
     [I] 802.10 SAID                     100003
     [M] MTU Size                        1500
     [L] Translational Bridge 1          0
     [J] Translational Bridge 2          0
     [T] VLAN State                      Enabled

     [S] Save and Exit                   [X] Cancel and Exit

Enter Selection:  N


This command selects the unique VLAN identifier of a VLAN.

Configuration change only takes effect when the VLAN SAVE command is executed.

Enter VLAN Number [2-1001]:

Current setting ===> 3

    New setting ===> 2
```

Enter **S** to save the configuration.

```
        Catalyst 1900 - Add Ethernet VLAN

    ----------------------- Settings ----------------------------------------
     [N] VLAN Number                     2
     [V] VLAN Name                       VLAN0003
```

```
     [I] 802.10 SAID                  100003
     [M] MTU Size                     1500
     [L] Translational Bridge 1       0
     [J] Translational Bridge 2       0
     [T] VLAN State                   Enabled

     [S] Save and Exit                [X] Cancel and Exit

Enter Selection:  S

Press any key to continue.
```

The next step is to place the port in the newly created VLAN. From the Main Menu, select **V** for VLAN:

```
        Catalyst 1900 - Main Menu

     [C] Console Settings
     [S] System
     [N] Network Management
     [P] Port Configuration
     [A] Port Addressing
     [D] Port Statistics Detail
     [M] Monitoring
     [V] Virtual LAN
     [R] Multicast Registration
     [F] Firmware
     [I] RS-232 Interface
     [U] Usage Summaries
     [H] Help
     [K] Command Line

     [X] Exit Management Console

Enter Selection:  V
```

From the Virtual LAN Configuration menu, select **E** for VLAN Membership:

```
 Catalyst 1900 - Virtual LAN Configuration

    ----------------------- Information ----------------------------------
     VTP version: 1
     Configuration revision: 3
     Maximum VLANs supported locally: 1005
     Number of existing VLANs: 7
     Configuration last modified by: 10.10.4.234 at 00-00-0000 00:00:00

    ----------------------- Settings -------------------------------------
     [N] Domain name
     [V] VTP mode control              Server
     [F] VTP pruning mode              Disabled
     [O] VTP traps                     Enabled

    ----------------------- Actions --------------------------------------
     [L] List VLANs                    [A] Add VLAN
     [M] Modify VLAN                   [D] Delete VLAN
     [E] VLAN Membership               [S] VLAN Membership Servers
```

```
     [T] Trunk Configuration              [W] VTP password
     [P] VTP Statistics                   [X] Exit to Main Menu

Enter Selection:  E
```

The VLAN membership configuration menu appears, listing all the ports and the VLANs of which they are members. Note that all the ports are in VLAN 1, which is the default setting on the Catalyst.

To change the VLAN membership on a port, select **V** for VLAN assignment:

```
        Catalyst 1900 - VLAN Membership Configuration

   Port  VLAN    Membership Type       Port  VLAN    Membership Type
   -----------------------------       -----------------------------
   1        1       Static             13       1       Static
   2        1       Static             14       1       Static
   3        1       Static             15       1       Static
   4        1       Static             16       1       Static
   5        1       Static             17       1       Static
   6        1       Static             18       1       Static
   7        1       Static             19       1       Static
   8        1       Static             20       1       Static
   9        1       Static             21       1       Static
   10       1       Static             22       1       Static
   11       1       Static             23       1       Static
   12       1       Static             24       1       Static
                                       AUI      1       Static
   A        1       Static
   B        1       Static

     [M] Membership type                  [V] VLAN assignment
     [R] Reconfirm dynamic membership     [X] Exit to previous menu

Enter Selection:  V
```

The next screen prompts you to enter the ports that you want to place in a particular VLAN. The command enables you to enter the ports one by one or as a group using commas and dashes. In the following example, you enter port 5 and 6, separating them with a dash.

```
This command assigns or re-assigns ports to a VLAN.  If the port is configured with
dynamic VLAN membership, then assigning VLAN 0 causes the discovery of VLAN membership to
start all over again.

Port numbers should be separated by commas or spaces.  A port number range may also be
specified. The word ALL indicates all ports. Example:  1, 7-11, AUI, 4, A

Enter port numbers:  5-6
```

You are then prompted to enter the VLAN in which to place the ports. This example uses VLAN 2. Repeat the same steps to place port 7 in VLAN 3.

```
Identify VLAN 0 - VLAN 1005:
Select [0-1005] 2
```

13. *Connect the rest of the routers serially as per Figure 2-6. Make sure that the correct ports are configured for DTE/DCE (the DCE clock is set to 64K on all routers).*

This step is straightforward. The only thing that needs to be done is to set the clock rate on the DCE side of each serial link.

To set the clock rate, use the following interface command:

```
R7(config-if)#clock rate 64000
```

Use the **show controller** interface command to verify that the DCE cable is connected to the correct serial port. The following is a truncated output from the command, note that the cable attached is a V35 DCE cable and the clockrate is set to 64K.

```
R7#show controllers s 0
HD unit 0, idb = 0x970D0, driver structure at 0x9AE28
buffer size 1524  HD unit 0, V.35 DCE cable, clockrate 64000
cpb = 0xE1, eda = 0x4940, cda = 0x4800
RX ring with 16 entries at 0xE14800
```

14. *R3 and R2 are connected via Frame Relay to R1. Use physical interfaces on R3 and R2, and a logical interface on R1. Assign addresses to each interface using subnet 152.1.10.0 and do not allocate more addresses than are needed. Use the first subnet available and assign the addresses in ascending order, starting with R1. Configure the interfaces for Frame Relay. Make sure that only DLCIs used by Routers are seen and that all routers can reach each other.*

The following is the physical connectivity of R1, R2, and R3 to the Frame Relay switch (R8):

Frame Switch (R8) interface S0	R1 Interface S0/0
Frame Switch (R8) interface S0	R2 Interface S0
Frame Switch (R8) interface S0	R3 Interface S0/1

This question can be a bit tricky unless you read the case study completely. Question 22 specifies that you should use address 152.1.10.3 /32 for the loopback interface on R3, 152.1.10.2 /32 for the loopback interface on R2, and address 152.1.10.1 /32 on R1.

You need three IP addresses, so you must use a 29-bit mask. This will give you eight total addresses, of which six are usable. The first available subnet is 152.1.10.8 because subnet 152.1.10.0 is used by the loopback interface.

The IP addressing is as follows:

152.1.10.8←**Network address reserved**

152.1.10.9←**R1**

152.1.10.10←**R2**

152.1.10.11←**R3**

152.1.10.12←**Unused**

152.1.10.13←**Unused**

152.1.10.14←**Unused**

152.1.10.15←**Broadcast address reserved**

Now that the IP addressing is laid out, you must configure the interfaces for Frame Relay. The question specifies that R3 and R2 use a physical interface and that R1 use a logical or subinterface. To create a logical subinterface on R1, you must first configure the physical interface with the Frame Relay encapsulation information:

```
R1(config)#interface serial 0/0
R1(config-if)#encapsulation frame-relay IETF
R1(config-if)#frame-relay lmi-type ansi
```

Then create the subinterface by specifying the physical interface followed by a dot and the logical interface. If the interface does not exist, it is created. Two types of Frame Relay subinterfaces exist: point-to-point and multipoint:

```
R1(config)#int s0/0.1 multipoint
R1(config-subif)#ip add 152.1.10.9 255.255.255.248
```

The question also states that the router should only see DLCIs that it is using:

```
R1(config-subif)#no frame-relay inverse-arp
```

Because you are disabling inverse arp, you must map the far-end layer-3 address to a layer-2 DLCI. To do this, perform the following:

```
R1(config-subif)#frame-relay map ip 152.1.10.11 200 broadcast
R1(config-subif)#frame-relay map ip 152.1.10.10 100 broadcast
```

Because you are not using logical interfaces on R2 and R3, all configurations are done under the physical interface:

```
R2(config)#interface s0
R2(config-if) ip address 152.1.10.10 255.255.255.248
R2(config-if) encapsulation frame-relay IETF
R2(config-if)frame-relay lmi-type ansi
R2(config-if)no frame-relay inverse-arp
R2(config-if)frame-relay map ip 152.1.10.9 100 broadcast
R2(config-if) frame-relay map ip 152.1.10.11 100 broadcast


R3(config)#interface serial 0/1
R3(config-if)#ip address 152.1.10.11 255.255.255.248
R3(config-if)#encapsulation frame-relay IETF
R3(config-if)#frame-relay lmi-type ansi
R3(config-if)#no frame-relay inverse-arp
R3(config-if)#frame-relay map ip 152.1.10.10 200 broadcast
R3(config-if)#frame-relay map ip 152.1.10.9 200 broadcast
```

15. *Assign IP addresses to Ethernet 0/0 of R3 and R1 using the next available subnet in 152.1.8.0. The subnet should contain 48 host addresses, R1 should use the first address in the subnet, and R3 should use the last one.*

This question involves simple subnetting. You need to pick a subnet that contains 48 usable IP addresses. A 26-bit mask provides 64 addresses, two of which can't be used, leaving 62. This is a bit more than you need, but if you choose a 27-bit mask, you would be short. The following is the IP address assignment:

```
152.1.8.0 /26←Network Address
152.1.8.1 /26←R1
 . . .  . . .  . . .  . . .  . . . .
 . . .  . . .  . . .  . . .  . . . .
152.1.8.62 /26←R3
152.1.8.63 /26←Broadcast Address
```

The following commands define IP addresses on R3 and R1:

```
R3(config)#interface ethernet 0/0
R3(config-if)#ip address 152.1.8.62 255.255.255.192

R1(config)#interface ethernet 0/0
R1(config-if)#ip address 152.1.8.1 255.255.255.192
```

16. *Assign IP addresses to Ethernet 0/1 of R1 and Ethernet 0 of R2 using the first available subnet in 152.1.9.0. The subnet should contain 128 host addresses, R1 should use the first address in the subnet, and R2 should use the last one.*

This question is also a review of subnetting. You need to pick a subnet that contains 128 usable addresses. At first glance, you might think to use a 25-bit mask that provides 128 addresses, but because two of these addresses cannot be used, you are forced to use a 24-bit mask. The following is the IP address assignment:

```
152.1.9.0/24     ←Network Address
152.1.9.1/24     ←R1
152.1.9.254 /24 ←R2
152.1.9.255 /24 ←Network Broadcast
```

The following commands define IP addresses on R1 and R2.

```
R1(config)#interface ethernet 0/1
R1(config-if)#ip add 152.1.9.1 255.255.255.0

R2(config)#interface ethernet 0/
R2(config-if)#ip address 152.1.9.254 255.255.255.0
```

17. *Assign IP addresses to Serial 0 on R7 and Serial S0/0 on R3 using the first available subnet in 152.1.20.0. The address should be assigned in ascending order, starting with R3.*

The serial connection between R7 and R3 is point-to-point and requires two addresses. For this use a 30-bit mask. The following is the IP address assignment:

```
152.1.20.0 /30←Network Address
152.1.20.1 /30←R3
152.1.20.2 /30←R7
152.1.20.3 /30←Broadcast Address
```

The following commands define IP addresses on R1 and R2:

```
R3(config)#interface serial  0/0
R3(config-if)#ip address 152.1.20.1 255.255.255.252

R7(config)#interface serial  0
R7(config-if)#ip address 152.1.20.2 255.255.255.252
```

18. *Assign IP addresses to the serial interface on R3 that connects to R4 using the first available subnet in 152.1.12.0. The subnet should contain the minimal hosts needed and the IP addresses should be assigned in ascending order, staring with R3.*

The serial connection between R3 and R4 is a point-to-point connection, so only two addresses are needed. After carefully reading the rest of the case study, determine if you can use the first subnet 152.1.12.0. For this, use a 30-bit mask. The following is the IP address assignment:

```
152.1.12.0 /30←Network Address
152.1.12.1 /30←R3
152.1.12.2 /30←R4
152.1.12.3 /30←Broadcast Address
```

The following commands define IP addresses on R3 and R4:

```
R3(config)#interface serial 1/0
R3(config-if)#ip address 152.1.12.1 255.255.255.252

R4(config)#interface serial  0
R4(config-if)#ip address 152.1.12.2 255.255.255.252
```

19. *Assign IP addresses to the serial interfaces on R3 that connect to R5 using the next available subnet in 152.1.12.0. The subnet should contain the minimal number of hosts needed and the IP addresses should be assigned in ascending order, staring with R3.*

Like the previous questions, the link between R3 and R5 is a point-to-point link, so you only need two addresses. The first subnet is being used by loopback interfaces on R4 and R5, the next available subnet is 152.1.12.8. Use a 30-bit mask because you only need two host IP addresses. The following is the IP address assignment:

```
152.1.12.8 /30←Network Address
152.1.12.9 /30←R3
152.1.12.10/30←R5
152.1.12.11/30←Broadcast Address
```

The following commands define IP addresses on R3 and R5:

```
R3(config)#interface serial  1/1
R3(config-if)#ip address 152.1.12.9 255.255.255.252

R5(config)#interface serial  0
R5(config-if)#ip address 152.1.12.10 255.255.255.252
```

20. *Assign an IP address to Serial 1 on R4 and R5 using the next available subnet in 152.1.12.0. The subnet should contain the minimal hosts needed and the IP addresses should be assigned in ascending order, starting with R4.*

Once again, because the link between R4 and R5 is point-to-point, you only need two addresses so use a 30-bit mask. The next available subnet is 152.1.12.12. The following is the IP address assignment:

```
152.1.12.12/30←Network Address
152.1.12.13/30←R3
152.1.12.14/30←R5
152.1.12.15/30←Broadcast Address
```

The following commands define IP addresses on R3 and R5:

```
R4(config)#interface serial  1
R4(config-if)#ip address 152.1.12.13 255.255.255.252

R5(config)#interface serial  1
R5(config-if)#ip address 152.1.12.14 255.255.255.252
```

21. *Assign an IP address to Serial 1 of R2 and Serial 0 of R6 using the first available subnet in 152.1.11.0. The subnet should contain the minimal number of hosts needed and the IP addresses should be assigned in ascending order, starting with R2.*

Again, the link between R2 and R6 is point-to-point, so you only need two addresses for this use a 30-bit mask. The next available subnet is 152.1.11.0. The following is the IP address assignment:

```
152.1.11.0 /30←Network Address
152.1.11.1 /30←R3
152.1.11.2 /30←R5
152.1.12.3 /30←Broadcast Address
```

The following commands define IP addresses on R3 and R5:

```
R2(config)#interface serial  1
R2(config-if)#ip address 152.1.11.1 255.255.255.252

R6(config)#interface serial  0
R6(config-if)#ip address 152.1.11.2 255.255.255.252
```

22. *Assign the following loopback addresses to the corresponding routers:*

R7 Loopback 0 152.1.20.7 /32

R7 Loopback 1 152.1.20.17 /28

R4 Loopback 0 152.1.12.4 /32

R4 Loopback 0 152.1.13.1 /24

R5 Loopback 0 152.1.12.5 /32

R5 Loopback 1 152.1.14.1 /29

R3 Loopback 0 152.1.10.3 /32

R2 Loopback 0 152.1.10.2 /32

R1 Loopback 0 152.1.10.1 /32

R6 Loopback 0 152.1.11.6 /32

R6 Loopback 1 152.1.11.129 /26

IGP Protocols

23. *Place the serial connections between R3, R2, and R1 in OSPF Area 0.*

Enabling OSPF on a router is done in two steps. First, an OSPF process is defined and then an interface is added to the process. The command to start an OSPF process is **Router OSPF [Process #]**. Multiple OSPF processes can be configured on one router.

The following command enables OSPF process 64 on the router and assigns the serial interfaces attached to the Frame Cloud on the routers to area 0:

```
R1(config)#router ospf 64
R1(config-router)#network 152.1.10.8 0.0.0.7 area 0

R2(config)#router ospf 64
R2(config-router)#network 152.1.10.8 0.0.0.7 area 0

R3(config)#router ospf 64
R3(config-router)#network 152.1.10.8 0.0.0.7 area 0
```

Use the command **show ip ospf interface** to verify that OSPF is configured on the interface. The following is the output from this command. Notice that on R2, interface S0 is configured for OSPF and is in area 0. The command also shows the network type, timer intervals, and adjacent neighbors.

```
R2#show ip ospf interface
Ethernet0 is up, line protocol is up
   OSPF not enabled on this interface
```

```
Loopback0 is up, line protocol is up
   OSPF not enabled on this interface
Serial0 is up, line protocol is up
  Internet Address 152.1.10.10/29 , Area 0
  Process ID 64, Router ID 152.1.10.2, Network Type NON_BROADCAST, Cost: 64
  Transmit Delay is 1 sec, State DOWN, Priority 1
  No designated router on this networkNo backup designated router on this network
  Timer intervals configured, Hello 30, Dead 120, Wait 120, Retransmit 5
Serial1 is up, line protocol is up
   OSPF not enabled on this interface
Serial0/1  is up, line protocol is up
  Internet Address 152.1.10.11/29 , Area 0
  Process ID 64, Router ID 152.1.10.3, Network Type NON_BROADCAST, Cost: 64
  Transmit Delay is 1 sec, State WAITING, Priority 1
  No designated router on this network
  No backup designated router on this network
  Timer intervals configured, Hello 30, Dead 120, Wait 120, Retransmit 5
    Hello due in 00:00:06
    Wait time before Designated router selection 00:01:36
  Neighbor Count is 0, Adjacent neighbor count is 0
```

Because the case study requires that you use a logical interface at the hub (R1) and a physical interface on the spokes (R3 andR4), the interfaces are all non-broadcast. By default, a Frame Relay physical interface and a point-to-point subinterface are non-broadcast, which means a *designated router* (DR) and *backup designated router* (BDR) is elected for the network. However, due to the lack of broadcast capabilities, the DR and BDR must have a static list of all routers attached to the Frame Relay cloud. Because the Hub (R1) is the DR, you need to define static neighbors on R1.

The following commands define OSPF neighbors on the hub router:

```
R1(config)#router ospf 64
R1(config-router)#neighbor 152.1.10.10
R1(config-router)#neighbor 152.1.10.11
```

When configuring OSPF on NBMA networks such as Frame Relay or ATM, care must be taken on which router becomes the DR and BDR for the network. The DR and BDR require full logical connectivity with all routers on the network.

All multi-access networks with two or more attached routers elect a DR. The DR concept enables a reduction in the number of adjacencies that need to be formed on a network. In order for OSPF-enabled routers to exchange routing information, they must form an adjacency with one another. If a DR is not used, then each router on a multi-access network would need to form an adjacency with every other router (because link state databases are synchronized across adjacencies). This would result in N-1 adjacencies.

Instead, all routers on a multi-access network form adjacencies only with the DR and BDR. Each router sends the DR and BDR routing information, and the DR is responsible for flooding this information to all adjacent routers and originating a network link advertisement on behalf of the network. The BDR is used in case the DR fails.

The reduction in adjacencies reduces the volume of routing protocol traffic as well as the size of the topological database.

The DR is elected using the Hello protocol, which will be described in detail in the IGP case study. The election of the DR is determined by the router priority, which is carried in the Hello packet. The router with the highest priority is elected the DR; if a tie occurs, the router with the highest router ID is selected.

The router ID is the IP address of the loopback interface. If no loopback is configured, the router ID is the highest IP address on the router. The router priority can be configured on the router interface with the **ip ospf priority** command.

When a router first becomes active on a multi-access network, the router checks to see if a DR currently exists for the network. If a DR is present, the router accepts the DR regardless of what its priority is. Once a DR is elected, no other router can become the DR unless the DR fails. If no DR is present on the network, then the routers negotiate the DR based on router priority.

For this question, use the **ip ospf priority** command to assure that R1 becomes the DR for the network. To do this, set the OSPF priority on R1 to 255 and the OSPF priority on R2 and R3 to 0. An OSPF priority of zero means that the router is ineligible in the DR election process.

```
R1(config)#router ospf 64
R1(config-subif)#ip ospf priority 255

R2(config)#router ospf 64
R2(config-if)#ip ospf priority 0

R3(config)#router ospf 64
R3(config-if)#ip ospf priority 0
```

Use the **show ip ospf neighbor** command on R1 to verify that an adjacency with the two spokes has formed. The following code is the output from the command. Notice that R1 has two OSPF neighbors, 152.1.10.3 and 152.1.10.2. These neighbor IDs are the router IDs of each router. OSPF uses the highest loopback interfaces as its router IDs. and if no loopback interface is configured on the router, the highest IP address is used.

```
R1#show ip ospf neighbor

Neighbor ID     Pri   State           Dead Time   Address         Interface
152.1.10.3        0   FULL/DROTHER    00:01:40    152.1.10.11     Serial0/0 .1
152.1.10.2        0   FULL/DROTHER    00:01:43    152.1.10.10     Serial0/0 .1
```

The state of each neighbor is full, meaning that R1 has formed an adjacency with each one of its neighbors. Only the DR and BDR form adjacencies with all the routers on the network. R1 is the DR for the network, and each of the neighbors are DROTHER, which means that they are neither the BDR or DR for the network.

24. *Place VLAN 2 in OSPF area 1 and VLAN 3 in OSPF area 2. R1 should be the DR for Area 1 and R2 should be the DR for Area 2.*

This question is similar to question 23, the only difference being the network type. The broadcast network type is the default type on LANs, such asToken Ring, Ethernet, or FDDI. The only difference between the two network types is that no neighbors need to be defined in the broadcast model and the Hello, dead, and wait intervals are different.

If either the Hello interval or the dead interval do not match, the router does not form an adjacency with its neighbor.

The following commands enable OSPF process 64 on the router and assign the Ethernet interfaces in VLAN 2 to area 1 and VLAN 3 to area 2:

```
R1(config)#router ospf 64
R1(config-router)#network 152.1.8.0 0.0.0.63 area 1
R1(config-router)#network 152.1.9.0 0.0.0.255 area 2

R2(config)#router ospf 64
R2(config-router)#network 152.1.9.0 0.0.0.255 area 2
R3(config-router)#network 152.1.8.0 0.0.0.63 area 1
```

The default priority of an Ethernet interface is 1, so in order to assure that R1 and R2 are the DRs for their respective VLANs, you need to set their priorities higher than 1. To achieve this, you could use a number from 2 to 255. The following commands set the OSPF priority:

```
R1(config)#interface ethernet 0/0
R1(config-if)#ip ospf priority 2

R2(config)#interface ethernet 0
R2(config-if)#ip ospf priority 2
```

Because this question requires that the other routers become the BDRs for the network, you cannot set their priorities to 0. This could pose a problem, depending on what order the configuration was done in. For example, if you enable OSPF on VLAN1 and then went in and set the priority on R1, R2 would have already been elected the DR because it has the highest router ID. Remember that, by default, the priority of an Ethernet interface is 1, so when the routers elect a DR/BDR, they first look at the priority. If the priorities are equal, the routers then look at the router ID.

25. *R1 should prefer the Frame Relay link over the Ethernet connection. All other connections should prefer the interface with the highest bandwidth.*

Cisco's implementation of OSPF calculates its route based on bandwidth. The higher the bandwidth, the lower the cost and the more preferable the route. The path cost is calculated using the following formula: cost = 100,000,000/bandwidth in bits per second. The cost is inversely proportional to the bandwidth of the link. The higher the bandwidth, the lower the cost.

This question requires that you set the cost of the path over the Frame Relay circuit so that it is lower or more preferable than the Ethernet link. By default, the OSPF cost of an Ethernet interface is 10 (100M/10M) and the cost of a serial interface is 64 (100M/1.54M). In order to get OSPF to prefer the route over the serial interface, you must manually set the cost lower than that of the Ethernet.

The following command sets the OSPF cost of the serial interface of R1 to 9, which, of course, is one less than the Ethernet interface. The problem is that by doing this R1 also prefers the serial link to reach R2. You must set the cost on the Ethernet interface attached to VLAN2 to a cost of 8:

```
R1(config)#interface serial 0/0 .1
R1(config-subif)#ip ospf cost 9

R1(config)#interface ethernet 1/0
R1(config-if)#ip ospf cost 8
```

26. *Change the Hello interval on R3 to two minutes.*

This question is tricky if you don't understand what the Hello interval is used for and the implications of changing it on any router attached to a multi-access network.

The OSPF Hello interval is the length of time in seconds the router waits between sending Hello packets on a particular interface. Once the adjacency is formed, the Hello packet is used to maintain the neighbor relationship. If a Hello packet is not received within the specified dead interval, the neighbor is declared down. The default dead interval is four times the Hello interval.

The following command sets the dead interval on interface S0/1 of R3 to 120 seconds:

```
R3(config)#interface serial 0/1
R3(config-if)#ip ospf hello-interval 120
```

Let's see what happens when you change the Hello interval on R3 to 120 seconds. Use the command **show ip ospf neighbor** to display the status of the neighbors. The following is the output from the command. Note that R3 has lost its adjacency over its serial link with R1, while the adjacency over the Ethernet still exists.

```
R3#show ip ospf neighbor

Neighbor ID     Pri   State          Dead Time   Address         Interface
152.1.10.1        2   FULL/DR        00:00:34    152.1.8.1       Ethernet0/0
```

Monitor the OSPF events on R3 with the command **debug ip ospf events.** Notice that R3 is receiving an OSPF packet from R1 and the Hello intervals do not match. If either the Hello interval or the dead interval do not match, then the router does not form an adjacency with its neighbor.

```
OSPF: Mismatched hello parameters from 152.1.10.9
Dead R 120 C 480, Hello R 30 C 120  Mask R 255.255.255.248 C 255.255.255.248
```

In order for this to work, all routers that are attached to the multi-access network must set the Hello interval to 120 seconds.

```
R1(config)#interface serial 0/0 .1
R1(config-subif)#ip ospf hello-interval 120

R2(config)#interface serial 0
R2(config-if)#ip ospf hello-interval 120
```

27. *Configure EIGRP on the interface connecting R3 to R7. Do not send advertisements out any other interface. Enable EIGRP on all interfaces on R7, but do not send updates out the interface connecting to VLAN1.*

Enabling EIGRP on a router consists of two steps. First, an EIGRP routing process is enabled on the router, and then the network command is added, specifying which interfaces will receive and send EIGRP routing updates. It also specifies which networks will be advertised. The following command enables EIGRP on R3:

```
R3(config)#router eigrp 64
R3(config-router)#network 152.1.0.0
```

When enabling EIGRP on a router, you specify which classful network the protocol will be run on. Because every interface on the router is part of the classful network 152.1.0.0, EIGRP sends updates out every interface.

To verify which interfaces EIGRP is running on, use the **show ip eigrp interface** command. The following is the output from the command. Note that EIGRP is running on all interfaces.

```
R3#show ip eigrp interfaces
IP-EIGRP interfaces for process 64

                    Xmit Queue   Mean   Pacing Time   Multicast    Pending
Interface    Peers  Un/Reliable  SRTT   Un/Reliable   Flow Timer   Routes
Se0/0          1        0/0       269       0/15         1343          0
Et0/0          0        0/0         0       0/10            0          0
Se0/1          0        0/0         0       0/10            0          0
Se1/0          0        0/0         0       0/10            0          0
Se1/1          0        0/0         0       0/10            0          0
Lo0            0        0/0         0       0/10            0          0
```

This question requires that EIGRP updates only be sent out the serial interface connecting R3 to R7. The passive-interface command enables EIGRP-enabled routers to listen to, but not send, routing updates out a particular interface. This command is typically used when the network router configuration command configures more interfaces than is desirable. To disable the sending of EIGRP updates out the other interfaces, perform the following commands:

```
R3(config)#router eigrp 64
R3(config-router)#passive-interface ethernet 0/0
R3(config-router)#passive-interface serial 0/1
R3(config-router)#passive-interface serial 1/0
R3(config-router)#passive-interface serial 1/1
R3(config-router)#passive-interface loopback 0
```

Once again, use the command **show ip eigrp interfaces** to verify which interfaces which EIGRP is sending updates out. The following is the output from the command. Note that EIGRP is only sending updates out of S0/0, which is the interface connected to R7.

```
R3#show ip eigrp interfaces
IP-EIGRP interfaces for process 64

                    Xmit Queue   Mean   Pacing Time   Multicast    Pending
Interface    Peers  Un/Reliable  SRTT   Un/Reliable   Flow Timer   Routes
Se0/0          1        0/0       269       0/15         1343          0
```

The same needs to be done on R7, but this time only the interface attached to VLAN1 is passive. The following are the commands:

```
R7(config)#router eigrp 64
R7(config-router)#network 152.1.0.0
R7(config-router)#passive-interface ethernet 0
```

28. *R7 should be able to see all the networks behind R3. R3 only needs to see network 152.1.20.16/28 via R7.*

R3 can see all the networks attached to R7, but R7 cannot see all the networks behind R3. In order to get the OSPF routes into EIGRP and the EIGRP routes into OSPF, you must perform mutual redistribution, which is when each routing protocol is redistributed into the other.

When redistributing from one protocol to another, you must take the metrics used into account. As discussed earlier, OSPF uses a metric based on the bandwidth of a link. EIGRP uses a metric based up to five variables.

The EIGRP metric is a 32-bit number, which is calculated using bandwidth, delay, reliability, loading, and the Maximum Transmission Unit (MTU). Calculating the metric for a route is a two-step process using the five different characteristics of the link and the K values. The K values are configurable, but this is not recommended. The default K values are K1=1, K2=0, K3=1, K4=0, and K5=0.

Metric= K1*Bandwidth +(K2 * Bandwidth)/(256-load) + K3*Delay

If K5 is not equal to zero, take the metric from step 1 and multiple it by [K5/(reliability +K4)]. If K5 is 0, ignore step 2.

Metric = Metric* [K5/(reliability +K4)]

As shown here, Cisco sets K2, K4, and K5 to 0. This leaves only two variables to compute the EIGRP metric (bandwidth and delay). Because three of the K values are 0, the formula is reduced to

Metric= Bandwidth + Delay

The bandwidth is derived by finding the smallest of all bandwidths in the path to the destination and dividing 10,000,000 by that number.

The delay is found by adding all the delays along the paths and dividing that number by 10. The sum of the two numbers is then multiplied by 256. This equation can be written as

Metric = [(10,000,000/min bandwidth) + (SUM(interface delay)/10)] * 256

In order to redistribute EIGRP into OSPF, you must tell the router what it should use for a metric. The following command is used to redistribute OSPF into EIGRP:

```
R3(config-router)#redistribute ospf 64 metric 1500 100 255 1 1500
```

Show the routing table on R7 to verify that you are receiving all the routes that are in AS100:

```
R7#show ip route
Codes: C - connected, S - static, I - IGRP, R - RIP, M - mobile, B - BGP
       D - EIGRP, EX - EIGRP external, O - OSPF, IA - OSPF inter area
       N1 - OSPF NSSA external type 1, N2 - OSPF NSSA external type 2
       E1 - OSPF external type 1, E2 - OSPF external type 2, E - EGP
       i - IS-IS, L1 - IS-IS level-1, L2 - IS-IS level-2, * - candidate default
       U - per-user static route, o - ODR

Gateway of last resort is not set

     152.1.0.0/16  is variably subnetted, 10 subnets, 6 masks
C    152.1.20.7/32  is directly connected, Loopback0
C    152.1.20.0/30  is directly connected, Serial0
D EX 152.1.9.0/24  [170/2244096 ] via 152.1.20.1, 00:01:37, Serial0
D    152.1.10.3/32  [90/2297856 ] via 152.1.20.1, 00:25:57, Serial0
D    152.1.8.0/26  [90/2195456 ] via 152.1.20.1, 00:25:57, Serial0
D    152.1.12.0/30  [90/2681856 ] via 152.1.20.1, 00:25:57, Serial0
C    152.1.1.0/29  is directly connected, Ethernet0
D    152.1.10.8/29  [90/2681856 ] via 152.1.20.1, 00:25:57, Serial0
D    152.1.12.8/30  [90/2681856 ] via 152.1.20.1, 00:25:57, Serial0
C    152.1.20.16/28  is directly connected, Loopback1
```

Now you need to redistribute the EIGRP-learned routes into OSPF. The following command is used to redistribute EIGRP into OSPF:

```
R3(config)#router ospf 64
R3(config-router)#redistribute eigrp 64 subnets metric 64
```

29. *Make sure R7 does not advertise any routes back to R3 that were learned from R3 and vice versa. Do not rely on split horizons!*

Care must be taken when using mutual redistribution in order to prevent routing loops. Metrics and spilt horizons will help to prevent routing loops, but it is a good idea to configure distribution lists so the router cannot advertise invalid routes. You need to configure a distribute list out on R3 and R7.

On R3, you must configure a distribute list to deny networks 152.1.20.7/32 , 152.1.1.0/29, and 152.1.20.16/28. This is a two-step process. First, an access list must be set up to identify and permit or deny a network based on the network address. Then the access list must be applied to routing updates through the use of a distribute list.

a. Define an access list on R3, denying networks 152.1.20.7/32 , 152.1.1.0/29 , and 152.1.20.16/28 :

```
R3(config)#access-list 1 deny 152.1.20.7 0.0.0.0
R3(config)#access-list 1 deny 152.1.1.0 0.0.0.7
R3(config)#access-list 1 deny 152.1.20.16  0.0.0.15
R3(config)#access-list 1 permit any
```

b. Apply the access list to routing updates through the use of a distribute list:

```
R3(config)#router eigrp 64
R3(config-router)#distribute-list 1 out eigrp 64
```

Because you have less routes on R7 to advertise, configure an access list to permit that subset of routes.

c. Define an access list on R7, permitting networks 152.1.20.7/32 , 152.1.1.0/29 , and 152.1.20.16/28 :

```
R7(config)#access-list 1 permit 152.1.20.7 0.0.0.0
R7(config)#access-list 1 permit 152.1.1.0 0.0.0.7
R7(config)#access-list 1 permit 152.1.20.16  0.0.0.15
R7(config)#access-list 1 deny any
```

d. Apply the access list to routing updates through the use of a distribute list:

```
R7(config)#router eigrp 64
R7(config-router)#distribute-list 1 out eigrp 64
```

30. *Enable EIGRP using process 56 on R4 and R5. Don't run EIGRP on the interfaces that are connected to other ASs.*

Enabling EIGRP on a router is performed in two steps. First, an EIGRP-routing process is enabled on the router, and then the network command specifies which interfaces will receive and send EIGRP-routing updates. It also specifies which networks will be advertised. The following command enables EIGRP on R4 and R5:

```
R4(config)#router eigrp 56
R4(config-router)#network 152.1.0.0

R5(config)#router eigrp 56
R5(config-router)#network 152.1.0.0
```

When enabling EIGRP on a router, you must specify which classful network the protocol will be run on. Because every interface on the router is part of the classful network 152.1.0.0, EIGRP sends updates out every interface.

The question states that EIGRP should not be run on any interface connecting to another AS. To prevent this, use the passive interface command on S0 of R4 and R5 and the Ethernet interface on R5:

```
R4(config)#router eigrp 56
R4(config-router)#passive-interface serial 0

R5(config)#router eigrp 56
R5(config-router)#passive-interface serial 0
R5(config-router)#passive-interface ethernet 0
```

To verify which interfaces EIGRP is running on, use the **show ip eigrp interfaces** command. The following is the output from the command. Note that EIGRP is not running on the Serial interface 0 or the Ethernet interface of R5:

```
R5#show ip eigrp interfaces
IP-EIGRP interfaces for process 56

                        Xmit Queue   Mean   Pacing Time   Multicast    Pending
Interface        Peers  Un/Reliable  SRTT   Un/Reliable   Flow Timer   Routes
Lo0                0        0/0        0       0/10           0            0
Lo1                0        0/0        0       0/10           0            0
Se1                0        0/0        0       0/15           0            0
```

31. *Make sure that the loopback interfaces on R2, R1, and R3 are being propagated via OSPF. OSPF should not be run on these interfaces, and the routes should not show up as OSPF external routes.*

In order to get any network advertised in OSPF, it must be part of the OSPF process or be redistributed into the process. The instructions state that the routes should not show up as OSPF external. This means that you cannot redistribute connected subnets into OSPF. You must add the network to the OSPF process, but the problem with this is that once the network is added to the OSPF process, OSPF will be enabled on that interface. To get around this, make the interface passive.

The passive interface command disables the sending and receiving of OSPF router information. The specified interface address appears as a stub network in the OSPF domain:

```
R3(config)#router ospf 64
R3(config-router)#network 152.1.10.3 0.0.0.0 area 0
```

```
R3(config-router)#passive-interface loopback 0

R1(config)#router ospf 64
R1(config-router)#network 152.1.10.1 0.0.0.0 area 0
R1(config-router)#passive-interface loopback 0

R2(config)#router ospf 64
R2(config-router)#network 152.1.10.2 0.0.0.0 area 0
R2(config-router)#passive-interface loopback 0
```

EBGP/IBGP

32. *Place R3, R2, and R1 in AS 100. R1 will IBGP peer with R2 and R3. R1 is the only router in AS100 that should have two IBGP neighbors.*

Enabling BGP is a two-step process. First, the BGP routing process is enabled on the router; then neighbors are defined. If the neighbors are in the same AS, it is IBGP. If the neighbors are in a different AS, it is EBGP.

```
R1(config)#router bgp 100

R2(config)#router bgp 100

R3(config)#router bgp 100
```

To prevent routing loops within an AS, BGP does not advertise to internal BGP peers' routes that it has learned about via other internal BGP peers. In Figure 2-7, R3 advertises all the routes it has learned via EBGP to R1, and these routes are not advertised to R2. This is because R1 does not pass IBGP routes between R3 and R2. In order for R2 to learn about these routes, an IBGP connection between R1 and R2 is needed.

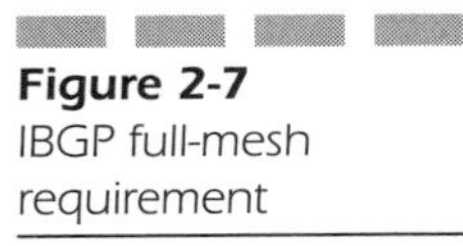

Figure 2-7
IBGP full-mesh requirement

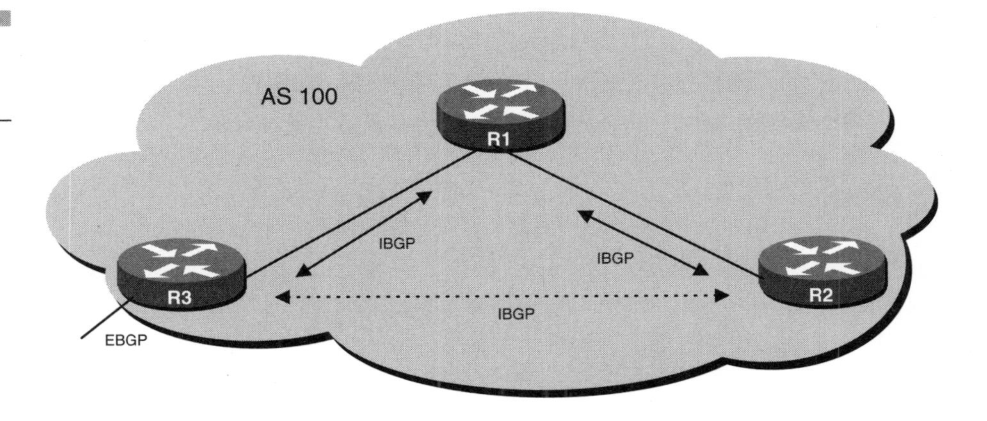

The full mesh requirement of IBGP creates the need for neighbor statements to be defined for each IBGP router. In an AS of 100 routers, this would require 100 neighbor statements to be defined. As you can see, this does not scale well.

To get around this, the concept of a route reflector has been defined. A route reflector acts as a concentration router or focal point for all IBGP sessions. Routers that peer with the route reflector are called *route reflector clients*. The clients peer with the route reflector and exchange routing information. The route reflector then exchanges or "reflects" this information to all clients, thereby eliminating the need for a fully meshed environment.

The instructions specify that R3 should not neighbor with R2, so you need to make R1 a route reflector for AS 100. R3 and R2 then only need to neighbor with R1. To do this, first define the neighbors:

```
R1(config)#router bgp 100
R1(config-router)#neighbor 152.1.10.10 remote-as 100
R1(config-router)#neighbor 152.1.10.11 remote-as 100
R1(config-router)# no synchronization

R2(config)#router bgp 100
R2(config-router)#neighbor 152.1.10.9 remote-as 100
R2(config-router)#neighbor 152.1.11.2 remote-as 300
R2(config-router)#no synchronization

R3(config)#router bgp 100
R3(config-router)#neighbor 152.1.10.9 remote-as 100
R3(config-router)#neighbor 152.1.12.2 remote-as 200
R3(config-router)#neighbor 152.1.12.10 remote-as 200
R3(config-router) no synchronization
```

NOTE: *In order for a router to advertise a network to another BGP speaker, the network to be advertised must be present in the IP routing table. Because you don't want to redistribute BGP into OSPF in AS100, disable synchronization. If this is not done, the routers will not load the BGP routes into the IP routing table and R1 will not reflect the routes to R2.*

Then specify R1 as a BGP route reflector and configure the specified neighbor as the route-reflector-client:

```
R1(config)#router bgp 100
R1(config-router)#neighbor 152.1.10.11 route-reflector-client
R1(config-router)#neighbor 152.1.10.10 route-reflector-client
```

To verify that BGP is working properly, display the BGP neighbors on R1 with the command **show ip bgp neighbors**. The following is the output from this command, which has been truncated. Notice that R1 has two IBGP neighbors: R2 (152.1.10.10) and R3

(152.1.10.11). The router ID of R2 is 152.1.10.2, which is the loopback address on R2, and the router ID of R3 is 152.1.10.3, which is the loopback on R3.

```
R1#show ip bgp neighbors
BGP neighbor is 152.1.10.10,  remote AS 100, internal link
 Index 0, Offset 0, Mask 0x0
  Route-Reflector Client
  BGP version 4, remote router ID 152.1.10.2
  BGP state = Established, table version = 3, up for 00:03:08
  Last read 00:00:08, hold time is 180, keepalive interval is 60 seconds
  Minimum time between advertisement runs is 5 seconds
  Received 6 messages, 0 notifications, 0 in queue
  Sent 6 messages, 0 notifications, 0 in queue
  Connections established 1; dropped 0
  Last reset 00:09:18, due to : User reset request
  No. of prefix received 0
Connection state is ESTAB, I/O status: 1, unread input bytes: 0
Local host: 152.1.10.9, Local port: 11006
Foreign host: 152.1.10.10, Foreign port: 179


BGP neighbor is 152.1.10.11,  remote AS 100, internal link
 Index 0, Offset 0, Mask 0x0
  Route-Reflector Client
  BGP version 4, remote router ID 152.1.10.3
  BGP state = Established, table version = 3, up for 00:03:50
  Last read 00:00:51, hold time is 180, keepalive interval is 60 seconds
  Minimum time between advertisement runs is 5 seconds
  Received 6 messages, 0 notifications, 0 in queue
  Sent 6 messages, 0 notifications, 0 in queue
  Connections established 1; dropped 0
  Last reset 00:09:26, due to : User reset request
  No. of prefix received 0
Connection state is ESTAB, I/O status: 1, unread input bytes: 0
Local host: 152.1.10.9, Local port: 179
Foreign host: 152.1.10.11, Foreign port: 11000
```

To identify itself to its neighbors, the BGP process uses a RouterID, which is an IP address. This address is either the loopback address or the highest IP address of an active interface on the router. The router ID is calculated at boot time and will remain constant until the BGP process is removed or the router is reloaded.

One of the most important things shown by the command **show ip bgp neighbors** is the line BGP state =. This shows the status of the BGP connection, which has six possible states.

Before BGP speakers can exchange network layer reachablity information (networks being advertised), a BGP session must be established. Figure 2-8 illustrates the states that BGP neighbor negotiation goes through before a connection becomes fully established.

Figure 2-8 BGP neighbor negotiation states

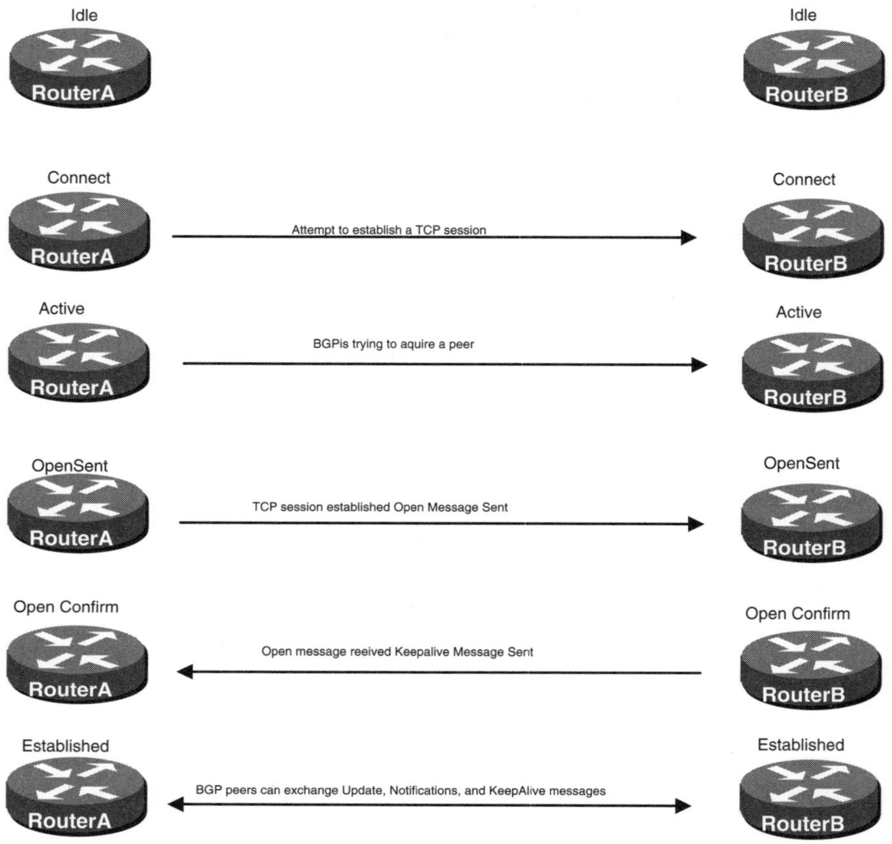

Idle Initially, BGP is in an idle state until an operator initiates a start event, which is usually caused by establishing or restarting a BGP session.

Connect In this state, BGP is waiting for the transport protocol connection to be completed. If the transport protocol connection succeeds, an Open message is sent to the peer router and the BGP state changes to OpenSent. If the connection fails, the local system changes to an active state and continues to listen for connections.

Active State In this state, BGP is trying to acquire a peer by initiating a transport protocol connection. If the connection is successful, an OPEN message is sent to the peer router. If the connection retry timer expires, the BGP state changes to connect and continues to listen for connections that may be initiated by the remote BGP peer.

OpenSent State In this state, BGP is waiting for an OPEN message from its peer. When an OPEN message is received, all fields are checked for correctness. If an error is detected, the local system sends a NOTIFICTION message and changes its state to Idle. If no errors occur, BGP starts sending keepalive messages to its peer.

OpenConfirm In this state, BGP waits for a keepalive or a notification message. If the local system receives a keepalive message, it changes its state to established. If the Hold timer expires before a keepalive message is received, the local system sends a notification message and changes its state to Idle.

Established This is the final stage of the neighbor negotiation. In the established state, BGP peers can exchange update, notifications, and keepalive messages.

33. *Place R4 and R5 in AS 200. R4 should peer with R5 and R3.*

This question requires that you place R4 and R5 in AS 100. To do this, define the BGP process with the following command:.

```
R4(config)#router bgp 200
R5(config)#router bgp 200
```

The question also requires that R4 peer with R5 and R3. To do this, define neighbors under the BGP process:

```
R4(config)#router bgp 200
R4(config-router)#neighbor 152.1.12.1 remote-as 100
R4(config-router)#neighbor 152.1.12.14 remote-as 200
```

34. *Place R6 in AS 300 and R7 in AS 400.*

This question requires that you place R6 in AS 300 and R7 in AS 400. This is the same process that you used in the last question. The only difference is that the AS number has changed.

```
R6(config)#router bgp 300
R7(config)#router bgp 400
```

35. *R7 should BGP peer with R6, R5, and R8.*

This question requires that R7 peer with R6, R5, and R8. To do this, simply define neighbor statements to each router with the appropriate remote AS:

```
R7(config)#router bgp 400
R7(config-router)# neighbor 152.1.1.1 remote-as 200
R7(config-router)# neighbor 152.1.1.2 remote-as 300
R7(config-router)# neighbor 152.1.1.4 remote-as 500
```

36. *R5 should BGP peer with R7, R8, R6, and R3.*

This question requires that R5 peer with R7, R8, R6, and R3. To do this, simply define neighbor statements to each router with the appropriate remote AS:

```
R5(config)#router bgp 200
R5(config-router)#neighbor 152.1.1.2 remote-as 300
R5(config-router)# neighbor 152.1.1.3 remote-as 400
R5(config-router)# neighbor 152.1.1.4 remote-as 500
R5(config-router)#neighbor 152.1.12.9 remote-as 100
R5(config-router)#neighbor 152.1.12.13 remote-as 200
```

37. *R6 should peer with R7, R8, R5, and R2.*

This question requires that R6 peer with R7, R8, R5, and R2. To do this, simply define neighbor statements to each router with the appropriate remote AS:

```
R6(config)#router bgp 300
R6(config-router)# neighbor 152.1.1.1 remote-as 200
R6(config-router)# neighbor 152.1.1.3 remote-as 400
R6(config-router)# neighbor 152.1.1.4 remote-as 500
R6(config-router)# neighbor 152.1.11.1 remote-as 100
```

38. *R7, R5, and R6 should all see the following networks via BGP from R8.*

network 148.1.0.0

network 148.2.0.0

network 148.3.0.0

network 148.4.0.0

Use the **show ip bgp** command to verify that R5, R7, and R6 all see the networks that are being advertised by R8. The following is the output from the command on R5. Notice that the routes are being learned from AS 400 (R7), AS 300 (R6), and AS 500 (R8).

```
R5#show ip bgp
BGP table version is 6, local router ID is 152.1.14.1
Status codes: s suppressed, d damped, h history, * valid, > best, i - internal
Origin codes: i - IGP, e - EGP, ? - incomplete

   Network          Next Hop            Metric LocPrf Weight Path
*> 148.1.0.0        152.1.1.4                0      0    500 i
*                            152.1.1.4                       0 300 500 i
*                            152.1.1.4                       0 400 500 i
*> 148.2.0.0        152.1.1.4                0      0    500 i
*                            152.1.1.4                       0 400 500 i
*                            152.1.1.4                       0 300 500 i
*> 148.3.0.0        152.1.1.4                0      0    500 i
*                            152.1.1.4                       0 300 500 i
*                            152.1.1.4                       0 400 500 i
*> 148.4.0.0        152.1.1.4                0      0    500 i
*                            152.1.1.4                       0 300 500 i
*                            152.1.1.4                       0 400 500 i
```

39. *AS400 should advertise local network 152.1.20.0 via BGP, and AS300 should advertise local network 152.1.11.128 via BGP.*

Now from R7, you will advertise network 152.1.20.16 /30 via BGP and from R6 network 152.1.11.128 via BGP. In order for a router to advertise a network to another BGP speaker, two conditions must be met:

a. The BGP process must be aware of the route, either through the use of the network command or by redistribution.

b. The network to be advertised must be present in the IP routing table.

For the purposes of this case study, you will be using the network command under the BGP process. This takes care of the first rule, making the BGP process aware of the route. The network command gives you better control of what is being redistributed from the IGP into BGP, allowing the user to individually list the prefixes that need to be advertised via BGP. The maximum number of network statements that can be configured on a Cisco router is 200. If you have more than 200 networks to advertise, dynamic redistribution must be used.

The second rule is met because networks 152.1.20.0 and 152.1.11.128 are directly connected networks. Therefore, they are in the IP routing table.

To advertise network 152.1.20.16 /30 via BGP, add the following command under the BGP process. The mask must be added because the network does not fall on a major network boundary (255.0.0.0, 255.255.0.0, or 255.255.255.0). For example, the statement 152.1.0.0 is sufficient to send the prefix 152.1.0.0 /16:

```
R7(config-router)#network 152.1.20.16 255.255.255.240

R6(config-router)#network 152.1.11.128 mask 255.255.255.192
```

To verify that the network is being advertised, use the command **show ip bgp.** The following is the output from the command on R6. Notice that network 152.1.11.128 /26 is now in the BGP table.

```
R6#show ip bgp
BGP table version is 7, local router ID is 152.1.11.129
Status codes: s suppressed, d damped, h history, * valid, > best, i - internal
Origin codes: i - IGP, e - EGP, ? - incomplete

   Network          Next Hop          Metric LocPrf Weight Path
*> 148.1.0.0        152.1.1.4           0             0 500 i
*                          152.1.1.4                              0 200 500 i
*                          152.1.1.4                              0 400 500 i
*> 148.2.0.0        152.1.1.4           0             0 500 i
*                          152.1.1.4                              0 400 500 i
*                          152.1.1.4                              0 200 500 i
*> 148.3.0.0        152.1.1.4           0             0 500 i
```

```
*                        152.1.1.4                        0 200 500 i
*                        152.1.1.4                        0 400 500 i
*> 148.4.0.0        152.1.1.4          0           0 500 i
*                        152.1.1.4                        0 200 500 i
*                        152.1.1.4                        0 400 500 i
*> 152.1.11.128/26  0.0.0.0            0           32768 i
*> 152.1.20.16/28   152.1.1.3          0           0 400 i
*                        152.1.1.3                        0 200 400 i
*                        152.1.1.3                        0 500 400 i
```

40. *AS 200 should advertise local network 152.1.13.0 /24 and 152.1.14.0 / 29.*

Now from R4, you will advertise network 152.1.13.0 /24 and from R5 network 152.1.14.0 /29 via BGP:

```
R4(config-router)#network 152.1.13.0 mask 255.255.255.0
R5(config-router)#network 152.1.14.0 mask 255.255.255.248
```

41. *AS100 should only accept local routes from AS300. All other routes should be accepted via AS200.*

AS100 should only accept local routes from AS300. In order to achieve this, you must apply an inbound route filter on R2, denying any route that originated outside of AS300.

In order to filter the routes, you must first identify them. To do this, use a regular expression to identify routes based on AS path information. A regular expression is a pattern to match against an input string. When a regular expression is created it specifies the pattern that a string must match. The following is a list of keyboard characters that have special meaning when used in regular expressions.

Character	Symbol	Meaning
Period	.	Match any character, including white space.
Asterisk	*	Match zero or more sequences of the pattern.
Plus sign	+	Match one or more sequences of the pattern.
Question Mark	?	Match zero or one occurrence of the pattern.
Caret	^	Begins with.
Dollar Sign	$	Ends with.
Underscore	_	Match the following.
Brackets	[ ]	Match a single value in range.
Hyphen	-	Separates the end points of a range.

R2 prevents any network that does not originate from AS 300 from being accepted via BGP from R6. Filtering routes based on AS path information can be very useful when all routes from a particular AS need to be filtered. If filtering based on an AS path is not used, the administrator would have to list each route one by one or potentially filter on a prefix. AS path filtering provides an efficient alternative to this.

In order to filter routes based on AS path information, you must identify the AS path based on the defined regular expression and apply this to a BGP neighbor through a filter list:

a. Define the regular expression to deny any route that does not originate from AS300:

```
R2#configure terminal
R2(config)#ip as-path access-list 1 permit ^300$←Permit any route that originates from
AS 300
```

Use the command **show ip bgp regexp** to see which routes the regular expression matches. The following is the output from the command. Note that network 152.1.11.128/26 is the only route that matches the regular expression (^300$). This command is very useful in verifying that the regular expression covers the routes that you intend it to.

```
R2#show ip bgp regexp ^300$
BGP table version is 9, local router ID is 152.1.10.2
Status codes: s suppressed, d damped, h history, * valid, > best, i - internal
Origin codes: i - IGP, e - EGP, ? - incomplete

   Network          Next Hop            Metric LocPrf Weight Path
*> 152.1.11.128/26   152.1.11.2              0             0 300 i
```

b. Define a route map named **Accept_Local** that accepts partial routes from AS300. The route map will try to match any path defined by the as-path access-list 1:

```
R2(config)#route-map Accept_Local permit 10
R2(config-route-map)#match as-path 1
```

c. Apply the route map to inbound traffic from neighbor (R6) 152.1.11.2:

```
R2(config-router)#neighbor 152.1.11.2 route-map Accept_Local in
```

In order for the changes to take effect, the BGP neighbor must be reset. To do this, use the command **clear ip bgp *.** This causes the TCP session between neighbors to be reset, the neighbor negotiations to be restarted, and the cache to be invalidated.

Verify that this is working properly by displaying the BGP table on R2. The following is the output from the command. Note that the only route that R2 is receiving from AS 300 is 152.1.11.128:

```
R2#show ip bgp
BGP table version is 25, local router ID is 152.1.10.2
Status codes: s suppressed, d damped, h history, * valid, > best, i - internal
Origin codes: i - IGP, e - EGP, ? - incomplete

   Network          Next Hop            Metric LocPrf Weight Path
*>i148.1.0.0        152.1.12.10                   100      0 200 500 i
*>i148.2.0.0        152.1.12.10                   100      0 200 500 i
*>i148.3.0.0        152.1.12.10                   100      0 200 500 i
*>i148.4.0.0        152.1.12.10                   100      0 200 500 i
* i152.1.11.128/26  152.1.12.10                   100      0 200 300 i
*>                  152.1.11.2                    0        0 300 i
*>i152.1.13.0/24    152.1.12.2               0    100      0 200 i
*>i152.1.14.0/29    152.1.12.2               0    100      0 200 i
*>i152.1.20.16/28   152.1.12.10                   100      0 200 400 i
```

Verify that R3 is seeing network 152.1.11.128 /26 via R2 with the **show ip bgp command.** The following code is the output from the command. Note that R3 is not receiving network 152.1.11.128/26 from R2; the route is being learned from R5.

```
R3#show ip bgp
BGP table version is 13, local router ID is 152.1.10.3
Status codes: s suppressed, d damped, h history, * valid, > best, i - internal
Origin codes: i - IGP, e - EGP, ? - incomplete

   Network          Next Hop            Metric LocPrf Weight Path
*> 148.1.0.0        152.1.12.10                            0 200 500 i
*> 148.2.0.0        152.1.12.10                            0 200 500 i
*> 148.3.0.0        152.1.12.10                            0 200 500 i
*> 148.4.0.0        152.1.12.10                            0 200 500 i
*> 152.1.11.128/26  152.1.12.10                            0 200 300 i
*> 152.1.13.0/24    152.1.12.2                             0     0 200 i
*                   152.1.12.10                                  0 200 i
*> 152.1.14.0/29    152.1.12.2                                   0 200 i
*                   152.1.12.10                            0     0 200 i
*> 152.1.20.16/28   152.1.12.10                            0 200 400 i
```

The reason that the route to network 152.1.11.128/26 is not in the BGP table is because R3 does not know how to reach network 152.1.11.2 (that is, the next hop address). When a route is learned via BGP, the next hop address that is advertised is the IP address of the neighbor that announced the route, which in this case is R6 (152.1.11.2). When routes are injected into the AS via EBGP, the next hop learned from EBGP is carried unaltered into IBGP.

To fix this problem, you either have to advertise network 152.1.1.0 in our IGP (which is OSPF) or use the next hop self command. This command forces the router to advertise itself, rather than the external peer, as the next hop to reach the route. The following command on R2 forces the router to advertise all routes to neighbor 152.1.10.9 (R1) using its own IP address:

```
R2(config-router)#neighbor 152.1.10.9 next-hop-self
```

In order for the changes to take effect, the BGP neighbor must be reset. To do this, use the command **clear ip bgp *.** This causes the TCP session between neighbors to be reset, the neighbor negotiations to be restarted from scratch, and the cache to be invalidated.

Verify that this is working properly by displaying the BGP table on R3. The following is the output from the command. Note that R3 now sees the network 152.1.11.128/26 via both R5 and R2:

```
R3#show ip bgp
BGP table version is 12, local router ID is 152.1.10.3
Status codes: s suppressed, d damped, h history, * valid, > best, i - internal
Origin codes: i - IGP, e - EGP, ? - incomplete

   Network          Next Hop          Metric LocPrf Weight Path
*> 148.1.0.0        152.1.12.10                          0       200 500 i
*> 148.2.0.0        152.1.12.10                          0       200 500 i
*> 148.3.0.0        152.1.12.10                          0       200 500 i
*> 148.4.0.0        152.1.12.10                          0       200 500 i
*  152.1.11.128/26  152.1.12.10                          0       200 300 i
*>i                 152.1.10.10            0     100     0       300 i
*  152.1.13.0/24    152.1.12.10                          0       200 i
*>                  152.1.12.2       0                   0       200 i
*  152.1.14.0/29    152.1.12.10                          0       200 i
*>                  152.1.12.2                           0       200 i
*> 152.1.20.16/28   152.1.12.10                          0       200 400 i
```

42. *AS100 should accept a default route from AS300 in case of a link failure between A200 and AS100. The default should not be advertised outside of AS100.*

Cisco provides a way to target a default route to a specific peer. This prevents R6 from advertising the default to all peers. The following command on R6 only advertises a default route to neighbor (R2) 152.1.11.1:

```
R6(config)#router bgp 300
R6(config-router)#neighbor 152.1.11.1 default-originate
```

In order for the changes to take effect, the BGP neighbor must be reset. To do this, use the command **clear ip bgp *.** This causes the TCP session between neighbors to be reset, the neighbor negotiations to be restarted from scratch, and the cache to be invalidated.

You must also prevent R3 from advertising the route to AS200, so apply a filter list on R3.

In order to filter routes based on a network address, you must identify network addresses through the use of an access list and apply that list to a BGP neighbor using a distribute list:

a. Define the access list on R3 to deny network 0.0.0.0:

```
R3(config)#access-list 2 deny 0.0.0.0
R3(config)#access-list 2 permit any
```

b. Apply the distribution list to both BGP neighbors:

```
R3(config)#router bgp 100
R3(config-router)#neighbor 152.1.12.2 filter-list 2 out
R3(config-router)#neighbor 152.1.12.10 filter-list 2 out
```

In order for the changes to take effect, the BGP neighbor's must be reset. To do this, use the command **clear ip bgp *.** This causes the TCP session between neighbors to be reset, the neighbor negotiations to be restarted from scratch, and the cache to be invalidated.

```
RouterA#clear ip bgp *
```

Display the BGP table on R5 and R4 with the command **show ip bgp**. The following is the output from the command on R5. Notice that the route to network 0.0.0.0 is no longer in the BGP table.

```
R5#show ip bgp
BGP table version is 15, local router ID is 152.1.14.1
Status codes: s suppressed, d damped, h history, * valid, > best, i - internal
Origin codes: i - IGP, e - EGP, ? - incomplete

   Network          Next Hop            Metric LocPrf Weight Path
*  148.1.0.0        152.1.1.4                              0 400 500 i
*>                  152.1.1.4                0             0 500 i
*                   152.1.1.4                              0 300 500 i
*  148.2.0.0        152.1.1.4                              0 400 500 i
*>                  152.1.1.4                0             0 500 i
*                   152.1.1.4                              0 300 500 i
*  148.3.0.0        152.1.1.4                              0 400 500 i
*>                  152.1.1.4                0             0 500 i
*                   152.1.1.4                              0 300 500 i
*  148.4.0.0        152.1.1.4                              0 400 500 i
*>                  152.1.1.4                0             0 500 i
*                   152.1.1.4                              0 300 500 i
*  152.1.11.128/26  152.1.12.9                             0 100 300 i
*                   152.1.1.2                              0 400 300 i
*                   152.1.1.2                              0 500 300 i
*>                  152.1.1.2                0             0 300 i
*>i152.1.13.0/24    152.1.12.13              0    100      0 i
*> 152.1.14.0/29     0.0.0.0                 0         32768 i
*> 152.1.20.16/28   152.1.1.3                0             0 400 i
*                   152.1.1.3                              0 500 400 i
*                   152.1.1.3                              0 300 400 i
```

43. *Advertise network 152.1.9.0 via IBGP on R1 and R2. Do not use the redistribute command.*

Routes can be injected dynamically or statically into BGP. Because the question states that the redistribute command cannot be used, you will need to statically inject routes into BGP. To do this, use the network command, which is a way to individually list prefixes that need to be sent in BGP.

To advertise network 152.1.9.0 /24 via BGP, add the following command under the BGP process. The mask must be added because the network does not fall on a major network boundary (255.0.0.0, 255.255.0.0, or 255.255.255.0). For example, the statement 152.1.0.0 would be sufficient to send the prefix 152.1.0.0 /16:

```
R1(config)#router bgp 100
R1(config-router)#network 152.1.9.0 mask 255.255.255.0
```

44. *To prevent AS 100 from being used as a transit network, AS300 should only accept local networks from AS100. To accomplish this, don't add any configurations to any router in AS 100*

.

In order to prevent AS100 from being used as a transit network, either block the sending or receiving of routes that do not originate in AS100. Because the question specifies that you can only add configuration commands to AS300, you must block the receiving of updates from R2 that do not originate in AS100. To do this, apply an inbound filter on BGP updates received from R2.

R6 prevents any network that does not originate on AS 100 from being received from AS100. Filtering routes based on AS path information can be very useful when all routes from a particular AS need to be filtered. If filtering based on an AS path is not used, the administrator would have to list each route one by one or potentially filter on a prefix. AS path filtering provides an efficient alternative to this.

In order to filter routes based on AS path information, you must identify the AS path based on the defined regular expression and apply this to a BGP neighbor through a filter list.

a. Define the regular expression to permit any route that originates from AS100:

```
R6#configure terminal
R6(config)#ip as-path access-list 2 permit ^100$←Permit only local routes from AS100
```

Use the **show ip bgp regexp** command to see which routes the regular expression matches. The following is the output from the command. Note that network 152.1.9.0/24 is the only route that matches the regular expression (^100$). This command is very useful in verifying that the regular expression covers the routes that you intend it to cover.

```
R6#show ip bgp regexp ^100$
BGP table version is 16, local router ID is 152.1.11.129
Status codes: s suppressed, d damped, h history, * valid, > best, i - internal
Origin codes: i - IGP, e - EGP, ? - incomplete

   Network          Next Hop          Metric LocPrf Weight Path
*> 152.1.9.0/24      152.1.11.1                             0 100 i
```

b. Define a route map named **Receive_Local** that permits only local networks routes from AS100. The route map will try to match any path defined by the as-path access-list 2:

```
R6(config)#route-map Receive_Local permit 10
R6(config-route-map)#match as-path 2
```

c. Apply the route map to inbound traffic from neighbor (R2) 152.1.11.1:

```
R2(config-router)#neighbor 152.1.11.1 route-map Receive_Local in
```

In order for the changes to take effect, the BGP neighbor must be reset. To do this, use the command **clear ip bgp *.** This causes the TCP session between neighbors to be reset. This also restarts the neighbor negotiations from scratch and invalidates the cache.

To quickly determine which routes are being learned via AS100, use the command **show ip bgp regexp 100.** This command shows any network that has AS100 in its path. The following is the output from the command. Note that only network 152.1.9.0 /24 is being advertised via AS100.

```
R6#show ip bgp regexp 100
BGP table version is 17, local router ID is 152.1.11.129
Status codes: s suppressed, d damped, h history, * valid, > best, i - internal
Origin codes: i - IGP, e - EGP, ? - incomplete

   Network          Next Hop            Metric LocPrf Weight Path
*  152.1.9.0/24     152.1.1.1                              0 200 100 100 100 i
*                   152.1.1.1                              0 500 200 100 100 100 i

*>                  152.1.11.1                             0 100 i
*                   152.1.1.1                              0 400 200 100 100 100 i
```

45. *AS100 prefers that network 152.1.9.0 be reachable by the outside world via the link between AS100 and AS300.*

For this question, R3 must be configured so that when it advertises network 152.1.9.0 /24 to AS200 it will not be as preferable as the route through R2. In order to do this, you must first understand how BGP selects its path.

BGP uses a set of parameters (attributes) that describe the characteristics of a route. The attributes are sent in the BGP update packets with each route. The router then uses these attributes to select the best route to the destination.

It is important to understand the BGP decision process in order to be able to correctly manipulate path selection. The following is the order of the BGP decision process used by the router in path selection:

1. If the next hop is unreachable, do not consider it.
2. Prefer the path that has the largest weight.
3. If the routes have the same weight, use the route with the highest local preference.
4. If the routes have the same local preference, prefer the route that was originated by BGP on this router.
5. If no route was originated, prefer the route with the shortest AS path.

6. If all paths are of the same AS length, prefer the route with lowest origin code (IGP < EGP < INCOMPLETE).
7. If the origin codes are the same, prefer the path with the lowest *Multi-Exit Discriminator* (MED).
8. If the MEDs are the same, prefer external paths over internal paths.
9. If still the same, prefer the path through the closest IGP neighbor.
10. If still the same, prefer the path with the lowest BGP Router ID

For this question, you will prepend an additional AS number to network 152.1.9.0 /24 before R3 advertises it to AS 200. This makes the AS_path length for this network shorter via AS300 (rule number 5)

In order to manipulate the AS path information, you need to identify which routes will be manipulated through the use of an access list, define a policy that will be applied to those routes through a route map, and then assign the route map to a BGP neighbor:

a. Add access-list 3 to R3, permitting network 152.1.9.0 /24:

```
R3#configure terminal
R3(config)#access-list 3 permit 152.1.9.0 0.0.0.255
```

b. Define a route map named **Add_Path** that prepends two additional AS path numbers (AS100 and AS100) to the route if it matches access list 3:

```
R3#configure terminal
R3(config)#route-map Add_Path permit 10
R3(config-route-map)#match ip address 3
R3(config-route-map)#set as-path prepend 100 100
R3(config)#route-map Add_Path permit 20
R3(config-route-map)#set as-path prepend
```

NOTE: *Instance 20 of the route map is added so that the other routes will be distributed . When an update does not meet the criteria of a route map instance, BGP applies the next instance. If the update does not meet any criteria, the update is not redistributed or controlled. So if instance 20 is left out, only network 152.1.9.0 would be advertised.*

Apply the route map to outbound routing updates to BGP neighbor 152.1.12.2 (R4) and neighbor 152.1.12.10 (R5):

```
R3#configure terminal
R3(config)#router bgp 100
R3(config-router)#neighbor 152.1.12.10 route-map Add_Path out
R3(config-router)#neighbor 152.1.12.2 route-map Add_Path out
```

In order for the changes to take effect, the BGP neighbors must be reset. To do this, use the command **clear ip bgp *.** This causes the TCP session between neighbors to be reset, the neighbor negotiations to be restarted from scratch, and the cache to be invalidated:

```
R3#clear ip bgp *
```

Display the BGP table entry for network 152.1.9.0 on R4 with the command **show ip bgp 152.1.9.0**. The following is the output from the command. Notice that the best route to network 152.1.9.0 is now via AS300.

```
R5#show ip bgp 152.1.9.0
BGP routing table entry for 152.1.9.0/24 , version 37
Paths: (4 available, best #2, advertised over IBGP, EBGP)
  500 300 100
    152.1.1.2 from 152.1.1.4 (148.4.0.1)
      Origin IGP, valid, external
  300 100
    152.1.1.2 from 152.1.1.2 (152.1.11.129)
      Origin IGP, valid, external, best
  400 300 100
    152.1.1.2 from 152.1.1.3 (152.1.20.17)
      Origin IGP, valid, external
  100 100 100
    152.1.12.9 from 152.1.12.9 (152.1.10.3)
      Origin IGP, valid, external
```

46. *Under no circumstances can AS100 be used as a transit network for AS200 to reach AS300. To accomplish this, no additional configurations should be made to any router except R2.*

The tricky part of this question is how do you prevent routes from being advertised to AS 200 without additional configurations. Because you already have an outbound filter in place, you simply need to remove instance 20 of route map Add_Path. It is added is to permit the rest of the networks to be advertised. By default, when an update does not meet the criteria of a route map instance, BGP applies the next instance. If the update does not meet any criteria, the update is not redistributed or controlled. So if instance 20 is left out, only network 152.1.9.0 would be advertised, which is exactly what you want. So simply remove instance 20 of the route map (no additional configuration changes are made):

```
R3(config)#no route-map Add_Path permit 20
```

In order for the changes to take effect, the BGP neighbor must be reset. To do this, use the command **clear ip bgp *.** This causes the TCP session between neighbors to be reset, the neighbor negotiations to be restarted, and the cache to be invalidated.

To quickly determine which routes are being learned via AS100, use the command **show ip bgp regexp 100** on R5. This command shows any network that has AS100 in its path. The following is the output from the command. Note that only network 152.1.9.0 /24 is being advertised via AS100.

```
R5#show ip bgp regexp 100
BGP table version is 26, local router ID is 152.1.14.1
Status codes: s suppressed, d damped, h history, * valid, > best, i - internal
Origin codes: i - IGP, e - EGP, ? - incomplete

   Network          Next Hop            Metric LocPrf Weight Path
* i152.1.9.0/24     152.1.12.1                    100      0 100 100 100 i
*>                  152.1.12.9                             0 100 100 100 i
```

47. *Network 152.1.9.0 should not be passed by AS200 or AS300. No additional configuration should be used to achieve this outside of AS100.*

In the past examples, you used filter lists and route maps to prevent updates from being advertised. Because this question requires that no changes be made outside of AS100, you must use community attributes.

To prevent network 152.1.9.0 /24 from being advertised outside AS200 and AS300, you need to use the no-export community value. A community is a group of destinations that shares some common property. They are used to simplify routing policies by identifying routes based on a logical property, rather than an IP prefix or AS number.

The community attribute (type code 8) is an optional transitive attribute that varies in length and consists of a set of four-byte values. Communities in the range of 0x00000000 through 0x0000FFFF and 0xFFFF0000 through 0xFFFFFFFF are reserved. These are well-known communities, which means they have global significance. The no-export community is an example of a well-known community. When a route is carrying this community, it is not advertised to peers outside the confederation. If only one AS is in the confederation, then it is not advertised outside the AS. By default, each AS is a confenderation.

In order to prevent routes from being advertised outside of an AS using the no-export community, you first need to identify the prefix using an access list, define the community that is assigned to that prefix with a route map, and apply the route map to a neighbor.

a. Define an access list to permit prefix 152.1.9.0 /24. This is already done on R3, so you only need to do this on R2:

```
R2(config)#access-list 3 permit 152.1.9.0 0.0.0.255
```

b. A route map named Set_Community is defined to match on prefix 152.1.9.0/24 and set its community attribute to no-export:

```
R3(config)#route-map Set_Community permit 10
R3(config-route-map)#match ip address 3
R3(config-route-map)#set community no-export

R2(config)#route-map Set_Community permit 10
R2(config-route-map)#match ip address 3
R2(config-route-map)#set community no-export
```

c. The last step is to apply the route map to a neighbor. The **send-community** keyword must be assigned to a neighbor session in order to enable the community attribute to be sent to a specified neighbor.

```
R2(config-router)#neighbor 152.1.11.2 route-map Set_Community out
R2(config-router)#neighbor 152.1.11.2 send-community

R3(config-router)#neighbor 152.1.12.10 route-map Set_Community out
R3(config-router)#neighbor 152.1.12.2 route-map Set_Community out
R3(config-router)#neighbor 152.1.12.10 send-community
R3(config-router)#neighbor 152.1.12.2 send-community
```

Use the **show ip bgp community no-export** on R5 to verify that network 152.1.9.0/24 has the community attribute set. The following is the output from the command:

```
R5#show ip bgp community no-export
BGP table version is 92, local router ID is 152.1.14.1
Status codes: s suppressed, d damped, h history, * valid, > best, i - internal
Origin codes: i - IGP, e - EGP, ? - incomplete

   Network          Next Hop            Metric LocPrf Weight Path
*> 152.1.9.0/24      152.1.12.9                            0 100 i
```

48. *R3 should use R5 to reach network 152.1.14.0 and R4 to reach network 152.1.13.0. No additional configuration should be used to achieve this outside of AS200.*

Display the BGP table entry for network 152.1.14.0 /29 on R3. The following is the output from the command. Note the best path is via R4 (152.1.12.2).

```
R3#Show ip bgp 152.1.14.0
BGP routing table entry for 152.1.14.0/29 , version 41
Paths: (2 available, best a1, advertised over IBGP)
  200
    152.1.12.2 from 152.1.12.2 (152.1.13.1)
      Origin IGP, valid, external, best
  200
    152.1.12.10 from 152.1.12.10 (152.1.14.1)
      Origin IGP, metric 0, valid, external
```

Display the BGP table entry for network 152.1.13.0 on R3. The following is the output from the command. Note the best path is via R4 (152.1.12.2).

```
R3#Show ip bgp 152.1.13.0
BGP routing table entry for 152.1.13.0/24 , version 12
Paths: (2 available, best a1, advertised over IBGP)
```

```
200
  152.1.12.2 from 152.1.12.2 (152.1.13.1)
    Origin IGP, metric 0, valid, external, best
200
  152.1.12.10 from 152.1.12.10 (152.1.14.1)
    Origin IGP, valid, external
```

NOTE: The reason that R4 is preferred over R5 is because the router ID is lower, and all other factors are the same. The router uses the route learned from the router with the lowest router ID, which in this case is R4.

Set the MED for prefix 152.1.14.0 advertised from R4 to 50 and prefix 152.1.13.0 to 100. On R5, set the MED for prefix 152.1.14.0 to 100 and prefix 152.1.13.0 to 50. The lower the MED, the more preferred the route.

The MED attribute is the external metric of a route. Unlike the local preference attribute, the MED is exchanged between ASs, but the MED that comes into an AS does not leave.

In order to manipulate the MED, you must identify which networks will be manipulated through the use of an access list, define a policy that will be applied to those routes through a route map, and then assign the route map to a BGP neighbor:

a. Add access-list 1 to R4 and R5, permitting network 152.1.14.0 /29:

```
R4(config)#access-list 1 permit 152.1.14.0 0.0.0.7
R5(config)#access-list 1 permit 152.1.14.0 0.0.0.7
```

Also add access-list 2 to R4 and R5, permitting network 152.1.13.0 /24:

```
R4(config)#access-list 2 permit 152.1.13.0 0.0.0.255
R5(config)#access-list 2 permit 152.1.13.0 0.0.0.255
```

b. Define a route map on R4, named Set_Med. The route map sets the MED attribute for network 152.1.14.0 to 50 and sets the MED attribute of network 152.1.13.0 to 100.

Define a route map on R5, named Set_Med. The first route map sets the MED attribute for network 152.1.13.0 to 50 and the MED attribute of network 152.1.12.0 to 100.

```
R5(config)#route-map Set_Med permit 10
R5(config-route-map)#match ip address 1
R5(config-route-map)#set metric 50
R5(config-route-map)#exit
R5(config)#route-map Set_Med permit 20
R5(config-route-map)#match ip address 2
R5(config-route-map)#set metric 100
R5(config-route-map)#exit
R5(config)#route-map Set_Med permit 30
R5(config-route-map)#set metric
```

```
R4(config)#route-map Set_Med permit 10
R4(config-route-map)#match ip address 1
R4(config-route-map)#set metric 100
R4(config-route-map)#exit
R4(config)#route-map Set_Med permit 20
R4(config-route-map)#match ip address 2
R4(config-route-map)#set metric 50
R4(config-route-map)#exit
R4(config)#route-map Set_Med permit 30
R4(config-route-map)#set metric
```

Apply the route map **Set_Med** on outbound routing updates to R3:

```
R5(config-router)#neighbor 152.1.12.9 route-map Set_Med out
R4(config-router)#neighbor 152.1.12.1 route-map Set_Med out
```

In order for the changes to take effect, the BGP neighbor's must be reset. To do this, use the command **clear ip bgp *.** This causes the TCP session between neighbors to be reset, the neighbor negotiations to be restarted from scratch, and the cache to be invalidated.

49. *R7 should originate and advertise network 148.0.0.0, 148.5.0.0, 148.6.0.0, and 148.7.0.0. Add four loopback interfaces to R7: Loopback 2 IP address 148.0.0.1/16 , Loopback 3 IP address 148.5.0.1/16 , Loopback 4 IP address 148.6.0.1/16 , and Loopback 5 IP address 148.7.0.1 /16.*

The first part of the question requires the user to create four loopback interfaces and assign IP addresses:

```
R7(config)#int loopback 2
R7(config-if)#ip address 148.0.0.1 255.255.0.0
R7(config)#int loopback 3
R7(config-if)#ip address 148.5.0.1 255.255.0.0
R7(config)#int loopback 4
R7(config-if)#ip address 148.6.0.1 255.255.0.0
R7(config)#int loopback 5
R7(config-if)#ip address 148.7.0.1 255.255.0.0
```

In order to advertise a route via BGP, two conditions must be met:

- The BGP process must be aware of the route, either through the use of the network command or by redistribution.
- The network to be advertised must be present in the IP routing table.

The second condition is met because the routes are directly connected. To meet the first condition, use the network command under BGP:

```
R7(config)#router bgp 400
R7(config-router)#network 148.0.0.0
R7(config-router)#network 148.5.0.0
R7(config-router)#network 148.6.0.0
R7(config-router)#network 148.7.0.0
```

50. *R5 should advertise as few prefixes for 148.x.0.0 as needed.*

All the 148.x.0.0 prefixes can be advertised as an aggregate using the aggregate address command. This router configuration command creates an aggregate entry in the BGP routing table if any more specific BGP routes are available that fall in the aggregate range. In this example, the router had routes to networks 148.0.0.0 /16, 148.1.0.0 /16, 148.2.0.0 /16, 148.3.0.0 /16, 148.4.0.0 /16, 148.5.0.0 /16, 148.6.0.0 /16, and 148.7.0.0 /16. All these networks could be advertised in one aggregate route 148.0.0.0/13 .

Figure 2-10 shows the concept of route aggregation, also referred to as supernetting. Eight Class B networks , each using a standard 16-bit mask, are shown.

- 148.0.0.0 /16
- 148.1.0.0 /16
- 148.2.0.0 /16
- 148.3.0.0 /16
- 148.4.0.0 /16
- 148.5.0.0 /16
- 148.6.0.0 /16
- 148.7.0.0 /16

Without supernetting, a router advertising these eight networks would need to send a separate route update for each one. Supernetting enables a router to advertise more than one network with a single advertisement. In the case of your eight networks, a router can advertise the 148.0.0.0 network with a 13-bit mask. Notice that the default 16-bit mask has been shortened to 13 bits. Supernetting works by reducing the number of subnet bits in a routing advertisement. This has the effect of matching several networks with a single routing update.

In Figure 2-9, all eight networks have an exact match for their first 13 bits. The remaining three bits of the original 16-bit mask are now part of the supernet. Notice that the .0, .1, .2, .3, .4, .5, and .7 networks use all eight combinations of the three remaining bits. Thus, a 13-bit subnet mask can be used to advertise eight Class B networks with a single advertisement of 148.0.0.0/13 .

If the aggregate command is used with no additional arguments, the aggregate is advertised as originating from the aggregating router and has the atomic aggregate attribute set to show that information might be missing. The more specific routes will also be advertised along with the aggregate. This command has several optional arguments:

as-set The as-set keyword creates an aggregate entry with a path that consists of all the elements contained in all the paths that are being summarized.

Figure 2-9
Supernetting

First 13 Bits Match	
148.0.0.0	**10010100.00000**000.00000000.00000000
148.1.0.0	**10010100.00000**001.00000000.00000000
148.2.0.0	**10010100.00000**010.00000000.00000000
148.3.0.0	**10010100.00000**011.00000000.00000000
148.4.0.0	**10010100.00000**100.00000000.00000000
148.5.0.0	**10010100.00000**101.00000000.00000000
148.6.0.0	**10010100.00000**110.00000000.00000000
148.7.0.0	**10010100.00000**111.00000000.00000000
255.248.0.0 Supernet Mask	**11111111.11111**000.00000000.00000000

summary-only The summary-only keyword suppresses advertisements of more specific routes to all neighbors.

suppress-map The suppress-map keyword creates the aggregate route but suppresses advertisements of specified routes. A route map can be used to selectively suppress some more specific routes of the aggregate and allow others.

For this question, you want to send only the summary. The following command aggregates the addresses:

```
R5(config-router)# aggregate-address 148.0.0.0 255.248.0.0  summary-only
```

51. *R3 is running an IGP (EIGRP) on the private link between it and AS400. Make sure that only the IGP route to network 152.1.20.16 is favored over the EBGP route.*

Network 152.1.20.16 is learned via both EIGRP and BGP. Because multiple routing protocols are present on most networks, there needs to be a way to select which route to use if a network is announced via multiple sources. Cisco uses a parameter called the administrative distance of a protocol to determine which one is used to route a packet.

The following is a list of routing protocols supported on a Cisco router and the default administrative distance of each one. The lower the distance, the more preferred the routing source.

The table above indicates that a directly connected route is generally preferred over a static route, which is preferred over an EBGP route, and so on. EBGP has a distance of 20, which is preferred over all IGP routes.

Protocol	Administrative Distance
Directly connected	0
Static	1
EBGP	20
EIGRP	90
IGRP	100
OSPF	110
ISIS	115
RIP	120
EGP	140
EIGRP (External)	170
IBGP	200
BGP Local	200
Unknown	255

Cisco provides a way to force IGP routes to take precedence over EBGP routes. The concept is called backdoor links. EBGP routes can be tagged as backdoor routes, which set the distances of these routes to the same as BGP local or 200. Because the distance is higher than the IGP route, the backdoor IGP route is preferred.

For this question, you will tag network 152.1.20.16 as a backdoor route, in which case R3 will prefer the route learned via EIGRP.

Display the routing table on R3. The following code is the output, and note that network 152.1.120.16 is in the routing table as an EBGP-learned route.

```
R3#show ip route
Codes: C - connected, S - static, I - IGRP, R - RIP, M - mobile, B - BGP
       D - EIGRP, EX - EIGRP external, O - OSPF, IA - OSPF inter area
       E1 - OSPF external type 1, E2 - OSPF external type 2, E - EGP
       i - IS-IS, L1 - IS-IS level-1, L2 - IS-IS level-2, * - candidate default
       U - per-user static route

Gateway of last resort is not set

     152.1.0.0/16  is variably subnetted, 14 subnets, 6 masks
B    152.1.11.128/26  [20/0 ] via 152.1.12.10, 00:21:27
D    152.1.20.7/32  [90/2297856 ] via 152.1.20.2, 02:00:30, Serial0/0
C    152.1.20.0/30  is directly connected, Serial0/0
O IA 152.1.9.0/24  [110/18 ] via 152.1.8.1, 00:16:29, Ethernet0/0
```

```
C     152.1.10.3/32  is directly connected, Loopback0
C     152.1.8.0/26  is directly connected, Ethernet0/0
O IA 152.1.10.1/32  [110/11 ] via 152.1.8.1, 00:16:29, Ethernet0/0
B     152.1.13.0/24  [20/50 ] via 152.1.12.2, 01:59:39
C     152.1.12.0/30  is directly connected, Serial1/0
B     152.1.14.0/29  [20/50 ] via 152.1.12.10, 01:59:43
D     152.1.1.0/29  [90/2195456 ] via 152.1.20.2, 02:00:31, Serial0/0
C     152.1.10.8/29  is directly connected, Serial0/1
B     152.1.20.16/28  [20/0 ] via 152.1.12.10, 01:57:07
C     152.1.12.8/30  is directly connected, Serial1/1
B     148.0.0.0/13  [20/0 ] via 152.1.12.10, 01:57:12
```

To tag the network prefix as a backdoor route, perform the following on R3:

```
R3(config)#router bgp 100
R3(config-router)#network 152.1.20.16 mask 255.255.255.240 backdoor
```

In order for the changes to take effect, the BGP neighbor must be reset. To do this, use the command **clear ip bgp *.** This causes the TCP session between neighbors to be reset, the neighbor negotiations to be restarted from scratch, and the cache to be invalidated.

Display the routing table on R3. The following code is the output. Note that network 152.1.120.16 is now in the routing table as an EIGRP learned route.

```
R3#show ip route
Codes: C - connected, S - static, I - IGRP, R - RIP, M - mobile, B - BGP
       D - EIGRP, EX - EIGRP external, O - OSPF, IA - OSPF inter area
       E1 - OSPF external type 1, E2 - OSPF external type 2, E - EGP
       i - IS-IS, L1 - IS-IS level-1, L2 - IS-IS level-2, * - candidate default
       U - per-user static route

Gateway of last resort is not set

     152.1.0.0/16  is variably subnetted, 14 subnets, 6 masks
B     152.1.11.128/26  [20/0 ] via 152.1.12.10, 00:01:59
D     152.1.20.7/32  [90/2297856 ] via 152.1.20.2, 02:07:46, Serial0/0
C     152.1.20.0/30  is directly connected, Serial0/0
O IA 152.1.9.0/24  [110/18 ] via 152.1.8.1, 00:23:44, Ethernet0/0
C     152.1.10.3/32  is directly connected, Loopback0
C     152.1.8.0/26  is directly connected, Ethernet0/0
O IA 152.1.10.1/32  [110/11 ] via 152.1.8.1, 00:23:44, Ethernet0/0
B     152.1.13.0/24  [20/50 ] via 152.1.12.2, 00:02:05
C     152.1.12.0/30  is directly connected, Serial1/0
B     152.1.14.0/29  [20/50 ] via 152.1.12.10, 00:02:00
D     152.1.1.0/29  [90/2195456 ] via 152.1.20.2, 02:07:46, Serial0/0
C     152.1.10.8/29  is directly connected, Serial0/1
D     152.1.20.16/28  [90/2297856 ] via 152.1.20.2, 00:03:06, Serial0/0
C     152.1.12.8/30  is directly connected, Serial1/1
B     148.0.0.0/13  [20/0 ] via 152.1.12.10, 00:02:01
```

52. *The configuration on R3 is growing extremely large and approaching the limitation of NVRAM. Reduce the size of the configuration on R3.*

When the NVRAM is not large enough to store the router configuration, Cisco provides an option that enables the configuration to be compressed. This option should only be used if the configuration file is bigger than the memory available in NVRAM. To compress the file, perform the following command:

```
R3(config)#service compress-config
```

53. *Configure R3 so that informational messages are logged to the internal buffer. Make sure the router can store up to 16K of information and the information is time-stamped.*

Message logging to the console port is enabled by default. In order to send messages to any other destination, you must configure the router.

To disable message logging, use the **no logging on** command. Enabling the logging process can slow down the router because a process must wait until the messages are written to the console before continuing.

To enable message logging to the buffer, perform the following task in global configuration mode:

```
R3(config)#logging buffered 16000
```

The logging buffered command copies logging messages to an internal buffer. The buffer is circular, so newer messages overwrite older messages after the buffer is full. The IOS enables the user to configure the buffer size, but the buffer size should not be too large because the router could run out of memory for other tasks.

*NOTE: The free processor memory can be viewed on the router with the **show memory EXEC** command. This number represents the maximum memory available and should not be changed.*

In order to correlate events, it is helpful to time-stamp each message with the date and time that it occurred. By default, log messages are not time-stamped. You can enable time-stamping of log messages by performing the following task in global configuration mode:

```
R3(config)#service timestamps log datetime localtime
```

54. *Configure R3 so that SNMP information can only be read by host 192.1.1.1.*

SNMP is made up of three components: an SNMP manager, an SNMP agent, and a *Management Information Base* (MIB). SNMP provides a message format for sending and receiving information between the manager and the agent. The manager is typically part of a NMS such as HP Open view or Netview 6K. The agent and MIB reside on the router or device that will be managed.

The agent contains the MIB variables, which the SNMP manager can request or change. The manager can request a value from an agent (Read\Get) or change a value on an agent (Write\Set).

The question specifies that you only want to permit server 192.1.1.1 SNMP read-access.

The first step is to specify a community string and permit read-only access. The community string is used as a password to permit access to the agent on the router. Optionally, the MIB view can be set to define a subset of MIBs that the given community can access. By default, the community will have access to all MIBs.

To enable SNMP read-only access to community cisco on R3, perform the following command:

```
R3(config)#snmp-server community cisco ro
```

Because you only want to permit access from server 192.1.1.1, you need to define an access list and apply it to the SNMP community string. Define access list 99 to permit IP address 192.1.1.1:

```
R3(config)#access-list 99 permit 192.1.1.1 0.0.0.0
```

Apply the access list to the community string:

```
R3(config)#snmp-server community cisco ro 99
```

Using the snmp-server community command, you specify a string and, optionally, a MIB view and an access list. The string is used as a password. The MIB view defines the subset of all MIB objects that the given community may access. The access list identifies the IP addresses of systems with SNMPv1 managers that might use the community string to gain access to the SNMPv1 agent.

55. *Network 152.1.13.0 is suspected of being unstable, causing BGP UPDATE and WITHDRAWN messages to be repeatedly propagated on the network. Configure R3 so that if the network continues to FLAP it will be withdrawn.*

Cisco provides a mechanism to control route instability or flapping, with a feature called route dampening. A route that is unstable, frequently going up and down, causes BGP UPDATE and WITHDRAWN messages to be repeatedly propagated on the network. This routing traffic can quickly use up link bandwidth and *central processing unit* (CPU) cycles on the router.

The router monitors a route and categorizes it as either well-behaved or ill-behaved. The recent history of the route is used to estimate future stability. That is, a route that has gone up and down two times in the last three minutes is ill-behaved and will be penalized in proportion to the expected future instability. Each time the route flaps, it is given a penalty. When the penalty reaches a predefined threshold, the route is suppressed. The more frequently a route flaps in a period of time, the faster it will be suppressed.

Once the route is suppressed, the router continues to monitor its stability. An algorithm is put in place to reduce (decay) the penalty exponentially. Once the route is deemed stable, it is then advertised.

The following set of terms and parameters applies to Cisco's implementation of route dampening:

Penalty An incremented numeric number that is assigned to the route each time it flaps.

Half-lifetime A configurable numeric value that describes the amount of time that must pass to reduce the penalty by half.

Suppress limit A numeric value that is compared with the penalty. If the penalty is greater than the suppress limit, the route is suppressed.

Reuse limit A numeric value that is compared with the penalty. If the penalty is less than the reuse limit, a suppressed route will no longer be suppressed.

Suppressed route A route that is not advertised even though it is up. If the penalty is greater than the suppress limit, the route is suppressed.

History entry An entry used to store flap information.

To apply route dampening to a specific network, use a route map to identify the match criteria and then apply the dampening using the route map.

The first step is to define an access list to identify the network address 152.1.13.0:

```
R3(config)#access-list 5 permit 152.1.13.0 0.0.0.255
```

Next, define a route-map named dampen_152_1_13_0 and apply the dampening parameters:

```
R3(config)#route-map dampen_152_1_13_0 permit 10
R3(config-route-map)#set dampening 20 950 2500 80
```

This first parameter sets the time that must pass before the penalty of a suppressed route is reduced by half (half-life) to 20 minutes. The next number sets the limit, after which a suppressed route will be readvertised (reuse limit) to 950. The next number sets the limit at which a route will be suppressed to 2500. If the penalty exceeds 2500, the route will be suppressed. The last value is the maximum time in minutes that the route can be suppressed for (max-suppress-time), which in this case is 80 minutes.

The final step is to apply the route map to the BGP process:

```
R3(config)#router bgp 100
R3(config-router)#bgp dampening route-map dampen_152_1_13_0
```

In order for the changes to take effect, the BGP neighbor must be reset. To do this, use the command **clear ip bgp ***. This causes the TCP session between neighbors to be reset, the neighbor negotiations to be restarted from scratch, and the cache to be invalidated.

Let's take a look and see what happens when the network 152.1.13.0 flaps. Shut down interface loopback 1 on R4, which will cause the route flap for network 152.1.13.0.

The following output shows how R3 treats the flapping route 152.1.13.0. Note that the route has flapped twice, because the router counts a flap as any time the path information changes for a route. Because the route went up and down, this is treated as two flaps.

```
R3#show ip bgp 152.1.13.0
BGP routing table entry for 152.1.13.0/24 , version 8
Paths: (1 available, best a1)
  200
    152.1.12.2 from 152.1.12.2 (152.1.13.1)
      Origin IGP, metric 50, valid, external, best
      Dampinfo: penalty 1472, flapped 2 times in 00:00:49
```

You can cause the network to flap again by shutting down the loopback interface on R4. The following output shows the route after four flaps. The penalty is now 2652, which is higher than the 2500 limit. The route has been suppressed and will not be passed on. The route will be unusable for 29 minutes and 40 seconds, at which time, if no other flaps occur, the penalty will be decayed to the reuse limit of 950.

```
R3#show ip bgp 152.1.13.0
BGP routing table entry for 152.1.13.0/24 , version 9
Paths: (1 available, no best path)
  200, (suppressed due to dampening)
    152.1.12.2 from 152.1.12.2 (152.1.13.1)
      Origin IGP, metric 50, valid, external
      Dampinfo: penalty 2652, flapped 4 times in 00:07:32, reuse in 00:29:40
```

56. *Enable R3 so that when its neighbors make a BGP change, the change will take effect without resetting the BGP TCP session. R3 should be able to initiate the change locally; no commands should have to be entered on the neighboring routers.*

One requirement of BGP is resetting the neighbors' TCP connection in order for a new policy to take effect. BGP soft configuration enables policies to be configured and activated without resetting the BGP and TCP sessions. This enables the new policy to take effect without significantly affecting the network. Without BGP soft configuration, BGP is required to reset the neighbor TCP connection in order for the new changes to take effect. This is accomplished using the clear ip bgp * command, which is used throughout this chapter.

Two types of BGP soft reconfiguration exist: outbound reconfiguration, which will make the new local outbound policy take effect without resetting the BGP session, and inbound soft reconfiguration, which enables the new inbound policy to take effect.

This question requires that you use inbound soft reconfiguration on R1. To accomplish this, an additional router command needs to be added before a soft reconfiguration can be issued. The reason is that this command tells the router to start storing the received updates.

Add the following configurations to the BGP process on R1:

```
R1(config)#router bgp 100
R1(config-router)#neighbor 152.1.10.10 soft-reconfiguration inbound
R1(config-router)#neighbor 152.1.10.11 soft-reconfiguration inbound
```

Inbound soft reconfiguration can than be triggered with the command:

```
clear ip bgp [*|address | peer-group ] [soft in]
```

The problem with inbound reconfiguration is that in order to generate new inbound updates without resetting the BGP session, all inbound updates (whether accepted or rejected) need to be stored by the router. This is memory-intensive and wherever possible it should be avoided.

To avoid the memory overhead needed for inbound soft reconfiguration, the same outcome could be achieved by doing an outbound soft reconfiguration at the other end of the connection.

An outbound soft reconfiguration can be triggered with the command:

```
clear ip bgp [*|address | peer-group ] [soft out]
```

Complete Network Diagram

Table 2-2 shows the IP address worksheet for this Case Study. Figure 2-9 shows the complete network diagram with IP addressing.

Table 2-2

IP Address Worksheet

IP Address	Mask	Subnet	Broadcast Address	Router	Interface
148.0.0.1	255.255.0.0	148.0.0.0		R7	L2
148.1.0.1	255.255.0.0	148.1.0.0		R8	L0
148.2.0.1	255.255.0.0	148.2.0.0		R8	L1
148.3.0.1	255.255.0.0	148.3.0.0		R8	L3
148.4.0.1	255.255.0.0	148.4.0.0		R8	L4
148.5.0.1	255.255.0.0	148.5.0.0		R7	L3
148.6.0.1	255.255.0.0	148.6.0.0		R7	L4
148.7.0.1	255.255.0.0	148.7.0.0		R7	L5
152.1.1.1	255.255.255.248	152.1.1.0	152.1.1.7	R5	E0
152.1.1.2	255.255.255.248	152.1.1.0	152.1.1.7	R6	E0
152.1.1.3	255.255.255.248	152.1.1.0	152.1.1.7	R7	E0
152.1.1.4	255.255.255.248	152.1.1.0	152.1.1.7	R8	E0
152.1.8.1	255.255.255.192	152.1.8.0	152.1.8.63	R1	E0/0
152.1.8.62	255.255.255.192	152.1.8.0	152.1.8.63	R3	E0/0
152.1.9.1	255.255.255.0	152.1.9.0	152.1.9.255	R1	E0/1
152.1.9.254	255.255.255.0	152.1.9.0	152.1.9.255	R2	E0
152.1.10.1	255.255.255.255	152.1.10.1		R1	L0
152.1.10.2	255.255.255.255	152.1.10.2		R2	L0
152.1.10.3	255.255.255.255	152.1.10.3		R3	L0
152.1.10.9	255.255.255.248	152.1.10.8	152.1.10.15	R1	S0/0.1
152.1.10.10	255.255.255.248	152.1.10.8	152.1.10.15	R2	S0
152.1.10.11	255.255.255.248	152.1.10.8	152.1.10.15	R3	S0/1
152.1.11.1	255.255.255.252	152.1.11.0	152.1.11.3	R2	S1
152.1.11.2	255.255.255.252	152.1.11.0	152.1.11.3	R6	S0
152.1.11.6	255.255.255.255	152.1.11.6		R6	L0
152.1.11.129	255.255.255.192	152.1.11.128	152.1.11.191	R6	L1
152.1.12.1	255.255.255.252	152.1.12.0	152.1.12.3	R3	S1/0
152.1.12.2	255.255.255.252	152.1.12.0	152.1.12.3	R4	S0
152.1.12.4	255.255.255.255	152.1.12.4		R4	L0
152.1.12.5	255.255.255.255	152.1.12.5		R5	L0
152.1.12.9	255.255.255.252	152.1.12.8	152.1.12.11	R3	S1/1
152.1.12.10	255.255.255.252	152.1.12.8	152.1.12.11	R5	S0
152.1.12.13	255.255.255.252	152.1.12.12	152.1.12.15	R4	S1
152.1.12.14	255.255.255.252	152.1.12.12	152.1.12.15	R5	S1
152.1.13.1	255.255.255.0	152.1.13.0	152.1.13.255	R4	L0
152.1.14.1	255.255.255.248	152.1.14.0	152.1.14.8	R5	L0
152.1.20.1	255.255.255.252	152.1.20.0	152.1.20.3	R3	S0/0
152.1.00.2	255.255.255.252	152.1.20.0	152.1.50.3	R7	S0
152.1.20.7	255.255.255.255	152.1.20.7		R7	L0
152.1.20.17	255.255.255.240	152.1.20.16	152.1.20.31	R7	L1

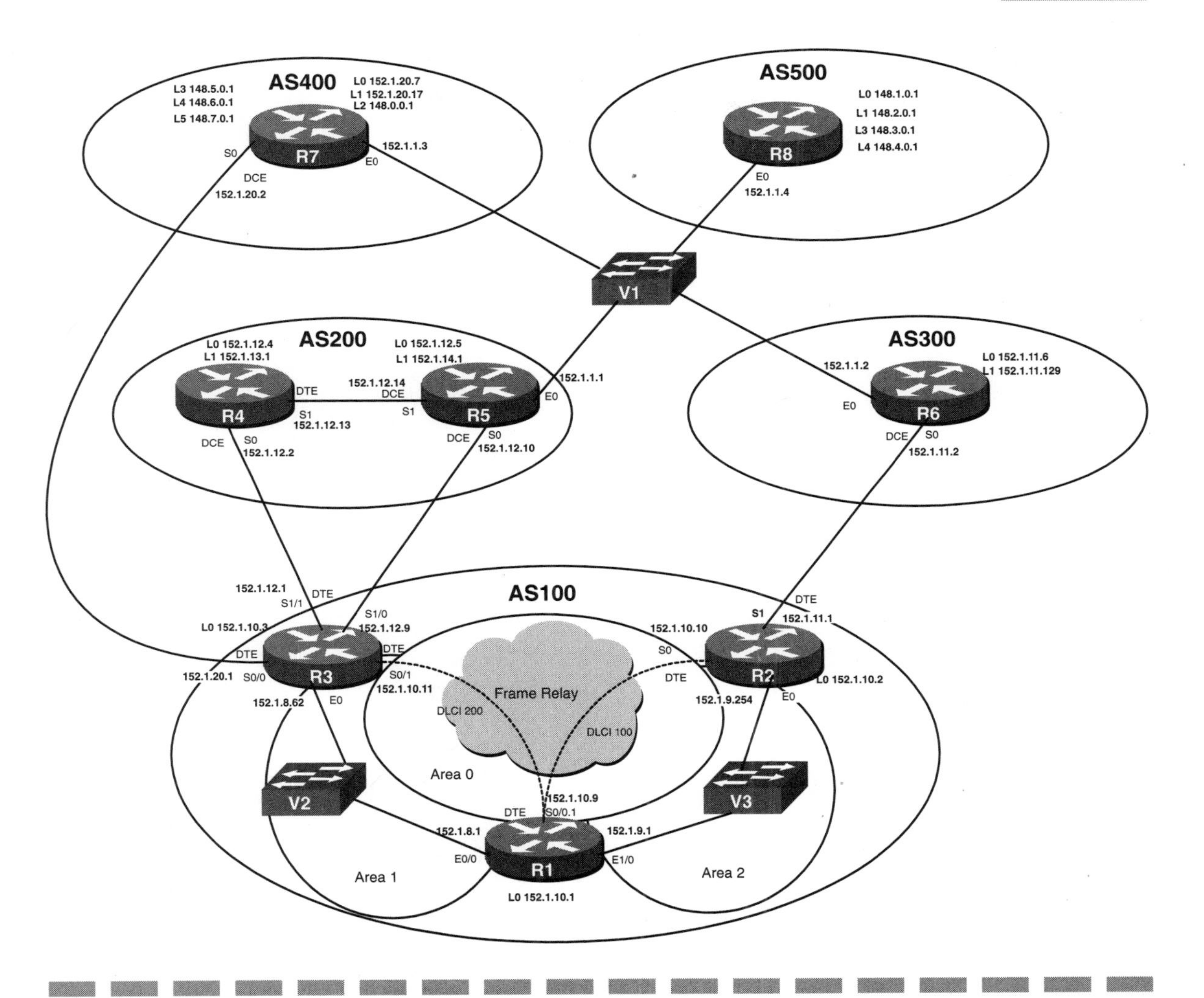

Figure 2-9
Network diagram with IP addressing

Router Configurations

R1

```
!
version 11.2
no service password-encryption
no service udp-small-servers
no service tcp-small-servers
!
hostname R1
!
!
!
interface Loopback0
 ip address 152.1.10.1 255.255.255.255
!
interface Ethernet0/0
 mac-address 0001.0100.0002
 ip address 152.1.8.1 255.255.255.192
 ip ospf priority 2
!
interface Serial0/0
 no ip address
 encapsulation frame-relay IETF
 no fair-queue
 frame-relay lmi-type ansi
!
interface Serial0/0 .1 multipoint
 ip address 152.1.10.9 255.255.255.248
 ip ospf cost 9
 ip ospf hello-interval 120
 ip ospf priority 255
 frame-relay map ip 152.1.10.10 100 broadcast
 frame-relay map ip 152.1.10.11 200 broadcast
 no frame-relay inverse-arp
!
interface Ethernet1/0
 ip address 152.1.9.1 255.255.255.0
 ip ospf cost 8
!
router ospf 64
 passive-interface Loopback0
 network 152.1.10.8 0.0.0.7 area 0
 network 152.1.8.0 0.0.0.63 area 1
 network 152.1.9.0 0.0.0.255 area 2
 network 152.1.10.1 0.0.0.0 area 0
 neighbor 152.1.10.11
 neighbor 152.1.10.10
!
router bgp 100
 no synchronization
 network 152.1.9.0 mask 255.255.255.0
 neighbor 152.1.10.10 remote-as 100
 neighbor 152.1.10.10 route-reflector-client
```

```
 neighbor 152.1.10.10 soft-reconfiguration inbound
 neighbor 152.1.10.11 remote-as 100
 neighbor 152.1.10.11 route-reflector-client
 neighbor 152.1.10.11 soft-reconfiguration inbound
!
no ip classless
!
!
line con 0
line aux 0
line vty 0 4
 login
!
end
```

R2

```
!
version 11.2
no service password-encryption
no service udp-small-servers
no service tcp-small-servers
!
hostname R2
!
interface Loopback0
 ip address 152.1.10.2 255.255.255.255
!
interface Ethernet0
 ip address 152.1.9.254 255.255.255.0
 ip ospf priority 2
!
interface Serial0
 ip address 152.1.10.10 255.255.255.248
 encapsulation frame-relay IETF
 ip ospf hello-interval 120
 ip ospf priority 0
 frame-relay map ip 152.1.10.9 100 broadcast
 frame-relay map ip 152.1.10.11 100 broadcast
 no frame-relay inverse-arp
 frame-relay lmi-type ansi
!
interface Serial1
 ip address 152.1.11.1 255.255.255.252
!
router ospf 64
 passive-interface Loopback0
 network 152.1.10.8 0.0.0.7 area 0
 network 152.1.9.0 0.0.0.255 area 2
 network 152.1.10.2 0.0.0.0 area 0
!
router bgp 100
 no synchronization
 neighbor 152.1.10.9 remote-as 100
 neighbor 152.1.10.9 next-hop-self
```

```
 neighbor 152.1.11.2 remote-as 300
 neighbor 152.1.11.2 send-community
 neighbor 152.1.11.2 route-map Accept_Local in
 neighbor 152.1.11.2 route-map Set_Community out
!
no ip classless
ip as-path access-list 1 permit ^300$
ip as-path access-list 2 permit ^$
access-list 3 permit 152.1.9.0 0.0.0.255
route-map Send_Local permit 10
 match as-path 2
!
route-map Accept_Local permit 10
 match as-path 1
!
route-map Set_Community permit 10
 match ip address 3
 set community no-export
!
!
line con 0
line 1 16
line aux 0
line vty 0 4
 login
!
end
```

R3

```
!
version 11.1
service timestamps log datetime localtime
service compress-config
service udp-small-servers
service tcp-small-servers
!
hostname R3
!
interface Loopback0
 ip address 152.1.10.3 255.255.255.255
!
!
interface Ethernet0/0
 ip address 152.1.8.62 255.255.255.192
!
interface Serial0/0
 ip address 152.1.20.1 255.255.255.252
!
interface Serial0/1
 ip address 152.1.10.11 255.255.255.248
 encapsulation frame-relay IETF
 ip ospf hello-interval 120
 ip ospf priority 0
 no frame-relay inverse-arp
```

```
 frame-relay lmi-type ansi
 frame-relay map ip 152.1.10.9 200 broadcast
 frame-relay map ip 152.1.10.10 200 broadcast
!
interface Serial1/0
 ip address 152.1.12.1 255.255.255.252
!
interface Serial1/1
 ip address 152.1.12.9 255.255.255.252
!
router eigrp 64
 redistribute ospf 64 metric 1500 100 255 1 1500
 passive-interface Ethernet0/0
 passive-interface Serial0/1
 passive-interface Serial1/0
 passive-interface Serial1/1
 passive-interface Loopback0
 network 152.1.0.0
 distribute-list 1 out eigrp 64
!
router ospf 64
 redistribute eigrp 64 metric 64 subnets
 passive-interface Loopback0
 network 152.1.10.8 0.0.0.7 area 0
 network 152.1.8.0 0.0.0.63 area 1
 network 152.1.10.3 0.0.0.0 area 0
!
router bgp 100
 no synchronization
 bgp dampening route-map dampen_152_1_13_0
 network 152.1.20.16 mask 255.255.255.240 backdoor
 neighbor 152.1.10.9 remote-as 100
 neighbor 152.1.12.2 remote-as 200
 neighbor 152.1.12.2 send-community
 neighbor 152.1.12.2 distribute-list 2 out
 neighbor 152.1.12.2 route-map Set_Community out
 neighbor 152.1.12.10 remote-as 200
 neighbor 152.1.12.10 send-community
 neighbor 152.1.12.10 distribute-list 2 out
 neighbor 152.1.12.10 route-map Set_Community out
!
no ip classless
logging buffered 16000
access-list 1 deny   152.1.20.7
access-list 1 deny   152.1.1.0 0.0.0.7
access-list 1 deny   152.1.20.16 0.0.0.15
access-list 1 permit any
access-list 2 deny   0.0.0.0
access-list 2 permit any
access-list 3 permit 152.1.9.0 0.0.0.255
access-list 5 permit 152.1.13.0 0.0.0.255
access-list 99 permit 192.1.1.1
route-map dampen_152_1_13_0 permit 10
 set dampening 20 950 2500 80
!
route-map Add_Path permit 10
 match ip address 3
 set as-path prepend 100 100
!
```

```
route-map Set_Community permit 10
 match ip address 3
 set community no-export
!
snmp-server community cisco RO 99
!
line con 0
line aux 0
line vty 0 4
 login
!
end
```

R4

```
!
version 11.2
service udp-small-servers
service tcp-small-servers
!
hostname R4
!
!
!
interface Loopback0
 ip address 152.1.12.4 255.255.255.255
!
interface Loopback1
 ip address 152.1.13.1 255.255.255.0
!
!
interface Serial0
 ip address 152.1.12.2 255.255.255.252
 clockrate 1000000
!
interface Serial1
 ip address 152.1.12.13 255.255.255.252
!
!
router eigrp 56
 passive-interface Serial0
 network 152.1.0.0
!
router bgp 200
 no synchronization
 network 152.1.13.0 mask 255.255.255.0
 neighbor 152.1.12.1 remote-as 100
 neighbor 152.1.12.1 route-map Set_Med out
 neighbor 152.1.12.14 remote-as 200
!
no ip classless
access-list 1 permit 152.1.14.0 0.0.0.7
access-list 2 permit 152.1.13.0 0.0.0.255
```

```
route-map Set_Med permit 10
 match ip address 1
 set metric 100
!
route-map Set_Med permit 20
 match ip address 2
 set metric 50
!
route-map Set_Med permit 30
!
!
line con 0
line aux 0
line vty 0 4
 login
!
end
```

R5

```
!
version 11.0
service udp-small-servers
service tcp-small-servers
!
hostname R5
!
!
!
interface Loopback0
 ip address 152.1.12.5 255.255.255.255
!
interface Loopback1
 ip address 152.1.14.1 255.255.255.248
!
interface Ethernet0
 mac-address 0005.0200.0000
 ip address 152.1.1.1 255.255.255.248
!
interface Serial0
 ip address 152.1.12.10 255.255.255.252
 clockrate 64000
!
interface Serial1
 ip address 152.1.12.14 255.255.255.252
 clockrate 64000
!
router eigrp 56
 passive-interface Ethernet0
 passive-interface Serial0
 network 152.1.0.0
!
router bgp 200
```

```
 no synchronization
 network 152.1.14.0 mask 255.255.255.248
 aggregate-address 148.0.0.0 255.248.0.0 summary-only
 neighbor 152.1.1.2 remote-as 300
 neighbor 152.1.1.3 remote-as 400
 neighbor 152.1.1.4 remote-as 500
 neighbor 152.1.12.9 remote-as 100
 neighbor 152.1.12.9 route-map Set_Med out
 neighbor 152.1.12.13 remote-as 200
!
access-list 1 permit 152.1.14.0 0.0.0.7
access-list 2 permit 152.1.13.0 0.0.0.255
route-map Set_Med permit 10
 match ip address 1
 set metric 50
!
route-map Set_Med permit 20
 match ip address 2
 set metric 100
!
route-map Set_Med permit 30
!
!
line con 0
 ip netmask-format bit-count
line 1 16
 transport input all
line aux 0
 transport input all
line vty 0 4
 login
!
end
```

R6

```
!
version 11.0
service udp-small-servers
service tcp-small-servers
!
hostname R6
!
enable password cisco
!
!
interface Loopback0
 ip address 152.1.11.6 255.255.255.255
!
interface Loopback1
 ip address 152.1.11.129 255.255.255.192
!
interface Ethernet0
 mac-address 0006.0300.0000
```

```
 ip address 152.1.1.2 255.255.255.248
!
interface Serial0
 ip address 152.1.11.2 255.255.255.252
 clockrate 64000
!
interface Serial1
 no ip address
 shutdown
!
router bgp 300
 network 152.1.11.128 mask 255.255.255.192
 neighbor 152.1.1.1 remote-as 200
 neighbor 152.1.1.3 remote-as 400
 neighbor 152.1.1.4 remote-as 500
 neighbor 152.1.11.1 remote-as 100
 neighbor 152.1.11.1 default-originate
 neighbor 152.1.11.1 route-map Receive_Local in
!
ip as-path access-list 2 permit ^100$
route-map Receive_Local permit 10
 match as-path 2
!
!
line con 0
line aux 0
 transport input all
 ip netmask-format bit-count
line vty 0 4
 exec-timeout 30 0
 password cisco
 login
!
end
```

R7

```
!
version 11.2
no service udp-small-servers
no service tcp-small-servers
!
hostname R7
!
enable password cisco
!
ip host r1 2001 152.1.1.3
ip host r2 2002 152.1.1.3
ip host r4 2004 152.1.1.3
ip host r5 2005 152.1.1.3
ip host r6 2006 152.1.1.3
ip host r7 2007 152.1.1.3
ip host r8 2008 152.1.1.3
ip host cat 2009 152.1.1.3
```

```
!
interface Loopback0
 ip address 152.1.20.7 255.255.255.255
!
interface Loopback1
 ip address 152.1.20.17 255.255.255.240
!
interface Loopback2
 ip address 148.0.0.1 255.255.0.0
!
interface Loopback3
 ip address 148.5.0.1 255.255.0.0
!
interface Loopback4
 ip address 148.6.0.1 255.255.0.0
!
interface Loopback5
 ip address 148.7.0.1 255.255.0.0
!
interface Ethernet0
 mac-address 0007.0300.0000
 ip address 152.1.1.3 255.255.255.248
!
interface Serial0
 ip address 152.1.20.2 255.255.255.252
 no fair-queue
 clockrate 64000
!
router eigrp 64
 passive-interface Ethernet0
 network 152.1.0.0
 distribute-list 1 out eigrp 64
!
router bgp 400
 network 152.1.20.16 mask 255.255.255.240
 network 148.0.0.0
 network 148.5.0.0
 network 148.7.0.0
 network 148.6.0.0
 neighbor 152.1.1.1 remote-as 200
 neighbor 152.1.1.2 remote-as 300
 neighbor 152.1.1.4 remote-as 500
!
no ip classless
access-list 1 permit 152.1.20.7
access-list 1 permit 152.1.1.0 0.0.0.7
access-list 1 permit 152.1.20.16 0.0.0.15
access-list 1 deny   any
!
line con 0
line 1 16
 no exec
 transport input all
line aux 0
line vty 0 4
 password cisco
 login
!
end
```

Frame Switch

```
!
version 11.1
service slave-log
service udp-small-servers
service tcp-small-servers
!
hostname R8
!
!
frame-relay switching
!
interface Loopback0
 ip address 148.1.0.1 255.255.0.0
!
interface Loopback1
 ip address 148.2.0.1 255.255.0.0
!
interface Loopback3
 ip address 148.3.0.1 255.255.0.0
!
interface Loopback4
 ip address 148.4.0.1 255.255.0.0
!
interface Ethernet0
 mac-address 0008.0500.0000
 ip address 152.1.1.4 255.255.255.248
 media-type 10BaseT
!
!
interface Serial0
 description Frame Relay connection to R1 S0
 no ip address
 encapsulation frame-relay IETF
 no fair-queue
 clockrate 64000
 frame-relay lmi-type ansi
 frame-relay intf-type dce
 frame-relay route 100 interface Serial1 100
 frame-relay route 200 interface Serial2 200
!
interface Serial1
 description Frame Relay connection to R2 s0
 no ip address
 encapsulation frame-relay IETF
 clockrate 64000
 frame-relay lmi-type ansi
 frame-relay intf-type dce
 frame-relay route 100 interface Serial0 100
!
interface Serial2
 description Frame Relay connection to R3 s 1
 no ip address
 encapsulation frame-relay IETF
 clockrate 64000
 frame-relay lmi-type ansi
 frame-relay intf-type dce
```

```
 frame-relay route 200 interface Serial0 200
!
!
!
!
!
!
router bgp 500
 network 148.2.0.0
 network 148.3.0.0
 network 148.4.0.0
 network 148.1.0.0
 neighbor 152.1.1.1 remote-as 200
 neighbor 152.1.1.2 remote-as 300
 neighbor 152.1.1.3 remote-as 400
!
no ip classless
!
line con 0
line aux 0
line vty 0 4
 login
!
end
```

BGP Technology Appendix

BGP Message Format

All BGP messages use the same fixed size header, which is shown here.

0	15	23	31
MARKER			
MARKER			
MARKER			
MARKER			
LENGTH		TYPE	

Marker This is a 16-byte field that is used to either authenticate incoming BGP messages or to detect a loss of synchronization between BGP peers. If the type of message is OPEN or if the OPEN message carries no authentication information, then the marker is all ones. Otherwise, the marker field is computed based on the authentication being used.

Length The length field is a two-byte field that indicates the total length of the message. The smallest permitted length is 19 bytes and the largest is 4,096.
Type: The type field is a one-byte field that indicates the type of BGP message. Four BGP message types are available:

- OPEN
- UPDATE
- KEEPALIVE
- NOTIFICATION

OPEN Message Format

BGP can not exchange routing information until the neighbor negotiation is complete. After the transport connection is established, the first message is an OPEN message. This message contains information on the BGP version number, the AS number, the hold time, the BGP identifier, and other optional parameters. If the peers disagree on any of the parameters, a notification error is sent and the peer connection does not get established.

If the OPEN message is acceptable, meaning the peer router agrees on the parameters, then a KEEPALIVE message is sent back to confirm the OPEN message.

In addition to the fixed-size BGP header, the OPEN message contains the following fields:

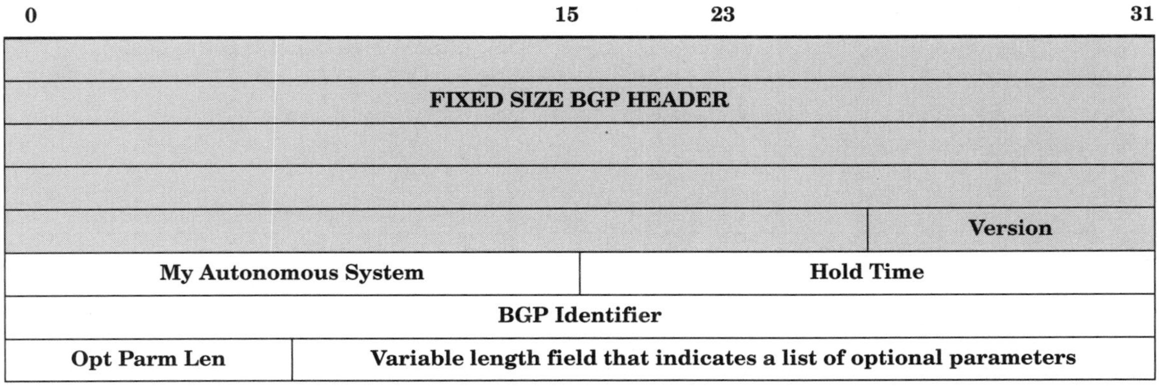

0	15	23	31
FIXED SIZE BGP HEADER			
			Version
My Autonomous System		Hold Time	
BGP Identifier			
Opt Parm Len	Variable length field that indicates a list of optional parameters		

Version The version field is a one-byte field that indicates the version of the BGP protocol. During the neighbor negotiation, peer routers agree on the BGP version number to be used. The highest version that both routers support is usually used.

My Autonomous System This field is two bytes long and indicates the AS number of the sending router.

Hold Time This field is two bytes long and indicates the maximum time in seconds that may elapse between the receipt of a successive KEEPALIVE and/or UPDATE message. If the hold time is exceeded, the neighbor is considered dead.

The hold time is negotiated between neighbors and is set to the lowest value. The router that receives the open message must calculate the hold time by using the smallest of its configured hold times and the hold time received in the open message.

BGP Identifier This field is four bytes long and indicates the BGP identifier of the sending router. This is the router ID, which is the highest loopback address or the highest IP address on the router at BGP session startup.

Optional Parameter Length This field is one byte long and indicates the total length in bytes of the optional parameter field. If no optional parameters are present, the field is set to 0.

Optional Parameters This is a variable-length field that indicates the list of optional parameters used in BGP neighbor session negotiation.

UPDATE Message Format

The UPDATE message is the primary message used to communicate information between BGP peers. When a BGP speaker advertises or withdraws a route from a peer router, an UPDATE message is used. The UPDATE message always includes the fixed-length BGP header and can optionally include the following:

Field	Length
Unfeasible Routes Length	Two bytes
Withdrawn Routes	Variable
Total Path Attribute Length	Two bytes
Path Attributes	Variable
Network Layer Reachability Information	Variable

Unfeasible Routes Length This two-byte field indicates the total length in octets of the Withdrawn Routes field. If the value is 0, it indicates that no routes are being withdrawn from service.

Withdrawn Routes This variable-length field contains a list of IP address prefixes of the routes that are being withdrawn from service.

Total Path Attribute Length This two-byte field indicates the total length in octets of the Path Attributes field.

Path Attributes This variable-length field contains a list of BGP attributes associated with the prefixes in the Network Layer Reachability field. The path attributes give information on the prefixes that are being advertised, such as the degree of preference or the origin of the prefix. This information is used for filtering and in the route decision process. The path attributes fall into four categories:

- ***Well-Known Mandatory:*** An attribute that has to exist in the BGP update and must be recognized by all BGP vendor implementations. ORIGIN, AS_PATH, and NEXT_HOP are three examples of well-know mandatory attributes.
  - **ORIGIN:** The ORIGIN attribute is an example of a well-known mandatory attribute that indicates the origin of the routing update with respect to the AS that originated it. This attribute tells how the original route was put into the BGP table. A route could be learned via an IGP such as OSPF, which was redistributed into BGP, learned through an external routing protocol (such as EGP), or learned through something other than an IGP or EGP, such as a static route.

    Three origins are possible: IGP, EGP, or INCOMPLETE. The router uses this information in its decision process when choosing between multiple routes. The router prefers the path with the lowest ORIGIN type, with an IGP lower than EGP, and with an EGP that is lower than INCOMPLETE.
  - **AS_PATH:** AS_PATH is a well-known mandatory attribute that indicates the ASs through which routing information contained in this UPDATE message has passed. In Figure 2-11, prefix 1.0.0.0 /8 is advertised to AS 300 and AS 200. When AS 300 passes the prefix to AS 200, it appends its AS number to the AS_PATH. When AS 200 receives the UPDATE from AS300, it knows that 1.0.0.0 originated in AS100 and then passed through AS300.
  - **NEXT_HOP:** This attribute defines the IP address of the border router that should be used as the next hop to the destinations listed in the UPDATE message. For example, in Figure 2-12, RouterB learns route 1.0.0.0 /8 via 192.1.1.1, which is the IP address of its EBGP peer. When RouterB advertises this route to RouterC, it includes the next hop information unaltered. RouterC then receives an IBGP update for network 1.0.0.0 /8 with a next hop of 192.1.1.1.
- ***Well-know discretionary***: An attribute that must be recognized by all BGP implementations but may or may not be sent in a BGP update.
  - **LOCAL_PREF:** The local preference attribute is a degree of preference given to a route to compare it with other routes to the same destination. The highest local preference is the

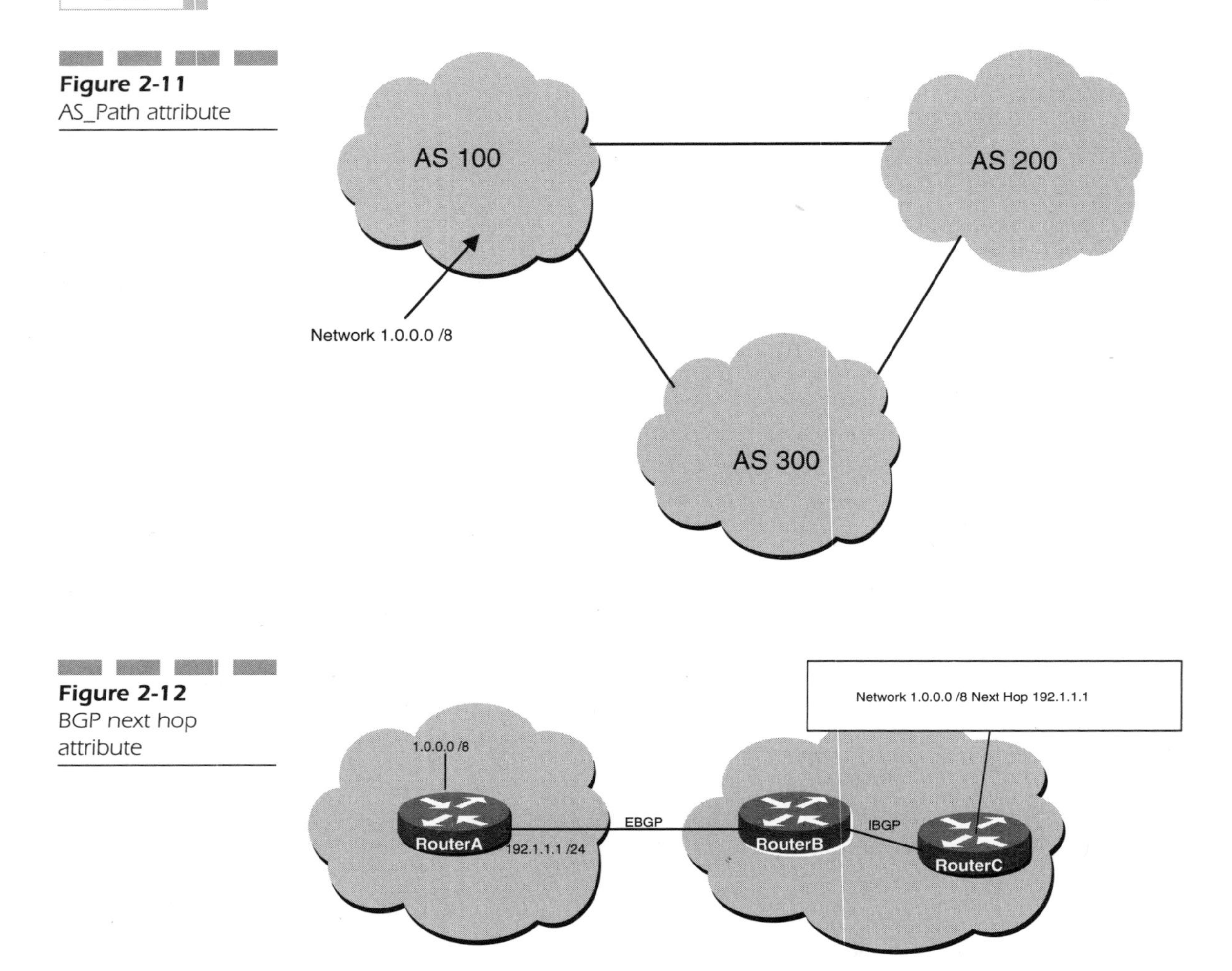

Figure 2-11 AS_Path attribute

Figure 2-12 BGP next hop attribute

most preferred route. The local preference is not included in update messages that are sent to BGP neighbors outside of the AS. If the attribute is contained in an update received from a BGP neighbor in a different AS, it is ignored.

In Figure 2-13, you see where the local preference attribute would be used. RouterB and RouterC are both advertising network 1.0.0.0 /8 into AS 400, but because the link between RouterB and RouterD is a high-speed link, you would like traffic destined for network 1.0.0.0 /8 to use this route. The local preference attribute will be used to manipulate the flow of traffic within AS 400. RouterD gives routes coming from RouterB a local preference of 150, and RouterE gives routes coming from RouterC a local preference of 100. Because RouterD and RouterE are exchanging routes via IBGP, they both will use the route with the highest local preference. In this case, all traffic

Figure 2-13
BGP local preference attribute

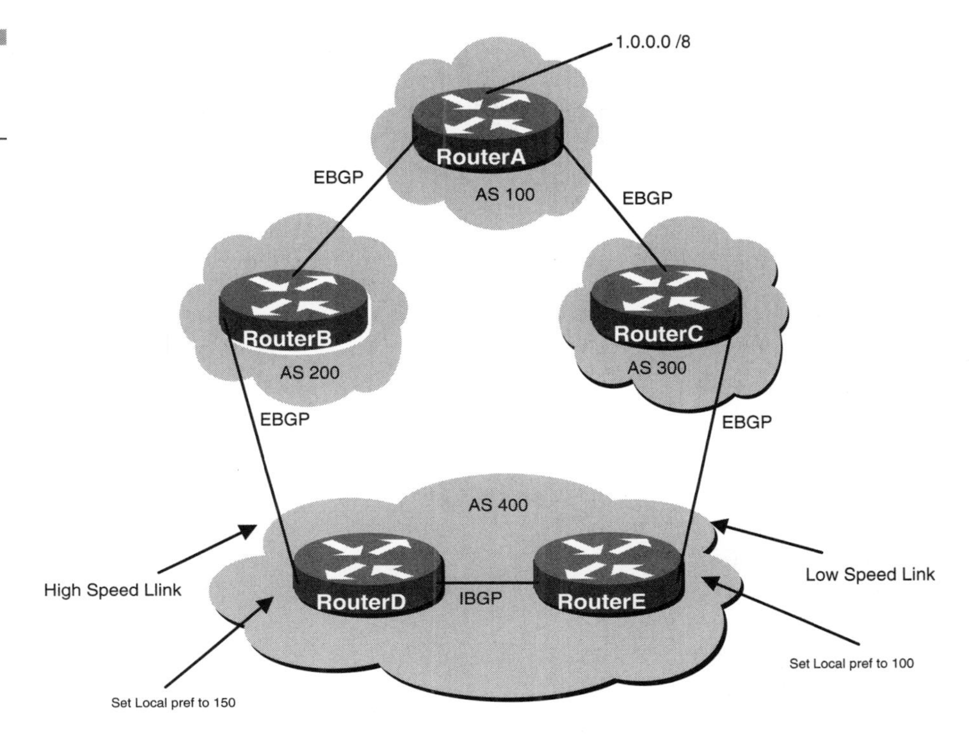

destined for network 1.0.0.0 /8 will be routed over the high-speed link between RouterB and RouterD.

- **ATOMIC_AGGREGATE:** The atomic aggregate attribute indicates that information has been lost. When routes are aggregated, this causes a loss of information because the aggregate is coming from different sources that have different attributes. If a router sends an aggregate that causes a loss of information, it is required to attach the atomic aggregate attribute to the route.

- ***Optional Transitive***: An optional transitive attribute is not required to be supported by all BGP implementations. However, if it is not recognized by the BGP process, it will look at the transitive flag. If the flag is set, the BGP implementation should accept the attribute and pass it along to other BGP speakers.
  - **AGGREGATOR:** This attribute identifies the BGP speaker (IP address) and AS number that performed the route aggregation.

- ***Optional nontransitive:*** An optional nontransitive attribute is not required to be supported by all BGP implementations. However, if the attribute is not recognized by the BGP process, it will look at the transitive flag. If the flag is not set, the attribute should be quietly ignored and not passed along to other BGP peers.

- **MULTI_EXIT_DISC (MED):** MED is used by the BGP speakers to discriminate between multiple exit points to a neighboring AS. A lower MED is preferred over a higher one. The MED attribute is exchanged between ASs, but a MED attribute that comes into an AS does not leave the AS (nontransitive). This differs from the local preference attribute. External routers can influence the path selection of another AS, whereas with local preference you can only influence the route selection within your own AS.

  In Figure 2-14, the MED attribute is used to influence the path that RouterA uses to reach network 1.0.0.0 /8. Both RouterB and RouterC are advertising network 1.0.0.0 /8 to RouterA. However, because RouterC is closer to this network, you would like all traffic destined to network 1.0.0.0 /8 from RouterA to be routed through RouterC. To achieve this, set the MED on route 1.0.0.0 /8 advertised from RouterC to 100 and the MED on route 1.0.0.0 /8 advertised from RouterB to 200. Because the MED coming from RouterC is lower, RouterA will prefer this route.
- **Network Layer Reachability:** This variable-length field contains a list of IP address prefixes that are reachable via the sender. All path attributes contained in a given update message apply to the destinations carried in the Network Layer Reachablity Information field of the UPDATE message.

Figure 2-14
BGP MED

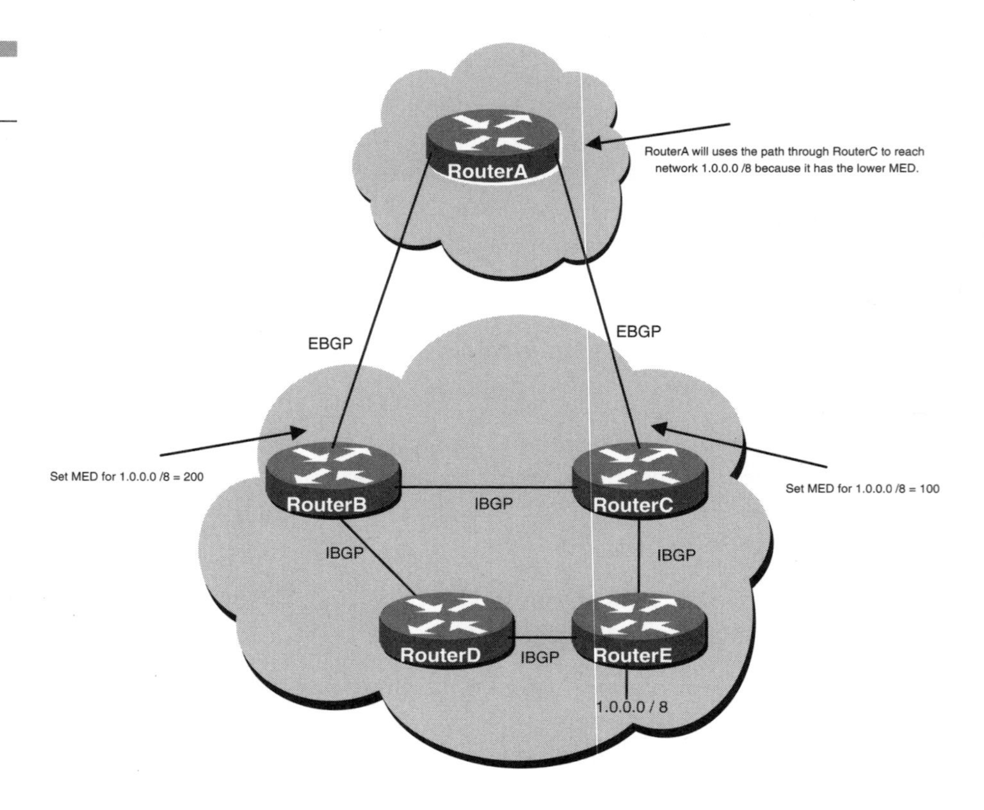

KEEPALIVE Message Format

KEEPALIVE messages are exchanged periodically between peers to determine whether peers are reachable. These messages are exchanged between peers often enough to prevent the hold timer from expiring.

NOTIFICATION Message Format

A NOTIFICATION message is sent whenever an error condition is detected. The BGP connection is closed immediately after sending it. In addition to the fixed-size BGP header, the NOTIFICATION message contains the following fields:

Field	Length
Error code	One byte
Error Subcode	One byte
Data	Variable

Error Code This one-byte field indicates the type of notification. The following are the possible types:

Error Code	Symbolic Name
1	Message Header Error
2	OPEN Message Error
3	UPDATE Message Error
4	Hold Timer Expired
5	Finite State Machine Error
6	Cease

Error Subcode This one-byte field provides more specific information about the nature of the reported error. Each error code may have one or more error subcodes associated with it. The following is a list of error subcodes:

Data This variable length field is used to diagnose the reason for the notification. The contents of the Data field depend upon the error code and error subcode.

Error Code	Error Subcode
1- Message Header Error	—Connection Not Synchronized —Bad Message Length — Bad Message Type
2- OPEN Message Error Subcodes	—Unsupported Version Number —Bad Peer AS —Bad BGP Identifier —Unsupported Optional Parameter —Authentication Failure —Unacceptable Hold Time
3-UPDATE Message Error Subcodes	—Malformed Attribute List —Unrecognized Well-Known Attribute —Missing Well-Known Attribute —Attribute Flags Error —Attribute Length Error —Invalid ORIGIN Attribute —AS Routing Loop —Invalid NEXT_HOP Attribute —Optional Attribute Error —Invalid Network Field —Malformed AS_Path
4- Hold Timer Expired	Not applicable
5- Finite State Machine Error	Not applicable
6- Cease	Not applicable

CHAPTER 3

A Case Study in Desktop Protocols

Introduction

TCP/IP has become the de facto networking protocol for the Internet as well as most corporate data networks. Although TCP/IP is now the dominant networking protocol, other protocols are still being used.

The three most popular routable protocols after TCP/IP are the IPX, AppleTalk, and DECnet. This chapter examines each of these protocols in detail. You will see that running multiple protocols on a router can be handled in one of two ways:

- Run "Ships in the Night" routing. This type of routing involves configuring and maintaining each individual desktop protocol independently of each other. When you run Ships in the Night routing, a separate routing table will be maintained for each network layer protocol that is configured on the router. For example, if you configure a router to run IPX, DECnet, and AppleTalk, your router will maintain three separate routing tables.
- Tunnel IPX, DECnet, or AppleTalk within TCP/IP. This has the advantage of not having to run a protocol besides TCP/IP on any of your routers, except for the edge routers where you take other protocols, such as IPX, DECnet, or AppleTalk, and tunnel them in TCP/IP for transport across the network.

Technology Overview

This section provides an overview of the three desktop technologies that will be discussed throughout this case study.

DECnet

DECnet was developed in 1975 by Digital Equipment Corporation as a way to provide communications between its computer systems. DECnet should not be confused with DEC LAT. DECnet is a routable protocol, while DEC LAT is not.

DECnet is configured differently from the other major routed protocols (TCP/IP, IPX, AppleTalk). Instead of assigning a DECnet address to each router interface, a single DECnet address is assigned to the entire router using a global configuration command. Each interface on a router running DECnet is assigned a DECnet cost, and the cost value is an arbitrary number determined by the network administrator. DECnet routing decisions are based on this cost value. The lower the cost, the more preferred the path.

Addressing

DECnet addressing uses an area.node notation. A DECnet area can range from 1 to 63, while a DECnet node can range from 1 to 1,023. A DECnet area can span multiple routers.

When DECnet is configured on a router, the MAC addresses of any LAN interfaces will be changed. DECnet uses an equation that multiplies the area number by 1,024 and then adds on the node number. This number is then converted to a hexadecimal value and appended in byte-swapped order to address AA00.0400.

As an example, the E0/0 interface on R1 has a burned-in MAC address of 00e0.1e5b.2601. During this case study, you will be assigning R1 to DECnet address 2.1. In DECnet addressing terminology, address 2.1 is Area 2 and Node 1. DECnet forms the new MAC address for E0/0 by taking the Area value of 2 and multiplying it by 1,024 and adding the Node value of 1. This becomes 2 * 1,024+1, which is 2,049. The 2049 value is then converted to hexadecimal, which becomes 0801(HEX). This number is then byte-swapped to become 0108(HEX). The 0108(HEX) value is then appended to the value AA00.0400. The new MAC address for the E0/0 interface of R1 then becomes AA00.0400.0108.

Routing Levels

A router running the DECnet protocol is referred to as either a Level 1 or Level 2 router. Level 1 routers communicate with DECnet end nodes and other Level 1 routers within a defined area address. Level 2 routers communicate with Level 1 routers within the same area address as well as Level 2 routers in different areas.

IPX

Novell NetWare is both an operating system and a networking protocol. Novell IPX is based on the *Xerox Network System* (XNS) protocols. When the term IPX is used, it refers to the entire Novell protocol stack, just as the term IP refers to the entire TCP/IP suite.

IPX Addressing

Addressing in a Novell IPX network is different than addressing in an IP network. An IPX address is in the format of network.node.socket. The three parts of the IPX address are as follows:

- **Network portion:** Every IPX network is assigned a globally unique 32-bit network number.
- **Node portion:** Each IPX device (usually a workstation, router, or server) is assigned a 48-bit node address. This 48-bit address is taken from the MAC address of the interface itself. This is an attractive feature of IPX. Having the node address of an IPX device use the device's MAC address means that IPX will not need the ARP. An IPX-enabled device wanting to send a datagram to another IPX device only needs to know the IPX address of the end station because the MAC address of the end station is embedded in the IPX address. The sending station has enough information to create an Ethernet or Token Ring frame because the destination MAC address is already known. With IP, you know an end device's IP address but do not know the end device's MAC address that is needed to build the Ethernet, Token Ring, or FDDI datalink frame. IP uses ARP to find the MAC address of a destination device or network with an IP address that is already known.
- **Socket portion**: The socket is a 16-bit number that identifies a software process using IPX in the end station. Some socket numbers are reserved and the remainder are available for use by the end station.

The IPX network address is 32 bits, the node address is 48 bits, and the socket number is 16 bits. A complete IPX address is a 96-bit number expressed as a 12-byte hexadecimal number.

IPX Protocol Stack

The key elements of the IPX protocol stack are described in the following sections.

IPX Network Layer IPX provides network-layer connectionless datagram delivery to support Novell NetWare. The minimum IPX packet size is 30 bytes, and the maximum packet size is 65,535 bytes. An IPX packet has a 30-byte header.

IPX Transport Layer Novell IPX uses SPX as its transport layer protocol. SPX is a connection oriented protocol. No data transfer can take place between two end stations using SPX until a connection has been built.

Service Advertising Protocol (SAP) SAP is used to advertise and distribute Novell server information. NetWare servers and routers broadcast a SAP message every 60 seconds. This message advertises what services they provide.

Three types of SAP packets exist:

- **Periodic updates**: A periodic update is used by a server when it has a service to advertise. The server sends a SAP broadcast with the service's name, service type, and full IPX address (network.node.socket). Routers listen to and store these broadcasts and periodically broadcast these updates to all directly connected neighbors.

- **Service queries**: A service query is used by a NetWare client to locate a server. This kind of query is often referred to as a *Get Nearest Server* (GNS) query. This service query is a broadcast and does not go off the local network. The query is answered by the local router that has stored the periodic updates it has received from servers on the network.
- **Service responses**: A service response is a response to a service query. This is usually a response from a router.

IPX Routing Protocols IPX uses three different routing protocols to propagate routing information:

- **IPX Routing Information Protocol (RIP)**: IPX RIP is a distance vector protocol and has many similarities to IP RIP. IPX RIP differs from IP RIP because the IPX end station requests route information, whereas with an IP network the end station has a default route to the nearest router. IPX RIP updates are broadcast every 60 seconds. (IP RIP sends updates every 30 seconds). IPX RIP uses two metrics. The first metric is delay, referred to as ticks. Each router interface running IPX is assigned a tick value. The higher the speed of the router interface, the lower the tick value. An Ethernet interface, for example, is assigned a tick value of 1, while a Serial interface is assigned a tick value of 6. The second metric is hop count. The route with the lowest delay is given the preference over the route with the lowest hop count. If two routes exist with the same tick value, then the router uses the hop count of each route to determine the best route.
- **NetWare Link Services Protocol (NLSP):** NLSP is a link state protocol. It provides load balancing across equal cost paths and features much faster convergence times than IPX RIP.
- **Enhanced Interior Gateway Routing Protocol (EIGRP):** This is the Cisco proprietary routing protocol that features automatic redistribution between RIP/SAP and EIGRP.

IPX Encapsulation Types When running on a LAN, IPX runs on Ethernet, Token Ring, and FDDI. IPX can use four different Ethernet encapsulation types on a LAN. This means that four different MAC frames can be used on an IPX network. If two workstations on a NetWare LAN use different Ethernet encapsulation types, they cannot talk to each other directly; their traffic has to go through a router.

The four Ethernet encapsulation types are

- **Ethernet II**: This encapsulation type is referred to by Cisco as ARPA encapsulation.
- **802.2**: This encapsulation type is referred to by Cisco as SAP encapsulation.
- **802.3**: This encapsulation type is referred to by Cisco as Novell-Ether encapsulation. This is the default encapsulation on a Cisco Ethernet interface.
- **Subnetwork Access Protocol (SNAP)**: This encapsulation type is referred to by Cisco as SNAP encapsulation.

IPX supports two different Token Ring encapsulation types:

- Cisco SAP, which is the default
- Cisco SNAP

IPX supports three different FDDI encapsulation types:

- Cisco SNAP
- Cisco SAP
- Novell-FDDI Raw

AppleTalk

AppleTalk is a networking protocol developed by Apple Computer to provide networking services for its Macintosh computers. AppleTalk is the most automatic of all the desktop protocols but it is also the chattiest. For example, the default routing protocol for AppleTalk is the *Routing Table Maintenance Protocol* (RTMP). RTMP sends routing updates every 10 seconds to all directly connected neighbors.

AppleTalk Terminology

An AppleTalk node can be any device that is connected to an AppleTalk network and is assigned an AppleTalk address. Nodes can be Macintosh computers, printers, or any other device that resides on the network and is addressable.

An AppleTalk network can be thought of as a physical LAN or WAN that contains one or more AppleTalk nodes.

An AppleTalk zone is a logical group of networks. A zone usually consists of AppleTalk nodes that reside in different physical locations. Zones are very similar in concept to a VLAN. Figure 3-1 shows an example of how AppleTalk zones can work. An AppleTalk network is displayed with three Ethernet segments. The Ethernet segment on RouterA and RouterB are both in zone Engineering. When a Macintosh user on the Ethernet LAN connected to RouterC wants to access resources in the Engineering zone, he is given access to the LAN on RouterA and RouterB. Zones enable you to functionally group network resources without any regard to their actual physical location.

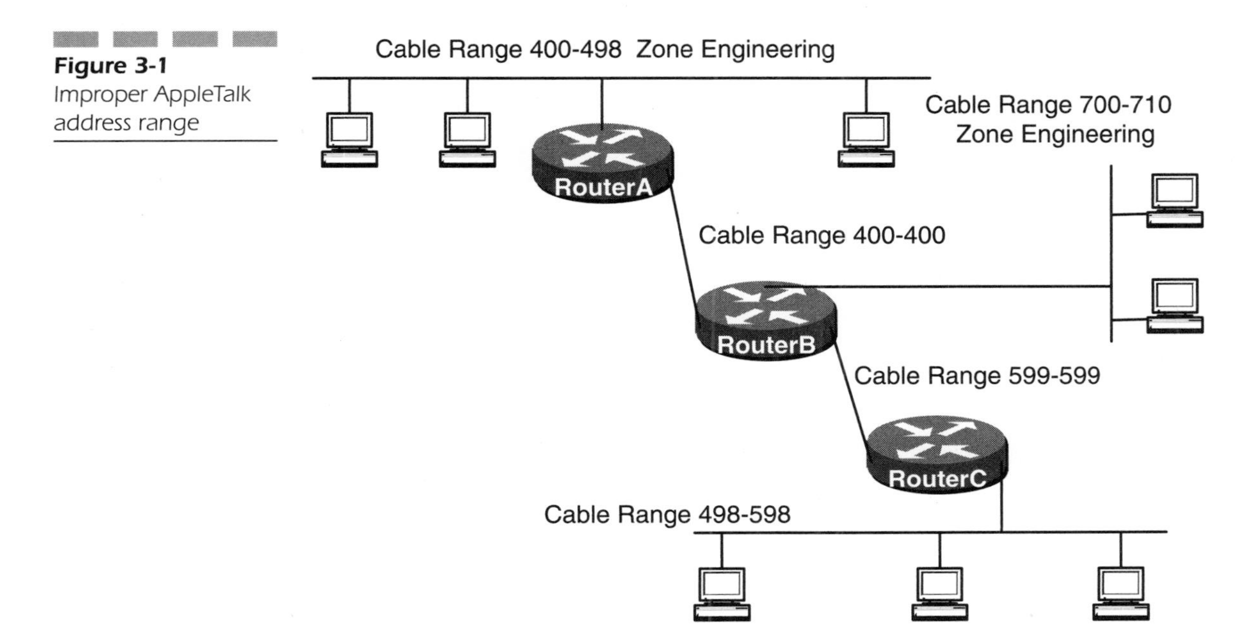

Figure 3-1 Improper AppleTalk address range

AppleTalk Addressing

Early AppleTalk networks were referred to as Phase I or non-extended networks. Phase I networks had a limited address space. Each LAN or WAN segment was allowed to contain up to 127 hosts and up to 127 servers. Each segment could also only be assigned a single AppleTalk network number.

AppleTalk Phase II networks are much more flexible in their network addressing. A Phase II network enables multiple network numbers to exist on each network segment. This means that a LAN can contain multiple AppleTalk networks. The range of network numbers that exist on a network segment is referred to as the cable range of the segment. The cable range must be unique and cannot overlap with other router interfaces. Figure 3-1 shows an example of an AppleTalk network with improperly assigned cable range numbers. In this case, an address conflict exists because network 498 has been assigned to both Ethernet LANs. Figure 3-2 shows a properly configured AppleTalk network, with no address overlaps.

AppleTalk node address assignment is designed to minimize the amount of configuration needed on a Macintosh computer. When a Macintosh is first powered on, it sends a broadcast

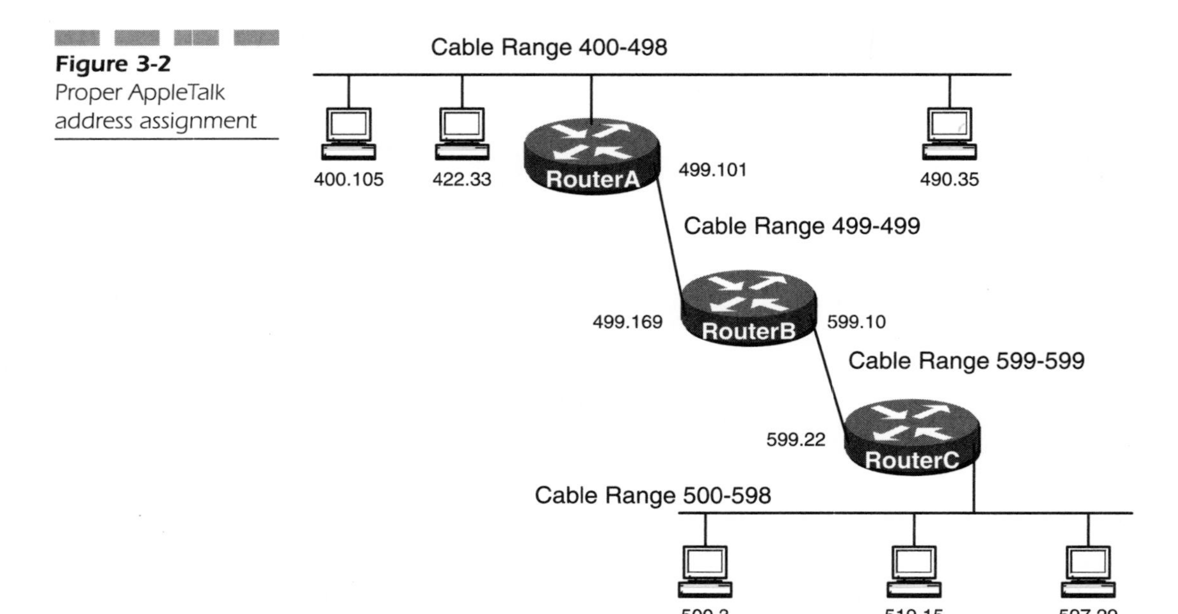

Figure 3-2 Proper AppleTalk address assignment

to any routers on the same network segment, asking what the cable range of the network segment is. Once a router responds, the Macintosh chooses a network number within the cable range and then picks a node number. Before the AppleTalk node uses the network.node combination it has picked, it queries the network to see if the network.node combination is already in use. If the address is already used, the AppleTalk node continues to choose new addresses until an unused address has been found.

AppleTalk Address Structure

An AppleTalk address is 24 bits long. The address is written in network.node format. The first 16 bits are the network number and the last eight bits are the node number. This means that all AppleTalk networks will be numbered less than 65,536 and all AppleTalk nodes will be numbered less than 256. Node numbers 0 and 255 are reserved (255 is used as a network broadcast address). An AppleTalk network can therefore have 254 nodes per network.

AppleTalk Protocol Stack

The AppleTalk Protocol Stack consists of several layers: the physical and datalink layers, the network layer, the transport layer, and the application/presentation/session layers.

Physical and Datalink Layers In addition to being supported on WAN links such as Frame Relay and ISDN, AppleTalk is supported on four major LAN platforms:

- **EtherTalk**: Apple's version of Ethernet
- **TokenTalk**: Apple's version of Token Ring
- **FDDITalk**: Apple's version of FDDI
- **AppleTalk**: An Apple proprietary serial link that runs at 230Kbits/second

Network Layer AppleTalk uses the *Datagram Delivery Protocol* (DDP) at the network layer for routing packets in a network. AppleTalk is a routable protocol because it has a network-layer address associated with each AppleTalk node. DDP is a connectionless network protocol.

Transport Layer Several protocols exist in the AppleTalk transport layer:

- **Routing Table Maintenance Protocol (RTMP):** This is a distance vector routing protocol that is similar to IP RIP. RTMP is very chatty; it sends out a routing update to all connected neighbors every 10 seconds.
- **AppleTalk Echo Protocol (AEP):** AEP is a simple protocol that generates packets that can be used to test the reachability of various network nodes.
- **AppleTalk Transaction Protocol (ATP):** ATP provides connection-based data transfers for AppleTalk traffic. It functions in a similar mode to TCP in an IP network. ATP provides for data acknowledgment, retransmission, packet sequencing, and fragmentation and reassembly.
- **Name Binding Protocol (NBP):** NBP associates an AppleTalk name with an addresses.

Application/Presentation/Session Layers AppleTalk supports several upper-layer protocols:

- **AppleTalk Session Protocol (ASP):** ASP establishes and maintains sessions between an AppleTalk client and a server.
- **Zone Information Protocol (ZIP):** ZIP maintains network-number-to-zone-name mappings in zone information tables. It uses RTMP routing tables to keep up with network topology changes. When ZIP finds a routing table entry that is not in the ZIP, it creates a new ZIP entry.
- **AppleTalk Printer Access Protocol (PAP):** PAP is a connection-oriented protocol that establishes and maintains connections between clients and printers.
- **AppleTalk Filing Protocol (AFP):** AFP helps clients share server files across a network.

AppleTalk Routing Protocols

Cisco supports three routing protocols for AppleTalk networks:

- **RTMP**: This is enabled by default on an AppleTalk network. RTMP is a distance vector routing protocol that uses the hop count as its metric. The update period for RTMP is every 10 seconds, regardless of whether or not a change occurs in the network. This frequent update has the effect of producing a large amount of routing traffic on an AppleTalk network.
- **AURP**: This routing protocol is similar to RTMP in that it is a distance vector routing protocol with a maximum hop count of 15 hops. AURP differs from RTMP in that it only sends routing updates when a change has occurred in the network, while RTMP sends updates every 10 seconds. AURP is also a tunneling protocol that enables AppleTalk to be tunneled within IP, thus allowing two AppleTalk networks to be connected over a TCP/IP network. The TCP/IP connection is called a tunnel and is counted as one network hop. The router that connects an AppleTalk network to a tunnel is referred to as an exterior router.
- **EIGRP**: AppleTalk EIGRP is used mainly for WAN links in an AppleTalk network. AppleTalk EIGRP uses the same composite metric that IP and IPX EIGRP use. It also uses the same *Diffusing Update Algorithm* (DUAL) routing algorithm, only sending out routing updates when a change has occurred in the network. AppleTalk EIGRP differs from IP and IPX EIGRP in that the AS number used to start the routing process must be unique for each router. AppleTalk EIGRP features automatic redistribution with the RTMP routing protocol.

AppleTalk Zones

An AppleTalk zone is a grouping of similar resources. It is very similar in concept to a VLAN. Each AppleTalk network must be defined as a member of one or more zones. The AppleTalk ZIP maintains a listing of all zone names and associated AppleTalk network numbers for the entire network. Members of a particular zone can be located anywhere in the entire network. Let's look at what will happen when an AppleTalk node such as an Apple Macintosh needs a service such as a printer:

1. The Macintosh chooser sends a request to the local router for a list of all the zones.
2. The Macintosh looks in the list of zones for the appropriate service.
3. If the appropriate service is found, the Macintosh sends a request to each of the cable numbers in the selected zone.
4. The local router sends this request as a multicast to the selected zones.
5. The services in the selected zones reply to the router that sent the multicast.
6. The router that sent the multicast forwards these replies to the originating Macintosh node.
7. The Macintosh node can now select the appropriate service.

Case Study 2: Desktop Protocols

IOS Requirements

IOS 11.2 is used in this case study.

Equipment Needed

The following equipment is needed to perform this case study:

- One Cisco router with two serial ports
- One Cisco router with two serial ports and an ISDN BRI port
- Two Cisco routers with one serial port and one Ethernet port
- Two Cisco routers with one Ethernet port
- One Cisco router with one serial, one Ethernet, and one ISDN BRI port
- One Cisco router with three serial ports. This router will be used as the Frame Relay switch
- One Cisco terminal server. (One of the other Cisco routers could also double as a terminal server. In this case study, R7 will be your terminal server.)
- A Cisco IOS desktop image loaded on each router.
- A PC running a terminal emulation program for connecting to the console port of the terminal server
- Three Ethernet cables
- An Ethernet hub
- Five Cisco DTE/DCE crossover cables
- Two ISDN BRI circuits and two ISDN BRI cables

Obtaining ISDN Circuits

This case study requires two ISDN BRI circuits that will be used to configure IPX DDR on Routers R1 and R6. You have two options for obtaining BRI circuits for this case study:

- Purchase the two ISDN circuits from your local telecommunications carrier.
- Obtain an ISDN simulator you can use to provision your own BRI circuits. This is the option that is used in this case study and an Adtran Atlas 800 desktop ISDN switch is utilized here.

Details on configuring this ISDN switch for this case study can be found at the end of this chapter.

Physical Connectivity Diagram

Figure 3-3 shows the physical connectivity for the routers in this case study.

Questions

DECnet

1. Cable the routers and make sure that the DTE/DCE cables are connected as per Figure 3-3.
2. Configure the terminal server so that all the routers can be accessed by name through it. The port number of the terminal server attaches to the corresponding router. That is, port 1 of the terminal server goes to the console port of R1, and the router acting as a Frame Relay switch goes to port 8 of the terminal server.
3. Set each router hostname as per Figure 3-3.
4. Set the enable password on all routers to be **cisco**.
5. Configure the router acting as a Frame Relay switch so that a full mesh of PVCs is used between the three serial ports on the Frame Relay switch. Configure the PVCs as follows:

Frame Switch Port In	DLCI In	Frame Switch Port Out	DLCI Out
S1/0	101	S1/1	102
S1/0	301	S1/2	302
S1/1	102	S1/0	101
S1/1	201	S1/2	202
S1/2	202	S1/1	201
S1/2	302	S1/0	301

As an example, the first line of the table would be interpreted as follows: Any traffic that enters port S1/0 of the Frame Relay switch with a DLCI value of 101 will be sent out port S1/1 of the Frame Relay switch with a DLCI value of 102.

Figure 3-3
Physical connectivity diagram

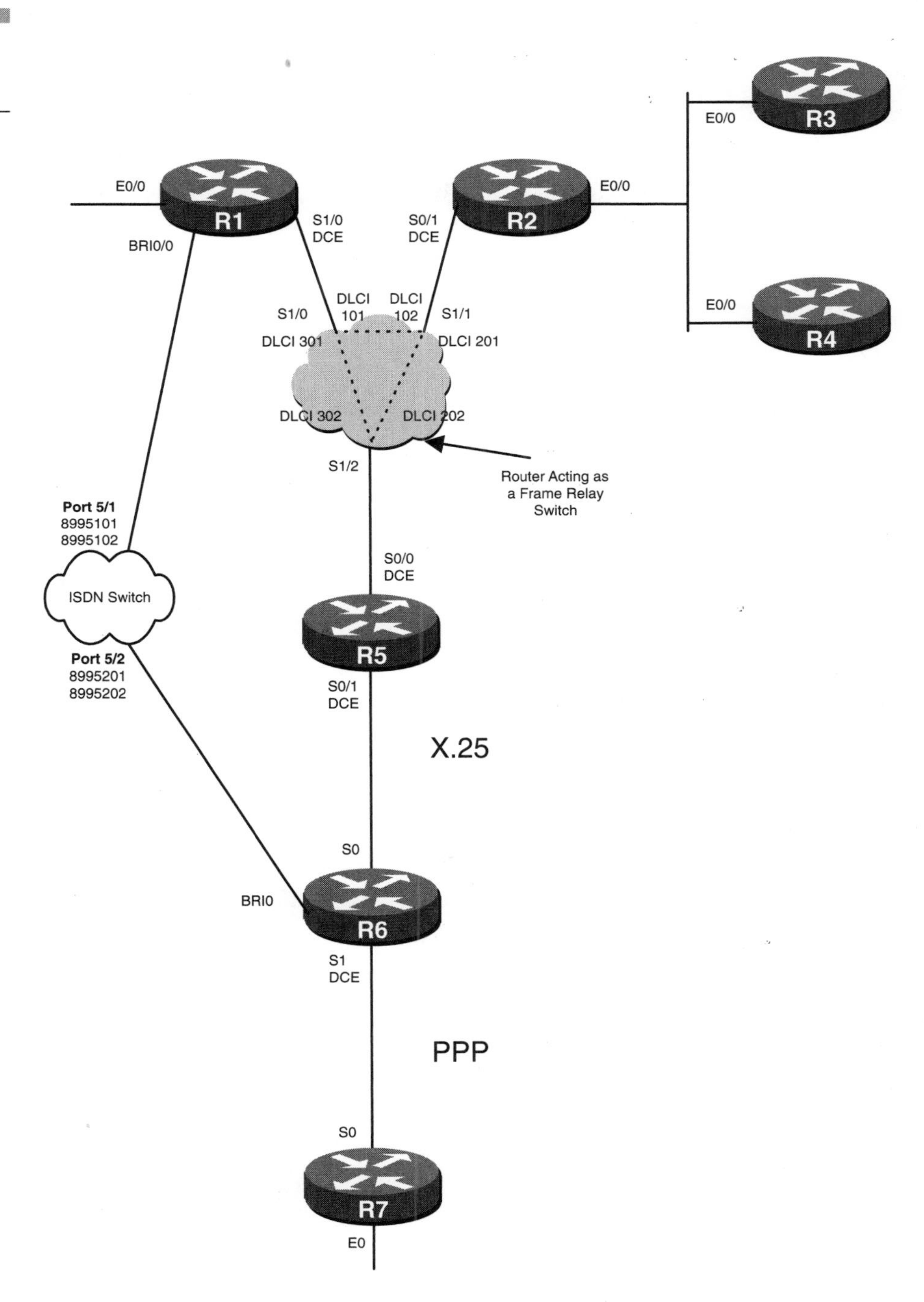

6. Configure R1 with a DECnet address of 2.1:
   a. Configure the router with an interface cost equal to the router number.
   b. R1 is connected to the Frame Relay network via interface S1/0. Configure R1 so that Frame Relay traffic only uses DLCI 101.
7. Configure R2 with a DECnet address of 2.2.
   a. Configure the router with an interface cost equal to the router number.
   b. R2 is connected to the Frame Relay network via interface S0/1. Configure R2 so that Frame Relay traffic to and from R1 uses DLCI 102, and Frame Relay traffic to and from R5 uses DLCI 201.
8. Configure R3 with a DECnet address of 2.3. Also configure it with an interface cost equal to the router number.
9. Configure R4 with a DECnet address of 2.4. Also configure it with an interface cost equal to the router number.
10. Configure R5 with a DECnet address of 2.5. Interface S0/1 of R5 should use X.25 encapsulation with an X.25 address of 5555. Interface S0/1 should act as an X.25 DCE.
    a. Configure the router with an interface cost equal to the router number.
    b. R5 is connected to the Frame Relay network via interface S0/0. Configure R5 so that Frame Relay traffic only uses DLCI 202.
11. Configure R6 with a DECnet address of 2.6. Interface S0 of R6 should use X.25 encapsulation with an X.25 address of 6666. Interface S0 should act as an X.25 DTE. Also configure the router with an interface cost equal to the router number.
12. Configure R7 with a DECnet address of 2.7. Also configure the router with an interface cost equal to the router number.
13. Configure R3 so that its routing table does not have an entry for R6.
14. Configure R7 for LAT translation. A LAT session request to a loopback interface on R7 should be translated into a telnet session into R7.
15. Verify that the configuration is working properly.

IPX

16. Configure R1 as follows:
    a. Interface E0/0 should be assigned IPX network number 12. It should use the default encapsulation for an IPX Ethernet interface.
    b. Create a loopback 0 interface with an IPX network number of 1.
    c. The IPX node number should be set to 0001.0001.0001.
    d. Interface S1/0 should be assigned IPX network number 8. Use a physical interface.

   e. Run IPX EIGRP on all interfaces except the loopback interface, which should run IPX RIP.

17. Configure R2 as follows:
   a. Interface E0/0 should be assigned two different IPX network numbers, 10 and 11.
   b. Create a loopback 0 interface with an IPX network number of two.
   c. The IPX node number should be set to 0002.0002.0002.
   d. Interface S0/1 should be assigned IPX network number 8. Use a physical interface.
   e. Run IPX EIGRP on all interfaces except the loopback interface, which should run IPX RIP.
18. Configure R3 as follows:
   a. Interface E0/0 should be assigned IPX network number 11.
   b. The IPX node number should be set to 0003.0003.0003.
   c. Create a loopback 0 interface with an IPX network number of 3.
   d. Run IPX EIGRP on all interfaces except the loopback interface, which should run IPX RIP.
19. Configure R4 as follows:
   a. Interface E0/0 should be assigned IPX network number 10.
   b. The IPX node number should be set to 0004.0004.0004.
   c. Create a loopback 0 interface with an IPX network number of 4.
   d. Run IPX EIGRP on all interfaces except the loopback interface that should run IPX RIP.
20. Configure R5 as follows:
   a. Interface S0/0 should be assigned IPX network number 8. Use a physical interface.
   b. The IPX node number should be set to 0005.0005.0005.
   c. Create a loopback 0 interface with an IPX network number of 5.
   d. Run IPX EIGRP on interface S0/0.
   e. Run IPX RIP on the loopback interface.
   f. Using internal network number 100, run IPX WAN with NLSP on interface S0/1.
21. Configure R6 as follows:
   a. Interface S1 should be assigned IPX network number 9.
   b. The IPX node number should be set to 0006.0006.0006.
   c. Create a loopback 0 interface with an IPX network number of 6.
   d. Run IPX EIGRP on interface S1.
   e. Run IPX RIP on the loopback interface.
   f. Using Internal network number 101, run IPX WAN with NLSP on interface S0.
22. Configure R7 as follows:
   a. Interface S0 should be assigned IPX network number 9.
   b. The IPX node number should be set to 0007.0007.0007.

c. Create a loopback 0 interface with an IPX network number of 7.
d. Run IPX EIGRP on all interfaces except the loopback interface that should run IPX RIP.
e. Interface E0 should be assigned IPX network number 13.

23. Configure two static SAP entries on R6 as follows:

a. The first static entry should be for a print server residing on IPX network 13 on R7.
b. The second static entry should be for a file server residing on IPX network 13 on R7.

24. Configure R1 so that it cannot see any print servers.

25. Configure R2 so that R3 and R4 do not have any entries for IPX network 7 in their routing tables.

26. Configure DDR between R1 and R6. Configure R1 so that it will call R6 when R1's Frame Relay connection is lost. Also configure DDR so that R6 will accept R1's call, authenticate R1, and then securely call R1 back. Use IPX network 14 for the ISDN BRI network.

27. Verify that the configuration is working properly.

AppleTalk

28. Configure R1 as follows:

a. Interface E0/0 should be assigned AppleTalk cable range 41-99, with an interface address of 69.1. The zone name for this interface should be LAN_1.
b. Interface S1/0 should be assigned AppleTalk cable range 100-199, with an interface address of 124.1. The zone name should be WAN_1_2_5.
c. Run AppleTalk EIGRP on interface S1/0 and RTMP on interface E0/0.

29. Configure R2 as follows:

a. Interface E0/0 should be assigned AppleTalk cable range 200-299, with an interface address of 234.2. The zone name should be set to LAN_2_3_4.
b. Interface S0/1 should be assigned AppleTalk cable range 100-199, with an interface address of 124.2. The zone name should be WAN_1_2_5.
c. Run AppleTalk EIGRP on interface S0/1 and RTMP on interface E0/0.

30. Configure R3 as follows:

a. Interface E0/0 should be assigned AppleTalk cable range 200-299, with an interface address of 234.3. The zone name should be set to LAN_2_3_4.
b. Run RTMP on this router.

31. Configure R4 as follows:

a. Interface E0/0 should be assigned AppleTalk cable range 200-299, with an interface address of 234.4. The zone name should be set to LAN_2_3_4.
b. Run RTMP on this router.

32. Configure R5 as follows:

a. Interface S0/0 should be assigned AppleTalk cable range 100-199, with an interface address of 124.5. The zone name should be set to WAN_1_2_5.

b. Run AppleTalk EIGRP on all AppleTalk interfaces on this router.

c. Interface S0/1 should be configured with IP address 142.2.1.17/29. Run IP EIGRP on this interface.

33. Configure R6 as follows:

a. Do not run any AppleTalk routing on this router.

b. Interface S0 should be assigned IP address 142.2.1.18/29.

c. Interface S1 should be assigned IP address 142.2.1.9/29.

d. Run IP EIGRP on this router.

34. Configure R7 as follows:

a. Interface E0 should be assigned AppleTalk cable range 701-799, with an interface address of 720.14. The zone name should be set to LAN_7 with a secondary zone name of TopSecret. Run RTMP on this interface.

b. Interface S0 should be configured with IP address 142.2.1.10/29. Run IP EIGRP on this interface.

35. Configure a GRE tunnel between R5 and R7.

a. Run AppleTalk EIGRP over this tunnel.

b. The tunnel cable range should be 500-700.

c. The tunnel zone should be WAN_5_7.

d. The tunnel endpoint on R5 should be AppleTalk address 545.5.

e. The tunnel endpoint on R7 should be AppleTalk address 660.7.

36. Configure R2 so that R3 and R4 cannot reach any addresses in AppleTalk cable range 701-799.

37. Configure R4 so that it can no longer see any routes to the 41-99 cable range.

38. Verify that the configuration is working properly.

Answer Guide

1. *Cable the routers per Figure 3-3. Make sure that the DTE/DCE cables are connected as in Figure 3-3.*

You can see from the diagram that this case study requires five serial connections and three Ethernet connections. Cable the configuration as in Figure 3-3, making sure to connect the DTE and DCE sides of the serial cables to the correct ports.

2. *Configure the terminal server so that all the routers can be accessed by name through it. The port number of the terminal server attaches to the corresponding router. That is, port 1 of the terminal server goes to the console port of R1, and the router acting as a Frame Relay switch goes to port 8 of the terminal server.*

   The terminal server provides access to all your routers via reverse telnet, which is the process of using telnet to make a connection out an asynchronous port.

   All the routers in the case study will have their console port connected directly to one of the 16 asynchronous interfaces on a Cisco 2511 terminal server router, using a standard Cisco rolled cable. The routers will be accessed using a reverse telnet connection. To make such a connection, you must telnet to any active IP address on the box, followed by 20xx where xx is the terminal server port number that you want to access. If you have defined a loopback on the router at IP address 1.1.1.1, then the following command would be used to reverse telnet to port 1 on the router: Telnet 1.1.1.1 2001.

   Use the following line commands to configure the terminal server:

```
R7(config)#line 1 16
R7(config-line)#transport input all
R7(config-line)#no exec
```

   The command **transport input all** specifies the input transport protocol. By default on IOS 11.1 and later, the transport input is set to none. Prior to 11.1, the default was all. If the transport input is left to none, you will receive an error stating that the connection is refused by the remote host. The command **no exec** enables only outgoing connections for the line. This prevents the device that you are attached to from sending out unsolicited data. If the port receives unsolicited data, an EXEC process starts, making the line unavailable.

   The next step is to configure hostnames on the terminal server so you can attach to any router simply by typing in the router name. The Cisco IOS software maintains a table of host names and their corresponding addresses. Similar to a DNS server, you can statically map host names to an IP address. This is very useful and saves a lot of keystrokes when you have multiple devices connected to the terminal server.

   The following global configuration commands are used to set the host names:

```
R7(config)#ip host R1 2001 1.1.1.1
R7(config)#ip host R2 2002 1.1.1.1
R7(config)#ip host R3 2002 1.1.1.1
R7(config)#ip host R4 2004 1.1.1.1
R7(config)#ip host R5 2005 1.1.1.1
R7(config)#ip host R6 2006 1.1.1.1
R7(config)#ip host R7 2007 1.1.1.1
```

3. *Set each router hostname as in Figure 3-3.*

   Connect to each of the routers and set their hostnames with the global command:

```
Router(config)#hostname R6
```

4. *Set the enable password on all routers to be* ***cisco****.*

   When a router is first configured, no default enable password exists. The enable password is used to prevent unauthorized users from changing the configuration. To set the enable password, use the following global configuration command:

```
R1(config)#enable password cisco
```

5. *Configure the router acting as a Frame Relay switch so that a full mesh of PVCs exists between the three serial ports on the Frame Relay switch. Configure the PVCs as follows:*

Frame Switch In	DLCI In	Frame Switch Out	DLCI Out
S1/0	101	S1/1	102
S1/0	301	S1/2	302
S1/1	102	S1/0	101
S1/1	201	S1/2	202
S1/2	202	S1/1	201
S1/2	302	S1/0	301

Frame Relay switching is enabled on a Cisco router by the **frame-relay switching** global command. In addition to the standard configuration commands defining Frame Relay encapsulation and LMI signaling, each interface on the router that supports switching needs to be defined as a Frame Relay DCE via the **frame-relay intf-type dce** command.

Frame Relay PVCs are defined with the **frame-relay route** command. As an example, let's look at interface S1/0 of your router that is acting as a Frame Relay switch. You want the Frame switch to take any traffic that comes into port S1/0 with a DLCI value of 101 to send that traffic to port S1/1 with a DLCI value of 102. This would be accomplished with the interface command:

```
FrameSwitch(config-if)# frame-relay route 101 interface Serial1/1 102
```

Similarly, any traffic entering port S1/0 with a DLCI value of 301 will be sent out port S1/2 with a DLCI value of 302. As shown in Figure 3-4, your Frame Relay switch will be configured with a full mesh of PVCs, meaning that a PVC exists between each port and every other port on the Frame Relay switch.

The router's Frame Relay-specific commands for the three interfaces on the router would be as follows:

```
FrameSwitch(config)# interface Serial1/0
FrameSwitch(config-if)# encapsulation frame-relay
FrameSwitch(config-if)# frame-relay lmi-type ansi
FrameSwitch(config-if)# frame-relay intf-type dce
FrameSwitch(config-if)# frame-relay route 101 interface Serial1/1 102
FrameSwitch(config-if)# frame-relay route 301 interface Serial1/2 302

FrameSwitch(config)# interface Serial1/1
FrameSwitch(config-if)# encapsulation frame-relay
FrameSwitch(config-if)# frame-relay lmi-type ansi
FrameSwitch(config-if)# frame-relay intf-type dce
FrameSwitch(config-if)# frame-relay route 102 interface Serial1/0 101
FrameSwitch(config-if)# frame-relay route 201 interface Serial1/2 202

FrameSwitch(config)# interface Serial1/2
FrameSwitch(config-if)# encapsulation frame-relay
FrameSwitch(config-if)# frame-relay lmi-type ansi
FrameSwitch(config-if)# frame-relay intf-type dce
FrameSwitch(config-if)# frame-relay route 202 interface Serial1/1 201
FrameSwitch(config-if)# frame-relay route 302 interface Serial1/0 301
```

6. *Configure R1 with a DECnet address of 2.1:*

a. Configure the router with an interface cost equal to the router number.

b. R1 is connected to the Frame Relay network via interface S1/0. Configure R1 so that Frame Relay traffic only uses DLCI 101.

A minimum DECnet configuration on a router consists of three commands, two global commands, and one interface command.

DECnet Global Commands

In order to enable DECnet routing on a router, you must define a global DECnet address on the router with the **Decnet Routing** command. DECnet addressing uses Area.Node notation. A DECnet area's value can range from 1 to 63, and a DECnet node address can be between 1 and 1,023. You can therefore have 64,449 nodes (63 * 1,023) in a DECnet network. DECnet differs from other major desktop protocols such as IP, IPX, and AppleTalk in that the entire router takes on the DECnet address. You do not configure individual DECnet addresses for each interface on a router running a DECnet protocol.

R1 needs to be configured with a DECnet Area.Node address of 2.1. The global command to accomplish this would be:

```
R1(config)#decnet routing 2.1
```

The second global task in enabling DECnet on a router is to define the node type of the router. DECnet routing nodes are referred to as either Level 1 or Level 2 routers. Level 1 routers exchange packets with other routers that are in the same area. Level 2 routers exchange packets to and from routers in other areas. Level 1 routers are referred to as intra-area routers and Level 2 routers are referred to as inter-area routers.

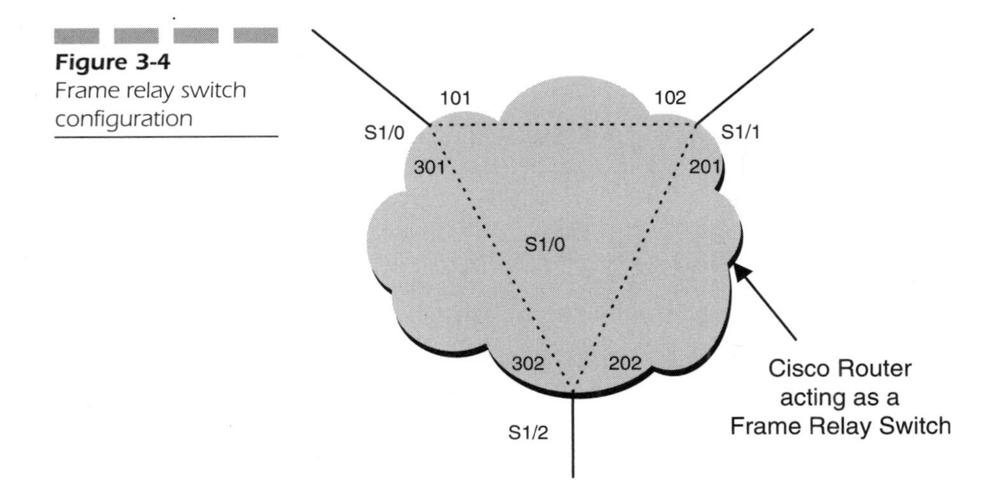

Figure 3-4
Frame relay switch configuration

Cisco defines the node type with the decnet node-type command. The keyword **routing-iv** is used for a Level 1 router and the keyword **area** is used for a Level 2 router.

R1 is acting as a Level 1 intra-area router so the configuration command to define the node type is

```
R1(config)#decnet node-type routing-iv
```

DECnet Interface Commands

At a minimum, you must configure an interface cost for each interface running DECnet on your router, and routing decisions are based on the interface cost. A Cisco router always prefers the lowest cost path to a specific destination. The DECnet cost is defined with the **decnet cost** command.

R1 needs a cost equal to the router number. You would therefore issue the following command under each interface of the router:

```
R1(config)# interface Ethernet0/0
R1(config-if)#decnet cost 1

R1(config)# interface Serial1/0
R1(config-if)#decnet cost 1
```

R1 is running Frame Relay encapsulation on its S1/0 interface. As depicted in Figure 3-4. a fully meshed Frame Relay network has been defined between R1, R2, and R5. Two DLCIs (101 and 301) are defined on interface S1/0 of R1. You want all traffic to only use DLCI 101. In order to do this, you must disable Frame Relay inverse arp on this interface with the **no frame-relay inverse-arp** command. This keeps R1 from trying to dynamically resolve network-layer addresses for far-end routers. Because Frame Relay inverse arp is disabled, you must manually define your Frame Relay neighbor next hop address information for each protocol. This is done by using the Frame Relay map command.

R1 has two Frame Relay neighbors, R2 and R5. Both of these routers must be reached via DLCI 101. Thus, define two Frame Relay map statements as follows:

```
R1(config-if)#frame-relay map decnet 2.2 101 broadcast
R1(config-if)#frame-relay map decnet 2.5 101 broadcast
```

The map statements can be interpreted as follows:

- If DECnet traffic has a next hop address of 2.2, then encapsulate the traffic in DLCI 101.
- If DECnet traffic has a next hop address of 2.5, then encapsulate the traffic in DLCI 101.

You can see that the two map statements have the effect of sending all traffic from R1 into the Frame Relay cloud over DLCI 101.

7. *Configure R2 with a DECnet address of 2.2:*

a. Configure the router with an interface cost equal to the router number.

b. R2 is connected to the Frame Relay network via interface S0/1. Configure R2 so that Frame Relay traffic to and from R1 uses DLCI 102 and Frame Relay traffic to and from R5 uses DLCI 201.

As with R1, you must configure DECnet routing globally along with the DECnet node type and interface costs for this router. The following commands will achieve this:

```
R2(config)#decnet routing 2.2
R2(config)#decnet node-type routing-iv

R2(config)#interface Ethernet0/0
R2(config-if)#decnet cost 2

R2(config)#interface Serial0/1
R2(config-if)#decnet cost 2
```

Notice that the cost of each interface has been set to 2.

R2 is running Frame Relay encapsulation on interface S0/1. As with R1, you first disable Frame Relay inverse arp with the **no frame-relay inverse-arp** command. You then must define two Frame Relay maps:

```
R2(config-if)#frame-relay map decnet 2.1 102 broadcast
R2(config-if)#frame-relay map decnet 2.5 201 broadcast
```

These two maps tell R2 that any DECnet traffic with a next hop DECnet address of 2.1 should be encapsulated in DLCI 102 and any DECnet traffic with a next hop DECnet address of 2.5 should be encapsulated in DLCI 201.

Two DLCIs are coming into R2 from the Frame Relay network, 102 and 201. DLCI 102 is a circuit from R2 to R1, and DLCI 201 is a circuit from R2 to R5. R2 is the hub router in this configuration. This means that R2 has a connection between itself and all other routers. R1 and R5 are referred to as spoke routers. A hub router running Frame Relay presents special issues in configuring a router. For example, traffic originating on R1 destined for R5 must be sent from R1 to R2 and from R2 to R5. Although routed traffic can take this path, routing protocol traffic is not normally allowed to take this path due to the rules of split horizon. To get around this issue, you must disable split horizon on this interface. Disabling this enables routing protocol traffic to exit from the same router interface on which it entered. DECnet split horizon is disabled with the following interface command:

```
R2(config)#interface Serial0/1
R2(config-if)#no decnet split-horizon
```

8. *Configure R3 with a DECnet address of 2.3. Also configure it with an interface cost equal to the router number.*

   R3 has a single Ethernet connection to the network through interface E0/0. You must configure DECnet Routing globally along with the DECnet node type and interface costs for this router. The following commands achieve this:

```
R3(config)#decnet routing 2.3
R3(config)#decnet node-type routing-iv

R3(config)#interface Ethernet0/0
R3(config-if)#decnet cost 3
```

9. *Configure R4 with a DECnet address of 2.4. Also configure it with an interface cost equal to the router number.*

   R4 has a single Ethernet connection to the network through interface E0/0. You must configure DECnet routing globally as well as the DECnet node type and interface costs for this router. The following commands achieve this:

```
R4(config)#decnet routing 2.4
R4(config)#decnet node-type routing-iv

R4(config)#interface Ethernet0/0
R4(config-if)#decnet cost 4
```

10. *Configure R5 with a DECnet address of 2.5. Interface S0/1 of R5 should use X.25 encapsulation with an X.25 address of 5555. Interface S0/1 should act as an X.25 DCE.*
    - **a.** Configure the router with an interface cost equal to the router number.
    - **b.** R5 is connected to the Frame Relay network via interface S0/0. Configure R5 so that Frame Relay traffic only uses DLCI 202.

Begin configuring R5 by entering the DECnet routing global commands needed to enable DECnet on the router:

```
R5(config)#decnet routing 2.5
R5(config)#decnet node-type routing-iv
```

Each of the two router interfaces must also have a DECnet cost defined:

```
R5(config)#interface Serial0/0
R5(config-if)#decnet cost 5

R5(config)#interface Serial0/1
R5(config-if)#decnet cost 5
```

R5 is running Frame Relay encapsulation on its S0/0 interface. As depicted in Figure 3-4, a fully meshed Frame Relay network has been defined between routers R1, R2, and R5. Two DLCIs (302 and 202) are defined on interface S0/0 of R5. You want all traffic to only use DLCI 202. In order to do this, disable Frame Relay inverse arp on this interface with the **no frame-relay inverse-arp** command. This will keep R5 from trying to dynamically resolve network-layer addresses for far-end routers. Because Frame Relay inverse arp is disabled, you must manually define your Frame Relay neighbor next hop address information for each protocol. This is done by using the Frame Relay map command.

R5 has two Frame Relay neighbors, R2 and R1. Both of these routers must be reached via DLCI 202. Thus, define two Frame Relay map statements under interface S0/0 as follows:

```
R5(config-if)#frame-relay map decnet 2.1 202 broadcast
R5(config-if)#frame-relay map decnet 2.2 202 broadcast
```

The map statements can be interpreted as follows:

- If DECnet traffic has a next hop address of 2.1, then encapsulate the traffic in DLCI 202.
- If DECnet traffic has a next hop address of 2.2, then encapsulate the traffic in DLCI 202.

You can see that the two map statements have the effect of sending all traffic from R5 into the Frame Relay cloud over DLCI 202.

Interface S0/1 needs to run X.25 encapsulation with an X.25 address of 5555. This interface acts as an X.25 DCE. This is configured with the following interface commands:

```
R5(config-if)#encapsulation x25 dce
R5(config-if)#x25 address 5555
```

X.25 requires a map statement that tells the router which X.25 address to use to get to the next hop. The following interface command is used to define your X.25 map:

```
R5(config-if)#x25 map decnet 2.6 6666 broadcast
```

11. *Configure R6 with a DECnet address of 2.6. Interface S0 of R6 should use X.25 encapsulation with an X.25 address of 6666. Interface S0 should act as an X.25 DTE. Also configure the router with an interface cost equal to the router number.*

R6 has two network connections, one through S0 and another through S1. You must configure DECnet routing globally as well as the DECnet node type and interface costs for each of these interfaces. The following commands achieve this:

```
R6(config)#decnet routing 2.6
R6(config)#decnet node-type routing-iv

R6(config)#interface Serial1
R6(config-if)#decnet cost 6

R6(config)#interface Serial0
R6(config-if)#decnet cost 6
```

In addition, you need to set the X.25 parameters under interface S0. Notice that you must configure an X.25 map telling the router which X.25 address to use to reach the next hop:

```
R6(config-if)#encapsulation x25
R6(config-if)#x25 address 6666
R6(config-if)#x25 map decnet 2.5 5555 broadcast
```

12. *Configure R7 with a DECnet address of 2.7. Also configure the router with an interface cost equal to the router number.*

R7 has two network connections, one through S0 and another through E0. You must configure DECnet Routing globally as well as the DECnet node type and interface costs for each of these interfaces. The following commands achieve this:

```
R7(config)#decnet routing 2.7
R7(config)#decnet node-type routing-iv

R7(config)#interface Ethernet0
R7(config-if)#decnet cost 7

R7(config)#interface Serial0
R7(config-if)#decnet cost 7
```

13. *Configure R3 so that its routing table does not have an entry for R6.*

You want to configure R3 so that no entries for the DECnet 2.6 network are in R3's routing table. This requires a DECnet access list. DECnet access list numbers range from 300 to 399. DECnet access lists use a DECnet source address followed by a mask. The mask has bits set wherever the corresponding bits in the address should be ignored. DECnet addresses and masks are written in decimal notation. As with IP access lists, an implied deny all is included at the end of a DECnet access list.

In order to keep a routing entry for the DECnet 2.6 network from appearing in R3's routing table, you must filter any routing information coming into R3. You can accomplish this with the **decnet in-routing-filter access-list number** command. You must also configure a DECnet access list. The following access list commands provide you with your desired results:

```
R3(config)#access-list 300 deny   2.6 0.0
R3(config)#access-list 300 permit 2.0 0.1023
```

The first line of the access list denies any DECnet traffic with an DECnet source address of 2.6. The second line permits all other DECnet traffic with an area address of 2. All nodes within area 2 are permitted due to the mask value on the second line of the access list being set to 0.1023. Recall from the discussion earlier that DECnet access lists are written in decimal notation. The 0.1023 mask can be written in binary as follows:

```
0.1111111111
```

Because a 1 in a mask causes the corresponding address bit position to be ignored, the 1023 portion of this mask has the effect of ignoring the entire node portion of the DECnet address. The 0 portion of the mask has the effect of only matching DECnet area 2.

Now let's verify that your access list is working properly. Display the routing table with the **show decnet route** command:

```
R3# show decnet route
  Node          Cost  Hops           Next Hop to Node         Expires   Prio
*2.1              5     2        Ethernet0/0   -> 2.2
*2.2              3     1        Ethernet0/0   -> 2.2              42       64   V
*2.3              0     0             (Local) -> 2.3
*2.4              3     1        Ethernet0/0   -> 2.4              40       64   V
*2.5              5     2        Ethernet0/0   -> 2.2
*2.7             16     4        Ethernet0/0   -> 2.2
```

You can see that no route exists to node 2.6 (R6). DECnet access lists running on a router can be viewed with the **show decnet access-lists** command:

```
R3# show decnet access-lists
DECnet access list 300
    deny   2.6 0.0
    permit 2.0 0.1023
```

14. *Configure R7 for LAT translation. A LAT session request to a loopback interface on R7 should be translated into a telnet session into R7.*

The Cisco IOS has the capability to perform translations between different types of protocols. For example, in this section you will be configuring the router to accept an incoming LAT session request and translate it to a Telnet session.

Your first step is to configure a loopback interface on R7 with IP address 1.1.1.1. You must also enter the **lat enabled** command under this interface:

```
R7(config)#int loop0
R7(config-if)#ip address 1.1.1.1 255.255.255.255
R7(config-if)#lat enabled
```

The **translate lat ServerB tcp 1.1.1.1** command causes the router to take a LAT request for ServerB and translate it to a telnet session for 1.1.1.1. Because 1.1.1.1 is an IP address on the same router that the LAT request is going to originate on, you will end up with a telnet session into your own router:

```
R7(config)#translate lat ServerB tcp 1.1.1.1
```

You need to configure a password on the vty port of your router so that you can telnet into it:

```
R7(config)#line vty 0 4
R7(config-line)#password cisco
R7(config-line)#login
```

The **show translate** command is used to display a listing of all defined protocol translations on the router. You can see that your LAT-to-Telnet translation is displayed:

```
R7#show translate

Translate From: LAT SERVERB
          To:   TCP 1.1.1.1 Port 23
          0/0  users active, 0 peak, 0 total, 0 failures
```

You can test your LAT configuration by trying to establish a LAT session to ServerB with the **lat serverb** command. Once youenter the login and enable passwords, you are presented with a telnet session into your router:

```
R7#lat serverb
Trying SERVERB . . . Open
Trying 1.1.1.1  . . .  Open


User Access Verification

Password:
R7>ena
Password:
R7a
```

Once your LAT-to-Telnet translated session had been established, you can use the **show lat sessions** command to view information about your LAT session:

```
R7#show lat sessions

tty0, connection 1 to service serverb

Session data:
  Name SERVERB, Remote Id 1, Local Id 1
  Remote credits 3, Local credits 0, Advertised Credits 2
  Flags: none
  Max Data Slot 255, Max Attn Slot 255, Stop Reason 0

Remote Node data:
Node "R7", usage 3, Local
  Timer 256,  sequence 2,  changes 159,  flags 0x0, protocol 5.2
  Facility 0,  Product code 234,  Product version 48
  Recv 143/101 /1012,  Xmit 143/101 /1012,  0 Dups, 0 ReXmit
  Bad messages: 0,  Bad slots: 0,  Solicits accepted: 0
  Solicits rejected: 0,  Multiple nodes: 0
  Groups:   0
  Service classes:   1

tty2, virtual tty from host R7

Session data:
  Name SERVERB, Remote Id 1, Local Id 1
  Remote credits 2, Local credits 1, Advertised Credits 3
  Flags: Send Credits
  Max Data Slot 255, Max Attn Slot 255, Stop Reason 0

Remote Node data:
Node "R7", usage 3, Local
  Timer 256,  sequence 2,  changes 159,  flags 0x0, protocol 5.2
  Facility 0,  Product code 234,  Product version 48
  Recv 151/106 /1731,  Xmit 151/106 /1731,  0 Dups, 0 ReXmit
  Bad messages: 0,  Bad slots: 0,  Solicits accepted: 0
  Solicits rejected: 0,  Multiple nodes: 0
  Groups:   0
  Groups:   0
```

The **show lat traffic** command gives you traffic and utilization statistics on active LAT circuits:

```
R7#sh lat traffic
Local host statistics:
  2/0  circuits, 2/0  sessions, 2/0  services
  255 sessions/circuit, circuit timer 80, keep-alive timer 20

Recv:   219 messages (0 duplicates),  151 slots,  2675 bytes
        0 bad circuit messages,  10 service messages (6 used)
Xmit:   219 messages (0 retransmit),  151 slots,  2675 bytes
        0 circuit timeouts,  5 service messages

Total:  6 circuits created,  6 sessions
```

Once you close your telnet session, you can verify with the **show lat sessions** command that the LAT session has been closed:

```
R7#exit

[Connection to serverb closed by foreign host]

R7#sh lat sessions
```

15. *Verify that DECnet is functioning properly on your network.*

At this point, you should have a proper DECnet configuration in all routers (R1 through R7) as well as a configured Frame Relay switch. The best place to start your verification is on your Frame switch, making sure that you have active PVCs to R1, R2, and R5.

Connect to the router acting as a Frame Relay switch. Several commands exist to test a Cisco router emulating a Frame Relay switch. The **show frame pvc** command is used to display all PVCs that are passing through the router. Several items in the output of the **show frame pvc** command indicate that the router is acting as a Frame Relay switch. The PVC statistics indicate that S1/0, S1/1, and S1/2 are acting as Frame Relay DCE interfaces. DLCI usage is indicated as Switched. Finally, the indication Num Pkts Switched shows how many frames have been switched through the interface:

```
FrameSwitch#show frame pvc

PVC Statistics for interface Serial1/0  (Frame Relay DCE)

              Active     Inactive      Deleted       Static
  Local         0           0             0             0
  Switched      2           0             0             0
  Unused        0           0             0             0

DLCI = 101, DLCI USAGE = SWITCHED, PVC STATUS = ACTIVE, INTERFACE = Serial1/0

  input pkts 290            output pkts 167           in bytes 120087
  out bytes 68541           dropped pkts 0            in FECN pkts 0
  in BECN pkts 0            out FECN pkts 0           out BECN pkts 0
  in DE pkts 0              out DE pkts 0
  out bcast pkts 12          out bcast bytes 3360          Num Pkts Switched 290

  pvc create time 00:12:58, last time pvc status changed 00:12:48

DLCI = 301, DLCI USAGE = SWITCHED, PVC STATUS = ACTIVE, INTERFACE = Serial1/0

  input pkts 0              output pkts 4             in bytes 0
  out bytes 1120            dropped pkts 0            in FECN pkts 0
  in BECN pkts 0            out FECN pkts 0           out BECN pkts 0
  in DE pkts 0              out DE pkts 0
  out bcast pkts 4           out bcast bytes 1120          Num Pkts Switched 0

  pvc create time 00:04:15, last time pvc status changed 00:04:10

PVC Statistics for interface Serial1/1  (Frame Relay DCE)

              Active     Inactive      Deleted       Static
  Local         0           0             0             0
```

```
  Switched       2           0            0            0
  Unused         0           0            0            0

DLCI = 102, DLCI USAGE = SWITCHED, PVC STATUS = ACTIVE, INTERFACE = Serial1/1

  input pkts 155           output pkts 303          in bytes 65181
  out bytes 123727         dropped pkts 0           in FECN pkts 0
  in BECN pkts 0           out FECN pkts 0          out BECN pkts 0
  in DE pkts 0             out DE pkts 0
  out bcast pkts 13         out bcast bytes 3640         Num Pkts Switched 155

  pvc create time 00:13:01, last time pvc status changed 00:12:52

DLCI = 201, DLCI USAGE = SWITCHED, PVC STATUS = ACTIVE, INTERFACE = Serial1/1

  input pkts 152           output pkts 319          in bytes 65051
  out bytes 136419         dropped pkts 0           in FECN pkts 0
  in BECN pkts 0           out FECN pkts 0          out BECN pkts 0
  in DE pkts 0             out DE pkts 0
  out bcast pkts 13         out bcast bytes 3640         Num Pkts Switched 152

  pvc create time 00:13:03, last time pvc status changed 00:12:53

PVC Statistics for interface Serial1/2  (Frame Relay DCE)

              Active     Inactive      Deleted       Static
  Local          0           0            0            0
  Switched       2           0            0            0
  Unused         0           0            0            0

DLCI = 202, DLCI USAGE = SWITCHED, PVC STATUS = ACTIVE, INTERFACE = Serial1/2

  input pkts 306           output pkts 165          in bytes 132779
  out bytes 68691          dropped pkts 0           in FECN pkts 0
  in BECN pkts 0           out FECN pkts 0          out BECN pkts 0
  in DE pkts 0             out DE pkts 0
  out bcast pkts 13         out bcast bytes 3640         Num Pkts Switched 306

  pvc create time 00:13:05, last time pvc status changed 00:12:55

DLCI = 302, DLCI USAGE = SWITCHED, PVC STATUS = ACTIVE, INTERFACE = Serial1/2

  input pkts 0             output pkts 5            in bytes 0
  out bytes 1400           dropped pkts 0           in FECN pkts 0
  in BECN pkts 0           out FECN pkts 0          out BECN pkts 0
  in DE pkts 0             out DE pkts 0
  out bcast pkts 5          out bcast bytes 1400         Num Pkts Switched 0

  pvc create time 00:04:48, last time pvc status changed 00:04:16
```

The **show frame lmi** command displays the status of the LMI interface between the router FrameSwitch and its directly connected routers, R1, R2, and R5. Again, you can see that the router FrameSwitch is acting as a Frame Relay switch. The indication is the Frame Relay DCE message on the command output. The key item to verify is that the number of status enquiries received is equal to the number of status messages sent. In this case, 103 status messages have been exchanged between the routers and the Frame Relay switch:

```
FrameSwitch#sh frame lmi

LMI Statistics for interface Serial1/0  (Frame Relay DCE) LMI TYPE = ANSI
  Invalid Unnumbered info 0               Invalid Prot Disc 0
  Invalid Unnumbered info 0               Invalid Prot Disc 0
  Invalid Status Message 0                Invalid Lock Shift 0
  Invalid Information ID 0                Invalid Report IE Len 0
  Invalid Report Request 0                Invalid Keep IE Len 0
  Num Status Enq. Rcvd 103                Num Status msgs Sent 103
  Num Update Status Sent 0                Num St Enq. Timeouts 0

LMI Statistics for interface Serial1/1  (Frame Relay DCE) LMI TYPE = ANSI
  Invalid Unnumbered info 0               Invalid Prot Disc 0
  Invalid dummy Call Ref 0                Invalid Msg Type 0
  Invalid Status Message 0                Invalid Lock Shift 0
  Invalid Information ID 0                Invalid Report IE Len 0
  Invalid Report Request 0                Invalid Keep IE Len 0
  Num Status Enq. Rcvd 103                Num Status msgs Sent 103
  Num Update Status Sent 0                Num St Enq. Timeouts 0

LMI Statistics for interface Serial1/2  (Frame Relay DCE) LMI TYPE = ANSI
  Invalid Unnumbered info 0               Invalid Prot Disc 0
  Invalid dummy Call Ref 0                Invalid Msg Type 0
  Invalid Status Message 0                Invalid Lock Shift 0
  Invalid Information ID 0                Invalid Report IE Len 0
  Invalid Report Request 0                Invalid Keep IE Len 0
  Num Status Enq. Rcvd 103                Num Status msgs Sent 103
  Num Update Status Sent 0                Num St Enq. Timeouts 0
```

The **show frame route** command is used when a router is configured as a Frame Relay switch. The command displays all DLCIs that are configured on the router. The following sample output from router FrameSwitch shows that six DLCIs are configured on the router. Each line of the output indicates the route of a different PVC. For example, the first line reads: Any traffic entering interface S1/0 with a DLCI value of 101 will be switched to interface S1/1 with a DLCI value of 102. The **show frame route** output also indicates that all of your DLCIs are currently in an active state:

```
FrameSwitch#show frame route
Input Intf      Input Dlci      Output Intf     Output Dlci      Status
Serial1/0        101             Serial1/1        102              active
Serial1/0        301             Serial1/2        302              active
Serial1/1        102             Serial1/0        101              active
Serial1/1        201             Serial1/2        202              active
Serial1/2        202             Serial1/1        201              active
Serial1/2        302             Serial1/0        301              active
```

Let's verify that your seven-router network is working by starting at the Frame Relay core. Connect to R1 and verify that this router is communicating with the Frame Relay switch. Interface S1/0 should be in an up/up state. The number of LMI enquiries sent should be approximately equal to the number of LMI status enquiries received.

```
R1#show interface s 1/0
Serial1/0  is up, line protocol is up
  Hardware is QUICC Serial
  MTU 1500 bytes, BW 1544 Kbit, DLY 20000 usec, rely 255/255 , load 1/255
```

```
  Encapsulation FRAME-RELAY, loopback not set, keepalive set (10 sec)
  LMI enq sent  189, LMI stat recvd 188, LMI upd recvd 0, DTE LMI up
  LMI enq recvd 0, LMI stat sent  0, LMI upd sent  0
  LMI DLCI 0  LMI type is ANSI Annex D  frame relay DTE
  Broadcast queue 0/64 , broadcasts sent/dropped 628/0 , interface broadcasts 322
  Last input 00:00:03, output 00:00:03, output hang never
  Last clearing of "show interface" counters never
  Queueing strategy: fifo
  Output queue 0/40 , 0 drops; input queue 0/75 , 0 drops
  5 minute input rate 0 bits/sec, 0 packets/sec
  5 minute output rate 0 bits/sec, 0 packets/sec
     573 packets input, 164770 bytes, 0 no buffer
     Received 0 broadcasts, 0 runts, 0 giants, 0 throttles
     1 input errors, 0 CRC, 1 frame, 0 overrun, 0 ignored, 0 abort
     837 packets output, 286705 bytes, 0 underruns
     0 output errors, 0 collisions, 19 interface resets
     0 output buffer failures, 0 output buffers swapped out
     6 carrier transitions
     DCD=up  DSR=up  DTR=up  RTS=up  CTS=up
```

The **show frame map** command should indicate that two static maps have been defined pointing to DECnet addresses 2.2 and 2.5. Both these maps should use DLCI 101 for their traffic. Notice that both of these maps are specific to the DECnet protocol:

```
R1#show frame map
Serial1/0  (up): decnet 2.2 dlci 101(0x65,0x1850), static,
              broadcast,
              CISCO, status defined, active
Serial1/0  (up): decnet 2.5 dlci 101(0x65,0x1850), static,
              broadcast,
              CISCO, status defined, active
```

The show frame pvc command verifies that you are only sending traffic to the network over DLCI 101, even though you have two DLCIs coming into R1:

```
R1#sh frame pvc

PVC Statistics for interface Serial1/0  (Frame Relay DTE)

DLCI = 101, DLCI USAGE = LOCAL, PVC STATUS = ACTIVE, INTERFACE = Serial1/0

  input pkts 414           output pkts 743          in bytes 178411
  out bytes 329343         dropped pkts 0           in FECN pkts 0
  in BECN pkts 0           out FECN pkts 0          out BECN pkts 0
  in DE pkts 0             out DE pkts 0
  out bcast pkts 720        out bcast bytes 327130
  pvc create time 00:36:30, last time pvc status changed 00:34:54

DLCI = 301, DLCI USAGE = UNUSED, PVC STATUS = ACTIVE, INTERFACE = Serial1/0

  input pkts 27            output pkts 0            in bytes 7560
  out bytes 0              dropped pkts 0           in FECN pkts 0
  in BECN pkts 0           out FECN pkts 0          out BECN pkts 0
  in DE pkts 0             out DE pkts 0
  out bcast pkts 0          out bcast bytes 0
```

```
  pvc create time 00:26:25, last time pvc status changed 00:26:15
  Num Pkts Switched 0
```

Let's try to ping the other two routers in the Frame Relay cloud with the **ping decnet** command. The successful pings verify that your Frame Relay cloud is functioning properly.

```
R1#ping decnet 2.2

Type escape sequence to abort.
Sending 5, 100-byte DECnet echos to 2.2, timeout is 5 seconds:
!!!!!
Success rate is 100 percent (5/5), round-trip min/avg/max = 4/4/8 ms

R1#ping decnet 2.5

Type escape sequence to abort.
Sending 5, 100-byte DECnet echos to 2.5, timeout is 5 seconds:
!!!!!
Success rate is 100 percent (5/5), round-trip min/avg/max = 8/8/12 ms
```

An important point to understand about DECnet is that when it is enabled on a router, it changes the MAC addresses of the router's LAN interfaces. It can be seen from the following output that the router's burned in address (BIA) is different from the MAC address currently in use.

Recall that when DECnet is configured on a router, the MAC addresses of any LAN interfaces are changed. DECnet uses an equation that multiplies the area number by 1,024 and then adds on the node number. This number is then converted to a hexadecimal value and appended to address AA00.0400. The number is appended in byte-swapped order, least significant byte first.

You can see from the following output that the new MAC address of interface E0/0 is AA00.0400.0108. DECnet forms the new MAC address for E0/0 by taking the Area value of 2, multiplying it by 1,024, and adding the Node value of 1. This becomes 2 * 1024 + 1, which is 2,049. The 2,049 value is then converted to hexadecimal, which becomes 0801. This number is then byte-swapped to become 0108. The 0108 value is then appended on to the value AA00.0400. The new MAC address for the E0/0 interface of R1 then becomes AA00.0400.0108.

```
R1#sh int e 0/0
Ethernet0/0  is up, line protocol is up
  Hardware is AmdP2, address is aa00.0400.0108 (bia 00e0.1e5b.2601)
  MTU 1500 bytes, BW 10000 Kbit, DLY 1000 usec, rely 128/255 , load 1/255
  Encapsulation ARPA, loopback not set, keepalive not set
  ARP type: ARPA, ARP Timeout 04:00:00
  Last input never, output 00:00:08, output hang never
  Last clearing of "show interface" counters never
  Queueing strategy: fifo
  Output queue 0/40 , 0 drops; input queue 0/75 , 0 drops
  5 minute input rate 0 bits/sec, 0 packets/sec
  5 minute output rate 0 bits/sec, 0 packets/sec
```

```
0 packets input, 0 bytes, 0 no buffer
Received 0 broadcasts, 0 runts, 0 giants, 0 throttles
0 input errors, 0 CRC, 0 frame, 0 overrun, 0 ignored, 0 abort
0 input packets with dribble condition detected
468 packets output, 41023 bytes, 0 underruns
468 output errors, 0 collisions, 3 interface resets
0 babbles, 0 late collision, 0 deferred
468 lost carrier, 0 no carrier
0 output buffer failures, 0 output buffers swapped out
```

You should be familiar with several other DECnet commands. Similar to IP and the other desktop protocols, the router maintains a separate routing table for DECnet. This routing table can be viewed with the **show decnet route** command. Notice that R1 has a route to every other DECnet network in its routing table:

```
R1# show decnet route
  Node         Cost  Hops          Next Hop to Node        Expires  Prio
*2.1             0     0            (Local) -> 2.1
*2.2             1     1           Serial1/0  -> 2.2            42      64  V
*2.3             3     2           Serial1/0  -> 2.2
*2.4             3     2           Serial1/0  -> 2.2
*2.5             3     2           Serial1/0  -> 2.2
*2.6             8     3           Serial1/0  -> 2.2
*2.7            14     4           Serial1/0  -> 2.2
```

Let's take some time to understand the meaning of this routing table:

- All seven of your DECnet addresses are visible in this routing table. This is good because it tells us that you have connectivity to your entire network.
- The next hop to every destination DECnet address is via R2, which is at DECnet address 2.2. This confirms that you are only using one of the two PVCs that are coming into R1.
- Let's understand how the cost column is calculated:

  a. First, you see that the router that you are on, 2.1, has a cost of 0.

  b. Router 2.2, which is your next hop, has a cost of 1. This is due to the fact that you have set the DECnet cost of your S1/0 interface to 1.

  c. The cost to addresses 2.3 (R3), 2.4 (R4), and 2.5 (R5) is 3. The DECnet cost to a specific destination is cumulative. To get to 2.2 (R2), the cost was 1. The DECnet cost on R2 was set to 2. Because R3, R4, and R5 are connected to R2, the cost will be 1 + 2, which is 3.

  d. The route to R6 (2.6) has a cost of 8. Because the cost to get to R5 (2.5) is 3 and the cost on R5 is 5, the total cost to R6 will be 5 + 3 = 8.

  e. The route to R7 (2.7) has a cost of 14. The cost to get to R6 is 8 and the cost on R6 is 6, so the total cost is 8 + 6 = 14.

The **show decnet interface** command is useful for displaying both global- and interface-specific DECnet parameters. In the following output, Ethernet0/0 is using routing input list 300, which is the access list that you defined to block any routes to R6:

```
R3# show decnet interface
Global DECnet parameters for network 0:
  Local address is 2.3, node type is routing-iv
  Nearest Level-2 router is  NONE
  Maximum node is 1023, maximum area is 63, maximum visits is 63
  Maximum paths is 1, path split mode is normal
  Local maximum cost is 1022, maximum hops is 30
  Area maximum cost is 1022, maximum hops is 30
  Static routes *NOT* being sent in routing updates
Ethernet0/0  is up, line protocol is up, encapsulation is ARPA
  Interface cost is 3, priority is 64, DECnet network: 0
  The designated router is 2.4
  Sending HELLOs every 15 seconds, routing updates 40 seconds
  Smallest router blocksize seen is 1498 bytes
  Routing input list is 300, output list is not set
  Access list is not set
  DECnet fast switching is enabled
  Number of L1 router adjacencies is : 2
  Number of non-PhaseIV+ router adjacencies is : 2
  Number of PhaseIV+ router adjacencies is : 0
Serial0/0  is administratively down, line protocol is down, encapsulation is HDLC
  DECnet protocol processing disabled
Serial0/1  is administratively down, line protocol is down, encapsulation is HDLC
  DECnet protocol processing disabled
Loopback0 is up, line protocol is up, encapsulation is LOOPBACK
  DECnet protocol processing disabled
```

The **show decnet interface** command can also be used to verify that the proper split horizon configuration is being used with NBMA networks, such as Frame Relay. In the following output, interface S0/1 is set for Frame Relay encapsulation and split horizon has been turned off:

```
R2# show decnet interface
Global DECnet parameters for network 0:
  Local address is 2.2, node type is routing-iv
  Nearest Level-2 router is  NONE
  Maximum node is 1023, maximum area is 63, maximum visits is 63
  Maximum paths is 1, path split mode is normal
  Local maximum cost is 1022, maximum hops is 30
  Area maximum cost is 1022, maximum hops is 30
  Static routes *NOT* being sent in routing updates
Ethernet0/0  is up, line protocol is up, encapsulation is ARPA
  Interface cost is 2, priority is 64, DECnet network: 0
  The designated router is 2.4
  Sending HELLOs every 15 seconds, routing updates 40 seconds
  Smallest router blocksize seen is 1498 bytes
  Routing input list is not set, output list is not set
  Access list is not set
  DECnet fast switching is enabled
  Number of L1 router adjacencies is : 2
  Number of non-PhaseIV+ router adjacencies is : 2
  Number of PhaseIV+ router adjacencies is : 0
Ethernet0/0 .1 is up, line protocol is up, encapsulation is ARPA
  DECnet protocol processing disabled
Ethernet0/0 .2 is up, line protocol is up, encapsulation is ARPA
```

```
  DECnet protocol processing disabled
Serial0/0  is administratively down, line protocol is down, encapsulation is HDLC
  DECnet protocol processing disabled
Serial0/1  is up, line protocol is up, encapsulation is FRAME-RELAY
  Interface cost is 2, priority is 64, DECnet network: 0
  The designated router is 2.5
  Sending HELLOs every 15 seconds, routing updates 40 seconds
  Smallest router blocksize seen is 1498 bytes
  Routing input list is not set, output list is not set
  Access list is not set
  DECnet fast switching is disabled
  Number of L1 router adjacencies is : 2
  Number of non-PhaseIV+ router adjacencies is : 2
  Number of PhaseIV+ router adjacencies is : 0
  Split horizon is OFF
Loopback0 is up, line protocol is up, encapsulation is LOOPBACK
  DECnet protocol processing disabled
```

Another useful command is the **show decnet traffic** command. This command displays data traffic and routing traffic information for the DECnet protocol:

```
R3# show decnet traffic
Total: 1816 received, 0 format errors, 0 unimplemented
       0 not a gateway, 0 no memory, 0 no routing vector
       0 non-empty queue encountered
Hellos: 1097 received, 0 bad, 0 other area, 371 sent
Level 1 routing: 719 received, 0 bad, 0 other area, 357 sent
Level 2 routing: 0 received, 0 not primary router, 1 sent
Data: 0 received, 0 not long format, 0 too many visits
      0 forwarded, 0 returned, 0 converted, 0 local destination
      0 access control failed, 0 no route, 0 encapsulation failed
      0 inactive network, 0 incomplete map
```

The **show decnet neighbors** command can be used to verify connectivity between adjacent routers. As you can see from the following output, R2 is neighbors with R1, R3, R4, and R5:

```
R2# show decnet neighbors
Net Node      Interface           MAC address    Flags
0   2.1       Serial0/1            0066.0000.0000 V
0   2.3       Ethernet0/0          aa00.0400.0308 V
0   2.4       Ethernet0/0          aa00.0400.0408 V
0   2.5       Serial0/1            00c9.0000.0000 V
```

Another troubleshooting tool is the **show interface** command. Let's connect to R7 and issue the show interface s 0 command. You can see that the NCP for DEC has been opened. DECnet does not come up on an interface until DECCP has been opened.

```
R7#sh int s 0
Serial0 is up, line protocol is up
  Hardware is HD64570
  Internet address is 142.2.1.10/29
  MTU 1500 bytes, BW 1544 Kbit, DLY 20000 usec, rely 255/255 , load 1/255
  Encapsulation PPP, loopback not set, keepalive set (10 sec)
  LCP Open
```

```
Open: DECCP, IPCP, CDPCP, IPXCP
Last input 00:00:01, output 00:00:01, output hang never
Last clearing of "show interface" counters never
Queueing strategy: fifo
Output queue 0/40 , 0 drops; input queue 0/75 , 0 drops
5 minute input rate 0 bits/sec, 1 packets/sec
5 minute output rate 0 bits/sec, 1 packets/sec
   146 packets input, 20299 bytes, 0 no buffer
   Received 146 broadcasts, 0 runts, 0 giants, 0 throttles
   0 input errors, 0 CRC, 0 frame, 0 overrun, 0 ignored, 0 abort
   130 packets output, 16519 bytes, 0 underruns
   0 output errors, 0 collisions, 5 interface resets
   0 output buffer failures, 0 output buffers swapped out
   0 carrier transitions
   DCD=up  DSR=up  DTR=up  RTS=up  CTS=up
```

IPX

16. *Configure R1 as follows:*

a. Interface E0/0 should be assigned IPX network number 12. It should use the default encapsulation for an IPX Ethernet interface.

b. Create a loopback 0 interface with an IPX network number of 1.

c. The IPX node number should be set to 0001.0001.0001.

d. Interface S1/0 should be assigned IPX network number 8. Use a physical interface.

e. Run IPX EIGRP on interface S1/0. All other interfaces should run IPX RIP.

IPX is enabled on a router with the **IPX routing *node*** global command. If you do not specify the node number, the router will use the MAC address of the first Ethernet, Token Ring, or FDDI interface card on the router. It is helpful to specify a known node number, as this will make it easier to manage your network. For example, with R1's node set to 1.1.1, interface S1/0 of R1 becomes IPX address 8.1.1.1, which is easier to remember than 8.aa00.0400.0108 (aa00.0400.0108 is the MAC address of interface E0/0 of R1). Therefore, enable IPX routing on R1 with the following command:

```
R1(config)#ipx routing 1. 1. 1
```

As with IP, loopback interfaces can be assigned an IPX address. IPX addresses are assigned with the IPX network command. Assign Interface Loopback0 IPX network number 1 with the following commands:

```
R1(config)#interface Loopback0
R1(config-if)#ipx network 1
```

Interface E0/0 needs to be assigned IPX network address 12. You'll want to use the default Ethernet encapsulation. Cisco supports four different Ethernet encapsulations. The default encapsulation that Cisco uses is NOVELL-ETHER. No interface-specific commands are needed to use this default encapsulation. Interface E0/0 would be configured with the following commands:

```
R1(config)#interface Ethernet0/0
R1(config-if)#ipx network 12
```

Interface S1/0 needs to be assigned IPX network address 8. It uses Frame Relay encapsulation. Recall from the earlier DECnet configuration steps on this interface that you have already disabled inverse arp on this interface with the command **no frame-relay inverse-arp**. With inverse arp disabled, R1 does not try to resolve neighbor router IPX addresses. Similar to your DECnet configuration, you will need to manually define Frame Relay map statements that tell R1 which DLCIs to use to reach its neighbors. The following commands configure IPX on interface S1/0:

```
R1(config)#interface Serial1/0
R1(config-if)#ipx network 8
R1(config-if)#frame-relay map ipx 8.0002.0002.0002 101 broadcast
R1(config-if)#frame-relay map ipx 8.0005.0005.0005 101 broadcast
```

The two Frame Relay map statements can be interpreted as follows:

- If IPX traffic has a next hop address of 8.2.2.2, then encapsulate the traffic in DLCI 101.
- If IPX traffic has a next hop address of 8.5.5.5, then encapsulate the traffic in DLCI 101.

These two maps cause all traffic leaving R1 going into the Frame Relay cloud to be encapsulated in DLCI 101. This ensures that the other DLCI defined on interface S1/0 (DLCI 301) will not be used.

Notice how the Frame Relay maps refer to node addresses 2.2.2 and 5.5.5. This shows why it is important to assign a node address when enabling IPX on a router. Frame Relay maps would be more complex to configure if you had to use an actual 48-bit MAC address for your neighbor's node, rather than the node address that you assign when enabling IPX routing.

IPX supports three different routing protocols: IPX RIP, IPX EIGRP, and NLSP. IPX RIP is enabled by default on any interface running IPX. EIGRP and NLSP have to be specifically configured. You want to run EIGRP on interface S1/0, which is IPX network 8. The following commands enable EIGRP on IPX network 8:

```
R1(config)#ipx router eigrp 1
R1(config-ipx-router)#network 8

R1(config)#ipx router rip
R1(config-ipx-router)#no network 8
```

Notice that you also have to disable IPX RIP on network 8 after enabling IPX EIGRP.

17. *Configure R2 as follows:*

a. Interface E0/0 should be assigned two different IPX network numbers, 10 and 11.

b. Create a loopback 0 interface with an IPX network number of 2.

c. The IPX node number should be set to 0002.0002.0002.

d. Interface S0/1 should be assigned IPX network number 8. Use a physical interface.

e. Run IPX EIGRP on all interfaces except the loopback interface, which should run IPX RIP.

Begin by enabling IPX routing on this router and define the node to be 0002.0002.0002. Cisco enables the leading zeros to be left out when defining the node, so the following command can be used:

```
R2(config)#ipx routing 2. 2. 2
```

Interface Loopback 0 needs to be configured as IPX network 2. The following commands will perform this task:

```
R2(config)#interface Loopback0
R2(config-if)#ipx network 2
```

You also want to have two different IPX networks assigned to Ethernet 0/0. Cisco supports two methods for achieving this:

- The first method is to define secondary addresses on a physical interface. Cisco does not recommend this method and has stated that they will not be supporting it in the future. Another drawback of this method is that you cannot run EIGRP on the interface because it contains secondary addresses.
- The second method is to use subinterfaces, which is the preferred method. Another advantage of this method is that it supports EIGRP. As previously mentioned, Cisco supports four types of encapsulation on an Ethernet interface. Each subinterface must use a different encapsulation. All other interfaces using the same network number must have the same encapsulation.

Let's use the standard NOVELL-ETHER encapsulation for IPX network 10 and the ARPA encapsulation for IPX network 11. Each of these networks will be defined on a subinterface because this is the preferred method, and this method supports EIGRP. The following commands will configure your two networks on E0/0:

```
R2(config)#interface Ethernet0/0

R2(config-if)#interface Ethernet0/0.1
R2(config-subif)#ipx network 10

R2(config-subif)#interface Ethernet0/0.2
R2(config-subif)#ipx network 11 encapsulation ARPA
```

Interface S0/1 needs to be assigned to IPX network 8. Frame Relay inverse arp has already been disabled when DECnet was configured on this interface. As with your DECnet Frame Relay map statements, you must now define Frame Relay map statements for IPX. The commands for configuring IPX on interface S0/1 should be entered as follows:

```
R2(config)#interface Serial0/1
R2(config-if)#ipx network 8
R2(config-if)#no ipx split-horizon eigrp 1
R2(config-if)#frame-relay map ipx 8.0001.0001.0001 102 broadcast
R2(config-if)#frame-relay map ipx 8.0005.0005.0005 201 broadcast
```

The two Frame Relay map statements define next hop addresses to Routers R1 and R5. The map statements can be read as follows:

- IPX traffic with a next hop address of 8.1.1.1 should be encapsulated in DLCI 102.
- IPX traffic with a next hop address of 8.5.5.5 should be encapsulated in DLCI 201.

Notice the statement **no ipx split-horizon eigrp 1** has been entered under your configuration for interface S0/1. R2 is a hub router and has a PVC from itself to all other routers (R1 and R5). A hub router running Frame Relay presents special issues in configuring a router. For example, traffic originating on R1 destined for R5 must be sent from R1 to R2 and from R2 to R5. Although routed traffic can take this path, routing protocol traffic is not normally allowed to take this path due to the rules of split horizon. To get around this issue, you need to disable split horizon on this interface. Disabling split horizon enables routing protocol traffic to exit from the same router interface where it entered.

You'll want to run EIGRP on all networks except the loopback network. This means that EIGRP should be run on S0/1 and both subinterfaces on E0/0. You also must specifically define EIGRP on these networks. Notice that the EIGRP autonomous-system-number matches on each EIGRP enabled router in your network.

```
R2(config)#ipx router eigrp 1
R2(config-ipx-router)#network 8
R2(config-ipx-router)#network 10
R2(config-ipx-router)#network 11
```

You should also disable IPX RIP on those networks that are running IPX EIGRP:

```
R2(config)#ipx router rip
R2(config-ipx-router)#no network 10
R2(config-ipx-router)#no network 11
R2(config-ipx-router)#no network 8
```

18. *Configure R3 as follows:*

a. Interface E0/0 should be assigned IPX network number 11.

b. The IPX node number should be set to 0003.0003.0003.

c. Create a loopback 0 interface with an IPX network number of 3.

d. Run IPX EIGRP on all interfaces except the loopback interface, which should run IPX RIP.

Begin your IPX configuration of R3 by globally enabling IPX routing with the **ipx routing 3.3.3** command:

```
R3(config)#ipx routing 3. 3. 3
```

Interface Loopback 0 needs to be defined with an IPX network number of 3. Enter the following commands to configure Loopback 0:

```
R3(config)#interface Loopback0
R3(config-if)#ipx network 3
```

Interface E0/0 has been assigned to IPX network 11. Recall from the previous configuration step that you had to configure a different Ethernet encapsulation for each different IPX network that you wanted on your LAN. Also, each host on the LAN that was on the same network had to be configured with the same encapsulation. You defined IPX network 11 on R2's E0/0 interface with ARPA encapsulation. You therefore need to define the E0/0 interface of R3 to have ARPA encapsulation as well. Because you only have a single IPX network defined on the E0/0 interface of R3, you can define your IPX network on the physical interface. The following commands configure E/0 on R3:

```
R3(config)#interface Ethernet0/0
R3(config-if)#ipx network 11 encapsulation ARPA
```

You'll want to run IPX EIGRP on network 11 and IPX RIP on the loopback network 2. IPX RIP is enabled on all interfaces by default, but IPX EIGRP must be explicitly enabled. The following commands enable IPX EIGRP on network 11:

```
R3(config)#ipx router eigrp 1
R3(config-ipx-router)#network 11
```

IPX RIP should be disabled on network 11 with the following commands:

```
R3(config)#ipx router rip
R3(config-ipx-router)#no network 11
```

19. *Configure R4 as follows:*

a. Interface E0/0 should be assigned IPX network number 10.

b. The IPX node number should be set to 0004.0004.0004.

c. Create a loopback 0 interface with an IPX network number of 4.

d. Run IPX EIGRP on all interfaces except the loopback interface, which should run IPX RIP.

Start your configuration of R4 by globally enabling IPX routing and setting the node to 4.4.4:

```
R4(config)#ipx routing 4. 4. 4
```

Interface Loopback0 needs to be assigned to IPX network 4:

```
R4(config)#interface Loopback0
R4(config-if)#ipx network 4
```

Interface Ethernet0/0 is assigned IPX network 10, which was assigned the default Ethernet encapsulation when you configured R2. In order for traffic to pass between Routers R2 and R4, you also need to set interface Ethernet0/0 on R4 to the default encapsulation:

```
R4(config)#interface Ethernet0/0
R4(config-if)#ipx network 10
```

IPX network 10 needs to be configured to run IPX EIGRP:

```
R4(config)#ipx router eigrp 1
R4(config-ipx-router)#network 10
```

IPX RIP should be disabled on IPX network 10:

```
R4(config)#ipx router rip
R4(config-ipx-router)#no network 10
```

20. *Configure R5 as follows:*

- **a.** Interface S0/0 should be assigned IPX network number 8. Use a physical interface.
- **b.** The IPX node number should be set to 0005.0005.0005.
- **c.** Create a loopback 0 interface with an IPX network number of 5.
- **d.** Run IPX EIGRP on interface S0/0.
- **e.** Run IPX RIP on the loopback interface.
- **f.** Using internal network number 100, run IPX WAN with NLSP on interface S0/1.

Start by globally enabling IPX on this router with the **ipx routing 5.5.5** command:

```
R5(config)#ipx routing 5.5.5
```

Next, define a unique internal IPX network number. This is a requirement when NLSP will be used on a router. Assign R5 the IPX internal network 100:

```
R5(config)#ipx internal-network 100
```

Interface Loopack0 needs to be assigned to IPX network 5. This is accomplished with the following commands:

```
R5(config)#interface Loopback0
R5(config-if)#ipx network 5
```

R5 is connected to the Frame Relay cloud via interface S0/0. R5 acts as a spoke router, and two DLCIs are coming into R5 from the Frame Relay switch. You want all traffic coming into and going out of R5 to use DLCI 202. You have already disabled Frame Relay inverse arp when you configured DECnet on this interface in a previous step. Interface S0/0 is assigned IPX network 8. You must also configure a Frame Relay map on this interface that defines the IPX address to DLCI mappings for other routers residing on the Frame Relay cloud. The following commands should be entered for interface S0/0:

```
R5(config)#interface Serial0/0
R5(config-if)#ipx network 8
R5(config-if)#frame-relay map ipx  8.0001.0001.0001 202 broadcast
R5(config-if)#frame-relay map ipx  8.0002.0002.0002 202 broadcast
```

The Frame Relay maps can be interpreted as follows: Traffic destined for either IPX network 8.1.1.1 or IPX network 8.2.2.2 should be encapsulated in DLCI 202.

Interface S0/1 needs to run IPX WAN with the NLSP routing protocol. The first configuration step is to enable IPX WAN on the interface. The command **ipx ipxwan 0 unnumbered R5** starts the IPX WAN processing and uses the defined internal network number on the router as the local node id. Next, you have to enable NLSP on the interface with the **ipx nlsp area1 enable**. The area1 argument is the tag value that is used to name the NLSP process. Because you are running X.25 encapsulation on this interface, you need to put the IPX configuration under a point-to-point subinterface. Notice that you also need to define an X.25 map statement that defines your next hop to R6. The **0.0000.0101.0000** IPX address in the X.25 map is the IPX address of Interface S0.2 on R6.

```
R5(config)#interface Serial0/1.2 point-to-point
R5(config-subif)#ipx ipxwan 0 unnumbered R5
R5(config-subif)#ipx nlsp area1 enable
R5(config-subif)#x25 map ipx 0.0000.0101.0000 6666 broadcast
```

You'll want to run EIGRP on network 8. An EIGRP AS number of 1 has been used for each of your router interfaces running EIGRP. Although IPX RIP and IPX EIGRP feature automatic redistribution, IPX EIGRP and IPX NLSP do not automatically redistribute routing information between routing processes. You must manually redistribute between IPX EIGRP and IPX NLSP. The **redistribute nlsp area1** command causes NLSP area1 routes to be redistributed into EIGRP AS number 1:

```
R5(config)#ipx router eigrp 1
R5(config-ipx-router)#redistribute nlsp area1
R5(config-ipx-router)#network 8
```

You must also define an NLSP process on this router. Currently, the tag identifier of area1 is being used to identify this process. The **area-address 0 0** command causes all networks to be part of the NLSP area. You must redistribute your EIGRP AS number 1 routes into NLSP by entering the command **redistribute eigrp 1:**

```
R5(config)#ipx router nlsp area1
R5(config-ipx-router)#area-address 0 0
R5(config-ipx-router)#redistribute eigrp 1
```

Because you are running IPX EIGRP on network 8, you should disable IPX RIP on this network:

```
R5(config)#ipx router rip
R5(config-ipx-router)#no network 8
```

21. *Configure R6 as follows:*

a. Interface S1 should be assigned IPX network number 9.

b. The IPX node number should be set to 0006.0006.0006.

c. Create a loopback 0 interface with an IPX network number of 6.

d. Run IPX EIGRP on interface S1.

e. Run IPX RIP on the loopback interface.

f. Using internal network number 101, run IPX WAN with NLSP on interface S0.

Begin configuring R6 by globally enabling IPX on the router with the **ipx routing 6.6.6** command. Because you will be running NLSP on this router, you need to define a unique internal IPX network number. You assigned an internal network number of 100 to R5, so you will assign an internal network number of 101 to R6:

```
R6(config)#ipx routing 6.6.6
R6(config)#ipx internal-network 101
```

Interface Loopback 0 is assigned IPX network 6 with the following commands:

```
R6(config)#interface Loopback0
R6(config-if)#ipx network 6
```

Interface S0 will run IPX WAN with the NLSP routing protocol. As in the previous step, you must first enable IPX WAN on the interface and then enable NLSP. Also use the same tag name, area1, that you previously did on R5. Because you are running X.25 encapsulation on this interface, you need to define the IPX parameters under a point-to-point subinterface. Notice that you also have to define an X.25 map to R5. The **0.0000.0100.0000** address in the X.25 map is the IPX address of Interface S0/1.2 on R5:

```
R6(config)#interface Serial0.2 point-to-point
R6(config-subif)#ipx ipxwan 0 unnumbered R6
R6(config-subif)#ipx nlsp area1 enable
R6(config-subif)#x25 map ipx 0.0000.0100.0000 5555 broadcast
```

Interface Serial1 is assigned IPX network 9:

```
R6(config)#interface Serial1
R6(config-if)#ipx network 9
```

You'll want to run EIGRP on IPX network 9. Use the same EIGRP AS number as you did on the other routers running EIGRP in your network. You must also redistribute any NLSP-learned routes into EIGRP:

```
R6(config)#ipx router eigrp 1
R6(config-ipx-router)#redistribute nlsp area1
R6(config-ipx-router)#network 9
```

Now define NLSP on R6. Use the same area1 tag that you used on interface S0. You also have to redistribute any EIGRP-learned routes into the NLSP routing process:

```
R6(config)#ipx router nlsp area1
R6(config-ipx-router)#area-address 0 0
R6(config-ipx-router)#redistribute eigrp 1
```

Because you are running EIGRP on IPX network 9, you should disable IPX RIP on this network:

```
R6(config)#ipx router rip
R6(config-ipx-router)#no network 9
```

22. *Configure R7 as follows:*

a. Interface S0 should be assigned IPX network number 9.

b. The IPX node number should be set to 0007.0007.0007.

c. Create a loopback 0 interface with an IPX network number of 7.

d. Run IPX EIGRP on all interfaces except the loopback interface, which should run IPX RIP.

e. Interface E0 should be assigned IPX network number 13.

R7 is assigned IPX node number 7.7.7:

```
R7(config)#ipx routing 7.7.7
```

Interface Loopback 0 is assigned IPX network 7:

```
R7(config)#interface Loopback0
R7(config-if)#ipx network 7
```

Interface E0 is assigned IPX network 13:

```
R7(config)#interface Ethernet0
R7(config-if)#ipx network 13
```

Interface S0 is assigned to IPX network 9:

```
R7(config)#interface Serial0
R7(config-if)#ipx network 9
```

You'll want to run EIGRP on IPX network 9. Notice that you use the same EIGRP AS number (1) as you have on the other routers in your network running EIGRP.

```
R7(config)#ipx router eigrp 1
R7(config-ipx-router)#network 9
```

Because you are running EIGRP on IPX network 9, you should disable IPX RIP processing:

```
R7(config)#ipx router rip
R7(config-ipx-router)#no network 9
```

23. *Configure two static SAP entries on R6 as follows:*

a. The first static entry should be for a print server residing on IPX network 13 on R7.

b. The second static entry should be for a file server residing on IPX network 13 on R7.

Normally, an IPX server sends out SAP broadcasts to advertise their services. Many SAP types are defined for functions such as print and file servers. The router receives these SAP advertisements and stores them in its SAP table. You can also add static SAP entries into the SAP table. A static SAP always takes precedence over any dynamically learned SAP updates. The predefined SAP value for a file server is 4 and the predefined value for a print server is 7. The last argument of the ipx sap command is the hop count to the server. Because R6 is one hop away from R7, set this value to a 1:

```
R6(config)#ipx sap 4 FileServer1 13.aa00.0400.0708 451 1
R6(config)#ipx sap 7 PrintServer1 13.aa00.0400.0708 451 1
```

24. *Configure R1 so that it cannot see any print servers:*

Once the two static SAP entries have been entered in the previous step, all routers, except R7, have a SAP entry to these two servers. You need to create an access list that will restrict R1 from seeing any print servers. Access lists 1000 through 1099 are used for IPX SAP access filtering. Because you want to restrict SAP information coming into R1, place an **ipx input-sap-filter** statement on interface S1/0. The access list, list 1000 in this case, denies any SAP service type of 7 and permits all other SAP service types:

```
R1(config)#interface Serial1/0
R1(config-if)#ipx input-sap-filter 1000

R1(config)#access-list 1000 deny FFFFFFFF 7
R1(config)#access-list 1000 permit FFFFFFFF 0
```

You can verify the access lists defined on R1 with the **show ipx access-list** command. You can see that the router indicates that this is a SAP access list. The router knows this from the access list number. Access lists in the range of 1000 to 1099 are defined as IPX SAP access lists.

```
R1# show ipx access-list
IPX SAP access list 1000
    deny FFFFFFFF 7
    permit FFFFFFFF 0
```

You enabled the IPX SAP filter by applying an IPX input SAP filter on interface S1/0. You can verify that the IPX filter has been defined on the S1/0 interface by using the **show ipx interface** command:

```
R1# show ipx interface s 1/0
Serial1/0  is up, line protocol is up
  IPX address is 8.0001.0001.0001 [up]
  Delay of this IPX network, in ticks is 6 throughput 0 link delay 0
  IPXWAN processing not enabled on this interface.
  IPX SAP update interval is 1 minute(s)
  IPX type 20 propagation packet forwarding is disabled
  Incoming access list is not set
  Outgoing access list is not set
  IPX helper access list is not set
  SAP GNS processing enabled, delay 0 ms, output filter list is not set
  SAP Input filter list is 1000
  SAP Output filter list is not set
  SAP Router filter list is not set
  Input filter list is not set
  Output filter list is not set
  Router filter list is not set
  Netbios Input host access list is not set
  Netbios Input bytes access list is not set
  Netbios Output host access list is not set
  Netbios Output bytes access list is not set
  Updates each 60 seconds, aging multiples RIP: 3 SAP: 3
  SAP interpacket delay is 55 ms, maximum size is 480 bytes
  RIP interpacket delay is 55 ms, maximum size is 432 bytes
  Watchdog processing is disabled, SPX spoofing is disabled, idle time 60
  IPX accounting is disabled
  IPX fast switching is configured (enabled)
  RIP packets received 0, RIP packets sent 1
  SAP packets received 0, SAP packets sent 1
```

The **show ipx servers** command can be used to verify that R1 only sees a single server, FileServer1. This server passes the input SAP filter because its SAP type is a 4. You are filtering on SAP types of 7.

```
R1# show ipx servers
Codes: S - Static, P - Periodic, E - EIGRP, N - NLSP, H - Holddown, + = detail
1 Total IPX Servers
```

```
Table ordering is based on routing and server info

   Type Name                        Net     Address    Port     Route Hops Itf
E     4 FileServer1              13.aa00.0400.0708:0451 287232000/03    4  Se1/0
```

25. *Configure R2 so that R3 and R4 do not have any entries for IPX network 7 in their routing tables.*

You'll want to create an access list on R2 so that no route updates reach R3 and R4, which contain routes to IPX network 7. Access lists 800 through 899 define standard IPX access lists. You must define access list 800, which denies network 7 and permits all other networks:

```
R2(config)#access-list 800 deny 7
R2(config)#access-list 800 permit FFFFFFFF
```

Apply this access list to the EIGRP routing process on R2. Apply it as an outbound distribute list:

```
R2(config)#ipx router eigrp 1
R2(config-ipx-router)#distribute-list 800 out
R2(config-ipx-router)#network 8
R2(config-ipx-router)#network 10
R2(config-ipx-router)#network 11
```

You can verify your access list by typing the **show ipx access-list** command. You used IPX access list 800. Access list numbers 800 to 899 are used for standard IPX access lists.

```
R2# show ipx access-list
IPX access list 800
    deny 7
    permit FFFFFFFF
```

Connect to R4 and verify with the **show ipx route** command that you do not see any route to IPX network 7:

```
R4# show ipx route
Codes: C - Connected primary network,    c - Connected secondary network
       S - Static, F - Floating static, L - Local (internal), W - IPXWAN
       R - RIP, E - EIGRP, N - NLSP, X - External, A - Aggregate
       s - seconds, u - uses

14 Total IPX routes. Up to 1 parallel paths and 16 hops allowed.

No default route known.

C          4 (UNKNOWN),        Lo0
C         10 (NOVELL-ETHER),   Et0/0
E          1 [2323456/1 ] via      10.aa00.0400.0208, age 00:46:28,
                        1u, Et0/0
E          2 [409600/1 ] via       10.aa00.0400.0208, age 00:47:07,
                        1u, Et0/0
E          3 [435200/1 ] via       10.aa00.0400.0208, age 00:47:07,
                        1u, Et0/0
```

```
E          5 [2323456/1 ] via      10.aa00.0400.0208, age 00:46:25,
                     1u, Et0/0
E          6 [268441600/2 ] via    10.aa00.0400.0208, age 00:46:25,
                     1u, Et0/0
E          8 [2195456/0 ] via      10.aa00.0400.0208, age 00:47:02,
                     1u, Et0/0
E          9 [268441600/2 ] via    10.aa00.0400.0208, age 00:46:25,
                     1u, Et0/0
E         11 [307200/0 ] via       10.aa00.0400.0208, age 00:47:08,
                     1u, Et0/0
E         12 [2221056/1 ] via      10.aa00.0400.0208, age 00:46:29,
                     1u, Et0/0
E         13 [286745600/3 ] via    10.aa00.0400.0208, age 00:46:26,
                     99u, Et0/0
E        100 [46379776/1 ] via     10.aa00.0400.0208, age 00:46:26,
                     1u, Et0/0
E        101 [269849600/2 ] via    10.aa00.0400.0208, age 00:46:26,
                     1u, Et0/0
```

Connect to R3 and verify with the **show ipx route** command that you do not see any route to IPX network 7:

```
R3# show ipx route
Codes: C - Connected primary network,    c - Connected secondary network
       S - Static, F - Floating static, L - Local (internal), W - IPXWAN
       R - RIP, E - EIGRP, N - NLSP, X - External, A - Aggregate
       s - seconds, u - uses

14 Total IPX routes. Up to 1 parallel paths and 16 hops allowed.

No default route known.

C          3 (UNKNOWN),        Lo0
C         11 (ARPA),           Et0/0
E          1 [2323456/1 ] via      11.aa00.0400.0208, age 00:45:43,
                     1u, Et0/0
E          2 [409600/1 ] via       11.aa00.0400.0208, age 00:46:24,
                     1u, Et0/0
E          4 [435200/1 ] via       11.aa00.0400.0208, age 00:46:24,
                     1u, Et0/0
E          5 [2323456/1 ] via      11.aa00.0400.0208, age 00:45:40,
                     1u, Et0/0
E          6 [268441600/2 ] via    11.aa00.0400.0208, age 00:45:40,
                     1u, Et0/0
E          8 [2195456/0 ] via      11.aa00.0400.0208, age 00:46:17,
                     1u, Et0/0
E          9 [268441600/2 ] via    11.aa00.0400.0208, age 00:45:40,
                     1u, Et0/0
E         10 [307200/0 ] via       11.aa00.0400.0208, age 00:46:24,
                     1u, Et0/0
E         12 [2221056/1 ] via      11.aa00.0400.0208, age 00:45:44,
                     1u, Et0/0
E         13 [286745600/3 ] via    11.aa00.0400.0208, age 00:45:41,
                     97u, Et0/0
E        100 [46379776/1 ] via     11.aa00.0400.0208, age 00:45:41,
                     1u, Et0/0
E        101 [269849600/2 ] via    11.aa00.0400.0208, age 00:45:41,
                     1u, Et0/0
```

26. *Configure Dial on Demand routing (DDR) between R1 and R6. R1 should call R6 when R1's Frame Relay connection is lost. Configure DDR so that R6 accepts R1's call, authenticate R1, and then securely call R1 back. Use IPX network 14 for the ISDN BRI network.*

This question requires a DDR configuration on R1 and R6. In addition, you will need to configure both routers for PPP callback. PPP callback is an IOS feature that enables secure callbacks in a DDR environment. In this case, R1 is the PPP callback client and R6 is the PPP callback server. When R1 places a call to R6, the two routers authenticate. If the authentication is successful, R6 disconnects the call and calls back R1.

Start your DDR configuration by configuring R1, but you must first define a username on R1 that will be used during your CHAP authentication. R1 and R6 must have a username statement for each other and must share the same common password.

```
R1(config)#username R6 password 0 cisco
```

The ISDN switch type of the Adtran Atlas 800 ISDN switch must be defined in global configuration mode:

```
R1(config)#isdn switch-type basic-ni1
```

Interface BRI0/0 needs to be configured for DDR. As shown in the following interface printout, define the following items on interface BRI0/0:

- Define an IPX network number of 14.
- Since you are using National ISDN, you need to define both your SPIDs on your BRI.
- Define a dialer map that maps your next hop IPX address of 14.6.6.6 to your dial number, which is 8995201.
- Define the load threshold that will determine under which load factor the second B channel of the BRI will be activated.
- Configure a dialer group that defines your interesting traffic.
- Because R1 is the PPP callback client, use the **ppp callback request** command.
- Configure the interface for CHAP authentication:

```
R1(config)#interface BRI0/0
R1(config-if)#no ip address
R1(config-if)#encapsulation ppp
R1(config-if)#ipx network 14
R1(config-if)#isdn spid1 5101 8995101
R1(config-if)#isdn spid2 5102 8995102
R1(config-if)#dialer map ipx 14.0006.0006.0006 name R6 broadcast 8995201
R1(config-if)#dialer load-threshold 255 either
R1(config-if)#dialer-group 1
R1(config-if)#ppp callback request
R1(config-if)#ppp authentication chap
```

Dialer list 1 is associated with IPX access list 900. Access list 900 defines your interesting traffic, which causes the ISDN circuit to be established:

```
R1(config)#dialer-list 1 protocol ipx list 900
```

Access list 900 defines the following criteria for interesting traffic:

- Cisco IPX pings are interesting.
- Cisco IPX RIP traffic is not interesting.
- Cisco IPX SAP traffic is not interesting.
- All other traffic is interesting.

```
R1(config)#access-list 900 permit any any cping
R1(config)#access-list 900 deny rip
R1(config)#access-list 900 deny sap
R1(config)#access-list 900 permit any any
```

Define an IPX floating static route on R1. This route defines R6 (IPX network address 14.6.6.6) as your next hop. Any traffic for IPX networks that are not listed in the IPX routing table is sent to this next hop address:

```
R1(config)#ipx route default 14.0006.0006.0006 floating-static
```

The configuration on R6 is similar to the configuration on R1, except for the PPP callback commands. Recall that R1 is the PPP callback client, while R6 is the PPP callback server.

Now globally define your CHAP username and ISDN call control for your Adtran Atlas 800 switch:

```
R6(config)#username R1 password 0 cisco

R6(config)#isdn switch-type basic-ni1
```

As with R1, you define your IPX network, your BRI SPID values, your dialer map, your dialer group, your load threshold, and your CHAP authentication. What is different from the configuration on R1 is the PPP callback commands. You need the **dialer callback secure** and the **ppp callback accept** commands under the BRI 0 interface. Also notice that a **class dial1** statement is added to your dialer map.

```
R6(config)#interface BRI0
R6(configif)#no ip address
R6(configif)#encapsulation ppp
R6(configif)#ipx network 14
R6(configif)#isdn spid1 5201 8995201
R6(configif)#isdn spid2 5202 8995202
R6(configif)#dialer callback-secure
R6(configif)#dialer map ipx 14.0001.0001.0001 name R1 class dial1 broadcast 8995101
```

```
R6(configif)#dialer load-threshold 255 either
R6(configif)#dialer-group 1
R6(configif)#ppp callback accept
R6(configif)#ppp authentication chap
```

You must configure a dialer map class for PPP callback that identifies which router to return the call to by looking up the authenticated hostname in the dialer map statement:

```
R6(config)#map-class dialer dial1
R6(config-map-class)#dialer callback-server username
```

Dialer list 1 is associated with IPX access list 900, which will define your interesting traffic:

```
R6(config)#dialer-list 1 protocol ipx list 900
```

Access list 900 only permits Cisco IPX pings to bring up the ISDN circuit:

```
R6(config)#access-list 900 permit any any cping
R6(config)#access-list 900 deny rip
R6(config)#access-list 900 deny sap
```

To verify that your DDR configuration is working, you first need to make sure that your ISDN circuits are functional. The **show isdn status** command verifies that your ISDN circuit is active. The output from this command on R1 is shown in the following output. The reader should verify that the ISDN circuit on R6 is also active.

```
R1#show isdn status
The current ISDN Switchtype = basic-ni1
ISDN BRI0/0  interface
    Layer 1 Status:
        ACTIVE
    Layer 2 Status:
        TEI = 64, State = MULTIPLE_FRAME_ESTABLISHED
        TEI = 65, State = MULTIPLE_FRAME_ESTABLISHED
    Spid Status:
        TEI 64, ces = 1, state = 5(init)
            spid1 configured, spid1 sent, spid1 valid
            Endpoint ID Info: epsf = 0, usid = 70, tid = 1
        TEI 65, ces = 2, state = 5(init)
            spid2 configured, spid2 sent, spid2 valid
            Endpoint ID Info: epsf = 0, usid = 70, tid = 2
    Layer 3 Status:
        0 Active Layer 3 Call(s)
    Activated dsl 0 CCBs = 1
        CCB: callid=0x0, sapi=0, ces=1, B-chan=0
ISDN BRI0/1  interface
    Layer 1 Status:
        DEACTIVATED
    Layer 2 Status:
        Layer 2 NOT Activated
    Layer 3 Status:
        0 Active Layer 3 Call(s)
```

```
Activated dsl 1 CCBs = 0
Total Allocated ISDN CCBs = 1
```

Now enable ISDN Q931 and PPP authentication debugging with the **debug isdn q931** and **debug ppp authentication** commands:

```
R1#debug isdn q931
ISDN Q931 packets debugging is on

R1#debug ppp authentication
PPP authentication debugging is on
```

You can bring up the ISDN circuit with a ping from R1 to R6. Recall that the interesting traffic on R1 includes IPX ping traffic. Notice how R1 places a call to R6, CHAP is exchanged, and then R6 disconnects the call:

```
R1#ping ipx 14.6.6.6

Type escape sequence to abort.
Sending 5, 100-byte IPX cisco Echoes to 14.0006.0006.0006, timeout is 2 seconds:
.
01:33:30: ISDN BR0/0 : TX ->  SETUP pd = 8  callref = 0x01←Router R1 placing the call
01:33:30:         Bearer Capability i = 0x8890
01:33:30:         Channel ID i = 0x83
01:33:30:         Keypad Facility i = '8995201'
01:33:30: ISDN BR0/0 : RX <-  CALL_PROC pd = 8  callref = 0x81
01:33:30:         Channel ID i = 0x89
01:33:30: ISDN BR0/0 : RX <-  CONNECT pd = 8  callref = 0x81
01:33:30:         Channel ID i = 0x89
01:33:30: %LINK-3-UPDOWN: Interface BRI0/0 :1, changed state to up
01:33:30: BR0/0 :1 PPP: Treating connection as. a callout
01:33:30: ISDN BR0/0 : TX ->  CONNECT_ACK pd = 8  callref = 0x01
01:33:30: BR0/0 :1 PPP: Phase is AUTHENTICATING, by both←CHAP Authentication
01:33:30: BR0/0 :1 CHAP: O CHALLENGE id 1 len 23 from "R1"
01:33:30: BR0/0 :1 CHAP: I CHALLENGE id 1 len 23 from "R6"
01:33:30: BR0/0 :1 CHAP: O RESPONSE id 1 len 23 from "R1"
01:33:30: BR0/0 :1 CHAP: I SUCCESS id 1 len 4
01:33:30: BR0/0 :1 CHAP: I RESPONSE id 1 len 23 from "R6"
01:33:30: BR0/0 :1 CHAP: O SUCCESS id 1 len 4
01:33:30: ISDN BR0/0 : RX <-  DISCONNECT pd = 8  callref = 0x81←R6 disconnects the ISDN
                                                             call
01:33:30:         Cause i = 0x8290 - Normal call clearing
01:33:30: %LINK-3-UPDOWN: Interface BRI0/0 :1, changed state to down
01:33:30: ISDN BR0/0 : TX ->  RELEASE pd = 8  callref = 0x01
01:33:30: ISDN BR0/0 : RX <-  RELEASE_COMP pd = 8  callref = 0x81 . . .
Success rate is 0 percent (0/5 )
```

After several seconds, R6 calls back R1:

```
01:33:46: ISDN BR0/0 : RX <-  SETUP pd = 8  callref = 0x01←R6 calls R1 back
01:33:46:         Bearer Capability i = 0x8890
01:33:46:         Channel ID i = 0x89
01:33:46:         Calling Party Number i = '!', 0x80, '0008995201'
01:33:46:         Called Party Number i = 0xC1, '8995101'
01:33:46: %LINK-3-UPDOWN: Interface BRI0/0 :1, changed state to up
01:33:46: BR0/0 :1 PPP: Treating connection as a callin
```

```
01:33:46: ISDN BR0/0 : TX ->  CONNECT pd = 8  callref = 0x81
01:33:46:         Channel ID i = 0x89
01:33:46: ISDN BR0/0 : RX <-  CONNECT_ACK pd = 8  callref = 0x01
01:33:49: BR0/0 :1 PPP: Phase is AUTHENTICATING, by both
01:33:49: BR0/0 :1 CHAP: O CHALLENGE id 2 len 23 from "R1"
01:33:49: BR0/0 :1 CHAP: I CHALLENGE id 2 len 23 from "R6"
01:33:49: BR0/0 :1 CHAP: Waiting for peer to authenticate first
01:33:49: BR0/0 :1 CHAP: I RESPONSE id 2 len 23 from "R6"
01:33:49: BR0/0 :1 CHAP: O SUCCESS id 2 len 4
01:33:49: BR0/0 :1 CHAP: Processing saved Challenge, id 2
01:33:49: BR0/0 :1 CHAP: O RESPONSE id 2 len 23 from "R1"
01:33:49: BR0/0 :1 CHAP: I SUCCESS id 2 len 4
01:33:50: %LINEPROTO-5-UPDOWN: Line protocol on Interface BRI0/0 :1, changed state to up
↑The circuit is now up
01:33:53: %ISDN-6-CONNECT: Interface BRI0/0 :1 is now connected to 8995201 R6
```

The interface is now up. An IPX ping from R1 to R2 should be successful:

```
R1#ping ipx 14.6.6.6

Type escape sequence to abort.
Sending 5, 100-byte IPX cisco Echoes to 14.0006.0006.0006, timeout is 2 seconds:
!!!!!
Success rate is 100 percent (5/5 ), round-trip min/avg/max = 36/36 /36 ms
R1#
```

The following output shows the IPX routing table when Interface S1/0 is down on R1. Notice that all your routes are now being learned with a next hop of 14.6.6.6 (R6).

```
R1#sh ipx route
Codes: C - Connected primary network,    c - Connected secondary network
       S - Static, F - Floating static, L - Local (internal), W - IPXWAN
       R - RIP, E - EIGRP, N - NLSP, X - External, A - Aggregate
       s - seconds, u - uses

18 Total IPX routes. Up to 1 parallel paths and 16 hops allowed.

Current default route is:

F   FFFFFFFE via       14.0006.0006.0006,       BR0/0

C          1 (UNKNOWN),        Lo0
C         12 (NOVELL-ETHER),   Et0/0
C         14 (PPP),            BR0/0
R          2 [20/03 ] via       14.0006.0006.0006,   21s, BR0/0
R          3 [21/03 ] via       14.0006.0006.0006,   21s, BR0/0
R          4 [21/03 ] via       14.0006.0006.0006,   21s, BR0/0
R          5 [07/02 ] via       14.0006.0006.0006,   21s, BR0/0
R          6 [07/01 ] via       14.0006.0006.0006,   21s, BR0/0
R          7 [13/02 ] via       14.0006.0006.0006,   21s, BR0/0
R          8 [07/02 ] via       14.0006.0006.0006,   21s, BR0/0
R          9 [07/01 ] via       14.0006.0006.0006,   21s, BR0/0
R         10 [13/03 ] via       14.0006.0006.0006,   22s, BR0/0
R         11 [13/03 ] via       14.0006.0006.0006,   23s, BR0/0
R         13 [13/02 ] via       14.0006.0006.0006,   23s, BR0/0
R        100 [08/02 ] via       14.0006.0006.0006,   23s, BR0/0
R        101 [07/01 ] via       14.0006.0006.0006,   23s, BR0/0
```

27. *Verify that the configuration is working properly.*

Let's start your verification of your IPX configuration by examining your Frame Relay core and working your way out. Connect to R1 and try to ping R2 with the **ping ipx 8.2.2.2** command. The results should be 100 percent successful.

```
R1#ping ipx 8.2.2.2

Type escape sequence to abort.
Sending 5, 100-byte IPX cisco Echoes to 8.0002.0002.0002, timeout is 2 seconds:
!!!!!
Success rate is 100 percent (5/5), round-trip min/avg/max = 4/7/8 ms
```

Now try to ping R5 with the ping ipx 8.5.5.5 command. The results should also be 100 percent successful.

```
R1#ping ipx 8.5.5.5

Type escape sequence to abort.
Sending 5, 100-byte IPX cisco Echoes to 8.0005.0005.0005, timeout is 2 seconds:
!!!!!
Success rate is 100 percent (5/5), round-trip min/avg/max = 12/12/16 ms
```

Recall from your previous steps that you have two DLCIs, 101 and 301, defined on interface S1/0 of R1. Your Frame Relay map configurations, displayed with the **show frame map** command, verify that you have defined R1 to limit all traffic, including IPX traffic to only use DLCI 101 and to not use DLCI 301.

```
R1#sh frame map
Serial1/0  (up): ipx 8.0002.0002.0002 dlci 101(0x65,0x1850), static,
              broadcast,
              CISCO, status defined, active
Serial1/0  (up): ipx 8.0005.0005.0005 dlci 101(0x65,0x1850), static,
              broadcast,
              CISCO, status defined, active
```

Use the **show frame pvc** command to verify that your Frame Relay traffic is taking the proper path. You can see that all traffic being sent out of R1 is using DLCI 101. DLCI 301 usage is in an unused state and the number of output packets on DLCI 301 is 0.

```
R1#show frame pvc

PVC Statistics for interface Serial1/0  (Frame Relay DTE)

DLCI = 101, DLCI USAGE = LOCAL, PVC STATUS = ACTIVE, INTERFACE = Serial1/0

  input pkts 359          output pkts 636          in bytes 150698
  out bytes 273319        dropped pkts 0           in FECN pkts 0
  in BECN pkts 0          out FECN pkts 0          out BECN pkts 0
  in DE pkts 0            out DE pkts 0
  out bcast pkts 600       out bcast bytes 269486
  pvc create time 00:30:11, last time pvc status changed 00:28:33
```

```
DLCI = 301, DLCI USAGE = UNUSED, PVC STATUS = ACTIVE, INTERFACE = Serial1/0

  input pkts 28            output pkts 0            in bytes 7840
  out bytes 0              dropped pkts 0           in FECN pkts 0
  in BECN pkts 0           out FECN pkts 0          out BECN pkts 0
  in DE pkts 0             out DE pkts 0
  out bcast pkts 0          out bcast bytes 0
  pvc create time 00:30:03, last time pvc status changed 00:28:34
  Num Pkts Switched 0
```

Your dynamic routing protocol that is running on the Frame Relay core is IPX EIGRP. You can verify proper IPX EIGRP operation with several commands. The **show ipx eigrp interface** command, for example, tells us that interface S1/0 on R1 is running IPX EIGRP. It also tells us that R1 has one EIGRP peer router.

```
R1# show ipx eigrp interface

IPX EIGRP Interfaces for process 1

                       Xmit Queue   Mean   Pacing Time   Multicast    Pending
Interface      Peers  Un/Reliable  SRTT   Un/Reliable   Flow Timer   Routes
Se1/0             1       0/0          9       0/15           55            0
```

If you want to know who the EIGRP peer is, you need to use the **show ipx eigrp neighbor** command. Your EIGRP neighbor is located at IPX address 8.2.2.2, which is R2. A key field in the output of the **show ipx eigrp neighbor** command is the SRRT, which stands for *Smooth Round Trip Time*. It is a measure of the amount of time in ms that it takes for an EIGRP packet from R1 to be sent to R2 and for R1 to receive an acknowledgement packet. Circuit problems involving high latency and high utilization can be detected by examining this parameter and seeing if it is much higher than its normal value.

```
R1# show ipx eigrp neighbor

IPX EIGRP Neighbors for process 1
H   Address                     Interface     Hold  Uptime     SRTT   RTO  Q   Seq
                                              (sec)            (ms)       Cnt  Num
0   8.0002.0002.0002            Se1/0          161  00:43:51      9   200   0   19
```

Now let's connect to R2 and make sure that you have connectivity to the other routers in your Frame Relay core. Use the **ping ipx 8.5.5.5** command to verify that you can reach R5:

```
R2#ping ipx 8.5.5.5

Type escape sequence to abort.
Sending 5, 100-byte IPX cisco Echoes to 8.0005.0005.0005, timeout is 2 seconds:
!!!!!
Success rate is 100 percent (5/5 ), round-trip min/avg/max = 4/6 /8 ms
```

Use the **ping ipx 8.1.1.1** command to verify that you can reach R1:

```
R2#ping ipx 8.1.1.1

Type escape sequence to abort.
Sending 5, 100-byte IPX cisco Echoes to 8.0001.0001.0001, timeout is 2 seconds:
!!!!!
Success rate is 100 percent (5/5 ), round-trip min/avg/max = 4/7 /8 ms
R2#
```

The **show ipx eigrp interface** command on R2 shows that you have a total of four IPX EIGRP neighbor peers on R2:

```
R2# show ipx eigrp interface

IPX EIGRP Interfaces for process 1

                    Xmit Queue   Mean   Pacing Time   Multicast    Pending
Interface    Peers  Un/Reliable  SRTT   Un/Reliable   Flow Timer   Routes
Se0/1                   0/0       164       0/15         850           0
Et0/0 .1       1        0/0         6       0/10          50           0
Et0/0 .2       1        0/0       380       0/10        1892           0
```

The **show ipx eigrp neighbor** command shows that your four EIGRP peers are R1 (8.1.1.1), R5 (8.5.5.5), R3 (11.aa00.0400.0308), and R4 (10.aa00.0400.0408). Notice how the IPX node addresses for R5 and R1 are 5.5.5 and 1.1.1 respectively. These numbers come from the node argument that you entered on these routers when you enabled IPX routing with the **ipx routing *node*** global command.

Also notice how the node addresses for R3 and R4 are not using the node numbers that you specified when you enabled IPX routing on those routers. IPX on R3 was enabled with the command **ipx routing 3.3.3**, while IPX on R4 was enabled with the command **ipx routing 4.4.4**. Your EIGRP neighbors on Routers R3 and R4 are Ethernet interfaces. These routers therefore use their Ethernet MAC addresses for their IPX node numbers, instead of using the node number that was specified when you enabled IPX routing on these routers.

```
R2# show ipx eigrp neighbor

IPX EIGRP Neighbors for process 1
H   Address                  Interface    Hold Uptime    SRTT    RTO  Q  Seq
                                          (sec)          (ms)        Cnt Num
3   8.0005.0005.0005         Se0/1         151 00:44:49   252   1512  0   6
2   8.0001.0001.0001         Se0/1         147 00:44:52    77    462  0   6
1   11.aa00.0400.0308        Et0/0 .2       13 00:45:32   380   2280  0   5
0   10.aa00.0400.0408        Et0/0 .1       11 00:45:35     6    200  0   6
```

You can verify that your IPX network is running properly by viewing your IPX routing table with the **show ipx route** command. You can see that your IPX routing table contains 16 routes. Examining the IPX network diagram will confirm that all your IPX networks are visible from R2, while examining the IPX routing tables on the other routers in your network should also confirm that all your IPX networks are available.

Notice that your routes are being learned by two different methods. The first group of routes consists of the directly connected routes, and the second group of routes is being learned via IPX EIGRP.

```
R2#sh ipx route
Codes: C - Connected primary network,    c - Connected secondary network
       S - Static, F - Floating static, L - Local (internal), W - IPXWAN
       R - RIP, E - EIGRP, N - NLSP, X - External, A - Aggregate
       s - seconds, u - uses

16 Total IPX routes. Up to 1 parallel paths and 16 hops allowed.

No default route known.

C          2 (UNKNOWN),        Lo0
C          8 (FRAME-RELAY),    Se0/1
C         10 (NOVELL-ETHER),   Et0/0 .1
C         11 (ARPA),           Et0/0 .2
E          1 [2297856/1 ] via     8.0001.0001.0001, age 00:07:46,
                          1u, Se0/1
E          3 [409600/1 ] via      11.aa00.0400.0308, age 00:08:36,
                          1u, Et0/0 .2
E          4 [409600/1 ] via      10.aa00.0400.0408, age 00:08:36,
                          1u, Et0/0 .1
E          5 [2297856/1 ] via     8.0005.0005.0005, age 00:07:51,
                          1u, Se0/1
E          6 [268416000/2 ] via   8.0005.0005.0005, age 00:07:51,
                          1u, Se0/1
E          7 [286720000/3 ] via   8.0005.0005.0005, age 00:07:52,
                          1u, Se0/1
E          9 [268416000/2 ] via   8.0005.0005.0005, age 00:07:52,
                          1u, Se0/1
E         12 [2195456/1 ] via     8.0001.0001.0001, age 00:07:47,
                          1u, Se0/1
E         13 [286720000/3 ] via   8.0005.0005.0005, age 00:07:52,
                          7u, Se0/1
E         14 [41024000/1 ] via    8.0001.0001.0001, age 00:07:47,
                          1u, Se0/1
E        100 [46354176/1 ] via    8.0005.0005.0005, age 00:07:53,
                          1u, Se0/1
E        101 [269824000/2 ] via   8.0005.0005.0005, age 00:07:53,
                          1u, Se0/1
```

The **show ipx servers** command lists all the IPX servers that R2 has learned about. Recall that you manually entered two static SAP entries on R6. One entry was for a print server and the other entry was for a file server.

The IPX servers are being shown as residing on IPX network 13, which is R7. This is due to the fact that the static SAP entries that were entered on R6 indicated that these servers were located on IPX network 13.

```
R2# show ipx servers
Codes: S - Static, P - Periodic, E - EIGRP, N - NLSP, H - Holddown, + = detail
2 Total IPX Servers
```

```
Table ordering is based on routing and server info

   Type Name                       Net     Address    Port     Route Hops Itf
E     4 FileServer1             13.aa00.0400.0708:0451 286720000/03    3  Se0/1
E    7 PrintServer1             13.aa00.0400.0708:0451 286720000/03    3  Se0/1
```

Now connect to R7 and display the IPX servers on R7 with the **show ipx servers** command. You can see that no output results, indicating that R7 has not learned of any IPX servers. You may ask, why is R7 indicating that no IPX servers exist when I have added two static SAP entries on Route R6 that point to a print server and a file server on R7? The answer requires an understanding of RIP/SAP split horizon. RIP/SAP split horizon says that a router never advertises RIP routing or SAP server information out of the same interface that it learned the information from. In this case, the static SAP entry on R6 points to two servers on R7. R6 will never send these SAP entries to R7 because R6 treats the static entries as if they were learned from R7.

```
R7#sh ipx servers
```

Now reconnect to R2. The **show ipx interface brief** command shows you a high-level overview of all the interfaces on the router as well as their corresponding IPX encapsulation types and IPX addresses (if IPX had been configured on the interface).

Recall that you did not configure IPX on the physical E0/0 interface, but you configured IPX on two subinterfaces, each with a different Ethernet encapsulation. This is shown in the output of the command:

```
R2#   show ipx interface brief
Interface             IPX Network Encapsulation Status                 IPX State
Ethernet0/0           unassigned  not config'd  up                     n/a
Ethernet0/0.1         10          NOVELL-ETHER  up                     [up]
Ethernet0/0.2         11          ARPA          up                     [up]
Serial0/0             unassigned  not config'd  administratively down  n/a
Serial0/1             8           FRAME-RELAY   up                     [up]
Loopback0             2           UNKNOWN       up                     [up]
```

More in-depth IPX information is available by using the **show ipx interface** command. This command displays detailed information on each of the router interfaces that have been configured for IPX routing:

```
R2# show ipx interface
Ethernet0/0 .1 is up, line protocol is up
  IPX address is 10.aa00.0400.0208, NOVELL-ETHER [up]
  Delay of this IPX network, in ticks is 1 throughput 0 link delay 0
  IPXWAN processing not enabled on this interface.
  IPX SAP update interval is 1 minute(s)
  IPX type 20 propagation packet forwarding is disabled
  Incoming access list is not set
  Outgoing access list is not set
```

```
  IPX helper access list is not set
  SAP GNS processing enabled, delay 0 ms, output filter list is not set
  SAP Input filter list is not set
  SAP Output filter list is not set
  SAP Router filter list is not set
  Input filter list is not set
  Output filter list is not set
  Router filter list is not set
  Netbios Input host access list is not set
  Netbios Input bytes access list is not set
  Netbios Output host access list is not set
  Netbios Output bytes access list is not set
  Updates each 60 seconds, aging multiples RIP: 3 SAP: 3
  SAP interpacket delay is 55 ms, maximum size is 480 bytes
  RIP interpacket delay is 55 ms, maximum size is 432 bytes
  IPX accounting is disabled
  IPX fast switching is configured (enabled)
  RIP packets received 0, RIP packets sent 1
  SAP packets received 0, SAP packets sent 49
Ethernet0/0 .2 is up, line protocol is up
  IPX address is 11.aa00.0400.0208, ARPA [up]
  Delay of this IPX network, in ticks is 1 throughput 0 link delay 0
  IPXWAN processing not enabled on this interface.
  IPX SAP update interval is 1 minute(s)
  IPX type 20 propagation packet forwarding is disabled
  Incoming access list is not set
  Outgoing access list is not set
  IPX helper access list is not set
  SAP GNS processing enabled, delay 0 ms, output filter list is not set
  SAP Input filter list is not set
  SAP Output filter list is not set
  SAP Router filter list is not set
  Input filter list is not set
  Output filter list is not set
  Router filter list is not set
  Netbios Input host access list is not set
  Netbios Input bytes access list is not set
  Netbios Output host access list is not set
  Netbios Output bytes access list is not set
  Updates each 60 seconds, aging multiples RIP: 3 SAP: 3
  SAP interpacket delay is 55 ms, maximum size is 480 bytes
  RIP interpacket delay is 55 ms, maximum size is 432 bytes
  IPX accounting is disabled
  IPX fast switching is configured (enabled)
  RIP packets received 0, RIP packets sent 1
  SAP packets received 0, SAP packets sent 49
```

The **show ipx traffic** command displays detailed traffic information for the router, including a breakdown of the amount of routing protocol traffic that has been sent and received. You can see from the following output that the majority of the IPX traffic on this router is EIGRP routing protocol traffic:

```
R2# show ipx traffic
System Traffic for 0.0000.0000.0001 System-Name: R2
Rcvd:   1537 total, 0 format errors, 0 checksum errors, 0 bad hop count,
        103 packets pitched, 1434 local destination, 0 multicast
Bcast:  1496 received, 1488 sent
```

```
Sent:    1524 generated, 0 forwarded
         3 encapsulation failed, 0 no route
SAP:     0 SAP requests, 0 SAP replies, 2 servers
         0 SAP Nearest Name requests, 0 replies
         0 SAP General Name requests, 0 replies
         0 SAP advertisements received, 135 sent
         12 SAP flash updates sent, 0 SAP format errors
RIP:     0 RIP requests, 0 RIP replies, 15 routes
         0 RIP advertisements received, 49 sent
         5 RIP flash updates sent, 0 RIP format errors
Echo:    Rcvd 0 requests, 0 replies
         Sent 0 requests, 0 replies
         0 unknown: 0 no socket, 0 filtered, 0 no helper
         0 SAPs throttled, freed NDB len 0
Watchdog:
         0 packets received, 0 replies spoofed
Queue lengths:
         IPX input: 0, SAP 0, RIP 0, GNS 0
         SAP throttling length: 0/(no limit), 0 nets pending lost route reply
         Delayed process creation: 0
EIGRP:   Total received 1434, sent 1317
         Updates received 16, sent 19
         Queries received 0, sent 0
         Replies received 0, sent 0
         SAPs received 7, sent 8
NLSP:    Level-1 Hellos received 0, sent 0
         PTP Hello received 0, sent 0
         Level-1 LSPs received 0, sent 0
         LSP Retransmissions: 0
         LSP checksum errors received: 0
         LSP HT=0 checksum errors received: 0
         Level-1 CSNPs received 0, sent 0
         Level-1 PSNPs received 0, sent 0
         Level-1 DR Elections: 0
         Level-1 SPF Calculations: 0
         Level-1 Partial Route Calculations: 0
```

Let's connect to R5, which is running the NLSP protocol on interface S0/1. This can be verfied with the **show ipx nlsp neighbors** command. You can then see that your NLSP neighbor is R6:

```
R5#sh ipx nlsp neighbors
NLSP Level-1 Neighbors: Tag Identifier = area1

System Id      Interface    State  Holdtime  Priority  Circuit Id
R6             Se0/1         Up      44         0         03
```

Connect to R6 and display the IPX routing table with the **show ipx route** command. Notice that some of your routes are being learned via IPX EIGRP and some are being learned via NLSP. The EIGRP routes are coming from IPX network 7 and IPX network 13. Both of these networks are on your neighbor router, R7. You are learning these routes via EIGRP because R6 and R7 are running EIGRP between themselves as their routing protocol. The NLSP routes are being learned from any IPX networks that are in the direction of Router 5. Recall that when you configured R5, the following commands were entered:

```
ipx router nlsp area1
area-address 0 0
redistribute eigrp 1
```

These three commands caused EIGRP routes on R5 to be redistributed into NLSP. Because R6 is learning all routes coming in from R5 via NLSP, your redistribution is working properly.

```
R6#show ipx route
Codes: C - Connected primary network,    c - Connected secondary network
       S - Static, F - Floating static, L - Local (internal), W - IPXWAN
       R - RIP, E - EIGRP, N - NLSP, X - External, A - Aggregate
       s - seconds, u - uses

15 Total IPX routes. Up to 1 parallel paths and 16 hops allowed.

No default route known.

L        101 is the internal network
C          6 (UNKNOWN),        Lo0
C          9 (PPP),            Se1
NX         1 [27][20/02 ][19/01 ] via      100.0000.0000.0001, 3145s, Se0
NX         2 [27][14/02 ][13/01 ] via      100.0000.0000.0001, 3146s, Se0
NX         3 [27][15/02 ][14/01 ] via      100.0000.0000.0001, 3146s, Se0
NX         4 [27][15/02 ][14/01 ] via      100.0000.0000.0001, 3146s, Se0
N          5 [27][01/01 ]         via      100.0000.0000.0001, 3183s, Se0
E          7 [2297856/1 ] via       9.0007.0007.0007, age 00:54:03,
                         1u, Se1
N          8 [27][01/01 ]         via      100.0000.0000.0001, 3146s, Se0
NX        10 [27][07/02 ][06/01 ] via      100.0000.0000.0001, 3146s, Se0
NX        11 [27][07/02 ][06/01 ] via      100.0000.0000.0001, 3146s, Se0
NX        12 [27][20/02 ][19/01 ] via      100.0000.0000.0001, 3146s, Se0
E         13 [2195456/1 ] via       9.0007.0007.0007, age 00:54:04,
                         325u, Se1
N        100 [27][02/01 ]         via      100.0000.0000.0001, 3184s, Se0
```

AppleTalk

28. *Configure R1 as follows:*

a. Interface E0/0 should be assigned AppleTalk cable range 41 to 99, with an interface address of 69.1. The zone name for this interface should be LAN_1.

b. Interface S1/0 should be assigned AppleTalk cable range 100 to 199, with an interface address of 124.1. The zone name should be WAN_1_2_5.

c. Run AppleTalk EIGRP on interface S1/0 and RTMP on interface E0/0.

AppleTalk must be enabled on a router with the **appletalk routing** global configuration command. If the AppleTalk EIGRP routing protocol is running on the router, then the EIGRP argument must be added to the appletalk routing command. AppleTalk is unique in that the EIGRP router number must be different on each router in the internetwork. The **appletalk route-redistribution** command is automatically added to the configuration.

```
R1(config)#appletalk routing eigrp 1
```

Configuring AppleTalk on an interface requires two mandatory tasks. The first task is to define an AppleTalk cable range and the second task is to define at least one zone. Interface E0/0 is in cable range 41 to 99 with an address of 69.1. Interface E0/0 is in zone LAN_1.

You'll want to run AppleTalk RTMP on interface E0/0. RTMP is enabled by default on an AppleTalk interface and you do not have to enter any special commands for it to run on this interface.

```
R1(config)#interface Ethernet0/0
R1(config-if)#appletalk cable-range 41-99 69.1
R1(config-if)#appletalk zone LAN_1
```

Interface S1/0 is in cable range 100 to 199 with an address of 124.1. It is in the zone WAN_1_2_5.

You do not want to run RTMP on this interface. Because this interface is running Frame Relay encapsulation and is part of an NBMA network, it is a good idea to run AppleTalk EIGRP. RTMP is disabled with the **no appletalk protocol rtmp** command and EIGRP is enabled with the **appletalk protocol eigrp** command.

Similar to your DECnet and IPX configurations, you must manually map your next hop addresses with a Frame Relay map statement. Recall that you have disabled Frame Relay inverse arp on this interface. The following output displays the commands required to configure AppleTalk on interface S1/0:

```
R1(config)#interface Serial1/0
R1(config-if)#appletalk cable-range 100-199 124.1
R1(config-if)#appletalk zone WAN_1_2_5
R1(config-if)#appletalk protocol eigrp
R1(config-if)#no appletalk protocol rtmp
R1(config-if)#frame-relay map appletalk 124.5 101 broadcast
R1(config-if)#frame-relay map appletalk 124.2 101 broadcast
```

The Frame Relay maps configure the interface so that all traffic with a next hop address of either 124.5 or 124.2 is encapsulated in DLCI 101.

29. *Configure R2 as follows:*

a. Interface E0/0 should be assigned AppleTalk cable range 200 to 299, with an interface address of 234.2. The zone name should be set to LAN_2_3_4.

b. Interface S0/1 should be assigned AppleTalk cable range 100 to 199, with an interface address of 124.2. The zone name should be WAN_1_2_5.

c. Run AppleTalk EIGRP on interface S0/1 and RTMP on interface E0/0.

Begin by configuring AppleTalk routing on R2. Notice that your EIGRP router number is different than the router number that is used for R1. AppleTalk EIGRP does not work if the router numbers are not unique throughout the internetwork.

```
R2(config)#appletalk routing eigrp 2
```

Interface E0/0 is in cable range 200 to 299 with an address of 234.2 and is in zone LAN_2_3_4.

```
R2(config)#interface Ethernet0/0
R2(config-if)#appletalk cable-range 200-299 234.2
R2(config-if)#appletalk zone LAN_2_3_4
```

Interface S0/1 is in cable range 100 to 199 with an address of 124.2. It is in zone WAN_1_2_5. You want to run AppleTalk EIGRP instead of RTMP on this interface.

Because interface S0/1 is acting as the hub router on your Frame Relay network, you must address AppleTalk split horizon issues. You need to disable EIGRP split horizon with the **no appletalk eigrp-splithorizon** command.

You also need to define two Frame Relay maps because Frame Relay inverse arp has been disabled on this interface. The Frame Relay maps tell the router that AppleTalk traffic with a next hop of 124.5 should be encapsulated in DLCI 201 and that AppleTalk traffic with a next hop of 124.1 should be encapsulated in DLCI 102.

```
R2(config)#interface Serial0/1
R2(config-if)#appletalk cable-range 100-199 124.2
R2(config-if)#appletalk zone WAN_1_2_5
R2(config-if)#appletalk protocol eigrp
R2(config-if)#no appletalk protocol rtmp
R2(config-if)#no appletalk eigrp-splithorizon
R2(config-if)#frame-relay map appletalk 124.5 201 broadcast
R2(config-if)#frame-relay map appletalk 124.1 102 broadcast
```

The **appletalk local-routing** global command is required on R2. It must be entered on a router with EIGRP split horizon disabled.

```
R2(config)#appletalk local-routing
```

30. *Configure R3 as follows:*

a. Interface E0/0 should be assigned AppleTalk cable range 200 to 299, with an interface address of 234.3. The zone name should be set to LAN_2_3_4.

b. Run RTMP on this router.

R3 will not be running AppleTalk EIGRP. The AppleTalk routing global command can therefore leave out the EIGRP option. Notice how you do not have to enter any RTMP-specific commands on this router. RTMP runs by default.

```
R3(config)#appletalk routing
```

Interface E0/0 is in cable range 200 to 299 with an address of 234.3. It is in AppleTalk zone LAN_2_3_4.

```
R3(config)#interface Ethernet0/0
R3(config-if)#appletalk cable-range 200-299 234.3
R3(config-if)#appletalk zone LAN_2_3_4
```

31. *Configure R4 as follows:*

a. Interface E0/0 should be assigned AppleTalk cable range 200 to 299, with an interface address of 234.4. The zone name should be set to LAN_2_3_4.

b. Run RTMP on this router.

R4's configuration is similar to R3's configuration. You are only running the AppleTalk RTMP routing protocol and must define an AppleTalk cable range and zone on the E0/0 interface.

```
R4(config)#appletalk routing

R4(config)#interface Ethernet0/0
R4(config-if)#appletalk cable-range 200-299 234.4
R4(config-if)#appletalk zone LAN_2_3_4
```

32. *Configure R5 as follows:*

a. Interface S0/0 should be assigned AppleTalk cable range 100 to 199, with an interface address of 124.5. The zone name should be set to WAN_1_2_5.

b. Run AppleTalk EIGRP on all AppleTalk interfaces on this router.

c. Interface S0/1 should be configured with IP address 142.2.1.17/29, so run IP EIGRP on this interface.

R5 will be running EIGRP, so you must define a unique router number for the EIGRP process:

```
R5(config)#appletalk routing eigrp 5
```

Interface S0/0 is part of your Frame Relay core. It is in AppleTalk cable range 100 to -199, with an address of 124.5. You'll want to run AppleTalk EIGRP on this interface, so you must specifically enable EIGRP and disable RTMP.

Because Frame Relay inverse arp has been disabled on this interface, you must define Frame Relay maps. These two maps tell the router that all AppleTalk traffic with a next hop address of either 124.1 or 124.2 should be encapsulated in DLCI 202.

```
R5(config)#interface Serial0/0
R5(config-if)#appletalk cable-range 100-199 124.5
R5(config-if)#appletalk zone WAN_1_2_5
R5(config-if)#appletalk protocol eigrp
R5(config-if)#no appletalk protocol rtmp
R5(config-if)#frame-relay map appletalk 124.1 202 broadcast
R5(config-if)#frame-relay map appletalk 124.2 202 broadcast
```

No AppleTalk commands are going to be configured on Interface S0/1 and you do want to add an IP address of 142.2.1.17/29. You also need to define an X.25 map that defines your next hop to R6.

```
R5(config)#interface Serial0/1
R5(config-if)#ip address 142.2.1.17 255.255.255.248
R5(config-if)#x25 map ip 142.2.1.18 6666 broadcast
```

You'll also want to run IP EIGRP on interface S0/1.

```
R5(config)#router eigrp 1
R5(config-router)#network 142.2.0.0
```

33. *Configure R6 as follows:*

a. Do not run any AppleTalk routing on this router.
b. Interface S0 should be assigned IP address 142.2.1.18/29.
c. Interface S1 should be assigned IP address 142.2.1.9/29.
d. Run IP EIGRP on this router.

R6 will not be running any AppleTalk routing process and you'll want to define an IP address on both interfaces S0 and S1. You'll also want to run IP EIGRP on interfaces S0 and S1. The following commands configure R6 with the proper IP and EIGRP configuration. Notice that you need to add an X.25 map that defines your next hop to R5:

```
R6(config)#interface Serial0
R6(config-if)#ip address 142.2.1.18 255.255.255.248
R6(config-if)# x25 map ip 142.2.1.17 5555 broadcast

R6(config)#interface Serial1
R6(config-if)#ip address 142.2.1.9 255.255.255.248
R6(config-if)#encapsulation ppp

R6(config)#router eigrp 1
R6(config-router)#network 142.2.0.0
```

34. *Configure R7 as follows:*

a. Interface E0 should be assigned AppleTalk cable range 701 to 799, with an interface address of 720.14. The zone name should be set to LAN_7 with a secondary zone name of TopSecret. Run RTMP on this interface.
b. Interface S0 should be configured with IP address 142.2.1.10/29. Run IP EIGRP on this interface.

R7 will be running AppleTalk EIGRP, so you have to define a unique EIGRP router number:

```
R7(config)#appletalk routing eigrp 7
```

Interface E0 will be in cable range 701 to 799 with an address of 720.14. Interface E0 will be in two different zones; the default zone will be named LAN_7 and the secondary zone will be named TopSecret. The first zone specified is always the default zone.

```
R7(config)#interface Ethernet0
R7(config-if)#appletalk cable-range 701-799 720.14
R7(config-if)#appletalk zone LAN_7
R7(config-if)#appletalk zone TopSecret
```

Interface S0 will not be configured for AppleTalk. Instead, you will define an IP address of 142.2.1.10/29 and enable IP EIGRP with the following commands:

```
R7(config)#interface Serial0
R7(config-if)#ip address 142.2.1.10 255.255.255.248
R7(config-if)#encapsulation ppp

R7(config)#router eigrp 1
R7(config-router)#network 142.2.0.0
```

35. *Configure a GRE tunnel between R5 and R7.*

a. Run AppleTalk EIGRP over this tunnel.
b. The tunnel cable range should be 500 through 700.
c. The tunnel zone should be WAN_5_7.
d. The tunnel endpoint on R5 should be AppleTalk address 545.5.
e. The tunnel endpoint on R7 should be AppleTalk address 660.7.

Because you are not running AppleTalk on R6, you must come up with a way to transport AppleTalk traffic between Routers R5 and R7. You will use a tunnel between R5 and R7 that will take all AppleTalk traffic on R5, tunnel the traffic within IP, and route it to R7. Traffic from R7 to R5 will be tunneled in a similar fashion.

R5 Configuration

Begin the tunnel process by defining a tunnel interface. The tunnel is configured similar to a standard interface running AppleTalk. First, define an AppleTalk cable range and zone name. You must also enable AppleTalk EIGRP and disable AppleTalk RTMP. The major difference between this tunnel interface and a standard interface is that you define a tunnel source and destination. The tunnel source is the router interface that receives the AppleTalk traffic for tunneling into IP. The tunnel endpoint is the address that receives the AppleTalk-tunneled traffic and regenerates the AppleTalk traffic. The tunnel endpoint must be reachable via IP routing. This is why you defined IP addressing and IP EIGRP routing on R5, R6, and R7, The following commands will configure an AppleTalk tunnel on R5:

```
R5(config)#interface Tunnel0
R5(config-if)#no ip address
R5(config-if)#appletalk cable-range 500-700 545.5
R5(config-if)#appletalk zone WAN_5_7
R5(config-if)#appletalk protocol eigrp
R5(config-if)#no appletalk protocol rtmp
R5(config-if)#tunnel source Serial0/1
R5(config-if)#tunnel destination 142.2.1.10
```

R7 Configuration

The tunnel configuration for R7 is similar to the configuration on R5. R7 will have the same cable range and zone as R5, but it will have a unique AppleTalk address.

```
R7(config)#interface Tunnel0
R7(config-if)#no ip address
R7(config-if)#appletalk cable-range 500-700 660.7
R7(config-if)#appletalk zone WAN_5_7
R7(config-if)#appletalk protocol eigrp
R7(config-if)#no appletalk protocol rtmp
R7(config-if)#tunnel source Serial0
R7(config-if)#tunnel destination 142.2.1.17
```

36. *Configure R2 so that R3 and R4 cannot reach any addresses in AppleTalk cable range 701 to 799.*

You'll want to restrict R3 and R4 from being able to send any traffic to addresses within the cable range of 701to 799. You can accomplish this by defining an AppleTalk access list denying cable range 701 to 799 and then applying that access list to the E0/0 interface of R2.

You will apply the access list to interface E0/0 with the **appletalk access-group** command. This command applies a data-packet filter to an interface. All AppleTalk data packets being sent out interface E0/0 will have their AppleTalk source network address checked. The packets are discarded if the source network address is denied in the access list. The following commands will configure your access list:

```
R2(config)#interface Ethernet0/0
R2(config-if)#appletalk cable-range 200-299 234.2
R2(config-if)#appletalk zone LAN_2_3_4
R2(config-if)#appletalk access-group 600

R2(config)#access-list 600 deny cable-range 701-799
R2(config)#access-list 600 permit other-access
```

The **show apple access-lists** command verifies that you have defined access list 600, which denies cable range 701 to 799:

```
R2#show apple access-lists
AppleTalk access list 600:
  deny cable-range 701-799
  permit other-access
```

The **show apple interface** command verifies that AppleTalk access list 600 has been applied to interface E0/0:

```
R2#sh apple interface
Ethernet0/0  is up, line protocol is up
  AppleTalk cable range is 200-299
  AppleTalk address is 234.2, Valid
```

```
  AppleTalk zone is "LAN_2_3_4"
  AppleTalk address gleaning is disabled
  AppleTalk route cache is disabled, port has access control
  AppleTalk access-group is 600
Serial0/1  is up, line protocol is up
  AppleTalk cable range is 100-199
  AppleTalk address is 124.2, Valid
  AppleTalk zone is "WAN_1_2_5"
  Routing protocols enabled: EIGRP
  AppleTalk port configuration verified by 124.1
  AppleTalk discarded 1 packet due to output errors
  AppleTalk address gleaning is not supported by hardware
  AppleTalk route cache is enabled
```

Now connect to R3. The **show apple route** command shows that AppleTalk cable range 701 to 799 is still listed in the routing table:

```
R3#show apple route
Codes: R - RTMP derived, E - EIGRP derived, C - connected, A - AURP
       S - static  P - proxy
5 routes in internet

The first zone listed for each entry is its default (primary) zone.

R Net 41-99 [2/G] via 234.2, 3 sec, Ethernet0/0 , zone LAN_1
R Net 100-199 [1/G] via 234.2, 3 sec, Ethernet0/0 , zone WAN_1_2_5
C Net 200-299 directly connected, Ethernet0/0 , zone LAN_2_3_4
R Net 500-700 [2/G] via 234.2, 3 sec, Ethernet0/0 , zone WAN_5_7
R Net 701-799 [3/G] via 234.2, 3 sec, Ethernet0/0, zone LAN_7
                    Additional zones: 'TopSecret'
```

A ping to interface E0 on R7 at AppleTalk address 720.14 will fail. This confirms that the access list is working properly:

```
R3#ping apple 720.14

Type escape sequence to abort.
Sending 5, 100-byte AppleTalk Echos to 720.14, timeout is 2 seconds:
 . . . . .
Success rate is 0 percent (0/5)
```

37. *Configure R4 so that it can no longer see any routes to the 41-to-99 cable range.*

You should configure an access list on R4 so that its routing table does not have any entries for the 41-to-99 cable range. This can be accomplished by defining an AppleTalk access list and applying it to interface E0/0 with an **appletalk distribute-list in** command. The following commands will configure your access list:

```
R4(config)#interface Ethernet0/0
R4(config-if)#appletalk cable-range 200-299 234.4
R4(config-if)#appletalk zone LAN_2_3_4
R4(config-if)#appletalk distribute-list 610 in

R4(config)#access-list 610 deny cable-range 41-99
R4(config)#access-list 610 permit other-access
```

The **show apple access** command confirms that access list 610 has been configured:

```
R4#show apple access
AppleTalk access list 610:
  deny cable-range 41-99
  permit other-access
```

The **show apple interface** command confirms that access list 610 has been applied to interface E0/0 as an incoming route filter:

```
R4#show apple interface
Ethernet0/0  is up, line protocol is up
  AppleTalk cable range is 200-299
  AppleTalk address is 234.4, Valid
  AppleTalk zone is "LAN_2_3_4"
  AppleTalk address gleaning is disabled
  AppleTalk route cache is enabled
  AppleTalk incoming route filter is 610
```

You can see from the output of the **show apple route** command that no entries are in the routing table for the 41-to-99 network. This confirms that your access list is functioning properly.

```
R4#sh apple route
Codes: R - RTMP derived, E - EIGRP derived, C - connected, A - AURP
       S - static  P - proxy
4 routes in internet

The first zone listed for each entry is its default (primary) zone.

R Net 100-199 [1/G] via 234.2, 8 sec, Ethernet0/0 , zone WAN_1_2_5
C Net 200-299 directly connected, Ethernet0/0 , zone LAN_2_3_4
R Net 500-700 [2/G] via 234.2, 8 sec, Ethernet0/0 , zone WAN_5_7
R Net 701-799 [3/G] via 234.2, 8 sec, Ethernet0/0 , zone LAN_7
                    Additional zones: 'TopSecret'
```

38. *Verify that the configuration is working properly.*

Let's start to verify your AppleTalk configuration by connecting to R1 and examining your Frame Relay core. The **show frame map** command tells you that your two static maps are active:

```
R1#sh frame map
Serial1/0  (up): appletalk 124.5 dlci 101(0x65,0x1850), static,
              broadcast,
              CISCO, status defined, active
Serial1/0  (up): appletalk 124.2 dlci 101(0x65,0x1850), static,
              broadcast,
              CISCO, status defined, active
```

A ping to R2 and R5 at AppleTalk addresses 124.2 and 124.5 should be 100 percent successful:

```
R1#ping apple 124.2

Type escape sequence to abort.
Sending 5, 100-byte AppleTalk Echos to 124.2, timeout is 2 seconds:
!!!!!
Success rate is 100 percent (5/5), round-trip min/avg/max = 8/8/8 ms


R1#ping apple 124.5

Type escape sequence to abort.
Sending 5, 100-byte AppleTalk Echos to 124.5, timeout is 2 seconds:
!!!!!
Success rate is 100 percent (5/5), round-trip min/avg/max = 12/14/16 ms
```

The output of the **show frame pvc** command should verify that all your Frame Relay traffic is being sent on DLCI 101 and that no data traffic is being sent on DLCI 301:

```
R1#show frame pvc

PVC Statistics for interface Serial1/0  (Frame Relay DTE)

DLCI = 101, DLCI USAGE = LOCAL, PVC STATUS = ACTIVE, INTERFACE = Serial1/0

  input pkts 173          output pkts 247          in bytes 65800
  out bytes 96069         dropped pkts 0           in FECN pkts 0
  in BECN pkts 0          out FECN pkts 0          out BECN pkts 0
  in DE pkts 0            out DE pkts 0
  out bcast pkts 222       out bcast bytes 93524
  pvc create time 00:23:19, last time pvc status changed 00:10:19

DLCI = 301, DLCI USAGE = UNUSED, PVC STATUS = ACTIVE, INTERFACE = Serial1/0

  input pkts 10           output pkts 0            in bytes 2800
  out bytes 0             dropped pkts 0           in FECN pkts 0
  in BECN pkts 0          out FECN pkts 0          out BECN pkts 0
  in DE pkts 0            out DE pkts 0
  out bcast pkts 0         out bcast bytes 0
  pvc create time 00:11:56, last time pvc status changed 00:10:20
  Num Pkts Switched 0
```

Because AppleTalk EIGRP is being run on your Frame Relay core, you can use the **show appletalk eigrp interfaces** command to verify that interface S1/0 is running AppleTalk EIGRP. You can see that you have one AppleTalk EIGRP peer:

```
R1# show appletalk eigrp interfaces
AT/EIGRP Neighbors for process 1, router id 1

                      Xmit Queue   Mean   Pacing Time   Multicast    Pending
Interface      Peers  Un/Reliable  SRTT   Un/Reliable   Flow Timer   Routes
Se1/0            1       0/0        377       0/10         1894           0
```

The AppleTalk EIGRP peer can be displayed with the **show appletalk eigrp neighbors** command. Notice that your EIGRP neighbor is AppleTalk address 124.2, which is R2:

```
R1# show appletalk eigrp neighbors
AT/EIGRP Neighbors for process 1, router id 1
H   Address                  Interface    Hold Uptime    SRTT    RTO  Q  Seq
                                           (sec)          (ms)        Cnt Num
0   124.2                    Se1/0          162 00:35:21  377   2262  0   3
```

The **show appletalk route** command displays the AppleTalk routing table for your internetwork. You can see that all your AppleTalk networks are visible in your routing table. Examination of the routing tables on the other routers in your internetwork should reveal that they too contain routes to all the networks. Notice how the AppleTalk routing table not only contains routes to AppleTalk networks, but also contains the zone names of each network:

```
R1# show appletalk route
Codes: R - RTMP derived, E - EIGRP derived, C - connected, A - AURP
       S - static  P - proxy
5 routes in internet

The first zone listed for each entry is its default (primary) zone.

C Net 41-99 directly connected, Ethernet0/0 , zone LAN_1
C Net 100-199 directly connected, Serial1/0 , zone WAN_1_2_5
E Net 200-299 [1/G] via 124.2, 2124 sec, Serial1/0 , zone LAN_2_3_4
E Net 500-700 [2/G] via 124.2, 2124 sec, Serial1/0 , zone WAN_5_7
E Net 701-799 [3/G] via 124.2, 2124 sec, Serial1/0 , zone LAN_7
                    Additional zones: 'TopSecret'
```

The **show appletalk zone** command displays zone names for your entire AppleTalk network. AppleTalk networks with more than one zone (such as AppleTalk cable range 701 to 799) will show all the zones defined. All the routers in your network should show an identical view of the zones on the network (assuming that no zone access lists have been defined).

```
R1# show appletalk zone
Name                                    Network(s)
TopSecret                               701-799
WAN_1_2_5                               100-199
LAN_2_3_4                               200-299
LAN_1                                   41-99
LAN_7                                   701-799
WAN_5_7                                 500-700
Total of 6 zones
```

Another helpful command is the **show appletalk globals** command. This command displays high-level route table, zone, RTMP, and EIGRP information for the router:

```
R1# show appletalk globals
AppleTalk global information:
  Internet is incompatible with older, AT Phase1, routers.
  There are 5 routes in the internet.
  There are 6 zones defined.
  Logging of significant AppleTalk events is disabled.
```

```
  ZIP resends queries every 10 seconds.
  RTMP updates are sent every 10 seconds.
  RTMP entries are considered BAD after 20 seconds.
  RTMP entries are discarded after 60 seconds.
  AARP probe retransmit count: 10, interval: 200 msec.
  AARP request retransmit count: 5, interval: 1000 msec.
  DDP datagrams will be checksummed.
  RTMP datagrams will be strictly checked.
  RTMP routes may not be propagated without zones.
  Routes will be distributed between routing protocols.
  Routing between local devices on an interface will not be performed.
  IPTalk uses the udp base port of 768 (Default).
  EIGRP router id is: 1
  EIGRP maximum active time is 3 minutes
  Alternate node address format will not be displayed.
  Access control of any networks of a zone hides the zone.
```

The **show appletalk interface** command displays addressing and zone information for each interface on the router:

```
R1# show appletalk interface
Ethernet0/0  is up, line protocol is up
  AppleTalk cable range is 41-99
  AppleTalk address is 69.1, Valid
  AppleTalk zone is "LAN_1"
  AppleTalk address gleaning is disabled
  AppleTalk route cache is enabled
Serial1/0  is up, line protocol is up
  AppleTalk cable range is 100-199
  AppleTalk address is 124.1, Valid
  AppleTalk zone is "WAN_1_2_5"
  Routing protocols enabled: EIGRP
  AppleTalk port configuration verified by 124.2
  AppleTalk discarded 1 packet due to output errors
  AppleTalk address gleaning is not supported by hardware
  AppleTalk route cache is enabled
Loopback0 is up, line protocol is up
  AppleTalk protocol processing disabled
```

The **show appletalk traffic** command displays AppleTalk traffic information broken down by AppleTalk protocol:

```
R1# show appletalk traffic
AppleTalk statistics:
  Rcvd:  59 total, 0 checksum errors, 0 bad hop count
         59 local destination, 0 access denied
         0 for MacIP, 0 bad MacIP, 0 no client
         5 port disabled, 0 no listener
         0 ignored, 0 martians
  Bcast: 0 received, 284 sent
  Sent:  290 generated, 0 forwarded, 0 fast forwarded, 0 loopback
         0 forwarded from MacIP, 0 MacIP failures
         1 encapsulation failed, 0 no route, 0 no source
  DDP:   59 long, 0 short, 0 macip, 0 bad size
  NBP:   15 received, 0 invalid, 0 proxies
         0 replies sent, 20 forwards, 15 lookups, 0 failures
```

```
  RTMP:   0 received, 0 requests, 0 invalid, 0 ignored
         224 sent, 0 replies
  ATP:    0 received
  ZIP:    8 received, 13 sent, 0 netinfo
  Echo:   0 received, 0 discarded, 0 illegal
         0 generated, 0 replies sent
  Responder:  0 received, 0 illegal, 0 unknown
         0 replies sent, 0 failures
  AARP:   0 requests, 0 replies, 0 probes
         0 martians, 0 bad encapsulation, 0 unknown
         10 sent, 0 failures, 0 delays, 0 drops
  Lost: 0 no buffers
  Unknown: 0 packets
  Discarded: 0 wrong encapsulation, 0 bad SNAP discriminator
  AURP: 0 Open Requests, 0 Router Downs
        0 Routing Information sent, 0 Routing Information received
        0 Zone Information sent, 0 Zone Information received
        0 Get Zone Nets sent, 0 Get Zone Nets received
        0 Get Domain Zone List sent, 0 Get Domain Zone List received
        0 bad sequence
  EIGRP: 92 received, 41 hellos, 2 updates, 0 replies, 0 queries
         44 sent,     1 hellos, 3 updates, 0 replies, 0 queries
         0 invalid, 0 ignored
```

The **AppleTalk neighbor** command shows all neighboring routers. Notice that two of your neighbors are EIGRP peers and two of your neighbors are RTMP peers:

```
R2#show apple neigh
AppleTalk neighbors:
  124.1          Serial0/1 , uptime 00:06:31, 53 secs
        Neighbor is reachable as a EIGRP peer
  124.5          Serial0/1 , uptime 00:05:42, 18 secs
        Neighbor is reachable as a EIGRP peer
  234.3          Ethernet0/0 , uptime 00:19:25, 5 secs
        Neighbor is reachable as a RTMP peer
  234.4          Ethernet0/0 , uptime 00:19:28, 8 secs
        Neighbor is reachable as a RTMP peer
```

The **show x25 map** command displays a list of all the defined X.25 maps on the router:

```
R6# sh x25 map
Serial0.2: X.121 5555 <--> ipx 0.0000.0100.0000
  PERMANENT, BROADCAST, 1 VC: 1023*
Serial0: X.121 5555 <--> decnet 2.5
  PERMANENT, BROADCAST, 1 VC: 1*
Serial0: X.121 5555 <--> ip 142.2.1.17
  PERMANENT, BROADCAST, 1 VC: 1024*
```

Completed Network Diagrams

Figures 3-5 through 3-7 show the final configurations for the three desktop protocols, DECnet, IPX, and AppleTalk, which you have configured during this case study.

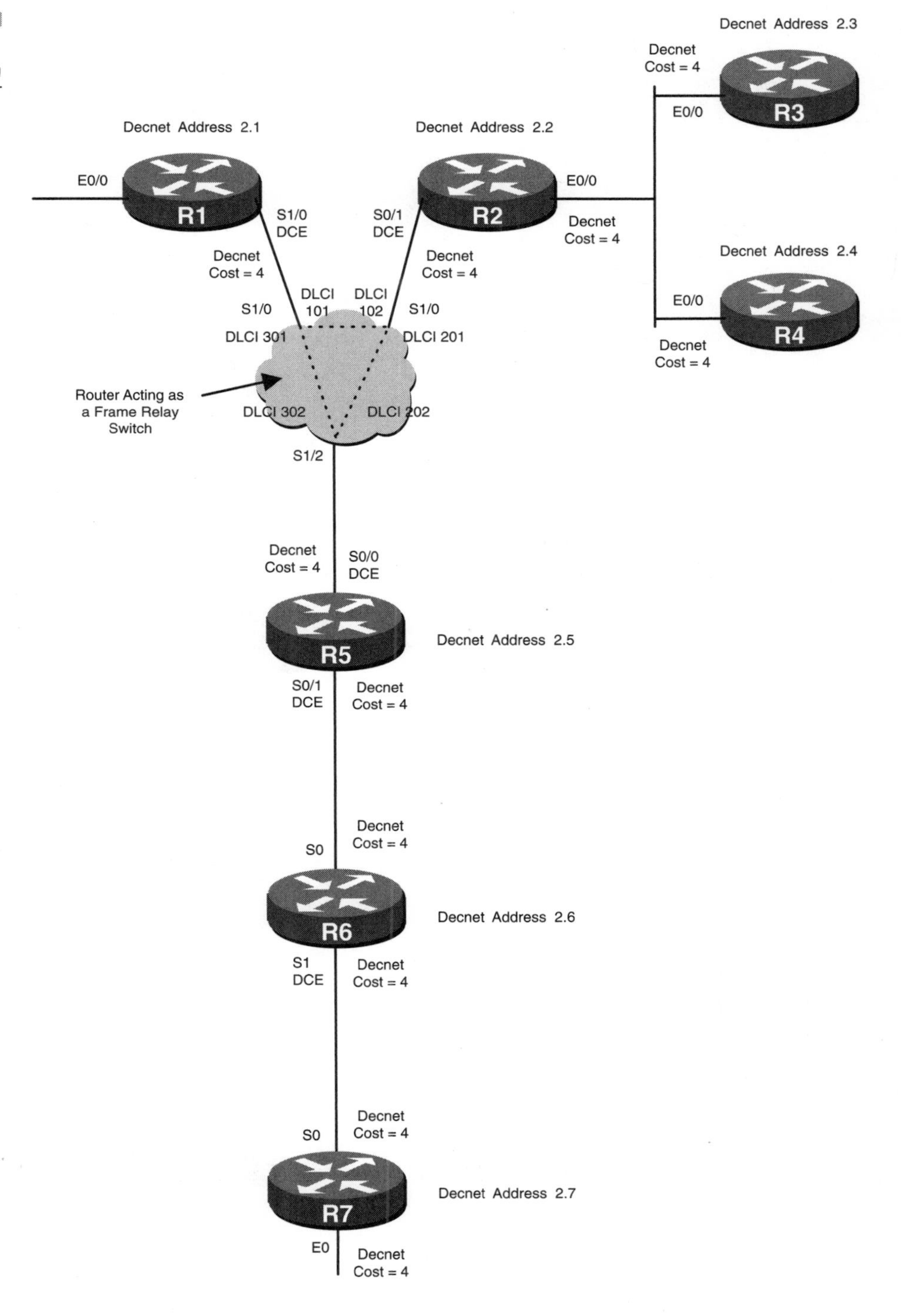

Figure 3-5
DECnet configuration

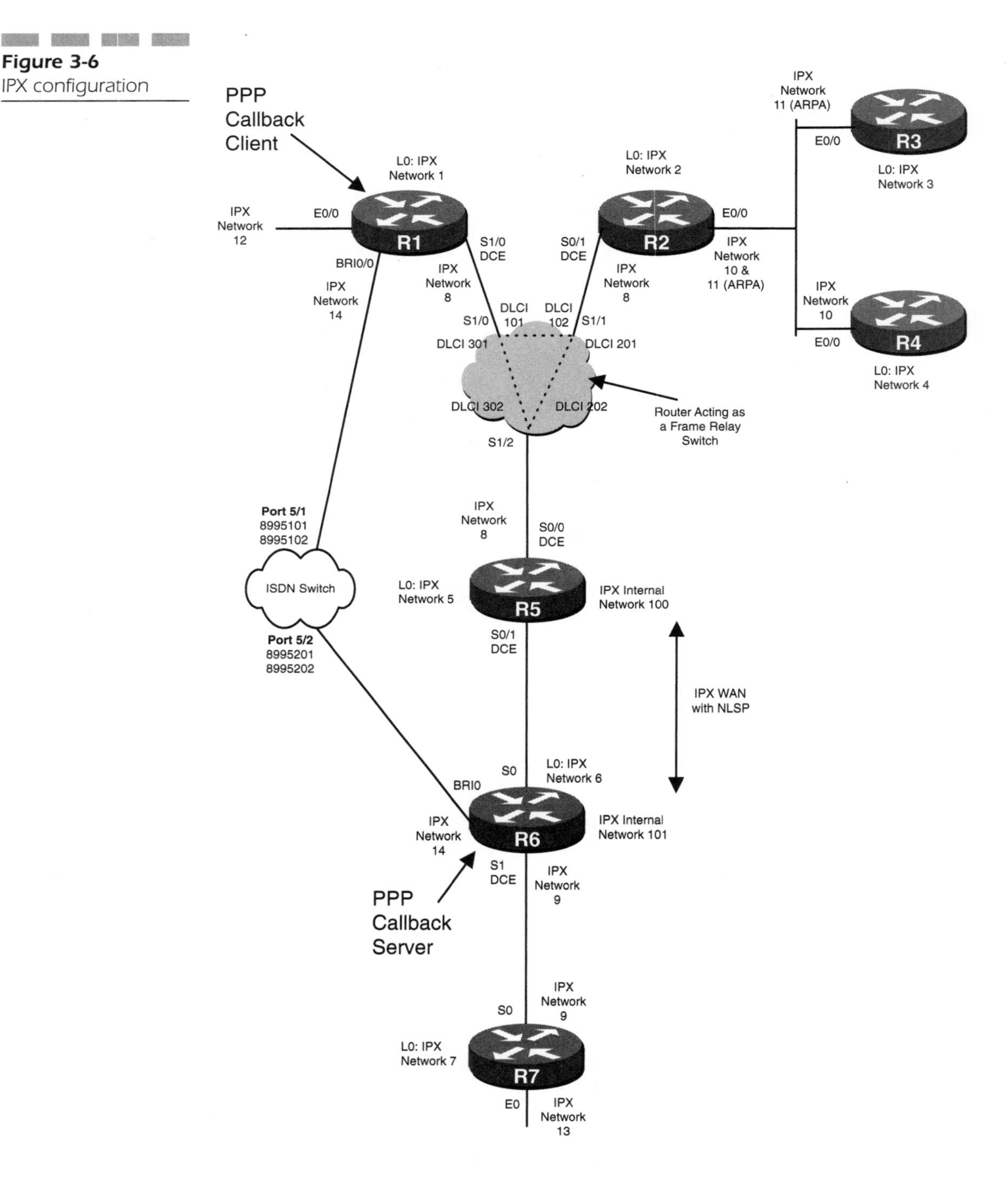

Figure 3-6
IPX configuration

Figure 3-7 AppleTalk configuration

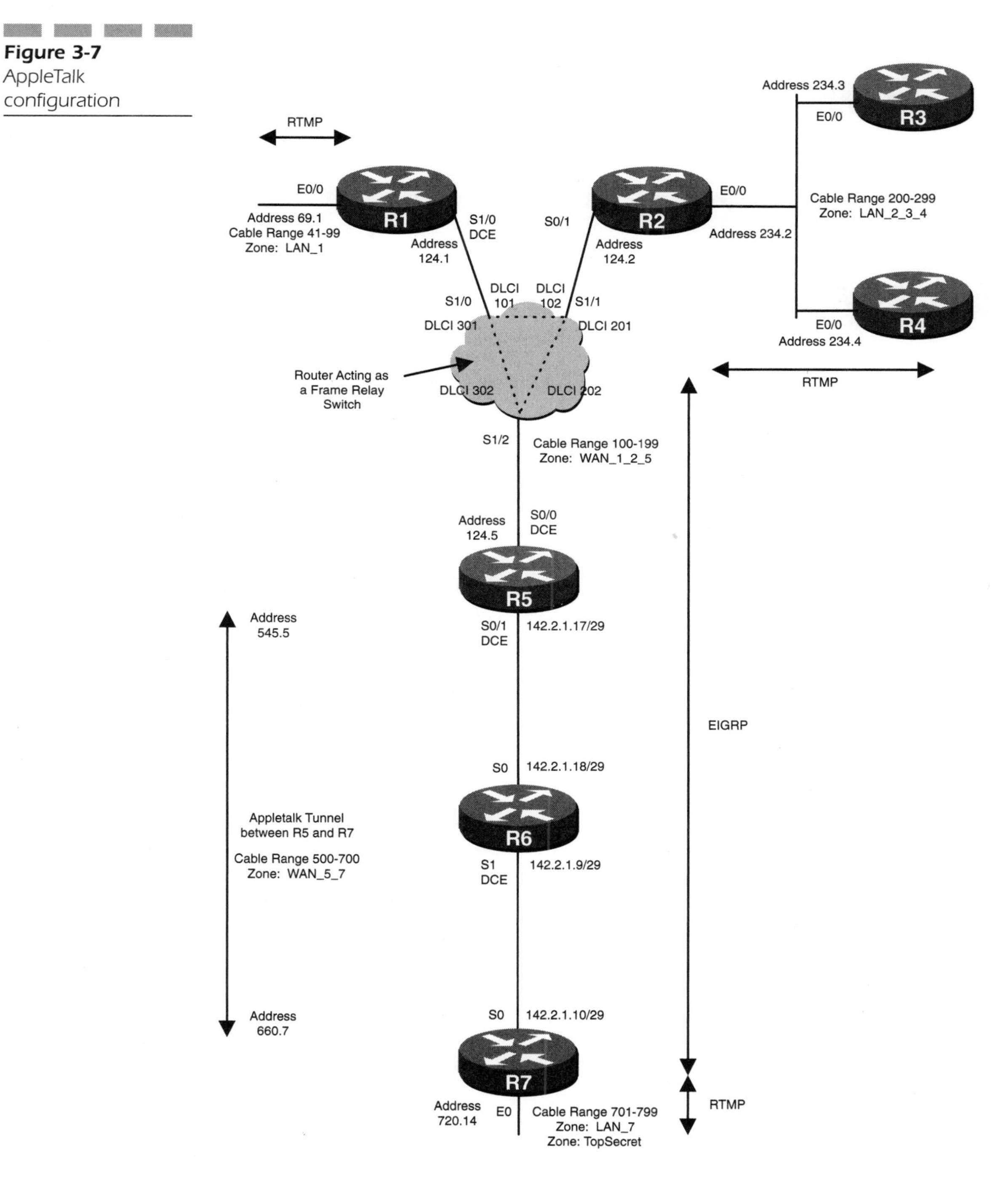

Router Configurations

The following section contains the complete router configurations for all routers used in this case study, including the router used to emulate a Frame Relay switch.

R1

```
!
version 11.2
service timestamps debug uptime
service timestamps log uptime
no service password-encryption
no service udp-small-servers
no service tcp-small-servers
!
hostname R1
!
boot system flash c3620-d-mz_112-16_p.bin
enable password cisco
!
username R6 password 0 cisco
!
decnet routing 2.1
decnet node-type routing-iv
!
appletalk routing eigrp 1
appletalk route-redistribution
ipx routing 0001.0001.0001
isdn switch-type basic-ni1
!
interface Loopback0
 no ip address
 ipx network 1
!
interface BRI0/0
 no ip address
 encapsulation ppp
 ipx network 14
 isdn spid1 5101 8995101
 isdn spid2 5102 8995102
 dialer map ipx 14.0006.0006.0006 name R6 broadcast 8995201
 dialer load-threshold 255 either
 dialer-group 1
 ppp callback request
 ppp authentication chap
!
interface Ethernet0/0
 no ip address
 no keepalive
 decnet cost 1
 appletalk cable-range 41-99 69.1
 appletalk zone LAN_1
 ipx network 12
```

```
!
interface Serial1/0
 no ip address
 encapsulation frame-relay
 decnet cost 1
 appletalk cable-range 100-199 124.1
 appletalk zone WAN_1_2_5
 appletalk protocol eigrp
 no appletalk protocol rtmp
 ipx input-sap-filter 1000
 ipx network 8
 no fair-queue
 clockrate 800000
 frame-relay map appletalk 124.5 101 broadcast
 frame-relay map decnet 2.2 101 broadcast
 frame-relay map decnet 2.5 101 broadcast
 frame-relay map ipx 8.0002.0002.0002 101 broadcast
 frame-relay map ipx 8.0005.0005.0005 101 broadcast
 frame-relay map appletalk 124.2 101 broadcast
 no frame-relay inverse-arp
 frame-relay lmi-type ansi
!
no ip classless
no logging buffered
access-list 900 permit any any cping
access-list 900 deny rip
access-list 900 deny sap
access-list 900 permit any any
access-list 1000 deny FFFFFFFF 7
access-list 1000 permit FFFFFFFF 0
!
!
ipx route default 14.0006.0006.0006 floating-static
!
ipx router eigrp 1
 network 8
!
!
ipx router rip
 no network 8
!
!
!
dialer-list 1 protocol ipx list 900
!
line con 0
line aux 0
line vty 0 4
 login
!
end
```

R2

```
!
version 11.2
```

```
no service password-encryption
no service udp-small-servers
no service tcp-small-servers
!
hostname R2
!
boot system flash c3620-d-mz_112-16_p.bin
enable password cisco
!
!
decnet routing 2.2
decnet node-type routing-iv
!
appletalk routing eigrp 2
appletalk route-redistribution
ipx routing 0002.0002.0002
!
interface Loopback0
 no ip address
 ipx network 2
!
interface Ethernet0/0
 no ip address
 decnet cost 2
 appletalk cable-range 200-299 234.2
 appletalk zone LAN_2_3_4
 appletalk access-group 600
!
interface Ethernet0/0.1
 arp timeout 0
 ipx network 10
!
interface Ethernet0/0.2
 arp timeout 0
 ipx network 11 encapsulation ARPA
!
interface Serial0/1
 no ip address
 encapsulation frame-relay
 decnet cost 2
 no decnet split-horizon
 appletalk cable-range 100-199 124.2
 appletalk zone WAN_1_2_5
 appletalk protocol eigrp
 no appletalk protocol rtmp
 no appletalk eigrp-splithorizon
 ipx network 8
 no ipx split-horizon eigrp 1
 clockrate 800000
 frame-relay map appletalk 124.5 201 broadcast
 frame-relay map decnet 2.1 102 broadcast
 frame-relay map ipx 8.0001.0001.0001 102 broadcast
 frame-relay map decnet 2.5 201 broadcast
 frame-relay map ipx 8.0005.0005.0005 201 broadcast
 frame-relay map appletalk 124.1 102 broadcast
 no frame-relay inverse-arp
 frame-relay lmi-type ansi
!
```

```
no ip classless
access-list 600 deny cable-range 701-799
access-list 600 permit other-access
access-list 800 deny 7
access-list 800 permit FFFFFFFF
!
appletalk local-routing
!
!
ipx router eigrp 1
 distribute-list 800 out
 network 8
 network 10
 network 11
!
!
ipx router rip
 no network 10
 no network 11
 no network 8
!
!
!
!
line con 0
line aux 0
line vty 0 4
 login
!
end
```

R3

```
!
version 11.2
no service password-encryption
no service udp-small-servers
no service tcp-small-servers
!
hostname R3
!
enable password cisco
!
!
decnet routing 2.3
decnet node-type routing-iv
!
appletalk routing
ipx routing 0003.0003.0003
!
interface Loopback0
 no ip address
 ipx network 3
!
interface Ethernet0/0
```

```
 no ip address
 decnet cost 3
 decnet in-routing-filter 300
 appletalk cable-range 200-299 234.3
 appletalk zone LAN_2_3_4
 ipx network 11 encapsulation ARPA
!
no ip classless
access-list 300 deny   2.6 0.0
access-list 300 permit 2.0 0.1023
!
!
!
ipx router eigrp 1
 network 11
!
!
ipx router rip
 no network 11
!
!
!
!
line con 0
line aux 0
line vty 0 4
 login
!
end
```

R4

```
!
version 11.2
no service password-encryption
no service udp-small-servers
no service tcp-small-servers
!
hostname R4
!
boot system flash c3620-d-mz_112-16_p.bin
enable password cisco
!
!
decnet routing 2.4
decnet node-type routing-iv
!
appletalk routing
ipx routing 0004.0004.0004
!
interface Loopback0
 no ip address
 ipx network 4
!
interface Ethernet0/0
```

```
 no ip address
 decnet cost 4
 appletalk cable-range 200-299 234.4
 appletalk zone LAN_2_3_4
 appletalk distribute-list 610 in
 ipx network 10
!
no ip classless
access-list 610 deny cable-range 41-99
access-list 610 permit other-access
!
!
!
ipx router eigrp 1
 network 10
!
!
ipx router rip
 no network 10
!
!
!
!
line con 0
line aux 0
line vty 0 4
 login
!
end
```

R5

```
!
version 11.2
no service password-encryption
no service udp-small-servers
no service tcp-small-servers
!
hostname R5
!
boot system flash c3620-d-mz_112-16_p.bin
enable password cisco
!
!
decnet routing 2.5
decnet node-type routing-iv
!
appletalk routing eigrp 5
appletalk route-redistribution
ipx routing 0005.0005.0005
ipx internal-network 100
!
interface Loopback0
 no ip address
 ipx network 5
```

```
!
interface Tunnel0
 no ip address
 appletalk cable-range 500-700 545.5
 appletalk zone WAN_5_7
 appletalk protocol eigrp
 no appletalk protocol rtmp
 tunnel source Serial0/1
 tunnel destination 142.2.1.10
!
interface Serial0/0
 no ip address
 encapsulation frame-relay
 decnet cost 5
 appletalk cable-range 100-199 124.5
 appletalk zone WAN_1_2_5
 appletalk protocol eigrp
 no appletalk protocol rtmp
 ipx network 8
 no fair-queue
 clockrate 800000
 frame-relay map decnet 2.1 202 broadcast
 frame-relay map decnet 2.2 202 broadcast
 frame-relay map ipx 8.0001.0001.0001 202 broadcast
 frame-relay map ipx 8.0002.0002.0002 202 broadcast
 frame-relay map appletalk 124.1 202 broadcast
 frame-relay map appletalk 124.2 202 broadcast
 no frame-relay inverse-arp
 frame-relay lmi-type ansi
!
interface Serial0/1
 ip address 142.2.1.17 255.255.255.248
 encapsulation x25 dce
 ip ospf network broadcast
 no ip mroute-cache
 decnet cost 5
 x25 address 5555
 x25 map decnet 2.6 6666 broadcast
 x25 map ip 142.2.1.18 6666 broadcast
 clockrate 800000
!
interface Serial0/1.2 point-to-point
 ipx ipxwan 0 unnumbered R5
 ipx nlsp area1 enable
 x25 map ipx 0.0000.0101.0000 6666 broadcast
!
router eigrp 1
 network 142.2.0.0
!
no ip classless
!
!
!
ipx router eigrp 1
 redistribute nlsp area1
 network 8
!
!
```

```
ipx router nlsp area1
 area-address 0 0
 redistribute eigrp 1
!
!
ipx router rip
 no network 8
!
!
!
!
line con 0
line aux 0
line vty 0 4
 login
!
end
```

R6

```
!
version 11.2
no service password-encryption
no service udp-small-servers
no service tcp-small-servers
!
hostname R6
!
enable password cisco
!
username R1 password 0 cisco
!
decnet routing 2.6
decnet node-type routing-iv
!
ipx routing 0006.0006.0006
ipx internal-network 101
isdn switch-type basic-ni1
!
interface Loopback0
 no ip address
 ipx network 6
!
interface Serial0
 ip address 142.2.1.18 255.255.255.248
 encapsulation x25
 ip ospf network broadcast
 no ip mroute-cache
 decnet cost 6
 x25 address 6666
 x25 map decnet 2.5 5555 broadcast
 x25 map ip 142.2.1.17 5555 broadcast
!
interface Serial0.2 point-to-point
```

```
 ipx ipxwan 0 unnumbered R6
 ipx nlsp area1 enable
 x25 map ipx 0.0000.0100.0000 5555 broadcast
!
interface Serial1
 ip address 142.2.1.9 255.255.255.248
 encapsulation ppp
 decnet cost 6
 ipx network 9
 clockrate 800000
!
interface BRI0
 no ip address
 encapsulation ppp
 ipx network 14
 isdn spid1 5201 8995201
 isdn spid2 5202 8995202
 dialer callback-secure
 dialer map ipx 14.0001.0001.0001 name R1 class dial1 broadcast 8995101
 dialer load-threshold 255 either
 dialer-group 1
 ppp callback accept
 ppp authentication chap
!
router eigrp 1
 network 142.2.0.0
!
no ip classless
!
map-class dialer dial1
 dialer callback-server username
access-list 900 permit any any cping
access-list 900 deny rip
access-list 900 deny sap
!
!
!
ipx router eigrp 1
 redistribute nlsp area1
 network 9
!
!
ipx router nlsp area1
 area-address 0 0
 redistribute eigrp 1
!
!
ipx router rip
 no network 9
!
!
ipx sap 4 FileServer1 13.aa00.0400.0708 451 1
ipx sap 7 PrintServer1 13.aa00.0400.0708 451 1
!
dialer-list 1 protocol ipx list 900
!
line con 0
line aux 0
```

```
line vty 0 4
 login
!
end
```

R7 (R7 was also our Terminal Server)

```
!
version 11.2
no service password-encryption
no service udp-small-servers
no service tcp-small-servers
!
hostname R7
!
enable password cisco
!
no ip domain-lookup
ip host R2 2002 1.1.1.1
ip host R3 2003 1.1.1.1
ip host R4 2004 1.1.1.1
ip host R5 2005 1.1.1.1
ip host R6 2006 1.1.1.1
ip host R8 2008 1.1.1.1
ip host R1 2001 1.1.1.1
!
decnet routing 2.7
decnet node-type routing-iv
!
appletalk routing eigrp 7
appletalk route-redistribution
ipx routing 0007.0007.0007
!
interface Loopback0
 ip address 1.1.1.1 255.255.255.255
 ipx network 7
 lat enabled
!
interface Tunnel0
 no ip address
 appletalk cable-range 500-700 660.7
 appletalk zone WAN_5_7
 appletalk protocol eigrp
 no appletalk protocol rtmp
 tunnel source Serial0
 tunnel destination 142.2.1.17
!
interface Ethernet0
 no ip address
 no keepalive
 decnet cost 7
 appletalk cable-range 701-799 720.14
 appletalk zone LAN_7
 appletalk zone TopSecret
```

```
 ipx network 13
!
interface Serial0
 ip address 142.2.1.10 255.255.255.248
 encapsulation ppp
 decnet cost 7
 ipx network 9
 no fair-queue
!
router eigrp 1
 network 142.2.0.0
!
no ip classless
!
!
!
ipx router eigrp 1
 network 9
!
!
ipx router rip
 no network 9
!
!
!
translate lat ServerB tcp 1.1.1.1
!
line con 0
line 1 16
 no exec
 transport input all
line aux 0
line vty 0 4
 password cisco
 login
!
end
```

FrameSwitch

```
!
version 12.0
service timestamps debug uptime
service timestamps log uptime
no service password-encryption
!
hostname FrameSwitch
!
!
!
!
!
!
ip subnet-zero
!
```

```
frame-relay switching
!
!
cns event-service server
!
!
!
process-max-time 200
!
interface Serial1/0
 no ip address
 no ip directed-broadcast
 encapsulation frame-relay
 no ip mroute-cache
 no fair-queue
 cdp enable
 frame-relay lmi-type ansi
 frame-relay intf-type dce
 frame-relay route 101 interface Serial1/1 102
 frame-relay route 301 interface Serial1/2 302
!
interface Serial1/1
 no ip address
 no ip directed-broadcast
 encapsulation frame-relay
 no ip mroute-cache
 no fair-queue
 cdp enable
 frame-relay lmi-type ansi
 frame-relay intf-type dce
 frame-relay route 102 interface Serial1/0 101
 frame-relay route 201 interface Serial1/2 202
!
interface Serial1/2
 no ip address
 no ip directed-broadcast
 encapsulation frame-relay
 no ip mroute-cache
 no fair-queue
 cdp enable
 frame-relay lmi-type ansi
 frame-relay intf-type dce
 frame-relay route 202 interface Serial1/1 201
 frame-relay route 302 interface Serial1/0 301
!
ip classless
no ip http server
!
!
!
line con 0
 transport input none
line aux 0
line vty 0 4
 login
!
!
end
```

Adtran Atlas 800 ISDN Switch Configuration

The Adtran Atlas 800 is being used in this chapter as the ISDN switch. The Atlas 800 can be populated with both ISDN BRI and ISDN PRI cards. The advantage of using the Atlas 800 is its ease of use and adherence to ISDN standards. In addition, the Atlas' small form factor makes it a portable unit that can be set up on a desktop. The base unit of the Atlas 800 comes with two ISDN PRI circuits. Additional BRI cards and PRI cards can be added. The Atlas 800 chassis can hold up to eight additional cards. Each BRI card contains eight ISDN BRI U interface circuits, while Each PRI card contains four ISDN PRI circuits.

The Atlas 800 chassis has a 10baseT Ethernet connection, which is used for system management. The Atlas 800 also has a control port that can be used for local terminal access. The Atlas 800 is managed via a simple menu system. The management screens are the same whether you telnet to the Atlas via the Ethernet port or directly connect to the unit via the rear control port. Additional information on the Atlas 800 can be found on the Adtran Web page at www.Adtran.com.

The Atlas 800 is simple to configure. The following section outlines the key configuration screens for setting up these labs.

The System Info screen is shown below. Key system information, such as firmware levels and system uptime, can be found on this screen.

```
CCIE LAB STUDY GUIDE/System Info
System Info    | System Name               CCIE LAB STUDY GUIDE
System Status  | System Location           Adtran ATLAS 800
System Config  | System Contact            Adtran ATLAS 800
System Utility | Firmware Revision         ATLAS 800  Rev. H 09/18/98 09:11:41
Modules        | System Uptime             0 days  1 hours  36 min  55 secs
Dedicated Maps | Startup Mode              Power cycle
Dial Plan      | Current Time/Date (24h)   Saturday March  6  16:55:12  1999
               | Installed Memory          Flash:1048576 bytes  DRAM:8388608 bytes
               | Serial Number             847B8304
               | Boot ROM Rev              C 11/18/97
```

The Modules screen displays all the active modules that are installed in the Atlas 800. You can see from the following screen print that your system has four occupied slots. Slot 0 is the system controller, which is part of the base chassis. Slot 1 contains a four-port ISDN PRI card, slot 5 contains an eight-port ISDN BRI card, and slot 8 contains a four-port V.35 card. For this case study, only the BRI card is used.

```
CCIE LAB STUDY GUIDE/Modules
System Info    | Slt   Type     Menu    Alarm    Test   State      Status       Rev
System Status  |  0  Sys Ctrl   [+]     [OK]     [OFF]  ONLINE   Online          T
System Config  |  1  T1/PRI-4   [+]     [OK]     [OFF]  ONLINE   Online          A
System Utility |  2  EMPTY                              ONLINE   Empty           -
Modules        |  3  EMPTY                              ONLINE   Empty           -
```

```
Dedicated Maps|  4  EMPTY                                ONLINE  Empty           -
Dial Plan     |  5  UBRI-8    [+]     [OK]      [OFF]   ONLINE  Online          C
              |  6  EMPTY                                ONLINE  Empty           -
              |  7  EMPTY                                ONLINE  Empty           -
              |  8  V35Nx-4   [+]     [OK]      [OFF]   ONLINE  Online          M
```

The Dial Plan screen is used to configure *directory numbers* (DNs) for each of the BRI circuits. Notice in the following screen print that BRI a1 in slot 5 has been assigned 8995101 as its number for channel B1, while BRI a2 in slot 5 has been assigned 8995201 as its number for channel B1:

```
CCIE LAB STUDY GUIDE/Dial Plan/User Term
Network Term|  #   Slot/Svc    Port/Link  Sig  In#Accept  Out#Rej   Ifce Config
User Term   |  1   5)UBRI-8    1)BRI 5/1       [8995101]    [+]      [8995101]
Global Param|  2   5)UBRI-8    2)BRI 5/2       [8995201]    [+]      [8995201]
```

The Interface Configuration screen is used to set the switch type for each circuit. You can see in the following screen that the National ISDN switch type has been selected for this circuit:

```
CCIE LAB STUDY GUIDE/Dial Plan/User Term[1]/Interface Configuration
Incoming Number Accept List| Switch Type          National ISDN
Outgoing Number Reject List| SPID list            [8995101]
Interface Configuration    | Strip MSD            None
                           | Source ID            0
                           | Outgoing Caller ID   Send as provided
```

The *Service Profile Identifier* (SPID) list screen is used to configure the SPID and call capability for each B channel. Notice in the following screen print that the SPID for channel B1 has been set to 5101, while the SPID for channel B2 has been set to 5102. Each of these two B channels is enabled for 64K, 56K, audio, and speech capabilities.

```
CCIE LAB STUDY GUIDE/Dial Plan/User Term[1]/Interface Configuration/SPID list
SPID list| #    Phone #          SPID #      Calls  D64    D56    Audio  Speech
         | 1    8995101           5101         2   Enable Enable Enable Enable
         | 2    8995102           5102         2   Enable Enable Enable Enable
```

CHAPTER 4

A Case Study in IGPs

Introduction

The following case study provides complex *Interior Gateway Protocol* (IGP) configuration scenarios. The chapter is broken into two case studies. The first will examine OSPF in detail because this is one of the most popular routing protocols used today. The second case study will focus on IS-IS and redistribution and filtering of routing protocols.

Technology Overview

A router operates at the network layer or layer three of the *Open Systems Interconnection* (OSI) model. Its primary job is to forward packets of data between nodes that are not connected to same network. In order to correctly route packets to their respective destinations, a router must acquire and store routing information for each destination network.

In order to understand the information that is required to route a packet, it is useful to examine what happens when a framed packet arrives at one of the router's interfaces (see Figure 4-1). When an Ethernet frames arrives at a router, the destination MAC address is examined. If the MAC address is that of the router's interfaces or the broadcast address, the router strips off the layer-two frame and passes the packet to the indicated network layer protocol in the Ethernet header. In the case of this scenario, the layer-three protocol is IP; it could have been IPX, AppleTalk, or any other routable protocol.

Once the packet is passed to the network layer, the router examines the destination address of the packet. If the destination address is either the IP address of one of the interfaces on the router or set to the broadcast address, the protocol field of the IP packet is examined and the packet is sent to the appropriate internal process. If the destination address is any other, the packet must be routed.

In order to route the packet, the router queries its routing table in order to determine the appropriate interface to which the packet should be forwarded. The routing table contains the destination addresses of all the networks that the router can reach. The table also contains a pointer that indicates the interface to forward the packet out of and, if the network is not directly connected, the IP address of another router on a directly connected interface. This router is called the next-hop router and is one router hop closer to the destination.

The routing table is searched for a routing entry that contains the most specific match to the destination network. This is referred to as longest match routing. The routing table can contain host-specific addresses, subnets, summary routes, major network numbers, supernets, or a default address (see Table 4-1). If the router cannot match the destination address to any entry in the routing table, the packet is dropped and an ICMP host unreachable message is sent to the source address.

The routing table can be populated in multiple ways. First the destination network is directly connected to the router. The destination network is automatically placed in the routing table, along with the interface of the router that connects to that network.

Figure 4-1 Static route in a single-homed environment

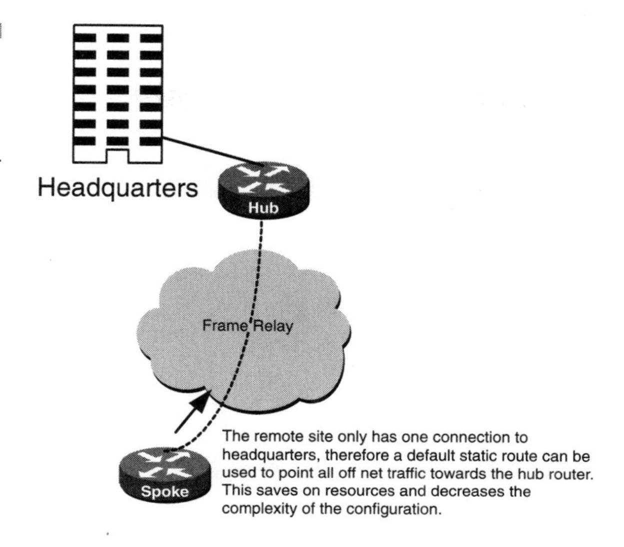

Table 4-1 Longest match routing

192.1.1.1 /32	Host-specific route
192.1.1.0 /30	Subnet
192.1.1.0 /25	Summary
192.1.1.0/24	Major network
192.1.0.0/16	Supernet
0.0.0.0	Default network

The other two ways that the router acquires information is through manual configuration (static routes) or dynamic routing protocols. This case study is focused on dynamic IP routing protocols, which can be further broken into two groups, distance vector and link state protocols.

Static Routes

Although static routes require manual configurations, they are very useful in certain situations. Static routes provide precise control over routing behavior that often cannot be obtained through dynamic protocols. The drawback, of course, is that these routes must be reconfigured any time a change in topology occurs.

Static routes are often used in hub and spoke situations where the spoke is singly homed to reduce routing overhead. In Figure 4-1, the spoke router only has one connection to the network. If that link were to go down, no alternative path would exist. Static routes are also useful in dial backup scenarios, where a floating static route is configured. A floating static route is unlike a regular static route in that it is not permanently entered in the routing table. It appears only after the failure of a more preferred route.

In Figure 4-2, the spoke router is learning its routes via RIP over its primary link from the hub. It also has a floating static route that points all traffic out of the ISDN interface. This floating static route has a higher administrative distance (admin distance of 130) than RIP (admin distance of 120). The administrative distance is the measure of preference that the router gives to all its routing protocols. The lower the administrative distance, the more preferred the route. A static route that is configured with a higher administrative distance is said to be floating because the only time that it will be invoked is if the more-preferred route, which in this case is the one learned by RIP, were to go away.

Dynamic Routing Protocols

Dynamic routing protocols enable routers to discover and share information automatically with other routers. A routing protocol can be though of as a language that routers use to communicate reachability information and network status. Routing protocols not only share information

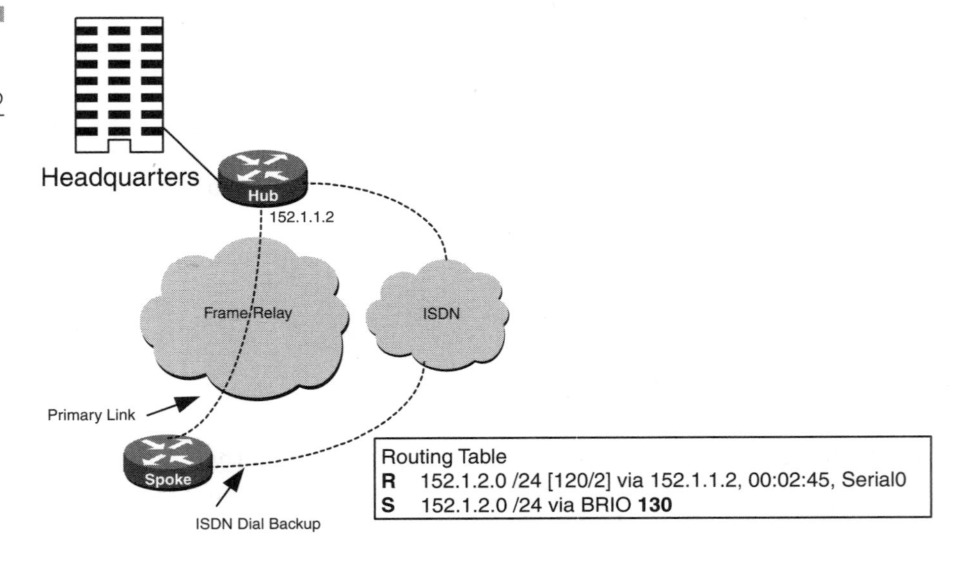

Figure 4-2 Using floating static routes for dial backup

between routers, but they also calculate the best paths to a destination. This capability to determine the best path to the destination during topology changes is the most important advantage dynamic protocols have over static routing.

Eight major IP routing protocols can be used to calculate and exchange routing information. Each of them have major advantages and disadvantages that will be explored throughout this case study. Table 4-2 lists all eight protocols along with their administrative distances.

The router uses the administrative distance to select the most preferred source of routing updates. If a router is running multiple protocols and learns a route to the same destination from each protocol, it needs a way to determine which one to use. This can be thought of as the measure of believability. The lower the administrative distance, the more believable the routing protocol. Administrative distances are assigned based on the routing protocol's capability to select the best path. For example, suppose that a router running RIP and the EIGRP learns of the same network from both protocols. Because RIP calculates the route based on the hop count and EIGRP uses a composite metric, EIGRP is most likely to have calculated the best path. As each protocol is explored in depth, the reasons that one protocol is preferred over another will become clear. As stated earlier, dynamic routing protocols are broken into two classes: distance vector and link state.

Distance Vector Protocols

The name *distance vector* is derived from the fact that routes are advertised as vectors of distance and direction. Distance is defined in terms of a metric, (hop count in the case of RIP) and

Table 4-2

Administrative distance

Routing Protocol	Administrative Distance
Static Route	1
BGP	20
EIGRP	90
IGRP	100
OSPF	110
IS-IS	115
RIP	120
EGP	140
IBGP	200

direction in terms of the next hop router. For example, destination X is a distance of three hops away in the direction of next hop router Y. Each router learns routes from its neighboring router perspective, adds a distance vector (its own distance value) to the route, adds the route to its routing table, and advertises it to its immediate neighbors.

This type of routing is often referred to as routing by rumor, because each router depends on its neighbor for routing information, which in turn may have been received from that router's neighbor. Most distance vector routing algorithms are based on the work done by Bellman, Ford, and Fulkerson; therefore, they are often referred to as Bellman-Ford algorithms.

For the most part, all distance vector protocols share the same characteristics and limitations. They exchange reachability information by sending periodic updates to all neighbors by broadcasting their entire routing table. In large networks, the routing table exchanged between routers becomes very large and hard to maintain, which leads to slow convergence. Convergence is the point in time when the entire network becomes updated to the fact that a route has appeared or disappeared.

Distance vector routers work on the basis of periodic updates and hold-down timers. If an update is not received in a given interval, the route goes into a hold-down state and is aged from the routing table. The hold-down aging process translates into minutes in convergence time before the whole network detects that a route has disappeared. This slow convergence problem creates inconsistencies, because routing update messages propagate slowly across the network. The larger the network, the greater the convergence time.

The other major drawback of distance vector protocols is their classfull nature. They do not support *Classless Interdomain Routing* (CIDR) or *Variable Length Subnet Masks* (VLSM) since no mask information is exchanged in their routing updates.

Finally, distance vector protocols are considered to be non-hierarchical or flat, which makes them incapable of scaling in large enterprise networks.

This case study explores the three most popular distance vector algorithms: RIP, IGRP, and EIGRP. Although EIGRP is occasionally described as a distance vector protocol that acts like a link state protocol (a hybrid), it still retains the routing characteristics of a distance vector protocol.

Routing Information Protocol (RIP)

RIP is a distance vector protocol used to exchange routing information among gateways (routers) and hosts. RIP is based on the Bellham-Ford (distance vector) algorithm, which was originally used in computer routing in 1969 by the *Advanced Research Projects Agency Network* (ARPANET). However, Xerox originally developed the RIP protocol as we know it today in the late 1970s as part of their *Xerox Networking Services* (XNS) protocol suite.

Despite its technical limitations, RIP is one of the most widely deployed IGPs designed for medium-sized homogeneous networks. RIP owes it widespread installed base to the fact that Berkley distributed routed software along with their popular 4BSD UNIX. Routed software used RIP to provide consistent routing and reachability information for machines on local networks. TCP/IP sites started using RIP to provide local area routing and eventually began using it in the wide area.

How RIP Works

RIP uses two packet types to convey information: updates and requests. Each RIP-enabled router on the network broadcasts update messages every 30 seconds using UDP port 520 to all directly connected neighbors. Update messages reflect the complete routing database, which currently exists in the router. Each entry in the database consists of two elements: the IP address of the network that can be reached and the distance to that network. Request messages are used by the router to discover other RIP-speaking devices on the network.

RIP uses the hop count as the metric to measure the distance to a network. Each router adds its internal distance (1) to the route before advertising the route to its neighbors. In Figure 4-3, R3 is directly connected to 152.1.0.0 (Network C). When it advertises this route to R2, it increments the metric by 1, and likewise R2 increases the metric to 2 and advertises the route to R1. R2 and R1 are said to be one and two hops, respectively, from Network C.

According to Figure 4-3, the number of hops to get to a given destination is the number of routers that a datagram must pass through to get to that destination network. Using hop count for path determination does not always provide the best path. For example, in Figure 4-4, to get from R1 to Network B, RIP prefers the shorter path over the 56Kbps link over the longer path

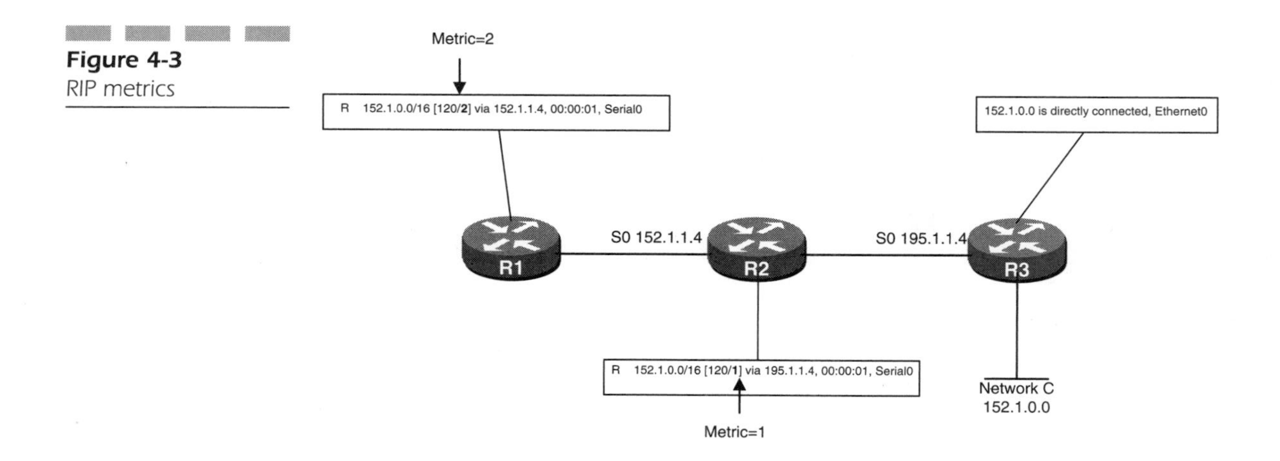

Figure 4-3
RIP metrics

Figure 4-4
Hop count

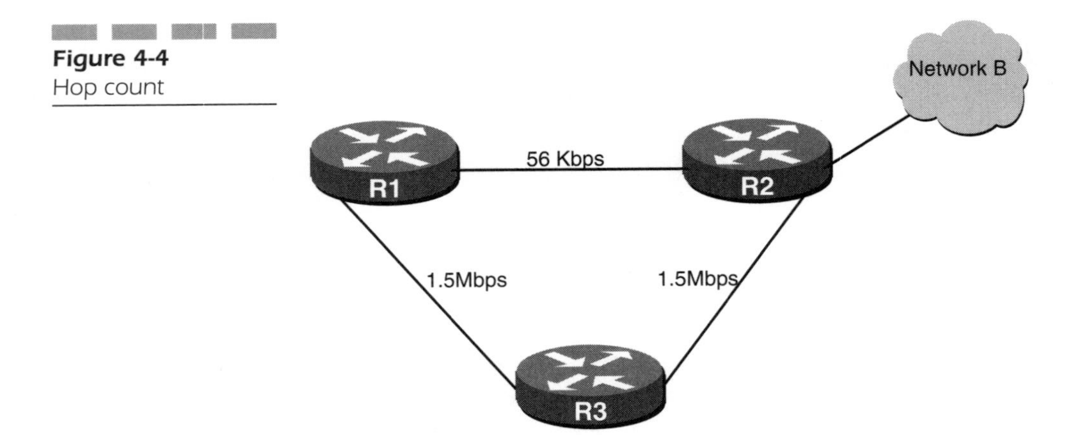

via the 1.5Mbp links. Even though the delay across a 56Kbps serial circuit is substantially greater than across two 1.5Mbps serial circuits, RIP chooses the least number of hops.

Routing Loops

The problem with any distance vector routing protocol is that each router does not have a complete view of the network. Routers must rely on the neighboring routers for network reachability information. The distance vector routing algorithm creates a slow convergence problem in which inconsistencies arise, because routing update messages propagate slowly across the network. To reduce the likelihood of routing loops caused by inconsistencies across the network, RIP uses the following mechanisms: count-to-infinity, split horizons, poison reverse updates, hold-down counters, and triggered updates.

Count-to-Infinity Problem

RIP permits a maximum hop count of 15. Any destination that is more than 15 hops away is considered unreachable. This number, while severely limiting the size of a network, prevents a problem called count-to-infinity (see Figure 4-5).

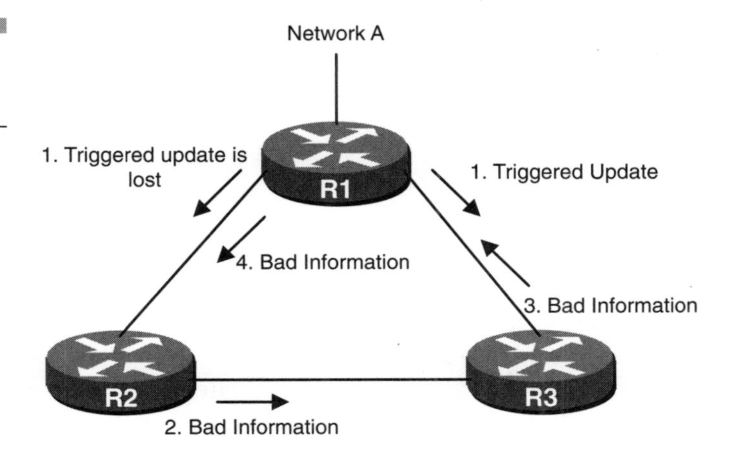

Figure 4-5 Count-to-infinity problem

Count-to-infinity works like this:

1. R1 loses its Ethernet interface (attached to Network A) and generates a triggered update, which is sent to R2 and R3. The triggered update tells R2 and R3 that R1 no longer can reach Network A. The update is delayed during transmission to R2 because of a CPU or a congested link, but it arrives at R3. R3 then removes the route to Network A from its routing table.
2. R2 still has not received the triggered update from R1 and sends its regular routing update advertising Network A as reachable with a hop count of 2. R3 receives the update and thinks a new route exits to Network A.
3. R3 then advertises to R1 that it can reach Network A with a hop count of 3.
4. R1 then advertises to R2 that that it can reach Network A with a hop count of 4.
5. This loop continues until the hop count reaches infinity, which is defined by RIP as 16. Once a route reaches infinity (16), it is declared unusable and deleted from the routing table.

With the count-to-infinity problem, the routing information continues to pass from router to router, incrementing the hop count by one. This problem, and the routing loop, will continue indefinitely, or until some limit is reached. That limit is RIP's maximum hop count. When the hop count of a route exceeds 15, the route is marked unreachable and, over time, eventually removed from the routing table.

Split Horizons

The rule of split horizon states that it is never useful for a router to advertise a route back in the direction from which it came. When split horizons is enabled on a router's interface, the router records the interface over which a route was received and does not propagate information about that route back out that interface.

The Cisco router enables you to disable split horizons on a per-interface basis. This is sometimes necessary in *Nonbroadcast Multiple Access* (NBMA) hub and spoke environments. In Figure 4-6, R2 is connected to R3 and R1 via Frame Relay, both PVCs are terminating on one physical interface on R2.

In Figure 4-6, if split horizon is not disabled on R2's serial interface, then R3 will not receive R1's routing advertisements and vice versa. The **no ip split-horizon** interface subcommand is used to disable split horizons.

Poison Reverse

Split horizon is a scheme used by the router to avoid problems caused by advertising routes back to the router from which they were learned. The split horizon scheme omits routes learned

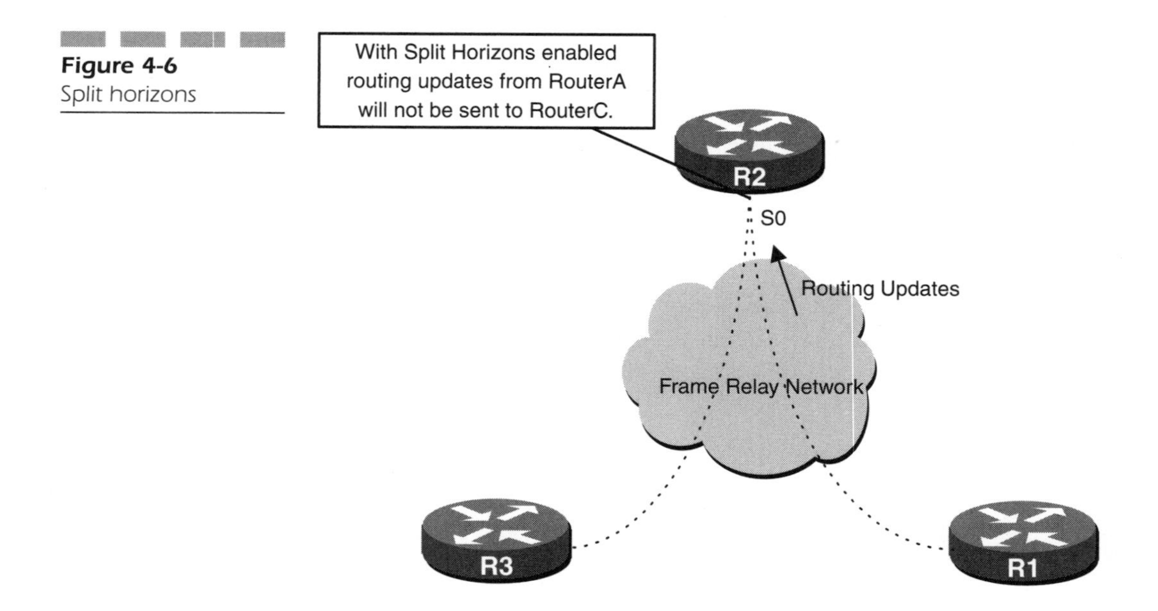

Figure 4-6 Split horizons

from one neighbor in updates sent to that neighbor. Split horizon with poisoned reverse includes the routes in updates, but sets their metrics to 16 (infinity).

By setting the hop count to infinity and advertising the route back to its source, it is possible to immediately break a routing loop. Otherwise, the inaccurate route stays in the routing table until it times out. The disadvantage to poison reverse is that it increases the size of the routing table.

Hold-down

Hold-down timers prevent the router from accepting routing information about a network for a fixed period of time after the route has been removed from the routing table. The idea is to make sure all routers have received the information, and no router sends out an invalid route. For example, back in Figure 4-5, R2 advertised bad information to R3 because of the delay in the routing update. With hold-down counters, this would not happen because R3 would not install a new route to Network A for 180 seconds. By then, R2 would have converged with the proper routing information.

Triggered Updates

Split horizons with poison reverse breaks any loop between two routers. Loops containing three or more routers can still occur, ending only when infinity (16) is reached. Triggered updates are an attempt to speed up convergence time. Whenever the metric of a route changes, a triggered update message is sent immediately, regardless of when the regular update message is scheduled to be sent.

RIP Message Format

Figure 4-7 shows the format of a RIP message. After the 32-bit header, the message contains a sequence of pairs. The pairs contain the network IP address and an integer distance to that network that can be reached.

The following list explains each of the RIP Header fields:

Command The command is generally either a RIP request (1) or a RIP response (2). Commands 3 and 4 are obsolete and command 5 is reserved for Sun Microsystems' internal use.

Version This field contains the protocol version number. Two versions of RIP exist.

Address Family Identifier RIP was designed to carry routing information for multiple protocols. This field specifies the family of the protocol that is being carried. The address family identifier for IP is 2.

Figure 4-7
RIP message format.

0	8	16	24	31
COMMAND (1-5)	VERSION (1)	MUST BE ZERO		
FAMILY OF NET 1		MUST BE ZERO		
IP ADDRESS OF NET 1				
MUST BE ZERO				
MUST BE ZERO				
DISTANCE TO NET 1				
FAMILY OF NET 2		MUST BE ZERO		
IP ADDRESS OF NET 2				
MUST BE ZERO				
MUST BE ZERO				
DISTANCE TO NET 2				

IP Address This field contains the IP address, which is stored as a four-octet number.

Must Be Zero: RIP can carry network addresses that are up to 12 octets long. Because an IP address only uses four of the 32 octets, the remaining eight octets are padded with zeros.

Distance to Net This field contains an integer count of the distance to the specified network. It contains a value of 16 if the network is unreachable.

RIP is the most widely used IGP in large organizations today, especially in organizations that have a large Unix-based routing environment. However, it is worth noting the limitations one faces when deploying a large RIP network:

- RIP uses a four-bit metric to count router hops to destinations. This limits the size of a RIP network, which cannot contain more than 15 hops to a destination. This is a severe limitation when trying to implement a typical modern large-scale network.
- RIP uses hop count as a routing metric, which does not always provide the most optimal path selection. More advanced protocols like IGRP use complex metrics to determine the optimal path.
- RIP was deployed prior to subnetting and has no subnet support. RIP assumes that all interfaces on the network have the same mask.

- RIP broadcasts a complete list of networks it can reach every 30 seconds. This can amount to a significant amount of traffic, especially on low-speed links.
- RIP has no security features built in. A RIP-enabled device accepts RIP updates from any other device on the network. More modern routing protocols such as OSPF enable the router to authenticate updates.

Interior Gateway Routing Protocol (IGRP)

IGRP is a Cisco proprietary distance vector routing protocol developed in 1986 to address the limitations of RIP. Unlike RIP, which uses UDP, IGRP accesses IP directly as protocol 9. IGRP also uses a concept of *autonomous systems* (AS) to separate IGRP process domains. IGRP allows multiple processes within the same domain, these processes are kept completely separate. No routes are exchanged between domains unless explicitly defined through the redistribution command. This enables traffic between IGRP processes to be closely regulated by redistribution and route filtering.

Although RIP works quite well in small homogenous internetworks, its small hop count (15) severely limits the size of the network, and its single metric (hop count) does not provide the routing flexibility needed in complex networks. IGRP addresses the shortcomings of RIP by allowing the network to grow up to 255 hops and by providing a wide range of metrics (link reliability, bandwidth, internetwork delay, and load) to provide routing flexibility for today's complex networks.

IGRP's other advantages over RIP are unequal-cost load balancing, an update period that is three times longer than RIP's, and a more efficient update packet format. The key disadvantage is that it is Cisco-proprietary, whereas RIP is an open standard.

Routing Loops

The problem with the first- and even second-generation distance vector routing protocols like IGRP is that each router does not have a complete view of the network. Routers must rely on the neighboring routers for network reachability information. This creates a slow convergence problem in which inconsistencies arise because routing update messages propagate slowly across the network. To reduce the likelihood of routing loops caused by inconsistencies across the network, IGRP uses split horizons, poison reverse updates, hold-down counters, and flash updates. The previous section discussed split horizons, poison reverse updates, and hold-down counters, so we will only cover flash updates in this section.

Flash Updates

Flash updates are an attempt to speed up convergence time. Whenever the metric of a route changes, the router must send an update message immediately. A flash update message is sent immediately, regardless of when the regular update message is scheduled to be sent.

IGRP Routes

IGRP advertises three types of routes: interior, system, and exterior. Interior routes are routes between subnets that are attached to the same router interface. System routes are routes to networks that are in the same AS, and exterior routes are routes to networks outside the AS (see Figure 4-8).

IGRP Metrics

The IGRP metric is a 32-bit number, which is calculated using bandwidth, delay, reliability, loading, and MTU. By default, IGRP chooses a route based on bandwidth and delay.

Bandwidth is expressed in units of kilobits per second. It is a static number that may not actually reflect the bandwidth of the link. For example, the default bandwidth of a serial link is 1544 regardless of the link's actual bandwidth. The bandwidth command should be used to set the interface to the actual bandwidth of the link.

Delay is expressed in 10-microsecond units and represents the end-to-end travel time. Like bandwidth, delay is also a static figure and may not actually reflect the delay of the link. For example, the default delay for a serial link is 1,000 microseconds regardless of the actual delay of the link. The delay command should be used to match the delay of the interface to the actual delay of the link.

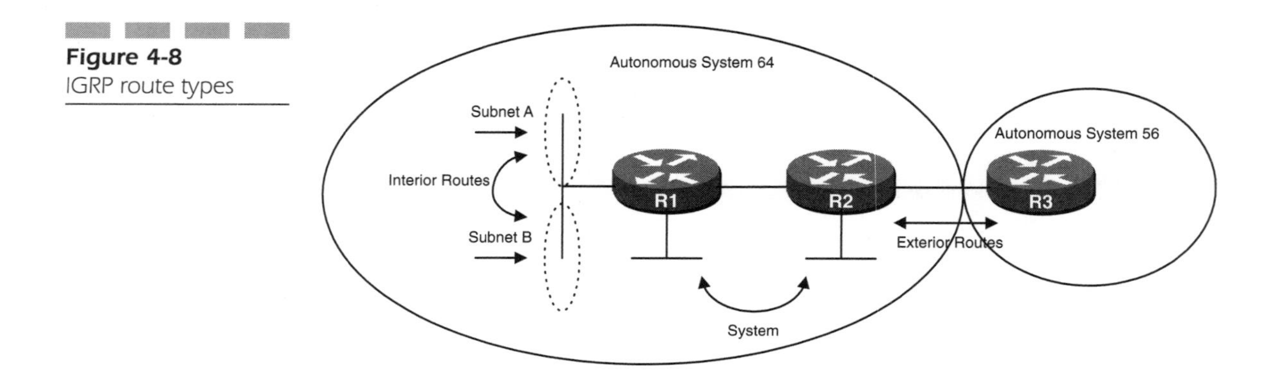

Figure 4-8
IGRP route types

Calculating the metric for a route is a two-step process using the five different characteristics of the link and the K values. The K values are configurable, but this is not recommended. The default K values are K1=1, K2=0, K3=1, K4=0, and K5=0. The two-step process is as follows:

1. **Metric** = K1 * Bandwidth + (K2 * Bandwidth)/(256 - load) + K3 * Delay
2. If K5 is not equal to zero, take the Metric from Step 1 and multiple it by [K5/(reliability + K4)]. If K5 is zero, ignore step 2:

Metric = Metric * [K5/(reliability + K4)]

As shown here, by default, Cisco sets K2, K4, and K5 to zero leaving only two variables to compute the IGRP metric (bandwidth and delay). Because three of the K values are zero, the formula reduces to:

Metric = Bandwidth + Delay

The IGRP bandwidth is derived by dividing 10,000,000 by the lowest bandwidth in the path to the destination

Delay is found by adding all the delays along the paths and dividing that number by 10. The equation is written as:

Metric = [(10,000,000/min bandwidth) + (SUM(interface delay)/10)]

IGRP Message Format

Figure 4-9 shows the format of an IGRP packet. The efficient design in comparison to RIP can be seen. IGRP updates provide much more information than RIP, which sends little more than a snapshot of the sender's routing table. Unlike RIP, no field is unused, and after the 12-byte header, individual route entries appear one after the other with no padding. Update packets can contain 104 network entries, versus the 25 for RIP. After the 32-bit header, the message contains a sequence of pairs. The pairs contain the network IP address and an integer distance to that network.

Version This field is always set to one.

Opcode This is set to one for an IGRP request packet and two for an IGRP update packet. A request packet consists of a header with no entries.

Figure 4-9
IGRP packet format

0	8	16	24 31
Version	OPCode	Edition	Autonomous System Number
Number of Interior Routes		Number of System Routes	
Number of Exterior Routes		Checksum	
Destination			Delay
Delay		Bandwidth	
Bandwidth	MTU		Reliability
Load	Hop Count	Destination	
Destination	Delay		
Bandwidth			MTU
MTU	Reliability	Load	Hop Count

Edition This is incremented by the sender of an update whenever a change in the routing information occurs. The edition number prevents the router from accepting old updates that arrive after newer updates.

Autonomous System Number The ID number of the IGRP process. This number enables multiple processes to exchange information over a common data link.

Number of Interior Routes The number of entries in the update that are subnets of a directly connected network. If none of the entries are subnets of a directly connected network, the field will be zero. This field along with the next two fields tells the router how many 14-octet entries are contained in the packet.

Number of System Routes This field indicates the number of entries in the update that are not directly connected networks, networks that have been summarized by a network border router.

Number of Exterior Routes This field indicates the number of routes to networks that have been identified as default networks.

Checksum This field contains a 16-bit ones complement sum of the packet that is calculated on the IGRP header and all entries.

Destination The address of the destination, which is only three octets long. At first, this might seem odd because an IP address is four octets long. This is possible because IGRP uses route categorization. If the entry is an interior route, at least the first octet of the IP address will always be known from the address of the interface on which the update was received. Thus, the destination field of an interior route entry only contains the last three octets of the address. If the entry is a system or an external route, the route will be summarized and at least the last octet will be all zeros. Therefore, the destination fields of system and external routes will contain only the first three octets of the address.

Delay This is a 24-bit number that is the sum of all the configured delays along the path and is expressed in units of 10 microseconds.

Bandwidth This is a 24-bit number, which is the result of 10,000,000 being divided by the lowest bandwidth along the path.

MTU This is the smallest of any link along the route to the destination.

Reliability This is a number between 0x01 and 0xFF that reflects the total outgoing error rate of the interface along the route and is calculated every five minutes.

Load This is a number between 0x01 and 0xff that reflects the total outgoing load of the interface along the route and is calculated every five minutes.

Hop Count This is a number between 0x01 and 0xFF indicating the number of hops to the destination.

Because IGRP is a distance vector protocol, it suffers from some of the same limitations as RIP, namely slow convergence. However, unlike RIP, IGRP can scale across large networks. IGRP's maximum hop count of 255 enables the protocol to be run in even the largest networks. Because IGRP uses five metrics to calculate route feasibility, it provides a more intuitive route selection process, providing optimal performance in even the most complex networks.

Enhanced Interior Gateway Routing Protocol (EIGRP)

EIGRP is a Cisco proprietary advanced distance vector routing protocol, which was first released in 1994 (IOS 9.21) to address the limitations of first- and second-generation distance vector protocols and link state protocols.

Traditional distance vector protocols, such as RIP and IGRP, forward routing updates to all attached neighbors, which in turn forward the updates to their neighbors. This hop-by-hop propagation of routing information creates large convergence times and looped topology problems.

Link state protocols such as OSPF have been offered as an alternative to the traditional distance vector protocols. The problem with link state protocols is that they solve the convergence problems of traditional distance vector protocols by replicating the topology information across the entire domain. This replication becomes undesirable in large networks and greatly affects CPU utilization due to the number of SPF calculations that need to be run.

EIGRP is different from the protocols discussed so far in that EIGRP uses a system of diffusing computations (route calculations that are performed in a coordinated fashion among multiple routers) to attain fast convergence while remaining loop-free at every instant. Unlike the protocols discussed earlier, EIGRP updates are non-periodic, partial, and bounded.

So what does all that mean? Non-periodic means that updates are not sent at regular intervals; EIGRP only sends updates in the event of a metric or topology change. Partial means that only the routes that have changed are sent, not all of the routes in the table. Bounded means that updates are only sent to the routers that are affected by the change. This means that EIGRP uses much less bandwidth than traditional distance vector protocols.

Some of the other major advantages of EIGRP are as follows:

- By default, EIGRP uses no more than 50 percent of the link's bandwidth. This is a major benefit when routing over low-speed or high-cost *Wide Area Network* (WAN) links.
- EIGRP is a classless protocol, and each route entry in an update includes a subnet mask.
- EIGRP supports MD5 authentication.
- EIGRP can also route IPX and AppleTalk traffic.

EIGRP Terminology

When dealing with EIGRP, it is important to understand the terminology being used.

Successor This is the directly connected neighboring router that has the best route to a particular destination. This route is used by the router to forward packets to a given destination. In order for a neighbor to become the successor for a particular destination, it must first meet the feasibility condition.

The feasibility condition states that the route must be advertised from a neighbor that is downstream with respect to the destination. The cost to reach the destination must be less than or equal to the cost of the route that is currently being used by the routing table.

For example, in Figure 4-10, R2's successor to reach Network A is R1, because the cost to reach Network A is 2, which is lower than going through R3, which is 3. However, if the metric of the link between R1 and R2 changed from 1 to 20, then R3 would meet the feasibility condition and become the successor.

Feasible Successor The feasible successor is a neighboring router that the destination can be reached through, but it is not used because the cost to reach the destination is higher than

Figure 4-10
EIGRP terminology

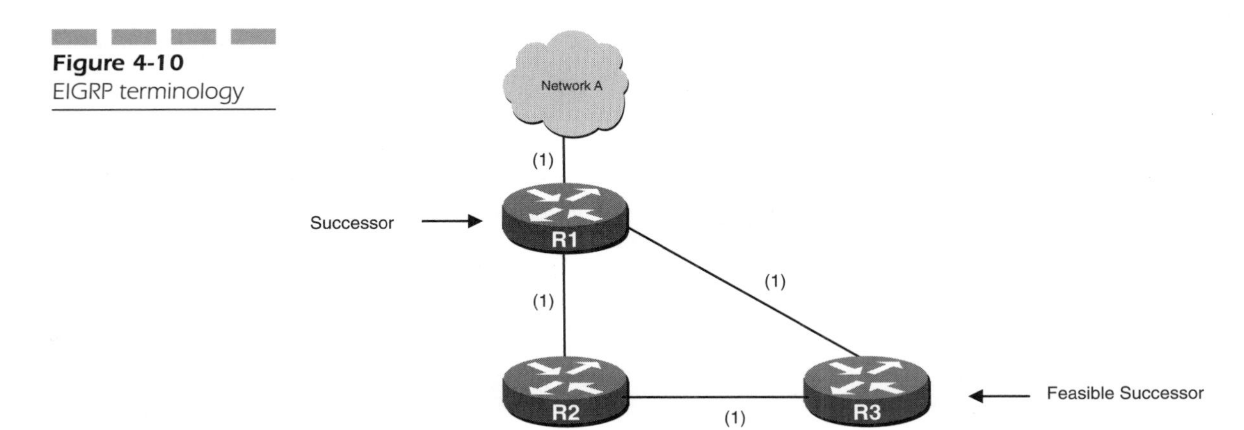

it would be using a different router. The feasible successor can be thought of as having the next best route to a destination.

Feasible successors are kept in the topology table and are used as backup routes. For example, in Figure 4-10, R2's feasible successor to reach Network A is R3. R3 has a route to Network A, but it is not the least-cost path and therefore is not used to forward data.

Feasibility Condition The feasibility condition is used to prevent routing loops. In order for the feasibility condition to be met, the route must be advertised from a neighbor that is downstream with respect to the destination. The cost to reach the destination must be less than or equal to the cost of the route that is currently being used in the routing table. If the feasibility condition is met, then the neighbor becomes the successor. For example, in Figure 4-10, if the link between R2 and R1 were to fail, R1 would no longer be the successor. R3 would move from being the feasible successor to the successor. If the link between R1 and R2 became active again, R1 would take over as successor because it meets the feasibility condition. It is downstream from Network A and its cost to reach network A is less than R3's cost.

Active State When the router loses its route to a destination and no feasible successor is available, the router goes into active state. While in an active state, the router sends out queries to all neighbors in order to find a route to the destination. At this time, the router must run the routing algorithm to recompute a new route to the destination.

Passive State When the router loses its successor but has a feasible successor, it goes into a passive state.

Hello Hello packets are exchanged between neighboring routers. As long as Hello packets are received, the router can determine that the neighbor is alive and functioning.

ACKs *Acknowledgement* (ACK) packets are sent by the router to acknowledge the receipt of update packets.

Update Update packets are used by the router to send routing information between neighbors. Update messages are sent if the metric of a route changes or when a router first comes up.

Query When the router loses its route to a destination and no feasible successor is available, the router goes into an active state. While in this active state, the router sends out query packets to all neighbors for a particular destination. The router waits for a response back from all neighbors before starting the computation for a new successor.

Replies Replies are sent in response to queries. The reply contains information on how to reach a destination. If the queried neighbor does not have the information requested, it sends queries to all its neighbors.

Technology Overview

When an EIGRP-enabled router first comes online, it sends Hello packets out all EIGRP-enabled interfaces using multicast address 224.0.0.10. The Hello packets are used for two things: discovering neighboring routers and, after the neighbors are discovered, determining if a neighbor has become unreachable or inoperative.

Once a new neighbor is discovered via a Hello packet, the router records the IP address and interface that the neighbor was discovered on. The router then sends an update to the neighbor containing all the routes that it knows about, and the neighbor does the same. This information is stored in the EIGRP topology table.

Subsequently, Hello packets are sent out every five seconds or every 60 seconds on low-speed NBMA networks. The Hello packets enable the router to discover the loss of its neighbor dynamically and quickly. If a Hello packet is not received from the neighbor router before the expiration of the hold timer, then the neighbor is declared down. At this point, the neighbor adjacency is deleted and all routes associated with that neighbor are removed.

The topology table includes the router's and neighbor's metrics for reaching the destination. The *Diffusing Update Algorithm* (DUAL) uses the topology table to find the lowest metric, loop-free path to each destination. The next hop router for the lowest cost path is referred to as the successor and is the next hop IP address that is loaded in the routing table. The DUAL also tries to find a feasible successor, or the next best route, which is kept in the topology database.

If the router loses its successor and a feasible successor is available, no route recomputation is necessary. The router simply makes the feasible successor the successor and adds the new route to the routing table remaining in a passive state. However, if no feasible successor is available, then the router goes into an active state for the destination network and recomputation for the route is necessary.

While the router is in active state, it sends a query packet out all EIGRP-enabled interfaces, except the interface the successor is on, inquiring if the neighbor has a route to the given destination. The neighbors respond and notify the sender if they have a route to the destination or not. Once all replies are received, the router can then calculate a new successor. If the neighbor receiving the query packet is using the sender to reach the destination network (as its successor), it queries all its neighbors for a route to the destination. The queried neighbors go through the same process, which creates a cascading of queries through the network, searching the network for a path to the destination.

As long as EIGRP has a feasible successor, no recomputation is necessary. This prevents the router from having to use CPU cycles and also speeds up convergence. Routers that are not affected by topology changes are not involved in recomputations

EIGRP Metrics

The EIGRP metric is a 32-bit number, which is calculated using bandwidth, delay, reliability, loading, and MTU. Calculating the metric for a route is a two-step process using the five different characteristics of the link and the K values. The K values are configurable, but this is not recommend. The default K values are K1=1, K2=0, K3=1, K4=0, and K5=0. The process is as follows:

1. **Metric** = K1 * Bandwidth + (K2 * Bandwidth)/(256 - load) + K3 * Delay
2. If K5 is not equal to zero, take the metric from Step 1 and multiply it by [K5/(reliability + K4)]. If K5 is zero, ignore step 2:

Metric = Metric * [K5/(reliability + K4)]

As shown, Cisco sets K2, K4, and K5 to zero. This leaves only two variables to compute the EIGRP metric (bandwidth and delay). Because three of the K values are zero, the formula reduces to:

Metric = Bandwidth + Delay

The bandwidth is derived by dividing 10,000,000 by the smallest of bandwidths in the path to the destination.

Delay is found by adding all the delays along the paths and dividing that number by 10. The sum of the two numbers is then multiplied by 256. This equation can be written as

Metric = [(10,000,000/min bandwidth) + (SUM(interface delay)/10)] * 256

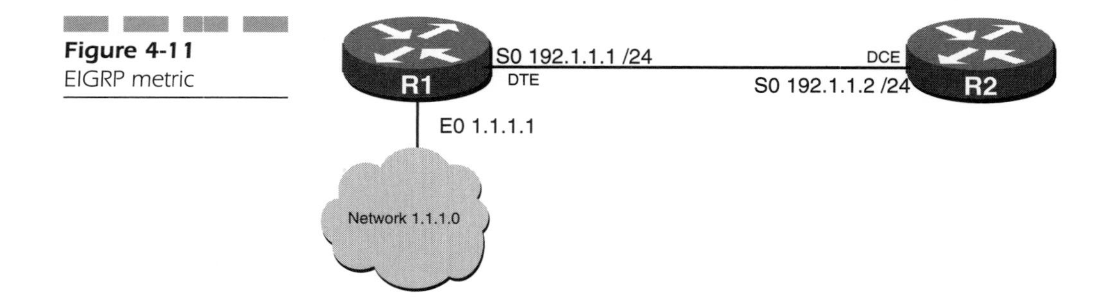

Figure 4-11 EIGRP metric

Let's look at Figure 4-11 and determine the metric to reach network 1.0.0.0 from R2.

Use the **show interface** command on each router to determine the bandwidth and delay for each interface. The following is the output from the command:

```
R2#show interfaces S0/0
Serial0/0 is up, line protocol is up
  Hardware is QUICC Serial
  Internet address is 192.1.1.1/24
  MTU 1500 bytes, BW 1544 Kbit, DLY 20000 usec, rely 255/255, load 1/255
  Encapsulation HDLC, loopback not set, keepalive set (10 sec)
  Last input 00:00:02, output 00:00:02, output hang never
  Last clearing of "show interface" counters never
  Input queue: 0/75/0 (size/max/drops); Total output drops: 0
  Queueing strategy: weighted fair
  Output queue: 0/64/0 (size/threshold/drops)
     Conversations  0/3 (active/max active)
     Reserved Conversations 0/0 (allocated/max allocated)
  5 minute input rate 0 bits/sec, 1 packets/sec
  5 minute output rate 0 bits/sec, 1 packets/sec
     155 packets input, 10368 bytes, 0 no buffer
     Received 80 broadcasts, 0 runts, 1 giants, 0 throttles
     5 input errors, 1 CRC, 2 frame, 0 overrun, 0 ignored, 1 abort
     246 packets output, 13455 bytes, 0 underruns
     0 output errors, 0 collisions, 910 interface resets
     0 output buffer failures, 0 output buffers swapped out
     154 carrier transitions
     DCD=up  DSR=up  DTR=up  RTS=up  CTS=up

R1#show interfaces e0/0
Ethernet0/0 is up, line protocol is up
  Hardware is AmdP2, address is 00e0.1e5b.25a1 (bia 00e0.1e5b.25a1)
  MTU 1500 bytes, BW 10000 Kbit, DLY 1000 usec, rely 243/255, load 1/255
  Encapsulation ARPA, loopback not set, keepalive not set
  ARP type: ARPA, ARP Timeout 04:00:00
  Last input never, output 00:00:08, output hang never
  Last clearing of "show interface" counters never
  Queueing strategy: fifo
  Output queue 0/40, 0 drops; input queue 0/75, 0 drops
  5 minute input rate 0 bits/sec, 0 packets/sec
  5 minute output rate 0 bits/sec, 0 packets/sec
```

```
0 packets input, 0 bytes, 0 no buffer
Received 0 broadcasts, 0 runts, 0 giants, 0 throttles
0 input errors, 0 CRC, 0 frame, 0 overrun, 0 ignored, 0 abort
0 input packets with dribble condition detected
6 packets output, 1071 bytes, 0 underruns
6 output errors, 0 collisions, 2 interface resets
0 babbles, 0 late collision, 0 deferred
6 lost carrier, 0 no carrier
0 output buffer failures, 0 output buffers swapped out
```

To reach network 1.0.0.0/8 from R2, a packet crosses the serial interface between R1 and R2 and the Ethernet interface on R1. Because the lowest bandwidth is used for the calculation, the bandwidth of the serial interface is used:

Metric = [(10,000,000/BW Serial link) + ((delay on serial link 1 delay on the Ethernet link)/10)] * 256

Metric = [(10,000,000/1544) + ((20000 1 1000)/10)] * 256

= Metric = 2195456

Let's take a look at the routing table on R2 and see if the calculations are correct.

```
R2#show ip route
Codes: C - connected, S - static, I - IGRP, R - RIP, M - mobile, B - BGP
       D - EIGRP, EX - EIGRP external, O - OSPF, IA - OSPF inter area
       N1 - OSPF NSSA external type 1, N2 - OSPF NSSA external type 2
       E1 - OSPF external type 1, E2 - OSPF external type 2, E - EGP
       i - IS-IS, L1 - IS-IS level-1, L2 - IS-IS level-2, * - candidate default
       U - per-user static route, o - ODR

Gateway of last resort is not set

D    1.0.0.0/8 [90/2195456] via 192.1.1.1, 00:21:50, Serial0/0
C    192.1.1.0/24 is directly connected, Serial0/0
```

EIGRP Packet Format

EIGRP accesses IP directly using protocol number 88. Figure 4-12 shows the EIGRP header.

The following is a list of EIGRP Packet details:

Version This field specifies the version of the originating EIGRP process, there are two releases of EIGRP software but they both use the same version number.

Opcode This field specifies the EIGRP packet type. Five possibilities are available, as shown in Table 4-3.

Checksum This is calculated for the entire EIGRP packet.

0	8	16	24	31
Version	Opcode	Checksum		
Flags				
Sequence				
ACK				
Autonomous System Number				
TLVs				

Figure 4-12
EIGRP packet format

Table 4-3
EIGRP opcodes

Opcode	Type
1	Update
3	Query
4	Reply
5	Hello
6	IPX SAP

Flags Only two possible flags can be set. The right-most bit is the *init bit*, which when set indicates that the enclosed route entries are the first new neighbor relationship. The send bit is the *conditional receive bit*, which is used in the proprietary reliable multicasting algorithm.

Sequence A 32-bit sequence number used by RTP.

ACK The 32-bit sequence number last heard for the neighbor to which the packet is being sent. A Hello packet with a nonzero sequence number is treated as an ACK packet rather than a Hello packet.

Autonomous System Number The number of the EIGRP domain.

TLV Each TLV contains a two-byte type number (the IP types are listed in Table 4-4; AppleTalk and IPX have specific TLV types that are not covered). The next two-byte field specifies the length of the TLV. The final field is the variable length and is determined by the type field.

Figure 4-13 is the format of the IP internal route and external routes TLV.

Table 4-4

EIGRP TLV types

Number	TLV Type
0x0001	EIGRP parameters that are used to convey metric weights and hold time
0x0003	Sequence number used by the reliable multicast algorithm
0x004	Software version used by the reliable multicast algorithm
0x005	Next multicast sequence used by the reliable multicast algorithm
0x0102	IP internal routes
0x0103	IP external routes

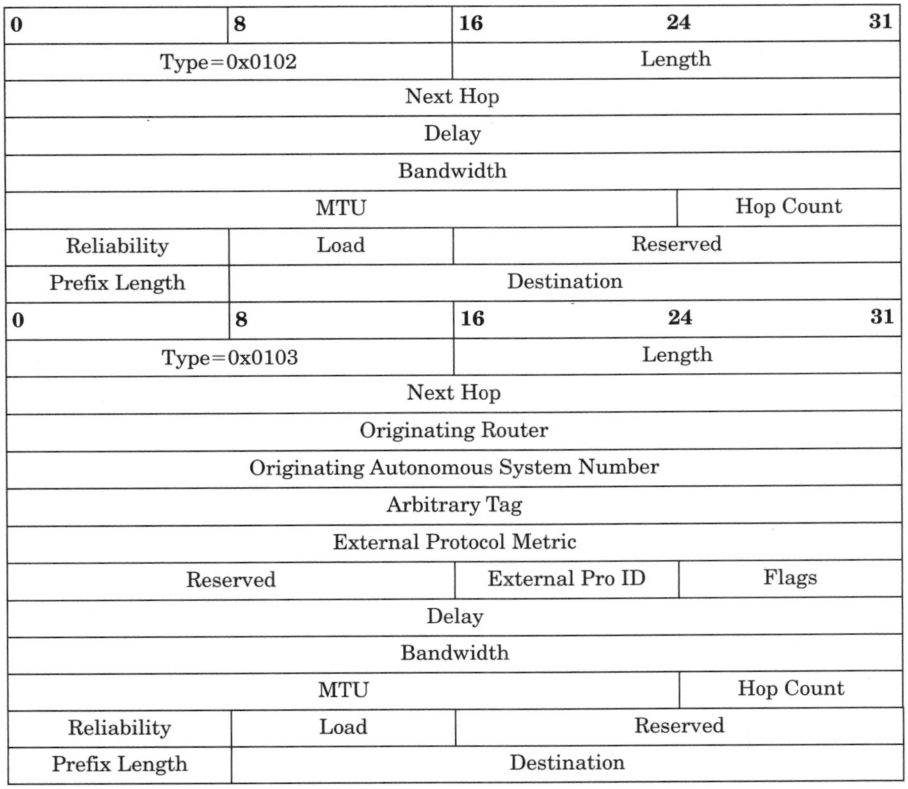

Figure 4-13
EIGRP IP internal route and external routes TLV

0	8	16	24	31
Type=0x0102		Length		
Next Hop				
Delay				
Bandwidth				
MTU			Hop Count	
Reliability	Load	Reserved		
Prefix Length	Destination			

0	8	16	24	31
Type=0x0103		Length		
Next Hop				
Originating Router				
Originating Autonomous System Number				
Arbitrary Tag				
External Protocol Metric				
Reserved		External Pro ID	Flags	
Delay				
Bandwidth				
MTU			Hop Count	
Reliability	Load	Reserved		
Prefix Length	Destination			

EIGRP is an enhanced version of IGRP and it uses the same distance vector algorithm and distance information as IGRP. EIGRP has been enhanced, making it converge faster and operate more efficiently than IGRP.

EIGRP also provides the following benefits:

- EIGRP provides fast convergence through the use of the DUAL.
- EIGRP only sends partial updates for routes that have changed, instead of sending the entire routing table.
- EIGRP supports *Variable Length Subnet Masking* (VLSM).
- The EIGRP metric is large enough to support thousands of hops.

Link State Protocols

Link state protocols derive their name from the fact that each router originates information about itself, its links, and the state of the links (hence, link state). This information is flooded (passed) by each router throughout the network. Link state protocols are often referred to as SPF protocols because they are built around the graph theory work done by E.W. Dijkstra on the shortest path algorithm.

Each router running a link state algorithm establishes a relationship called an adjacency with each neighbor (a router connected to the same network). Each router describes its local environment (its interfaces, the cost to send data over the interface, and what the interfaces connect to) in *Link State Advertisements* (LSAs). These LSAs are distributed to all the routers within a domain by a procedure called flooding. When a router receives a link state update packet, it makes a copy of it and forwards it unchanged to its neighbor. The object is for each router in the domain to have an identical view of the network, allowing each router to independently calculate its own best path.

The LSAs are stored, forming a topological database, also called a link state database. Using the Dijkstra algorithm, each router calculates the shortest path to each network. This information is stored in the routing table used to route packets. Two popular link state algorithms used today are OSPF and IS-IS.

Open Shortest Path First (OSPF)

OSPF is the most widely deployed link state routing protocol developed for IP networks. It was developed for use within a single AS to distribute routing information. The following section discusses the terminology and key concepts of OSPF.

OSPF Terminology

When dealing with OSPF, it is important to understand the terminology being used.

Autonomous System Abbreviated as AS, this is a group of routers under the control of a single administrative entity. For example, all the routers belonging to a particular corporation.

LSA Used to describe the local state of the router. The LSAs contain information about the state of the router's interfaces and the state of any adjacencies that are formed. The LSAs are flooded throughout the network.

The information contained in an LSA sent by each router in the domain is used to form the router's topological database. From this information, a shortest path is calculated to each destination.

Area An area is a collection of routers that have an identical topological database. OSPF uses areas to break an AS into multiple link state domains. Because the topology of one area is invisible to another, no flooding leaves an area, which greatly reduces the amount of routing traffic within an AS. Areas are used to contain link state updates and enable administrators to build hierarchical networks.

Cost This is the metric that the router uses to compare routes to the same destination. The lower the cost, the more preferred the route. OSPF calculates the cost of using a link based on bandwidth. The higher the bandwidth, the lower the cost and the more preferable the route.

Router ID This is a 32-bit number assigned to each OSPF-enabled router, which is used to uniquely identify the router within an AS. The router ID which is calculated at boot time is the highest loopback address on the router. If no loopback interfaces are configured, the highest IP address on the router is used.

Adjacency OSPF forms adjacencies between neighboring routers in order to exchange routing information. On a multi-access network, each router forms an adjacency with the DR.

Designated Router (DR) This is used to reduce the number of adjacencies that need to be formed on a multi-access network such as Ethernet, Token Ring, or Frame Relay. The reduction in the amount of adjacencies formed greatly reduces the size of the topological database.

The DR becomes adjacent with all other routers on the multi-access network. The routers send their LSAs to the DR, and the DR is responsible for forwarding them throughout the network. The idea behind a DR is that routers have a central point to send information to, versus every router exchanging information with every other router on the network.

Backup Designated Router (BDR Used on a multi-access network and is responsible for taking over for the DR if it should fail.

Inter-Area Route This is a route that is generated in an area other than the local one, inside the current OSPF routing domain.

Intra-Area Route This is a route that is within one area.

Neighbors Routers that share a common network. For example, two routers on an Ethernet interface are said to be neighbors.

Flooding A technique used to distribute LSAs between routers.

Hello Packet This is used to establish and maintain neighbor relationships. A Hello packet is also used to elect a DR for the network.

Technology Overview

Let's start with a brief introduction to OSPF before going into more detail. OSPF uses a link state algorithm to calculate the shortest path to all destinations in each area. When a router is first enabled or if any routing changes occur, the router configured for OSPF floods LSAs to all routers in the same hierarchical area. The LSAs contain information on the state of the router's links and the router's relationship to its neighboring routers. From the collection of LSAs, the router forms what is called a link state database. All routers in an area have an identical database describing the area's topology.

The router then runs the Dijkstra algorithm using the link state database to form a shortest path tree to all destinations inside the area. From this shortest path tree, the IP routing table is formed. Any changes that occur on the network are flooded via link state packets and cause the router to recalculate the shortest path tree using the new information.

Link State Routing Protocol

OSPF uses a link state algorithm to calculate the shortest path to all known destinations. Link state refers to the state of a router's interface (up, down, IP address, type of network, and so on) and the router's relationship to its neighbors (how the routers are connected on the network). The LSAs are flooded to each router and are used to create a topological database.

The Dijkstra algorithm is run on each router using the topological database, created by all LSAs received from all the routers in the area. The algorithm places each router at the root of the tree and calculates the shortest path to each destination based on the cost to reach that network.

Flooding

Flooding is the process of distributing LSAs between adjacent routers. The flooding procedure carries the LSA one hop further from its point of origin.

Because all routers in an OSPF domain are interconnected via adjacencies, the information disseminates throughout the network. To make this process reliable, each LSA must be acknowledged.

Dijkstra Algorithm

The Dijkstra algorithm is the heart of OSPF. Once a router receives all LSAs, it then uses the Dijkstra algorithm to calculate the shortest path to each destination inside the area based on the cumulative cost to reach that destination. Each router will have a complete view of the network topology inside the area. It builds a tree with itself as the route and has the entire path to any destination network or host.

However, the view of the topology from one router will be different from that of another because each router uses itself as the root of the tree. The Dijkstra algorithm is run any time a router receives a new LSA.

Areas

OSPF uses areas to segment the AS and contain link state updates. LSAs are only flooded within an area, so separating the areas reduces the amount of routing traffic on a network.

Each router within an area has an identical topological database, as do all other routers in the same area. A router in multiple areas has a separate topological database for each area it is connected to.

Routers that have all their interfaces within the same area are called *internal routers* (IR). Routers that connect areas within the same AS are called *Area Border Routers* (ABRs), and routers that act as gateways redistributing routing information from one AS to another are called *Autonomous System Border Routers* (ASBRs) (see Figure 4-14).

Backbone Area 0

OSPF has a concept of a backbone area referred to as area 0. If multiple areas are configured, one of these areas must be configured as area 0. The backbone (area 0) is the center for all areas; that is, all areas must have a connection to the backbone. All areas inject routing information into area 0 and the backbone propagates routing information back to each area.

In cases when an area does not have direct physical connectivity to the backbone, a virtual link must be configured. Virtual links will be discussed later in the chapter.

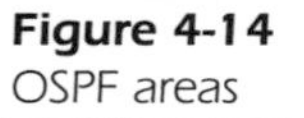

Figure 4-14
OSPF areas

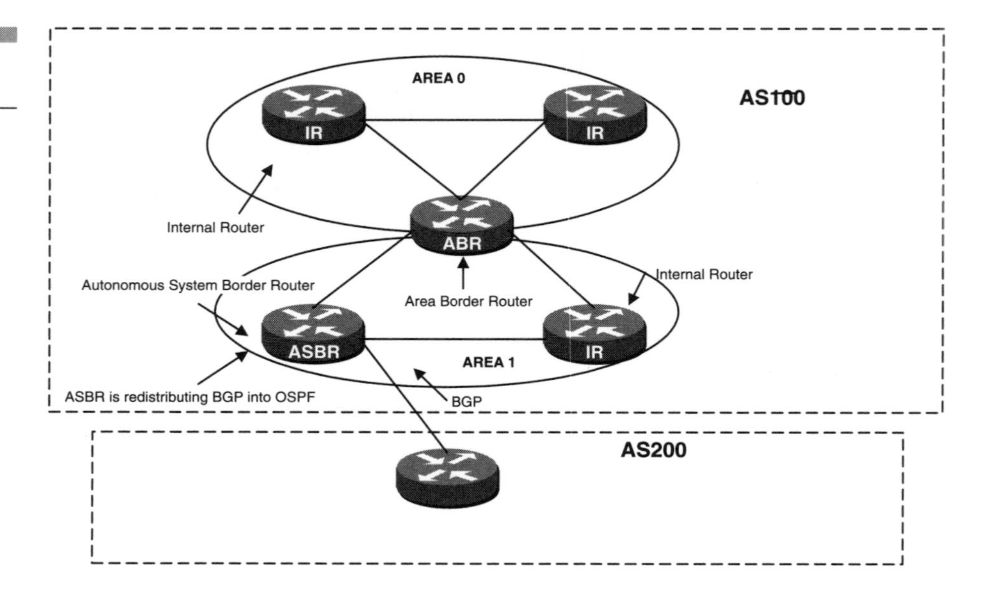

Designated Router (DR)

All multi-access networks with two or more attached routers elect a DR. The DR concept enables a reduction in the number of adjacencies that need to be formed on a network. In order for OSPF-enabled routers to exchange routing information, they must form an adjacency with one another. If a DR is not used, then each router on a multi-access network would need to form an adjacency with every other router (since link state databases are synchronized across adjacencies). This would result in N-1 adjacencies.

Instead, all routers on a multi-access network form adjacencies only with the DR and BDR. Each router sends the DR and BDR routing information and the DR is responsible for flooding this information to all adjacent routers and originating a network link advertisement on behalf of the network. The BDR is used in case the DR fails.

The reduction in adjacencies reduces the volume of routing protocol traffic as well as the size of the topological database.

The DR is elected using the Hello protocol, which was described earlier in this chapter. The election of the DR is determined by the router priority, which is carried in the Hello packet. The router with the highest priority will be elected the DR. If a tie occurs, the router with the highest router ID is selected.

The router ID is the IP address of the highest addressed loopback interface. If no loopback is configured, the router ID is the highest IP address on the router. The router priority can be configured on the router interface with the **ip ospf priority** command.

When a router first becomes active on a multi-access network, the router checks to see if a DR currently exists for the network. If a DR is present, the router accepts the DR regardless of what its priority is. Once a DR is elected, no other router can become the DR unless the DR fails. If no DR is present on the network, then the routers negotiate the DR based on router priority.

OSPF Protocol Packets

The OSPF protocol runs directly over IP protocol 89 and begins with the same 24-byte header, as shown in Figure 4-15.

There are five OSPF packet types, as shown in Table 4-5.

This section describes the OSPF packet types:

Hello Packets The Hello protocol is responsible for discovering neighbors and maintaining the neighbor relationship. Hello packets are sent periodically out the router's interface, depending on the network type. The Hello protocol is also responsible for electing a DR on multi-access networks. The role of the DR is discussed later in the chapter.

Figure 4-15
OSPF header

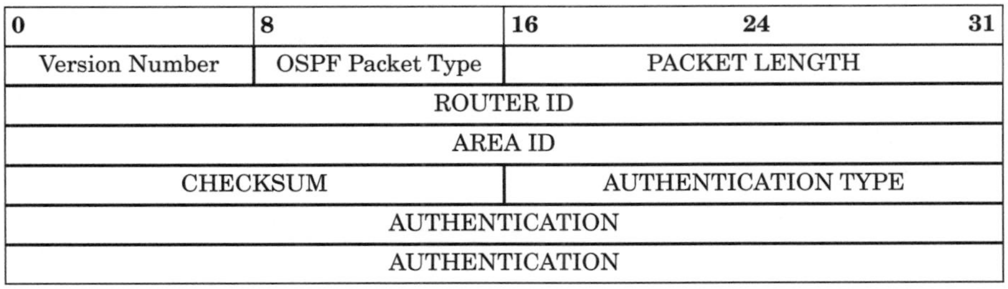

0	8	16	24	31
Version Number	OSPF Packet Type	PACKET LENGTH		
ROUTER ID				
AREA ID				
CHECKSUM		AUTHENTICATION TYPE		
AUTHENTICATION				
AUTHENTICATION				

Table 4-5
OSPF packet types

Type	Packet Name	Protocol Function
1	Hello	Discover and maintain neighbors
2	Database description	Summarize database contents
3	Link state request	Request for database information
4	Link state update	Database update
5	Link state acknowledgement	Acknowledgement

Database Description Packets Database description packets are OSPF type 2 packets. These packets are responsible for describing the contents of the Link State database of the router, which is one of the first steps to forming an adjacency.

Database descriptor packets are sent in a poll response manner. One router is designated the master and the other the slave. The master sends database polls, which are acknowledged by the database descriptor packets, sent by the slave.

Link State Request Packets Link state request packets are OSPF type 3 packets. Once the complete databases are exchanged between routers using the data base description packets, the routers compare the database of their neighbor with their own database. At this point, the router may find that parts of the neighbor's database may be more up to date than its own. If so, the router requests these pieces using the link state request packet.

Link State Update Packet Link state update packets are OSPF packet type 4. The router uses a flooding technique to pass LSAs. Multiple LSA types (router, network, summary, and external) are described in detail later in this chapter.

Link State Acknowledgement Packet Link state acknowledgements are OSPF type 4 packets, which are used to acknowledge the receipt of LSAs. This acknowledgement makes the OSPF flooding procedure reliable.

Link State Advertisements (LSAs)

All LSAs begin with the same 20-byte header, as shown in Figure 4-16. The CSA header contains the following fields:

LS age The time in seconds since the LSA was originated.

Options The optional capabilities supported by the router.

LS type The type of LSA.

Figure 4-16
OSPF LSA header

0	8	16	24	31
LS age		Options	LS Type	
Link State ID				
Advertising Router				
LS sequence number				
LS checksum		Length		

Link State ID This field identifies the portion of the internet environment that is being described by the advertisement.

Advertising Router The router ID of the router that originated the packet.

LS sequence number This is used to detect old or duplicate LSAs.

LS checksum The checksum of the complete contents of the LSA.

Length The length in bytes of the LSA, including the 20-byte header.

Router Link

Each router in the area generates a router LSA (type 1 LSA). This advertisement describes the state and cost of the router's interfaces to that area. All of the router's links to the area must be described in a single router LSA. The router LSAs are only flooded throughout a single area.

Network Link

Network link advertisements are type 2 LSAs. The DR for each multi-access network that has more than one attached router originates a network advertisement. The advertisement describes all the routers attached to the network as well as the DR itself.

Summary Link

Summary LSAs are type 3 and 4 LSAs. The ABR generates summary LSAs, which describe a route to a single destination. The summary LSA is advertised within the single area and the destination described is external to the area yet part of the same AS. Only intra-area routes are advertised in the backbone.

External Link

The ASBR generates an external type 5 LSA, which advertises each destination known to the router that is external to the AS. AS external type 5 LSAs are used to advertise default routes into the AS.

Two types of external routes exist: external type 1 and external type 2. The difference between the two is the way the cost or metric of the route is calculated. External type 1 routes use the external cost plus the internal cost of reaching a route. External type 2 routes only use the external cost of reaching the route. Type 2 routes are always preferred over type 1 routes and are the default type for any route that is redistributed into OSPF.

How OSPF Works

When an OSPF-enabled router first comes online, it sends out a Hello packet to the multicast address 224.0.0.5. This packet is sent out to all OSPF-enabled interfaces periodically, depending on the interface type. For broadcast media such as Ethernet or Token Ring or Point-to-Point interfaces, the Hello packet is sent every 10 seconds. On a NBMA such as Frame Relay or ATM, the Hello packet is sent out every 30 seconds.

The Hello packets are not only used to build neighbor relationships and discover which neighbors are on the same wire, but they are also used to describe any optional capabilities of the router, such as whether the router is in a regular or a stub area. The Hello packet is also used to elect the designated router on multi-access networks.

After the neighbor is discovered, bidirectional communication is assured and a designated router is elected (on a multi-access media). The router then attempts to form an adjacency with the neighboring router.

To form an adjacency, the routers must synchronize their databases. To do this, each router describes its databases to the others by sending a sequence of database description packets. This process is called the Database Exchange process and will be covered in more detail later in the chapter.

During the Database Exchange Process, the two routers form a master/slave relationship. Each database description packet sent by the master contains a sequence number. The slave acknowledges receipt of the packet by echoing back the sequence number.

During the database exchange process, each router checks its own database to see if any of the LSAs received by its neighbor are more recent than its own database copy. If any are, the router makes note of this and after the database exchange process is over the router requests updated LSAs using request packets. Each router responds to the link state request using a link state update. When the requesting router receives the updated LSA, it acknowledges the packet. When the database description process is complete and all link state requests have been updated, the databases are synchronized.

How an Adjacency Is Formed

In order for a router to exchange link state database information with another router, an adjacency must be formed. This is a key part of OSPF and therefore needs to be completely understood.

On a Cisco router, you can check the status of the adjacency using the **show ip ospf neighbor** command. The following is the output from the command. Notice that the state of the adjacency is full, which means that R1's database is synchronized with neighbor 1.1.1.1, which is R2.

```
R1#show ip ospf neighbor
Neighbor ID     Pri   State           Dead Time   Address         Interface
1.1.1.1           1   FULL/BDR        0:00:37     10.10.3.1       Ethernet0
```

Neighbor routers must go through five states before fully forming an adjacency or having a full neighbor state. Figure 4-17 shows an example of how an adjacency is formed between two neighboring routers on a broadcast media. R1 and R2 both connect to an Ethernet network and R3 is configured with a higher DR priority.

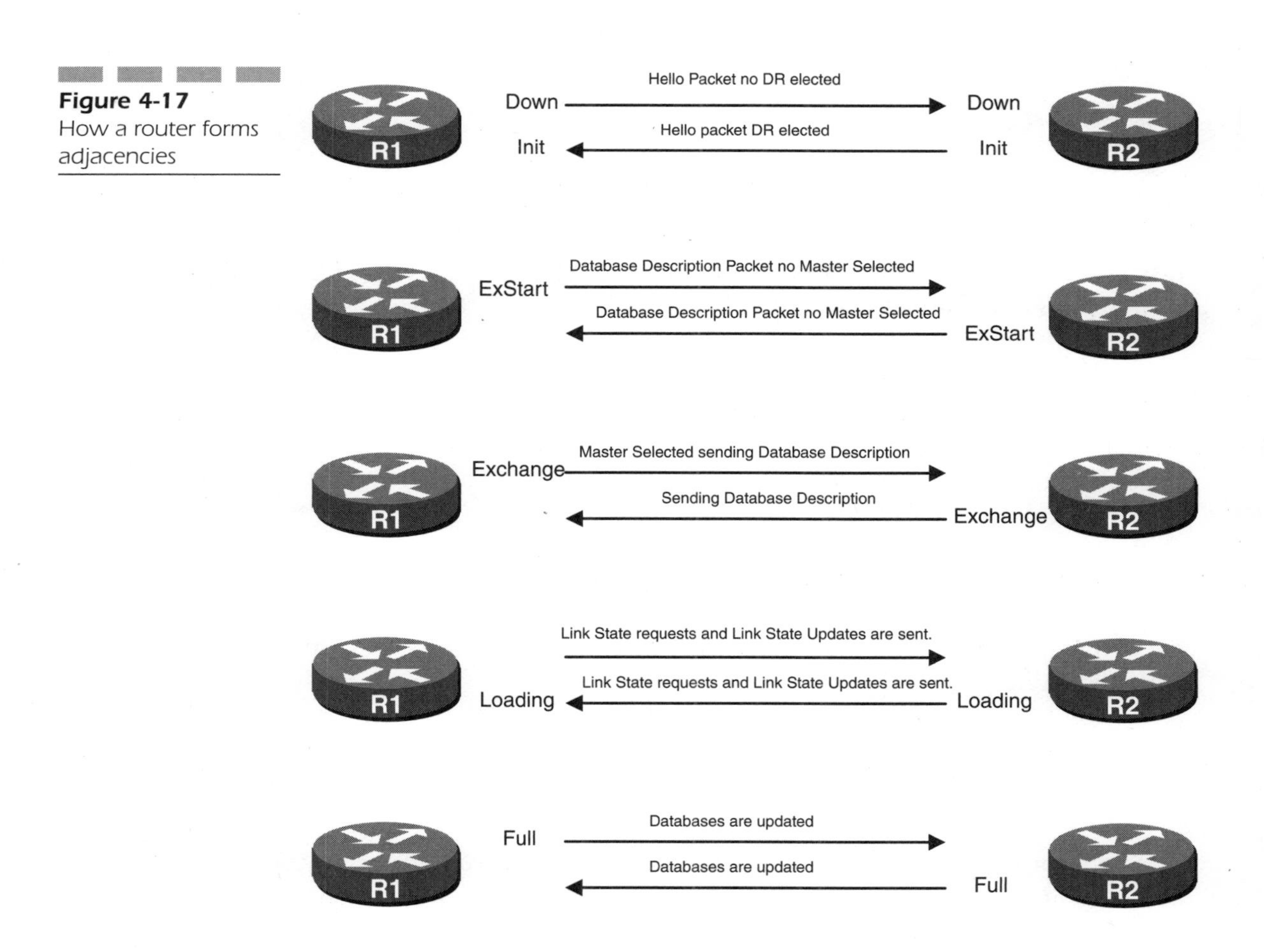

Figure 4-17 How a router forms adjacencies

When R1 and R2 first come online, they both initialize and begin sending Hello packets. At this point in time, neither router knows of the presence of the other router on the network, and no DR is elected. R2 hears the hello from R1 changing the state of the adjacency from down to Initializing (Init). This can be seen from the **show ip ospf neighbor** command on R2:

```
R2#show ip ospf neighbor

Neighbor ID     Pri   State           Dead Time   Address         Interface
1.1.1.1           1   INIT/DROTHER    0:00:39     10.10.3.1       Ethernet0
```

At this point, the routers have seen themselves in the Hello packet from their neighbors, and bidirectional communication is established. The adjacency changes from Initializing to 2way. This can be seen from the **show ip ospf neighbor** command on R2.

At the end of this stage, the DR and BDR is elected for the network and the router then decides whether or not to form an adjacency with its neighbor. On a multi-access network, routers only form adjacencies with the DR and BDR on the network.

```
R2#show ip ospf neighbor

Neighbor ID     Pri   State           Dead Time   Address         Interface
1.1.1.1         1     2WAY/DROTHER    0:00:36     10.10.3.1       Ethernet0
```

R2 in the next Hello packet indicates to R1 that it is the DR for the link. At this point, the state of the adjacency changes from Initializing to Exchange (Exstart). This can be seen from the **show ip ospf neighbor** command on R2. During the Exstart state, a master and slave relationship is formed between the two routers and the slave router adopts the master's *Database Description* (DD) sequence number.

```
R2#show ip ospf neighbor

Neighbor ID     Pri   State           Dead Time   Address         Interface
1.1.1.1         1     EXSTART/BDR     0:00:32     10.10.3.1       Ethernet0
                                                  ↑ R1 is the Backup DR
```

After the master slave relationship is formed and the routers agree on a common DD sequence number, the routers begin to exchange DD packets. At this point, the state of the adjacency changes from Exstart to Exchange. This can be seen from the **show ip ospf neighbor** command on R2:

```
R2#show ip ospf neighbor

Neighbor ID     Pri   State           Dead Time   Address         Interface
1.1.1.1         1     EXCHANGE/DR     0:00:38     10.10.3.1       Ethernet0
```

After the complete databases are exchanged between routers using the DD packets, the routers compare the database of their neighbor with their own database. The router may find that parts of the neighbor's database may be more up to date than its own. If so, the router requests these pieces using a link state request packet. At this point, the state of the adjacency is loading. This can be seen from the **show ip ospf neighbor** command on R2:

```
R2#show ip ospf neighbor

Neighbor ID     Pri   State           Dead Time   Address         Interface
1.1.1.1         1     LOADING/DR      0:00:38     10.10.3.1       Ethernet0
```

After the link state update requests have all been satisfied, R1 and R2 databases are deemed synchronized and the routers are fully adjacent. This can be seen from the **show ip ospf neighbor** command on R2:

```
R2#show ip ospf neighbor

Neighbor ID     Pri   State           Dead Time   Address         Interface
1.1.1.1         1     FULL/BDR        0:00:37     10.10.3.1       Ethernet0
```

OSPF Network Types

OSPF has four network types or models: broadcast, non-broadcast, point-to-point, and point-to-multipoint. Depending on the network type, OSPF works differently. Understanding how OSPF works on each network model is essential in designing a stable and robust OSPF network.

Broadcast The broadcast network type is the default type on LANs, such as Token Ring, Ethernet, and the FDDI. However, any interface can be configured as broadcast using the **ip ospf network** interface command. Here are some points to keep in mind when using the broadcast network type:

- On a broadcast model, a DR and a BDR are elected and all routers form adjacencies with the DR and BDR. This achieves optimal flooding because all LSAs are sent to the DR, which floods them to each individual router on the network.
- Neighbors do not need to be defined.
- All routers are on the same subnet.
- Care must be taken if the broadcast model is used on NMBA networks such as Frame Relay or ATM. Because a DR is elected, all routers must have physical connectivity to the DR. A full meshed environment should be used or the DR should be statically configured using the priority command to assure physical connectivity.
- The Hello timer is 10 seconds, the Dead interval is 40 seconds, and the Wait interval is 40 seconds.

In Figure 4-18, R1 and R3 are connected via Frame Relay to R2. The network is a hub and spoke environment configured as an OSPF network type broadcast. Because R2 is the only router that has logical connectivity to each router on the network, it must be elected the DR.

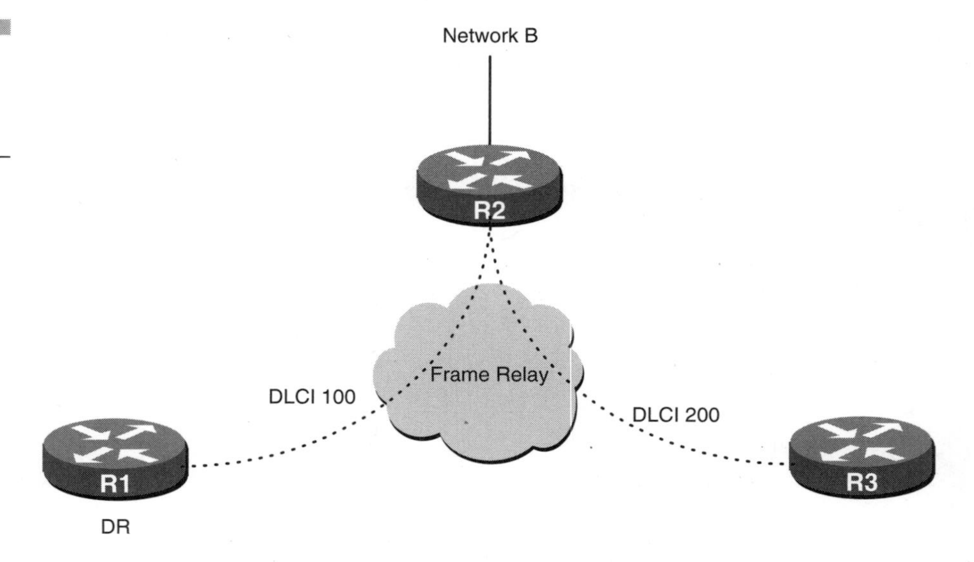

Figure 4-18 *An NBMA using the broadcast network type*

If a broadcast model is used on an NBMA network, all routers should be fully meshed or care should be taken on which router is elected the DR. In a hub and spoke environment, the hub should be configured as the DR.

The following is the output from the command **show ip ospf interface ethernet 0**. Note that the command shows the network type along with other key OSPF parameters.

```
R2#show ip ospf interface e0
Ethernet0 is up, line protocol is up
  Internet Address 10.10.3.2 255.255.255.0, Area 0
  Process ID 64, Router ID 2.2.3.2, Network Type BROADCAST, Cost: 10
  Transmit Delay is 1 sec, State BDR, Priority 1
  Designated Router (ID) 9.9.21.9, Interface address 10.10.3.1
  Backup Designated router (ID) 2.2.3.2, Interface address 10.10.3.2
  Timer intervals configured, Hello 10, Dead 40, Wait 40, Retransmit 5
    Hello due in 0:00:07
  Neighbor Count is 1, Adjacent neighbor count is 1
    Adjacent with neighbor 9.9.21.9  (Designated Router)
```

Non-Broadcast

The non-broadcast network type is the default type on serial interfaces configured for Frame Relay encapsulation. However, any interface can be configured as non-broadcast using the **ip ospf network** interface command. The following are some key points of the non-broadcast network type:

- With the non-broadcast model, a DR and a BDR are elected and all routers form adjacencies with them. This achieves optimized flooding because all LSAs are sent to the DR and the DR floods them to each individual router on the network.
- Due to the lack of broadcast capabilities, neighbors must be defined using the neighbor command.
- All routers are on the same subnet.
- Similar to the broadcast model, a DR is elected. Care must be taken to assure that the DR has logical connectivity to all routers on the network.
- The Hello timer is 30 seconds, the Dead interval is 120 seconds, and the Wait interval is 120 seconds.

The following is the output from the command **show ip ospf interface serial 0,** which is configured for Frame Relay encapsulation. Note that the command shows the network type along with other key OSPF parameters.

```
Serial0.2 is up, line protocol is down
  Internet Address 193.1.1.1 255.255.255.0, Area 0
  Process ID 64, Router ID 2.2.3.2, Network Type NON_BROADCAST, Cost: 64
  Transmit Delay is 1 sec, State DOWN, Priority 1
  No designated router on this network
  No backup designated router on this network
  Timer intervals configured, Hello 30, Dead 120, Wait 120, Retransmit 5
```

Point-to-Point The network type point-to-point is the default type on serial interfaces that are not using Frame Relay encapsulation or can be selected as a point-to-point sub-interface type. A sub-interface is a logical way of defining an interface. The same physical interface can be split into multiple logical interfaces, which was originally created to deal with issues caused by split horizons on NBMA networks.

The point-to-point model can be configured on any interface using the **ip ospf network point-to-point** interface command. Here are some significant point-to-point characteristics to be aware of:

- With a point-to-point model, no DR or BDR is elected and directly connected routers form adjacencies.
- Each point-to-point link requires a separate subnet.
- The Hello timer is 10 seconds, the Dead interval is 40 seconds, and the Wait interval is 40 seconds.

The following is the output from the command **show ip ospf interface serial 0,** which is not configured for Frame Relay encapsulation. Note that the command shows the network type along with other key OSPF parameters.

```
R2#show ip ospf interface s0
Serial0 is up, line protocol is down
```

```
  Internet Address 193.1.1.1 255.255.255.0, Area 0
  Process ID 64, Router ID 2.2.3.2, Network Type POINT_TO_POINT, Cost: 64
  Transmit Delay is 1 sec, State DOWN,
  Timer intervals configured, Hello 10, Dead 40, Wait 40, Retransmit 5
```

Point-to-Multipoint The network type point-to-multipoint can be configured on any interface using the **ip ospf network point-to-multipoint** interface command. The following list is some point-to-multipoint characteristics to be familiar with:

- No DR is elected.
- Neighbors do not need to be defined, because additional LSAs are used to convey neighbor router connectivity.
- One subnet is used for the whole network.
- The Hello timer is 30 seconds, the Dead interval is 120 seconds, and the Wait interval is 120 seconds.

The following is the output from the command **show ip ospf interface serial 0.** Notice that the command shows the network type along with other key OSPF parameters.

```
R2#show ip ospf interface s0
Serial0 is up, line protocol is down
  Internet Address 193.1.1.1 255.255.255.0, Area 0
  Process ID 64, Router ID 2.2.3.2, Network Type POINT_TO_MULTIPOINT, Cost: 64
  Transmit Delay is 1 sec, State DOWN,
  Timer intervals configured, Hello 30, Dead 120, Wait 120, Retransmit 5
```

As you can see from this section, a network that is based on a link state protocol such as OSPF is considerably more complex to configure, design, and troubleshoot than a network based on a distance vector protocol such as RIP. However, OSPF provides the following advantages over RIP version 1:

- A higher limitation on hop count.
- A faster convergence than RIP because routing changes are flooded throughout the network instantly.
- OSPF supports router authentication; RIP does not.
- OSPF has a concept of route tagging external routes that are injected into the AS. This enables the protocol to keep track of external routes that are injected by other protocols such as BGP.
- OSPF is classless; RIP is classful.
- OSPF uses the available bandwidth more effectively by only sending routing updates when a change occurs.
- OSPF uses multicast packets versus broadcast packets to send LSAs. This assures that routers that are not configured for OSPF do not have to process the packet.

Integrated IS-IS

Integrated *Intermediate System to Intermediate System* (IS-IS) is a intra-domain routing protocol that supports both IP and CLNS. IS-IS and OSPF have many features in common, which is not surprising since the ISO was working on IS-IS around the same time the *Internet Architecture Board* (IAB) was working on OSPF.

Like OSPF, IS-IS is based on two levels of hierarchy: Level 1 and Level 2 routing. Routing within an area is referred to as Level 1 routing, and routing between areas is called Level 2 routing. Level 2 routers connect multiple Level 1 routers similar to area0 in OSPF.

Both OSPF and IS-IS use Hello packets to form and maintain adjacencies and maintain a link state database from which an SPF algorithm computes the shortest path. Like OSPF, IS-IS is a classless protocol that provides address summarization between areas.

IS-IS Terminology

When dealing with IS-IS, it is important to understand the terminology being used.

Intermediate System (IS) A router running IS-IS.

Designated Intermediate System (DIS) Similar to the DR in OSPF, this is a router on a LAN that is responsible for flooding information for the broadcast network.

End System (ES) A host.

Network Service Access Point (NSAP) The address used to identify an intermediate systems.

Partial Sequence Number Protocol (PSNP) Used for database synchronization.

Complete Sequence Number Protocol (CSNP) Used for database synchronization.

IS-IS Hello (IIH) Used to discover other Intermediate Systems.

IS-IS Addressing

For routing IP, the NSAP address is divided into three parts: the area address, the system ID, and the N-selector.

Area Number	System ID	N Sel
←Variable Length→	6 Bytes	1 Byte

The area address is a variable length field that identifies the routing domain length of the area field and is associated with a single area within a routing domain. The system ID is six Bytes long and defines the ES or IS in an area. The NSAP selector is one Byte long and identifies a particular network service at the network layer of a node. A network service user can be a transport entity or the IS network itself.

IS-IS Areas

Like OSPF, IS-IS uses areas to create a two-level hierarchical topology, but a fundamental difference exists in the way that IS-IS defines these areas. In OSPF, area borders are marked by routers, while in IS-IS, they are marked by links (see Figures 4-19 and 4-20).

An intermediate system can be a Level 1 router, a Level 2 router, or a Level 1/Level 2 router. Similar to an OSPF non-backbone internal router, a Level 1 router has no direct connectivity to another area. Level 2 routers are similar to OSPF backbone routers and L1/L2 routers are analogous to OSPF ABRs. The L1/L2 routers must maintain both a Level 1 link state database and a Level 2 link state database, similar to how an OSPF ABR maintains separate databases for each area.

Figure 4-19
OSPF terminology

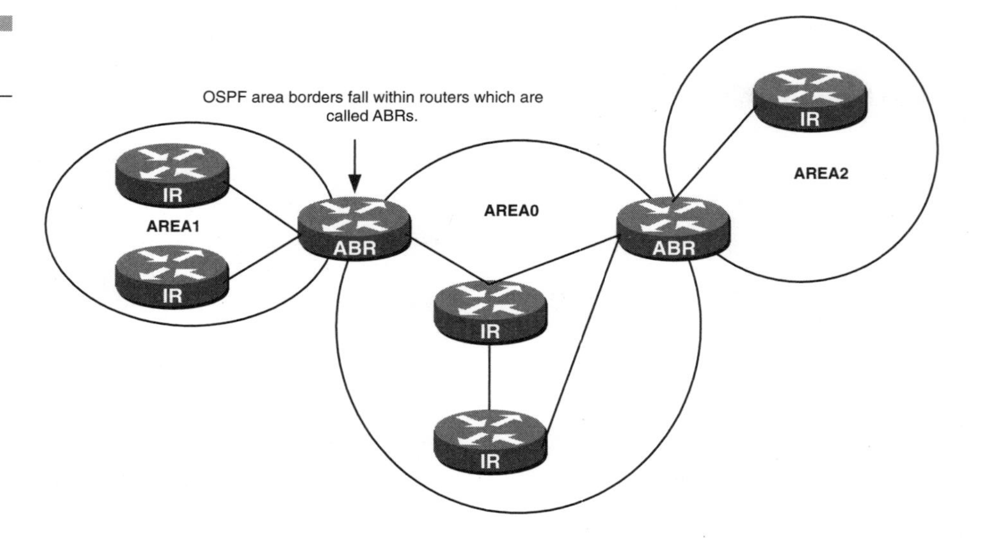

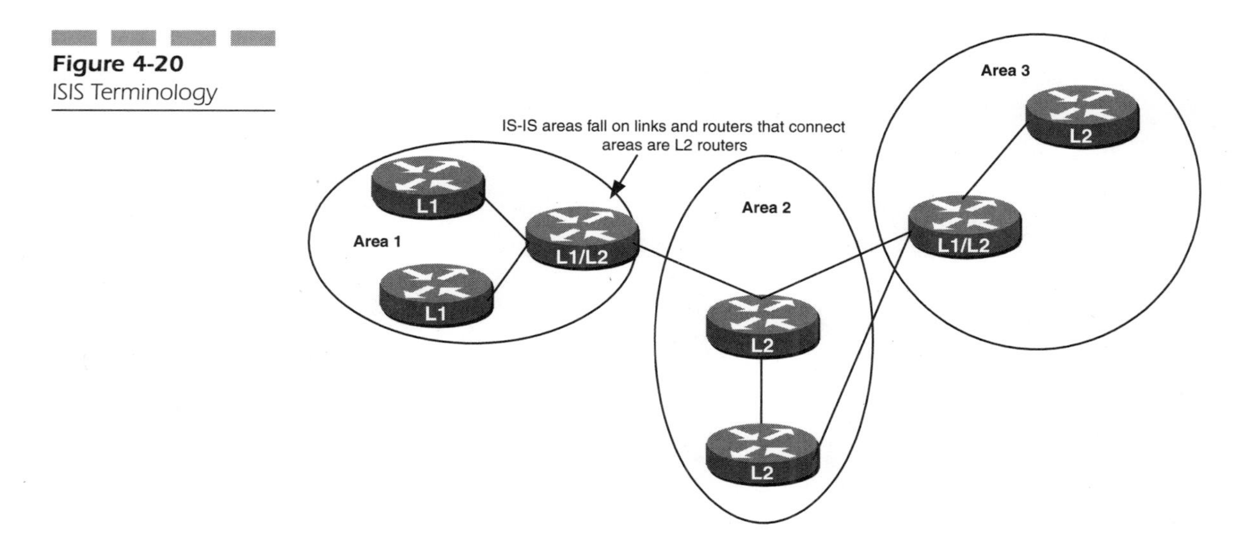

Figure 4-20
ISIS Terminology

How IS-IS Works

IS-IS routers discover and form adjacencies by exchanging IS-IS Hello packets, which are transmitted every 10 seconds. Like OSPF, an IS-IS Hello packet is used to identify a router and its capabilities and describe the interface that it was sent over. If both routers agree on the capabilities and interface parameters, they become adjacent. Once an adjacency is formed, the Hellos act as keepalives. If the router does not receive a keepalive within the dead interval, the router declares its neighbor dead. The default dead interval is three times the keepalive interval or 30 seconds.

Once an adjacency is formed, the routers use an update process to construct the Level 1 and Level 2 link state databases. Level 1 LSPs are flooded throughout the area and Level 2 LSPs are flooded over all Level 2 adjacencies. Like OSPF, IS-IS ages each LSP in the database, and the LSA must be refreshed every 15 minutes.

On point-to-point networks, routers send LSPs directly to the neighboring router. On broadcast networks, the LSPs are multicast to all neighbors. Unlike OSPF, IS-IS routers on a broadcast network form adjacencies with every router on the broadcast network, not just the DR.

IS-IS uses SNPs to acknowledge the receipt of LSPs and to maintain link state database synchronization. On point-to-point links, if an SNP is not received for a transmitted LSP in five seconds, a new LSP is generated. The IS-IS retransmission interval can be set on a per-interface basis with the command **isis retransmit-interval**. On a broadcast network, LSPs are not acknowledged by each router; instead, the DR periodically multicasts a CSNP that describes every LSP in the link state database. When a router receives a CSNP, it compares it with the LSP in its database. If the router's database does not contain a copy of all the LSPs listed in the CSNP update, the router multicasts a SNP listing the LSP that it requires. The DR then responds with the appropriate LSP.

Once the update process has built the link state database, it is used to calculate the shortest path tree. From this tree, a forwarding database is constructed and is what is used to route packets.

IS-IS can use four metrics to calculate the shortest path. Each metric can be an integer between 0 and 63. The four metrics are as follows:

- **Default,** which must be supported and understood by every router.
- **Delay,** which reflects the delay of the transit network.
- **Expense,** which reflects the cost of the link.
- **Error,** which is similar to IGRP/EIGRP and measures the error probability of the network.

For each metric, a separate route is calculated, resulting in an SPF calculation. So if a system is using all four metrics to calculate the cost, it runs the SPF algorithm four times for each destination network.

Cisco routers only support the default metric. As you can see, if all interfaces are left at their default settings, the IS-IS metric becomes a simple measure of the hop count. The default metric for every interface is set to 10 regardless of the interface type. This can be changed with the interface command **isis metric**.

Similar to OSPF, IS-IS classifies routes as internal and external as well as Level 1 and Level 2. Internal routes are routes that originate within the same domain, whereas external routes are routes that originate outside of the domain. Level 2 routes can be internal or external, and Level 1 routes are always internal. Level 1 routes are also always preferred over Level 2 paths to the same destination. Like OSPF, IS-IS has the capability to load balance over equal cost paths, and Cisco's IS-IS implementation supports up to six equal cost paths.

Case Study 3: OSPF

IOS Requirements

Most of the routing protocols discussed became available in IOS 10.0, however, all labs were performed using IOS 11.2.

Equipment Needed

The following equipment is needed to perform this case study. Table 4-6 outlines the interface requirements for each router.

- One Catalyst 5500 Ethernet switch.
- One terminal server with one Ethernet and one serial port.
- One Cisco router with two serial ports.
- Two Cisco routers with an Ethernet port, one serial port, and one ISDN BRI interface.
- One Cisco router with two serial ports and one Ethernet port.
- One peripheral router with one Ethernet port and three serial ports. This router (R8) will be used as the Frame Relay switch as well as the backbone router. **(The student does not configure this router and should use the configuration provided on the CD).**
- Cisco IOS 11.2.
- A PC running a terminal emulation program for connecting to the console port of the terminal server.
- Five Ethernet cables.
- Five Cisco DTE/DCE crossover cables.
- One Cisco Rolled cable.

Table 4-6

Interface requirements for each router

R1	1 Serial, 1 Ethernet, 1 BRI
R2	1 Serial, 1 Ethernet
R3	1 Serial, 1 Ethernet
R4	1 Serial, 1 Ethernet (terminal server)
R5	2 Serial
R6	1 Serial, 2 Ethernet, 1 BRI
R8	3 Serial, 1 Ethernet (Frame switch)

Physical Connectivity Diagram

Figure 4-21 shows the physical connecting network diagram for the routers in this case study.

Figure 4-21 Network diagram

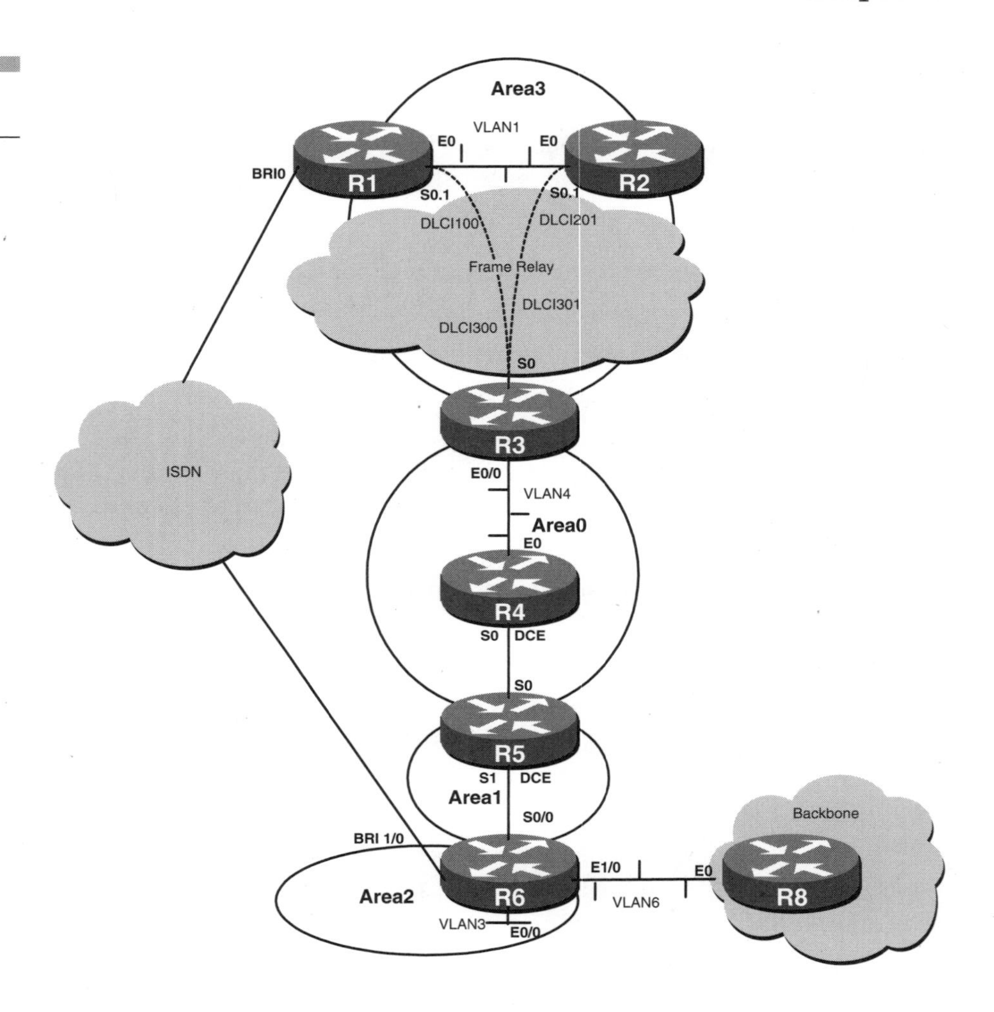

Notes

- All IP addresses will be assigned from the class B network 152.1.0.0.
- No static or default routes should be used in this case study unless explicitly called for.
- Use Process ID 64 for all OSPF configurations.
- Use AS 64 for all IGRP configurations.
- Do not use the network point-to-multipoint command under OSPF.
- No configuration is needed on R8; load the configuration that is provided off the CD.

- Read through the entire case study before beginning.
- Make sure you save your configurations regularly.
- The VTP domain name on the Catalyst should be CCIE_Lab.

Questions

1. Cable the routers per Figure 4-21 and load the configuration provided on the CD to R8. The configuration provided is from a 4500 with an Ethernet interface and two four-port serial network modules. If you plan on using a router other than a 4500 or one that has a different way of numbering its interfaces, such as a 3620, the interfaces provided in the configuration will need to be renamed.
2. Erase any existing configuration from the routers and set each router's name. The name will be the corresponding router number; for example, router 1 will be R1.
3. Set the enable secret on each router to be **cisco**.
4. Configure the terminal server (R4) so that all routers can be accessed by name through it. The port number of the terminal server attaches to the corresponding router; that is, port one of the terminal server goes to the console port of R1, while the Catalyst switch goes to port number 9. (Use the Ethernet IP address as the IP address for reverse telneting.)
5. Place Ethernet interface 0 of R1 and R2 in VLAN 1. The routers should be attached in ascending order staring with port 1 on the Catalyst switch.
6. A DNS server is located at 153.1.1.1. Configure R1 so that this server is used for address resolution.
7. Disable the CDP on all routers.
8. Assign IP addresses to Ethernet 0 of R1, and R2 from network 152.1.1.0. Do not allocate more addresses than needed. Use the first available subnet and assign the addresses in ascending order starting with R1.
9. VLAN 1 contains 28 servers in a server farm. Configure R1 and R2 so that in the event of a router failure the other router will take over routing for the subnet. Under normal conditions, the load should be split evenly between both routers. Fourteen of the servers are defined with a default gateway of 152.1.1.3 and the other fourteen are defined with a default gateway of 152.1.1.4.
10. R1 and R2 are connected via Frame Relay to R3. Use physical interfaces on R3 and R2 and a logical interface on R1. Assign addresses to each interface using the next available subnet of 152.1.1.0. Do not allocate more addresses than are needed and assign the addresses

in ascending order starting with R1. Configure the interfaces for Frame Relay. Make sure that the only DLCIs that are used are the ones provided and that all routers can reach each other.

The following is the physical connectivity of R1, R2, and R3 to the Frame Relay switch (R8).

Frame Switch (R8) interface S0	R1 Interface S0
Frame Switch (R8) interface S1	R2 Interface S0
Frame Switch (R8) interface S3	R3 Interface S0/0 (3600)

11. Assign IP addresses to BRI 0 of R1 and R6 from network 152.1.1.0. Do not allocate more addresses than needed. Use the next available subnet and assign the addresses in ascending order starting with R1.
12. Place Ethernet 0/0 of R3 and Ethernet 0 of R4. In VLAN 4, place Ethernet 0/0 of R6 in VLAN 3.
13. Assign IP addresses to Ethernet 0 of R3 and R4 from network 152.1.1.0. The subnet will contain 64 hosts. Assign the addresses in ascending order starting with R3.
14. Connect the rest of the routers serially per Figure 4-21. Make sure that the correct ports are configured for DTE/DCE (the DCE clock will be set to 64K on all routers unless otherwise specified).
15. Assign an IP address to S0 of R4 and R5 from network 152.1.2.0. Do not allocate more addresses than needed. Use the first available subnet and assign the addresses in ascending order starting with R4.
16. Assign IP addresses to S1 of R5 and S0/0 of R6 from network 152.1.10.0. Use an address that will support up to 80 hosts. Do not allocate more addresses than needed. Use the first available subnet and assign the addresses in ascending order starting with R5.
17. Set the maximum transmission unit on Serial 0 of R1 to 500.
18. Configure the Catalyst so that only hosts 152.1.1.1 and 152.1.1.2 have telnet access. Assign an address to the SCO interface of the catalyst switch, use the next available address on subnet 152.1.1.0/26.
19. Assign an IP address to Ethernet 0/0 of R6 from network 152.1.11.0. Use a subnet that supports up to 200 hosts.
20. Assign IP address 152.1.0.1 /24 to Ethernet 1/0 of R6, and place R6 E 1/0 in VLAN 6. The Ethernet interface from the backbone router R8 should also be placed in VLAN6.
21. Assign the following loopback addresses to the corresponding routers:

```
R1 Loopback 0    152.1.1.65 /32
R1 Loopback 1    152.1.1.97 /27
R2 Loopback 0    152.1.1.72 /32
R3 Loopback 0    152.1.1.69 /32
R4 Loopback 0    152.1.2.4  /32
```

```
R5 Loopback 0   152.1.2.5  /32
R3 Loopback 1   152.1.3.1  /24
R6 Loopback 0   152.1.12.1  /24
```

OSPF

22. Place VLAN 1 in OSPF area 3, and make sure that R1 is the DR for the area.

23. Place the Frame Relay cloud between R1, R2, and R3 in OSPF area 3.

24. Place VLAN4 and the serial interface connecting R4 and R5 in OSPF area 0.

25. Place the BRI interface on R6 and VLAN 3 in OSPF area 2.

26. Configure area 2 so that no external LSAs are flooded into the area from the backbone, however, area 2 must be able to send routes learned from the IGRP backbone into area 0.

27. Change the OSPF cost of using an Ethernet interface to 90 and the serial interface cost to 580 across the entire network. Do not use the OSPF cost command.

28. R4 should only see a single advertisement in its routing table from area 3.

29. Advertise all the loopback interfaces on R1 and R2, as well as loopback 1 on R3 via OSPF. Don't configure or run OSPF on any of these loopback interfaces, however, and don't propagate these networks outside of the area.

30. Configure the routers so that the source of each routing update is authenticated. Area 0 should use MD5 authentication, while area 3 and area 1 should use simple authentication. The key should be cisco, plus the area number. For example, the key for area 0 is cisco0.

31. Configure DDR between R1 and R6. R6 should call R1 when R6's serial connection is lost to R5. The backup should be invoked five seconds after the primary is lost. Once the primary is back, the router should wait 20 seconds before switching to the primary. Configure DDR so that R1 accepts R6's call, authenticates R6, and then securely calls R6 back.

32. Run OSPF on the ISDN circuit between R1 and R6. Make sure that all the routes are learned via R6 during backup.

33. Make sure that periodic OSPF Hello advertisements and LSAs are suppressed and that the link is only brought up by OSPF when a topology change occurs.

34. Because R6 is attached to the backbone, it should advertise a default route into the domain.

35. Configure R6 to run IGRP to the backbone. Only send IGRP messages out interface Ethernet 1/0.

36. R6 should only take in the following networks from the backbone:

```
152.12.0.0
152.13.0.0
152.14.0.0
152.15.0.0
```

37. Advertise the routes learned from the backbone in the OSPF domain. Also advertise as few of the networks as needed.

38. In addition to 152.1.11.0 and 152.1.12.0, which are already being advertised in the IGRP backbone, R6 should advertise the following networks.

```
152.1.1.0
152.1.2.0
```

Answer Guide

1. *Cable the routers per network Figure 4-21 and load the configuration provided on the CD to R8 (R8 will act as the backbone router and the Frame switch). The configuration provided is from a 4500 with an Ethernet interface and a four-port serial network module. If you plan on using a router other than a 4500 or one that has a different way of numbering its interfaces, such as a 3620, the interfaces provided in the configuration will need to be renamed.*

2. *Erase any existing configuration from the routers and set each router's name. The name will be the corresponding router number. For example, router 1 will be R1.*

   The first step is to erase the configuration for all the routers as well as the Catalyst switch. To erase the configuration on the router, use the following command:

```
R1#write erase or R1#erase startup-config
```

   This command erases the startup configuration stored in NVRAM (on all platforms except the Cisco 7000 series, Cisco 7200 series, and Cisco 7500 series). On the Cisco 7000 series, Cisco 7200 series, and the Cisco 7500 series, the router erases or deletes the configuration pointed to by the CONFIG_FILE environment variable. If the CONFIG_FILE variable points to NVRAM, then the router erases NVRAM. If the CONFIG_FILE environment variable specifies a flash memory device and a configuration filename, then the command deletes the named file.

   Erasing the configuration on the Catalyst 5500 is a bit different. You would use this command:

```
CAT#clear config all
```

3. *Set the enable secret on each router to be cisco.*

   When a router is first configured, no default enable password or enable secret exists. The enable secret is used to keep unauthorized users from changing the configuration. Both the enable password and the enable secret accomplish the same thing; that is, they enable you to establish an encrypted password that users must enter to access the enable mode (the default) or any privilege level you specify. The enable secret uses a one-way hash algorithm to encrypt the password, while the enable password uses a much weaker algorithm that can be reversed. Tools exist on the Internet such as GetPass V1.1, which enables you to enter the encrypted password and provides the decrypted password.

Cisco strongly recommends that you use the enable secret command because of the improved encryption algorithm.

To set the enable secret, use the following global configuration command:

```
R5(config)#enable secret cisco
```

4. *Configure the terminal server (R4) so that all routers can be accessed by name through it. The port number of the terminal server attaches to the corresponding router; that is, port 1 of the terminal server goes to the console port of R1, and the Catalyst switch goes to port 9 use the Ethernet IP address as the IP address for reverse telneting).*

The terminal server provides access to all your test routers via reverse telnet. Reverse telnet is the process of using telnet to make a connection out an asynchronous port.

All the routers in the case study, as well as the Catalyst switch, have their console port connected directly to one of the 16 asynchronous interfaces on the 2511RJ terminal server using a standard Cisco console rolled cable. The routers will be accessed using a reverse telnet connection. To make such a connection, telnet to any active IP address on the box followed by 200x where x is the port number that you want to access (**Telnet 1.1.1.1 2001**).

Use the following line commands to configure the terminal server:

```
R7(config)#line 1 16
R7(config-line)#transport input all
R7(config-line)#no exec
```

The command **transport input all** specifies the input transport protocol. By default, on IOS 11.1 and later, the transport input is set to none; prior to 11.1, the default was all. If the transport input is left to none, you will receive an error stating that the connection is refused by the remote host. The command **no exec** enables only outgoing connections from the line. This prevents the device that you are attached to from sending out unsolicited data. If the port receives unsolicited data, an EXEC process will start that makes the line unavailable.

The next step is to configure hostnames on R4 so you can attach to any router simply by typing in the router name. The Cisco IOS software maintains a table of host names and their corresponding addresses. Similar to a DNS server, you can statically map host names to IP addresses. This is very useful and saves a lot of keystrokes when you have multiple devices connected to the terminal server.

The following global configuration commands are used to set the host names:

```
R4(config)#ip host r1 2001 152.1.2.4
R4(config)#ip host r2 2002 152.1.2.4
R4(config)#ip host r4 2004 152.1.2.4
R4(config)#ip host r5 2005 152.1.2.4
R4(config)#ip host r6 2006 152.1.2.4
R4(config)#ip host r7 2007 152.1.2.4
R4(config)#ip host r8 2008 152.1.2.4
R4(config)#ip host cat 2009 152.1.2.4
```

5. *Place Ethernet interface 0 of R1 and R2 in VLAN 1. The routers should be attached in ascending order staring with port 1 on the Catalyst switch.*

Because, by default, all ports on a Catalyst are in VLAN1, no configuration is needed.

6. *A DNS server is located at 153.1.1.1. Configure R1 so that this server is used for address resolution.*

The Cisco IOS enables you to specify one or more hosts (up to six) that can function as a name server to supply name information. By default, the IOS will try to resolve a name to an IP address, so all that is needed is to define the name server. Use the following global configuration command to configure a name server:

```
R1(config)#ip name-server 153.1.1.1
```

7. *Disable CDP on all routers.*

CDP runs on all Cisco routers and switches. It can run over any physical media and over any protocol. Unlike a routing protocol that shows all known networks, CDP only shows information for directly connected neighbors. It is most useful for verifying that a router is connected to the proper port of its neighbor.

A CDP-enabled router will be able to learn directly connected neighbor port and hostname information. Additional information such as the neighbor's hardware model number and capabilities are also reported.

A CDP-enabled router sends out a periodic multicast packet containing a CDP update. The time between these CDP updates is determined by the **cdp timer** command; the timer value default is 60 seconds. To disable the sending of CDP updates, use the **no cdp run** global command on all of the routers.

8. *Assign IP addresses to Ethernet 0 of R1 and R2 from network 152.1.1.0. Do not allocate more addresses than needed. Use the first available subnet and assign the addresses in ascending order starting with R1.*

The next step is to assign IP addresses to the Ethernet interface of R1 and R2. The question specifies to use the first available subnet from 152.1.1.0 and not to allocate more address space than needed. After carefully reading through the whole case study, you can see that you need a subnet that contains 32 addresses. Twenty-eight for the servers on the

LAN, two for the Ethernet interfaces on the routers, and two for the *Hot Standby Routing Protocol* (HSRP) addresses.

Because 32 addresses are needed, you must use a 26-bit mask. This gives you a range of addresses from 152.1.1.0 to 152.1.1.63. This is a bit more than you need, but if you went with a 27-bit mask you would be short. The following is the IP address assignment:

```
152.1.1.0←Network Address reserved
152.1.1.1←R1
152.1.1.2←R2
152.1.1.3←HSRP Group 1 Address
152.1.1.4←HSRP Group 2 Address
152.1.1.6
..........
..........
152.1.1.63←Broadcast Address reserved
```

9. *VLAN 1 contains 28 servers in a server farm. Configure R1 and R2 so that in the event of a router failure the other router will take over routing for the subnet. Under normal conditions, the load should be split evenly between both routers. Fourteen of the servers are defined with a default gateway of 152.1.1.3 and the other 14 are defined with a default gateway of 152.1.1.4.*

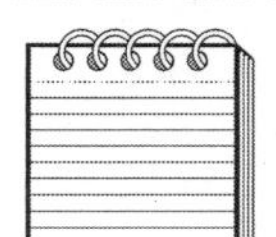

NOTE: *The Cisco 1000, 2500, 3000, and 4000 series routers that use Lance Ethernet hardware do not support multiple Hot Standby groups on a single Ethernet interface.*

The majority of today's TCP/IP LAN networks rely on the use of a default gateway (which is statically configured in the host) in order to route packets to hosts on other networks. The default gateway is usually a router connected to the Internet or a company's intranet. Each host on the LAN is configured to forward packets to this destination if the host it is trying to reach is not on the same network. This provides for a single point of failure on the network. If the gateway is down, all the hosts on the LAN are isolated from the rest of the network. To combat this, many companies install redundant gateways. The problem with this is that the user host is pointed at only one gateway. If this router should fail, the user must change their statically configured default gateway.

HSRP resolves this problem by enabling the network administrator to configure a set of routers to work together to present the appearance of a single default gateway. The routers in an HSRP group share a virtual MAC address and IP address, which is used by hosts on the LAN as the default gateway. The HSRP protocol selects which router is active, and the active router receives and routes packets that are destined for the group's MAC address.

HSRP uses multicast UDP-based hello packets to communicate with other routers that are part of the same HSRP group. Each router in the group watches for Hello packets from the active and standby routers. If the active router becomes unavailable, the standby assumes the active role and routes the packets for the network.

In order to load balance across both routers, you need to use multiple HSRP groups. R1 will be configured to act as the active router for HSRP group 1 (IP address 152.1.1.3) and the standby for HSRP group 2 (IP address 152.1.1.4). R2 will be configured to be the active router for HSRP group 2 and the standby for HSRP group 1.

Enabling HSRP on a router is done in two steps. First, an IP standby group is defined and then an HSRP priority is assigned for that interface and group. The HSRP member with the highest priority (assuming preemption is enabled) becomes the active router.

The following command enables multi-group HSRP on R1 and R2. In order to assure that R1 is the active router for group 1 and R2 is the active router for group 2, you need to set the HSRP priority. The router with the higher priority becomes active for the group, the default priority is 100.

```
R1(config-if)#standby 1 ip 152.1.1.3
R1(config-if)#standby 2 ip 152.1.1.4
R1(config-if)#standby 1 priority 150 ←Sets the priority to assure that R1 becomes the
                                      active router for group1

R2(config-if)#standby 1 ip 152.1.1.3
R2(config-if)#standby 2 ip 152.1.1.4
R2(config-if)#standby 2 priority 150 ←Sets the priority to assure that R2 becomes the
                                      active router for group2
```

10. *R1 and R2 are connected via Frame Relay to R3. Use physical interfaces on R3 and R2 and a logical interface on R1. Assign addresses to each interface using the next available subnet of 152.1.1.0. Do not allocate more addresses than are needed and assign the addresses in ascending order starting with R1. Configure the interfaces for Frame Relay. Make sure that the only DLCIs that are used are the ones provided and that all routers can reach each other.*

The following is the physical connectivity of R1, R2, and R3 to the Frame Relay switch (R8):

Frame Switch (R8) interface S0	R1 Interface S0
Frame Switch (R8) interface S1	R2 Interface S0
Frame Switch (R8) interface S3	R3 Interface S0/0 (3600)

Three IP addresses are needed, so use a 29-bit subnet mask. This gives you eight addresses of which six are usable. This question can be a bit tricky unless you read the case study completely.

You already used 152.1.1.0 /26 in question 9, so normally the next available subnet would be 152.1.1.64 /29. However, question 22 specifies to use address 152.1.1.65 /32 for the loopback interface on R1. The next subnet would be 152.1.1.72 /29, but loopback 0 on R2 will be assigned address 152.1.1.72 /32. So after carefully reading through the entire case study, you can see that the next available subnet is 152.1.1.80 /29.

The IP addressing is as follows:

```
152.1.1.80 ←Network Address reserved
152.1.1.81 ←R1
152.1.1.82 ←R2
152.1.1.83 ←R3
152.1.1.84 ←Unused
152.1.1.85 ←Unused
152.1.1.86 ←Unused
152.1.1.87 ←Broadcast Address reserved
```

Now that the IP addressing is laid out you must configure the interfaces for Frame Relay. The question specifies that R1 and R2 use point-to-point sub-interfaces and R3 use the physical interface. To create a logical sub-interface on R1, first configure the physical interface with the Frame Relay encapsulation information:

```
R1(config)#interface serial 0
R1(config-if)#encapsulation frame-relay IETF
R1(config-if)#frame-relay lmi-type ansi
```

Then create the sub-interface by specifying the physical interface followed by a dot and the logical interface. If the interface does not exist, it is created. Two types of Frame Relay sub-interfaces are available: point-to-point and multipoint. The case study specifies that point-to-point sub-interfaces must be used:

```
R1(config)#int s0.1 point-to-point
R1(config-subif)#ip add 152.1.1.81 255.255.255.248
```

The question also states that the router should only see the DLCI that it is using. Normally, you would use the command **no frame-relay inverse-arp**, but this is the default on point-to-point sub-interfaces. You still need to provide the router with a mapping between the layer three ip address and the layer two DLCI. To do this, use the frame-relay interface DLCI command:

```
R2(config-subif)#frame-relay interface-dlci 201
```

The same needs to be performed on R2:

```
R2(config)#interface serial 0
R2(config-if)#encapsulation frame-relay IETF
R2(config-if)#frame-relay lmi-type ansi
R1(config)#int s0.1 point-to-point
```

```
R1(config-subif)#ip add 152.1.1.82 255.255.255.248
R1(config-subif)#frame-relay interface-dlci 100
```

Because R3 is using the physical interface, no sub-interfaces need to be defined:

```
R3(config)#interface serial 0/0
R3(config-if)#encapsulation frame-relay IETF
R3(config-if)#frame-relay lmi-type ansi
R3(config-if)#ip address 152.1.1.83 255.255.255.248
```

The question also states that the router should only see the DLCI that it is using:

```
R3(config-subif)#no frame-relay inverse-arp
```

Because you are disabling inverse arp, you need to map the far-end layer three address to a layer two DLCI. To do this, perform the following:

```
R3(config-subif)#frame-relay map ip 152.1.1.81 300 broadcast
R3(config-subif)#frame-relay map ip 152.1.1.82 301 broadcast
```

11. *Assign IP addresses to BRI 1 of R1 and BRI 1/0 R6 from network 152.1.1.0. Do not allocate more addresses than needed. Use the next available subnet and assign the addresses in ascending order starting with R1.*

You need two IP addresses, so use a 30-bit subnet mask. This will give you four addresses of which two are usable. Once again, if you don't read the case study completely, this question can be tricky. Addresses 152.1.1.0 to 63 are used in VLAN 1. The next available three subnets are being used by R1, R2, and R3 loopback 0, leaving 152.1.1.76 /30 as the next available subnet. The IP addressing is as follows:

```
152.1.1.76   ←Network Address reserved
152.1.1.77   ←R1
152.1.1.78   ←R6
152.1.1.79   ←Broadcast Address reserved
```

The following commands define IP addresses on R1 and R6:

```
R1(config)#interface Bri 0
R1(config-if)#ip address 152.1.1.77 255.255.255.252

R6(config)#interface Bri 1/0
R6(config-if)#ip address 152.1.1.78 255.255.255.252
```

12. *Place Ethernet 0 of R3 and R4 in VLAN 4, place Ethernet 0/0 of R6 in VLAN 3, and place E1/0 of R6 in VLAN6 .*

Remember when the Catalyst first comes up it is in a state where all ports are in VLAN 1. The Catalyst switch must have a domain name before it can use VLAN numbers other than 1. You can see in the **show vtp domain** output that the domain name has not been set on this switch:

```
Console> (enable) show vtp domain
Domain Name                      Domain Index VTP Version Local Mode  Password
-------------------------------- ------------ ----------- ----------- ----------
                                 1            2           server      -

Vlan-count Max-vlan-storage Config Revision Notifications
---------- ---------------- --------------- -------------
5          1023             0               disabled

Last Updater    V2 Mode  Pruning  PruneEligible on Vlans
--------------- -------- -------- -------------------------
0.0.0.0         disabled disabled 2-1000
```

Set the vtp domain name with the command **set vtp domain:**

```
Console> (enable) set vtp domain CCIE_LAB
VTP domain CCIE_LAB modified

Console> (enable) show vtp domain
Domain Name                      Domain Index VTP Version Local Mode  Password
-------------------------------- ------------ ----------- ----------- ----------
CCIE_LAB                 1            2           server      -

Vlan-count Max-vlan-storage Config Revision Notifications
---------- ---------------- --------------- -------------
5          1023             0               disabled

Last Updater    V2 Mode  Pruning  PruneEligible on Vlans
--------------- -------- -------- -------------------------
0.0.0.0         disabled disabled 2-1000
```

Use the **set vlan 4 5/3** command to move port 5/3 to VLAN 4. Notice that the switch automatically modifies VLAN 4 and removes port 5/3 from VLAN 1:

```
Console> (enable) set vlan 4 5/3
Vlan 4 configuration successful
VLAN 4 modified.
VLAN 1 modified.
VLAN  Mod/Ports
---- -----------------------
4     5/3
```

Use the **set vlan 4 5/4** command to move port 5/4 to VLAN 4:

```
Console> (enable) set vlan 4 5/4
VLAN 4 modified.
VLAN 1 modified.
VLAN  Mod/Ports
---- -----------------------
4     5/3 -4
```

Activate the VLAN with the command **set vlan 4**:

```
Console> (enable) set vlan 4
Vlan 4 configuration successful
```

The **show vlan 4** command now indicates that VLAN4 is active and contains two ports, 5/3 and 5/4:

```
Console> (enable) show vlan 4
VLAN Name                             Status    Mod/Ports, Vlans
---- -------------------------------- --------- ----------------------------
4    VLAN0004                         active    5/3 -4

VLAN Type  SAID       MTU   Parent RingNo BrdgNo Stp  BrdgMode Trans1 Trans2
---- ----- ---------- ----- ------ ------ ------ ---- -------- ------ ------
4    enet  100002     1500   -      -      -      -    -        0      0

VLAN AREHops STEHops Backup CRF
---- ------- ------- ----------
```

Use the **set vlan 3 5/6** command to move port 5/6 to VLAN 3. Notice that the switch automatically modifies VLAN 3 and removes port 5/6 from VLAN 1:

```
Console> (enable) set vlan 3 5/6
Vlan 3 configuration successful
VLAN 3 modified.
VLAN 1 modified.
VLAN  Mod/Ports
---- -----------------------
3     5/6
```

Activate the VLAN with the command **set vlan 3**:

```
Console> (enable) set vlan 3
Vlan 3 configuration successful
```

Use the **set vlan** command to move port 5/7 and 5/8 to VLAN 6. Notice that the switch automatically modifies VLAN 5 and removes port 5/7 from VLAN 1:

```
Console> (enable) set vlan 6 5/7
Vlan 5 configuration successful
VLAN 5 modified.
VLAN 1 modified.
VLAN  Mod/Ports
---- -----------------------
6     5/7

Console> (enable) set vlan 6 5/8
Vlan 5 configuration successful
VLAN 5 modified.
VLAN 1 modified.
VLAN  Mod/Ports
---- -----------------------
6     5/8
```

Activate the VLAN with the command **set vlan 6**:

```
Console> (enable) set vlan 6
Vlan 5 configuration successful
```

13. *Assign IP addresses to Ethernet 0 of R3 and R4 from network 152.1.1.0. The subnet contains 64 hosts. R3 should use the first address in the subnet and R4 should use the last one.*

This question involves simple subnetting. You need to pick a subnet that contains 64 usable IP addresses. For this, use a 25-bit mask giving 128 addresses, two of which can't be used, leaving 126. This is a bit more than you need, but if you went with a 26-bit mask you would be short. The following is the IP address assignment:

```
152.1.1.128 /25←Network Address
152.1.1.129 /25←R3
 . . .  . . .  . . .  . . .  . . . .
 . . .  . . .  . . .  . . .  . . . .
152.1.1.254 /25←R4
152.1.8.255 /25←Broadcast Address
```

The following commands define IP addresses on R3 and R4:

```
R3(config)#interface ethernet 0/0
R3(config-if)#ip address 152.1.1.129 255.255.255.128

R4(config)#interface ethernet 0
R4(config-if)#ip address 152.1.1.254 255.255.255.128
```

14. *Connect the rest of the routers serially as in Figure 4-21. Make sure that the correct ports are configured for DTE/DCE (the DCE clock will be set to 64K on all routers unless otherwise specified).*

This question is straightforward; the only thing that needs to be done is to set the clock rate on the DCE side of each serial link.

To set the clock rate, use the following interface command:

```
R4(config-if)#clock rate 64000
```

Use the **show controller** interface command to verify that the DCE cable is connected to the correct serial port. The following is truncated output from the command. Note that the cable attached is a V35 DCE cable and the clock rate is set to 64K.

```
R4#show controllers s0
HD unit 0, idb = 0x970D0, driver structure at 0x9AE28
buffer size 1524  HD unit 0, V.35 DCE cable, clockrate 64000
cpb = 0xE1, eda = 0x4940, cda = 0x4800
RX ring with 16 entries at 0xE14800
```

15. *Assign an IP address to S0 of R4 and R5 from network 152.1.2.0. Do not allocate more addresses than needed. Use the first available subnet and assign the addresses in ascending order starting with R4.*

The serial connection between R4 and R5 is a point-to-point connection, so only two addresses are needed. For this, use a 30-bit mask. The following is the IP address assignment:

```
152.1.2.0 /30 ←Network Address
152.1.2.1 /30 ←R4
152.1.2.2 /30 ←R5
152.1.2.3 /30 ←Broadcast Address
```

The following commands define IP addresses on R4 and R5:

```
R4(config)#interface serial 0
R4(config-if)#ip address 152.1.2.1 255.255.255.252

R5(config)#interface serial s0
R5(config-if)#ip address 152.1.2.2 255.255.255.252
```

16. *Assign IP addresses to S1 of R5 and S0/0 of R6 from network 152.1.10.0. Use an address that supports up to 80 hosts. Do not allocate more addresses than needed. Use the first available subnet and assign the addresses in ascending order starting with R5.*

The subnet between R5 and R6 requires 80 hosts. For this, you use a 25-bit mask. The following is the IP address assignment:

```
152.1.2.128 /25←Network Address
152.1.2.129 /25←R5
152.1.2.130 /25←R6
 . . .  . . . .  .
 . . .  . . . .  .
152.1.2.255 /25←Broadcast Address
```

The following commands define IP addresses on R5 and R6:

```
R5(config)#interface serial 1
R5(config-if)#ip address 152.1.1.77 255.255.255.128

R6(config)#interface serial s0/0
R6(config-if)#ip address 152.1.1.78 255.255.255.128
```

17. *Set the maximum transmission unit on Serial 0 of R1 to 500.*

Each interface has a default maximum packet size or MTU. This number generally defaults to the largest size possible for that type interface. On serial interfaces, the MTU size varies but cannot be set smaller than 64 bytes. To set the MTU, use the following interface command:

```
R1(config-if)#mtu 500
```

The MTU can only be set on a subinterface for ATM. For Frame Relay, set the MTU on the main interface. This also will be applied to each subinterface and can be verified with the **show int s0.1** command. The following is the output:

```
R1#show interfaces s0.1
Serial0.1 is up, line protocol is up
  Hardware is PowerQUICC Serial
  Internet address is 152.1.1.81/29
  MTU 500 bytes, BW 64 Kbit, DLY 20000 usec,
  reliability 255/255 , txload 1/255 , rxload 1/255
  Encapsulation FRAME-RELAY IETF
```

NOTE: *Changing the MTU value with the* ***mtu interface*** *configuration command can affect values for the protocol-specific versions of the command (ip mtu, for example). If the values specified with the* ***ip mtu*** *interface configuration command are the same as the value specified with the mtu interface configuration command, and you change the value for the mtu interface configuration command, the ip mtu value automatically matches the new mtu interface configuration command value. However, changing the values for the ip mtu configuration commands has no effect on the value for the mtu interface configuration command.*

18. *Configure the Catalyst so that only hosts 152.1.1.1 and 152.1.1.2 have telnet access. Assign an address to the SCO interface of the catalyst switch. Use the next available address on subnet 152.1.1.0/26.*

The Catalyst 5500 switches have two management interfaces. The Sc0 interface is an internal management interface and uses the TCP/IP protocol to communicate to any IP-capable station on the network. The sl0 interface is the console port and supports SLIP.

In order to access the Catalyst switch inband through SNMP or telnet, an internal IP address needs to be assigned to the Sc0. The IP address of the Catalyst can be verified with the **show interface** command. You can see that no IP addresses have been defined on the switch:

```
Console> (enable) show interface
sl0: flags=51<UP,POINTOPOINT,RUNNING>
        slip 0.0.0.0 dest 0.0.0.0
sc0: flags=63<UP,BROADCAST,RUNNING>
        vlan 1 inet 0.0.0.0 netmask 0.0.0.0 broadcast 0.0.0.0
```

The case study specifies to use the next available IP address on subnetwork 152.1.1.0/26, which is 152.1.1.5. To assign this address to the catalyst, use the **set interface sc0** command, as shown here:

```
Console> (enable) set interface sc0 152.1.1.5 255.255.255.192
Interface sc0 IP address set.
```

Use the **show interface** command to verify that the sc0 IP address has been set to 152.1.1.5:

```
Console> (enable) show interface
sl0: flags=51<UP,POINTOPOINT,RUNNING>
        slip 0.0.0.0 dest 0.0.0.0
sc0: flags=63<UP,BROADCAST,RUNNING>
        vlan 1 inet 152.1.1.5 netmask 255.255.255.0 broadcast 192.1.1.255
```

Once the sc0 address has been set, verify that it is active by pinging R1 (152.1.1.1):

```
Console> (enable) ping 152.1.1.1
152.1.1.1 is alive
```

The next part of the question requires that you configure the Catalyst so that it can only be accessed from host 152.1.1.1 or 152.1.1.2. The Catalyst switch, through the use of the permit command, enables the user to define up to 10 IP addresses that are allowed inbound SNMP and telnet access to the switch. The permit list can be displayed with the **show ip permit** command. You can see in the following output that, by default, no IP addresses are in the permit and the IP permit list feature is disabled:

```
Console> (enable) show ip permit
IP permit list feature disabled.
Permit List        Mask
----------------   ----------------

Denied IP Address   Last Accessed Time    Type
-----------------   ------------------    ------
```

Two IP addresses need to be added to the permit list, 152.1.1.1 and 152.1.1.2. To do this, use the following commands:

```
Console> (enable) set ip permit 152.1.1.1
152.1.1.1 added to IP permit list.

Console> (enable) set ip permit 152.1.1.2
152.1.1.2 added to IP permit list
```

Use the **show ip permit** command to verify that 152.1.1.1 and 152.1.1.2 have been added to the permit list. Notice that the IP permit list feature is still disabled, which is the default state of the IP permit list.

```
Console> (enable) show ip permit
IP permit list feature disabled.
Permit List        Mask
----------------   ----------------
152.1.1.1
152.1.1.2

Denied IP Address   Last Accessed Time    Type
-----------------   ------------------    ------
```

After the IP permit list has been defined, it must be enabled with the **set ip permit enable** command:

```
Console> (enable) set ip permit enable
IP permit list enabled.
```

Verify that R1 and R2 can telnet to the Catalyst switch:

```
R1#telnet 152.1.1.5
Trying 152.1.1.5  . . .  Open

Cisco Systems Console

Enter password:
```

19. *Assign an IP address to Ethernet 0/0 of R6 from network 152.1.11.0. Use an address that will support up to 200 hosts.*

The subnet on Ethernet 0/0 requires 200 hosts and for this use a 24-bit mask. The following is the IP address assignment:

```
152.1.11.0 /24   ←Network Address
152.1.11.1 /24   ←R6
152.1.11.2 /24
 . . .  . . .  . . .  . . .  . . . .
 . . .  . . .  . . .  . . .  . . . .
152.1.11.255 /24←Broadcast Address
```

The following commands define IP addresses on R6:

```
R6(config)#interface Ethernet 0/0
R6(config-if)#ip address 152.1.11.1 255.255.255.0
```

20. *Assign IP address 152.1.0.1 /24 to Ethernet 1/0 of R6, and place R6 E 1/0 in VLAN 6. The Ethernet interface from the backbone router R8 should also be placed in VLAN 6.*

When you try to enter the IP address under the interface, you will receive an error message: **bad mask /24 for address 152.1.0.1**. The reason for this is when 152.2.0.0 is subnetted as 255.255.255.0, subnet zero is written as 152.2.0.0, which is identical to the network address.

```
R6(config)#int ethernet 1/0
R6(config-if)#ip address 152.1.0.1 255.255.255.0
Bad mask /24 for address 152.1.0.1
```

To use the all-zeros and all-ones subnet (152.1.255.0), use the global configuration command **ip subnet zero**. The following is the configuration for allowing the use of subnet zero on R6:

```
R6(config)#ip subnet-zero
```

Now you can add the IP address to Ethernet 1/0 on R6:

```
R6(config)#int ethernet 1/0
R6(config-if)#ip address 152.1.0.1 255.255.255.0
```

NOTE: Although used in this case study for teaching purposes, subnetting with a subnet address of zero is illegal and strongly discouraged (as stated in RFC 791) because of the confusion that can arise between a network and a subnet that have the same addresses.

21. *Assign the following loopback addresses to the corresponding routers:*

```
R1 Loopback 0   152.1.1.65 /32
R1 Loopback 1   152.1.1.97 /27
R2 Loopback 0   152.1.1.72 /32
R3 Loopback 0   152.1.1.69 /32
R4 Loopback 0   152.1.2.4  /32
R5 Loopback 0   152.1.2.5  /32
R3 Loopback 1   152.1.3.1  /24
```

22. *Place VLAN 1 in OSPF area 3 and make sure that R2 is always selected as the DR router.*

Enabling OSPF on a router is performed in two steps. First, an OSPF process is defined and then an interface is added to the process. The command to start an OSPF process is **Router OSPF [Process #]**, and the process number is used internally by the router. Multiple OSPF routing processes can be configured on one router.

The following command enables OSPF process 64 on the router and assigns the Ethernet interface attached to VLAN 1 on the routers to area 3:

```
R1(config)#router ospf 64
R1(config-router)#network 152.1.1.0 0.0.0.63 area 3

R2(config)#router ospf 64
R2(config-router)#network 152.1.1.0 0.0.0.63 area 3
```

Use the command **show ip ospf interface** to verify that OSPF is configured on the interface. The following is the output from this command. Notice that on R2 interface E0 is configured for OSPF and is in area 3. The command also shows the network type, timer intervals, and adjacent neighbors:

```
R2a#show ip ospf interface ethernet 0
Ethernet0 is up, line protocol is up
  Internet Address 152.1.1.2/27 , Area 3
  Process ID 64, Router ID 152.1.1.72, Network Type BROADCAST, Cost: 10
  Transmit Delay is 1 sec, State DR, Priority 1
  Designated Router (ID) 152.1.1.72, Interface address 152.1.1.2
  No backup designated router on this network
```

```
  Timer intervals configured, Hello 10, Dead 40, Wait 40, Retransmit 5
    Hello due in 00:00:08
  Neighbor Count is 0, Adjacent neighbor count is 0
  Suppress hello for 0 neighbor(s)
```

The case study requires that R2 is always elected as the DR for the broadcast network. In order to accomplish this, you must first understand how the DR/BDR election process works.

The DR is elected using the Hello protocol, which was described earlier in this chapter. The election of the DR is determined by the router priority, which is carried in the Hello packet. The router with the highest priority is elected the DR, but if a tie occurs, the router with the highest router ID is selected. The router ID is the IP address of the highest loopback interface on the router. If no loopback is configured, the router ID is the highest IP address on the router.

The router priority can be configured on the router interface with the **ip ospf priority** command. If the ospf priority is set to zero, the router is not eligible to become the DR or BDR.

In order to prevent R1 from becoming the DR, you need to set its priority to zero. To do this, use the **ip osfp priority interface** command. The following command sets the ospf priority of R1 Ethernet 0 to zero:

```
R1(config)#interface Ethernet 0
R1(config-if)#ip ospf priority 0
```

Once the priority has been set, you need to verify that R2 is the DR. To do this, use the command **show ip ospf int e 0**. The following is the output from the command (note that the DR is router ID 152.1.1.72, which is the loopback address of R2). Also note that no BDR exists for the network. When you set the ospf priority to zero, the router is ineligible to participate in the DR/BDR election process.

```
R2#show ip ospf interface ethernet 0
Ethernet0 is up, line protocol is up
  Internet Address 152.1.1.2/27, Area 3
  Process ID 64, Router ID 152.1.1.72, Network Type BROADCAST, Cost: 10
  Transmit Delay is 1 sec, State DR, Priority 1
  Designated Router (ID) 152.1.1.72, Interface address 152.1.1.2
  No backup designated router on this network
  Timer intervals configured, Hello 10, Dead 40, Wait 40, Retransmit 5
    Hello due in 00:00:08
  Neighbor Count is 0, Adjacent neighbor count is 0
  Suppress hello for 0 neighbor(s)
```

23. *Place the Frame Relay cloud between R1, R2, and R3 in OSPF area 3.*

The following command enables OSPF process 64 on the router and assigns the serial interfaces attached to the Frame cloud on the routers to area 3:

```
R1(config)#router ospf 64
R1(config-router)#network 152.1.1.80 0.0.0.7 area 3

R2(config)#router ospf 64
R2(config-router)#network 152.1.1.80 0.0.0.7 area 3

R3(config)#router ospf 64
R3(config-router)#network 152.1.1.80 0.0.0.7 area 3
```

Because the case study requires that you use point-to-point subinterfaces on R1 and R2, and a physical interface on the R3, the OSPF network types will be mismatched. Because the Hello and dead intervals are not the same, the routers will not form an adjacency.

Because R1 and R2 are point-to-point subinterfaces, the OSPF network type is point-to-point, which, if you recall from earlier in the case study, does not elect a DR and has a Hello interval of 10 seconds and dead interval of 40 seconds.

R3 uses the main interface, which, when configured for Frame Relay, defaults to an OSPF network type non-broadcast. The non-broadcast network type elects a DR, and because it lacks broadcast capabilities, neighbors must manually be defined. The Hello interval is set to 30 seconds and a dead interval is set to 120.

To verify the network type, use the **show ip ospf interface** command. From the following output, you can see that interface s0/0 on R3 is non-broadcast and interface s0.1 on R1 is point-to-point. Also note the difference in Hello and dead intervals:

```
R3#show ip ospf interface s0/0
Serial0/0  is up, line protocol is up
  Internet Address 152.1.1.83/29 , Area 3
  Process ID 64, Router ID 152.1.3.1, Network Type NON_BROADCAST, Cost:64
  Transmit Delay is 1 sec, State DR, Priority 1
  Designated Router (ID) 152.1.3.1, Interface address 152.1.1.83
  No backup designated router on this network
  Timer intervals configured, Hello 30, Dead 120, Wait 120, Retransmit 5
    Hello due in 00:00:08
  Neighbor Count is 0, Adjacent neighbor count is 0

R1#show ip ospf int s0.1
Serial0.1 is up, line protocol is up
  Internet Address 152.1.1.81/29, Area 3
  Process ID 64, Router ID 152.1.1.97, Network Type POINT_TO_POINT, Cost: 64
  Transmit Delay is 1 sec, State POINT_TO_POINT,
  Timer intervals configured, Hello 10, Dead 40, Wait 40, Retransmit 5
    Hello due in 00:00:07
  Index 2/2, flood queue length 0
  Next 0x0(0)/0x0(0)
  Last flood scan length is 0, maximum is 0
  Last flood scan time is 0 msec, maximum is 0 msec
  Neighbor Count is 0, Adjacent neighbor count is 0
  Suppress hello for 0 neighbor(s)
```

The easiest way to fix this is to configure each interface as a point-to-multipoint network, in which case the Hello and dead timers would match and no neighbors would need to be defined. However, the case study does not permit the use of the point-to-multipoint command. This leaves you with two options: you can either configure all the interfaces as non-broadcast or broadcast. Because R3 is already non-broadcast, you will configure R1 and R2 to be non-broadcast. The following command changes the network type of point-to-point to non-broadcast:

```
R1(config-subif)#ip ospf network non-broadcast
R2(config-subif)#ip ospf network non-broadcast
```

The non-broadcast network type elects a DR, but because it lacks broadcast capabilities, neighbors cannot be discovered. Therefore, they must manually be defined. Also, because a DR is elected, care must be taken when choosing which router becomes the DR and BDR for the network. The DR and BDR require full logical connectivity with all routers on the network.

Because R3 is the hub router and has logical connectivity to each router, it will be the DR for this NBMA network. Because it is the DR, R3 is the only place neighbors need to be defined. The following commands define OSPF neighbors on R3:

```
R3(config-router)#neighbor 152.1.1.81
R3(config-router)#neighbor 152.1.1.82
```

To insure that R3 becomes the DR on the network, you need to set the OSPF priority on R1 and R2 to zero. The following commands set the OSPF priority on R2 and R1 to zero:

```
R2(config-subif)#ip ospf priority 0
R1(config-subif)#ip ospf priority 0
```

24. *Place VLAN 4 and the serial interface connecting R4 and R5 in OSPF area 0. Place the serial interface between R5 and R6 in area 1.*

The following commands enable OSPF process 64 on R4 and assign both the Ethernet and serial interfaces to area 0:

```
R4(config)#router ospf 64
R4(config-router)#network 152.1.1.128 0.0.0.127 area 0
R4(config-router)#network 152.1.2.0 0.0.0.3 area 0
```

The following command places Ethernet interface of R3 in OSPF area 0:

```
R3(config-router)#network 152.1.1.128 0.0.0.127 area 0
```

The following commands enable OSPF process 64 on R5 and R6. They also assign serial 0 of R5 to area 0, serial 1 of R5 to area 1, and serial 0/0 of R6 to area 1.

```
R5(config)#router ospf 64
R5(config-router)#network 152.1.2.129 0.0.0.127 area 1
R5(config-router)#network 152.1.2.0 0.0.0.3 area 0

R6(config)#router ospf 64
R6(config-router)#network 152.1.2.128 0.0.0.127 area 1
```

25. *Place the BRI interface on R6 and VLAN 3 in OSPF area 2.*

The following commands place the BRI and Ethernet interface of R6 in area 2:

```
R6(config-router)#network 152.1.1.76 0.0.0.3 area 2
R6(config-router)#network 152.1.11.0 0.0.0.255 area 2
```

As mentioned earlier in this case study, area 0 has to be at the center of all areas. Because area 2 is not physically connected to the backbone area (area 0), you must use a virtual link to connect the two areas. A virtual link provides disconnected areas with a logical path to the backbone. Virtual links are used for the purposes of linking an area that does not have a physical connection to the backbone and for patching the backbone in case of discontinuity.

A virtual link is established between two ABRs that share a common area, one of which has a connection to the backbone area. For this example, area 2 does not have a physical connection to the backbone, so a virtual link must be configured between R6, which is the ABR for area 2, and R5, which has physical connectivity to area 0. Area 1 will be used as a transit area and R5 will be used as the entry point into area 0, providing R6 with a logical connection to the backbone.

When configuring a virtual link, the router command **area <area-id> virtual-link <RID>** is used. The area-id is the transit area that is used to reach area 0. In your case, this is area 1. The RID is the router id of the router that you are forming the virtual link with. This is usually the highest loopback interface on the router or, if no loopback interfaces exist, the highest IP address. The router ID can be found with the **show ip ospf neighbor** command, which shows the neighbor's router ID, along with the interface IP address that it is connected to. The following is the output from the command on R5 and R6. Notice that the RID for R5 is 152.1.2.5 and the router ID for R6 is 152.1.12.1.

```
R5#show ip ospf neighbor

Neighbor ID     Pri   State           Dead Time   Address         Interface
152.1.2.4         1   FULL/  -        0:00:33     152.1.2.1       Serial0
152.1.12.1        1   FULL/  -        0:00:38     152.1.2.130

R6#show ip ospf neighbor

Neighbor ID     Pri   State           Dead Time   Address         Interface
                                                                  Serial0/0
152.1.2.5         1   FULL/  -        00:00:34    152.1.2.129
```

To configure a virtual link between R5 and R6, use the following router configuration commands:

```
R5(config-router)#area 1 virtual-link 152.1.12.1
R6(config-router)#area 1 virtual-link 152.1.2.5
```

Once the virtual link is configured, the **show ip osfp virtual-link** command can be used to determine the state of the link. The following is the output from the command on R5. Notice that the virtual link is up and the adjacency state is full.

```
R5#show ip ospf virtual-links
Virtual Link to router 152.1.2.130 is up
  Transit area 1, via interface Serial1, Cost of using 64
  Transmit Delay is 1 sec, State POINT_TO_POINT,
  Timer intervals configured, Hello 10, Dead 40, Wait 40, Retransmit 5
    Hello due in never
    Adjacency State FULL
    Link State retransmission due in 0:00:04
         LSA in retransmission queue 1
```

26. *Configure area 2 so that no external LSAs are flooded into the area from Area 0. Area 2 must be able to send routes learned from the IGRP backbone into the OSPF domain.*

To accomplish this, you must configure area 2 as a *Not-so-Stubby area* (NSSA). An NSSA is similar to an OSPF stub area in that Type 5 external LSAs from the core are not flooded into the area, but NSSA has the capability to import AS external routes in a limited fashion within the area. NSSA enables the importing of Type 7 AS external routes within NSSA areas by redistribution. These Type 7 LSAs are translated into Type 5 LSAs by the NSSA ABR, and they are flooded throughout the whole routing domain. Summarization and filtering are also supported during the translation.

To configure area 2 as a NSSA area, use the following command:

```
R6(config-router)#area 2 nssa
```

27. *Change the OSPF cost of using an Ethernet interface to 90 and a serial interface to 580 across the entire network. Do not use the OSPF cost command.*

OSPF calculates the cost of an interface by dividing the reference bandwidth by the speed of the interface. By default, the reference bandwidth is set to 100M, so normally the OSPF cost of a 10M Ethernet is 10:

$$(\text{Reference bandwidth}/\text{-}10\text{M}) = 100\text{M}/10\text{M} = 10$$

Changing the reference bandwidth can then change the OSPF cost of the interface. The question states that the cost of an Ethernet interface should be 90 and the cost of a serial interface should be 580. To accomplish this, you need to set the reference bandwidth to 900M:

```
OSPF cost of an Ethernet Interface = 900M/10M= 90
OSPF cost of a Serial Interface =900M /1.5M= 580
```

To set the OSPF reference bandwidth for the router, use the following router command:

```
R2(config-router)#ospf auto-cost reference-bandwidth 900
```

Once the reference bandwidth is set, use the **show ip ospf interface** command to verify that the OSPF cost has been changed. The following is the output from the command on R2. Note that the OSPF cost is now 90 for the Ethernet interface on R2:

```
R2#show ip ospf int e0
Ethernet0 is up, line protocol is up
  Internet Address 152.1.1.2/26, Area 3
  Process ID 64, Router ID 152.1.1.72, Network Type BROADCAST, Cost: 90
  Transmit Delay is 1 sec, State DR, Priority 1
  Designated Router (ID) 152.1.1.72, Interface address 152.1.1.2
  No backup designated router on this network
  Timer intervals configured, Hello 10, Dead 40, Wait 40, Retransmit 5
    Hello due in 00:00:07
  Neighbor Count is 1, Adjacent neighbor count is 1
    Adjacent with neighbor 152.1.1.97
  Suppress hello for 0 neighbor(s)
```

28. *R4 should only see a single advertisement in its routing table from area 3.*

Cisco enables you to summarize addresses in order to conserve resources by limiting the number of routes that need to be advertised between areas. Two types of address summarization are supported on a Cisco router: inter-area summarization and external route summarization. Inter-area summarization is used to summarize addresses between areas, while external summarization is used to summarize a set of external routes that have been injected into the domain.

The following is the output from the **show ip route** command on R4. Notice that R4 has two routes from area 3 in its routing table:

```
R4#show ip route
Codes: C - connected, S - static, I - IGRP, R - RIP, M - mobile, B - BGP
       D - EIGRP, EX - EIGRP external, O - OSPF, IA - OSPF inter area
       N1 - OSPF NSSA external type 1, N2 - OSPF NSSA external type 2
       E1 - OSPF external type 1, E2 - OSPF external type 2, E - EGP
       i - IS-IS, L1 - IS-IS level-1, L2 - IS-IS level-2, * - candidate default
       U - per-user static route, o - ODR

Gateway of last resort is not set

     152.1.0.0/16  is variably subnetted, 9 subnets, 6 masks
C       152.1.1.128/25  is directly connected, Ethernet0
O IA    152.1.2.128/25 [110/710] via 152.1.2.2, 00:02:45, Serial0
O IA    152.1.11.0/24  [110/656] via 152.1.2.2, 00:02:45, Serial0
O IA    152.1.1.0/26   [110/163] via 152.1.1.129, 00:00:09, Ethernet0
C       152.1.0.1/32  is directly connected, Loopback1
C       152.1.2.0/30  is directly connected, Serial0
C       152.1.2.4/32  is directly connected, Loopback0
O IA    152.1.1.80/29  [110/154] via 152.1.1.129, 00:00:09, Ethernet0
```

Because area 3 contains all the addresses from 152.1.1.0 to 152.1.1.128, they can all be summarized by R3 in one update, 152.1.1.0 /25. To do this, the area range command is used, which specifies the area that the summary belongs to: the summary address and the mask. The following is the command used to summarize network 152.1.1.0 on R3:

```
R3(config-router)#area 3 range 152.1.1.0 255.255.255.128
```

To verify that the summarization is working properly, display the routing table on R4. The following is routing table on R4. Notice that only the summary address is being learned from area 3.

```
R4#show ip route
Codes: C - connected, S - static, I - IGRP, R - RIP, M - mobile, B - BGP
       D - EIGRP, EX - EIGRP external, O - OSPF, IA - OSPF inter area
       N1 - OSPF NSSA external type 1, N2 - OSPF NSSA external type 2
       E1 - OSPF external type 1, E2 - OSPF external type 2, E - EGP
       i - IS-IS, L1 - IS-IS level-1, L2 - IS-IS level-2, * - candidate default
       U - per-user static route, o - ODR

Gateway of last resort is not set

     152.1.0.0/16  is variably subnetted, 8 subnets, 4 masks
C       152.1.1.128/25  is directly connected, Ethernet0
O IA    152.1.2.128/25 [110/710] via 152.1.2.2, 00:03:47, Serial0
O IA    152.1.11.0/24  [110/656] via 152.1.2.2, 00:03:47, Serial0
O IA    152.1.1.0/25   [110/154] via 152.1.1.129, 00:00:25, Ethernet0
C       152.1.0.1/32  is directly connected, Loopback1
C       152.1.2.0/30  is directly connected, Serial0
C       152.1.2.4/32  is directly connected, Loopback0
```

29. *Advertise all the loopback interfaces on R1 and R2, as well as loopback 1 on R3 via OSPF. Don't configure or run OSPF on any of these loopback interfaces, however, and don't propagate these networks outside of the area.*

The first part of the question requires that the loopback interfaces be advertised via OSPF. This can be done in multiple ways, but the easiest ,of course, is to run OSPF on the interface. However, the question specifies that OSPF cannot be run on any of the loopback interfaces.

In order to get these networks into OSPF, you must use the **redistribute connected** command. Because you want to limit the interfaces that are going to be redistributed into OSPF, a route map is needed. The first step is to define a route map that specifies the interfaces that will be redistributed into OPSF. This map is then applied to the redistribute command under the OSPF process.

The first step is to define a route-map on each router that permits the specified interface:

```
R1(config)#route-map only_loopbacks permit 10
R1(config-route-map)#match interface loopback 0
```

```
R1(config-route-map)#match interface loopback 1

R2(config)#route-map only_loopbacks permit 10
R2(config-route-map)#match interface loopback 0

R3(config-router)#route-map only_loopback0 permit 10
R3(config-route-map)#match interface loopback 0
```

The second step is to apply the route map to the distribute list under the routing process:

```
R1(config)#router ospf 64
R1(config-router)#redistribute connected subnets route-map only_loopbacks

R2(config)#router ospf 64
R2(config-router)#redistribute connected subnets route-map only_loopbacks

R2(config)#router ospf 64
R3(config-router)#redistribute connected subnets route-map only_loopback0
```

Now verify that R1 sees all the loopback interfaces being advertised from R3 and R2. The following is the output from the routing table on R1. Note that it has two OSPF external routes. Any time a route is redistributed into OSPF it is considered an external route:

```
R1#show ip route
Codes: C - connected, S - static, I - IGRP, R - RIP, M - mobile, B - BGP
       D - EIGRP, EX - EIGRP external, O - OSPF, IA - OSPF inter area
       N1 - OSPF NSSA external type 1, N2 - OSPF NSSA external type 2
       E1 - OSPF external type 1, E2 - OSPF external type 2, E - EGP
       i - IS-IS, L1 - IS-IS level-1, L2 - IS-IS level-2, * - candidate default
       U - per-user static route, o - ODR, P - periodic downloaded static route
       T - traffic engineered route

Gateway of last resort is not set

     152.1.0.0/16  is variably subnetted, 11 subnets, 7 masks
O IA    152.1.1.128/25  [110/592] via 152.1.1.83, Serial0.1
O IA    152.1.2.128/25  [110/1302] via 152.1.1.83, Serial0.1
O IA    152.1.11.0/24  [110/1248] via 152.1.1.83, Serial0.1
C       152.1.1.0/26  is directly connected, FastEthernet0
O IA    152.1.2.0/30  [110/1174] via 152.1.1.83, Serial0.1
C       152.1.1.80/29  is directly connected, Serial0.1
O E2    152.1.1.72/32  [110/20] via 152.1.1.2, FastEthernet0
O IA    152.1.1.76/30  [110/2800] via 152.1.1.83, Serial0.1
C       152.1.1.65/32  is directly connected, Loopback0
O E2    152.1.1.69/32  [110/20] via 152.1.1.83, Serial0.1
C       152.1.1.96/27  is directly connected, Loopback1
```

Now verify that R2 sees all the loopback interfaces being advertised from R3 and R1. The following is the output from the routing table on R2. Note that it has three OSPF external routes:

```
R2#show ip route
Codes: C - connected, S - static, I - IGRP, R - RIP, M - mobile, B - BGP
       D - EIGRP, EX - EIGRP external, O - OSPF, IA - OSPF inter area
       N1 - OSPF NSSA external type 1, N2 - OSPF NSSA external type 2
```

```
       E1 - OSPF external type 1, E2 - OSPF external type 2, E - EGP
       i - IS-IS, L1 - IS-IS level-1, L2 - IS-IS level-2, * - candidate default
       U - per-user static route, o - ODR

Gateway of last resort is not set

     152.1.0.0/16  is variably subnetted, 11 subnets, 7 masks
O IA    152.1.1.128/25 [110/592] via 152.1.1.83, 00:06:25, Serial0.1
O IA    152.1.2.128/25 [110/1302] via 152.1.1.83, 00:06:25, Serial0.1
O IA    152.1.11.0/24  [110/1248] via 152.1.1.83, 00:06:25, Serial0.1
C       152.1.1.0/26   is directly connected, Ethernet0
O IA    152.1.2.0/30   [110/1174] via 152.1.1.83, 00:06:25, Serial0.1
C       152.1.1.80/29  is directly connected, Serial0.1
C       152.1.1.72/32  is directly connected, Loopback0
O IA    152.1.1.76/30  [110/2800] via 152.1.1.83, 00:06:25, Serial0.1
O E2    152.1.1.65/32  [110/20] via 152.1.1.1, 00:06:25, Ethernet0
O E2    152.1.1.69/32  [110/20] via 152.1.1.83, 00:02:50, Serial0.1
O E2    152.1.1.96/27  [110/20] via 152.1.1.1, 00:06:26, Ethernet0
```

Verify that R3 sees all the loopback interfaces being advertised from R2 and R1. The following is the output from the routing table on R2. Note that it has three OSPF external routes:

```
R3#show ip route
Codes: C - connected, S - static, I - IGRP, R - RIP, M - mobile, B - BGP
       D - EIGRP, EX - EIGRP external, O - OSPF, IA - OSPF inter area
       E1 - OSPF external type 1, E2 - OSPF external type 2, E - EGP
       i - IS-IS, L1 - IS-IS level-1, L2 - IS-IS level-2, * - candidate default
       U - per-user static route

Gateway of last resort is not set

     152.1.0.0/16  is variably subnetted, 12 subnets, 7 masks
C       152.1.1.128/25  is directly connected, Ethernet0/0
O IA    152.1.2.128/25  [110/720] via 152.1.1.254, 00:03:28, Ethernet0/0
O IA    152.1.11.0/24   [110/666] via 152.1.1.254, 00:03:28, Ethernet0/0
O       152.1.1.0/26    [110/73] via 152.1.1.81, 00:03:28, Serial0/0
C       152.1.3.0/24    is directly connected, Loopback1
O       152.1.2.0/30    [110/592] via 152.1.1.254, 00:03:28, Ethernet0/0
C       152.1.1.80/29   is directly connected, Serial0/0
O E2    152.1.1.72/32   [110/20] via 152.1.1.82, 00:03:28, Serial0/0
O IA    152.1.1.76/30   [110/2218] via 152.1.1.254, 00:03:28, Ethernet0/0
O E2    152.1.1.65/32   [110/20] via 152.1.1.81, 00:03:29, Serial0/0
C       152.1.1.69/32   is directly connected, Loopback0
O E2    152.1.1.96/27   [110/20] via 152.1.1.81, 00:03:29, Serial0/0
```

The second part of the question states that the loopback interfaces should not be seen outside of the area. The following is the output from the **show ip route** command on R4. Note that the four loopback interfaces are in the routing table as External Type 2 routes, because you used the redistribute command:

```
R4#show ip route
Codes: C - connected, S - static, I - IGRP, R - RIP, M - mobile, B - BGP
       D - EIGRP, EX - EIGRP external, O - OSPF, IA - OSPF inter area
       N1 - OSPF NSSA external type 1, N2 - OSPF NSSA external type 2
```

```
       E1 - OSPF external type 1, E2 - OSPF external type 2, E - EGP
       i - IS-IS, L1 - IS-IS level-1, L2 - IS-IS level-2, * - candidate default
       U - per-user static route, o - ODR

Gateway of last resort is not set

     152.1.0.0/16  is variably subnetted, 12 subnets, 5 masks
C       152.1.1.128/25  is directly connected, Ethernet0
O IA    152.1.2.128/25  [110/710] via 152.1.2.2, 00:08:40, Serial0
O IA    152.1.11.0/24   [110/656] via 152.1.2.2, 00:08:40, Serial0
O IA    152.1.1.0/25    [110/154] via 152.1.1.129, 00:08:40, Ethernet0
C       152.1.0.1/32    is directly connected, Loopback1
C       152.1.2.0/30    is directly connected, Serial0
C       152.1.2.4/32    is directly connected, Loopback0
O E2    152.1.1.72/32   [110/20] via 152.1.1.129, 00:08:40, Ethernet0
O IA    152.1.1.76/30   [110/2208] via 152.1.2.2, 00:08:40, Serial0
O E2    152.1.1.65/32   [110/20] via 152.1.1.129, 00:08:41, Ethernet0
O E2    152.1.1.69/32   [110/20] via 152.1.1.129, 00:05:05, Ethernet0
O E2    152.1.1.96/27   [110/20] via 152.1.1.129, 00:08:41, Ethernet0
```

In order to remove these routes from the routing table, you must use the distribute list command. Two arguments for this command are available: in and out. **Distribute** list out can only be used on a router doing redistribution (ASBR), which in your case would be R1, R2, and R3. Because you want these routes advertised within area 3, you must use the **distribute** list in command on R4. One common misconception is that a distribute list filters the routing updates. This is not true. The command prevents the filtered routes from being placed in the routing table. The OSPF LSA database still contains the routes and therefore floods them to each neighbor. Thus, each downstream router will still see all the routes.

To prevent the routes from being placed in the routing table of R4, use the **distribute list in** command on R4.

In order to filter routes from the routing table, you need to identify network addresses that need to be filtered through the use of an access list and apply that list to an OSPF interface using a distribute list.

a. Define the access list on R4 to deny the loopback networks:

```
R4(config)#access-list 1 deny 152.1.1.72 0.0.0.0
R4(config)#access-list 1 deny 152.1.1.65 0.0.0.0
R4(config)#access-list 1 deny 152.1.1.69 0.0.0.0
R4(config)#access-list 1 deny 152.1.1.96 0.0.0.31
R4(config)#access-list 1 permit any
```

When defining the access list, a wildcard mask is used after the IP address. This mask is used to specify which bits in an IP address should be ignored when comparing that address with another IP address. A one in the wildcard mask means that the bit position can be ignored when being compared to another IP address, and a zero specifies that the bit position must match.

For example, the second to last entry of the previous access list specifies that in order for traffic to be denied the first 27 bits must match. This access list denies network 152.1.1.96 /27:

```
access-list 1 deny 152.1.1.96 0.0.0.31

   152          1            1           96      ←IP address
10011000     00000001     00000001    01100000   ←Binary representation of IP address
00000000     00000000     00000000    00011111   ←Binary representation of Wildcard mask
10011000     00000001     00000001    011xxxxx
---------------------------------------------
    ↑            ↑            ↑           ↑
   152           1            1          When comparing IP addresses match the first
                                         27 bits and ignore the last 5.
```

A permit any is needed at the end because, by default, if no match is found in the access list, the packet is denied. So for access list 1 if the permit any was left off, all traffic would be denied.

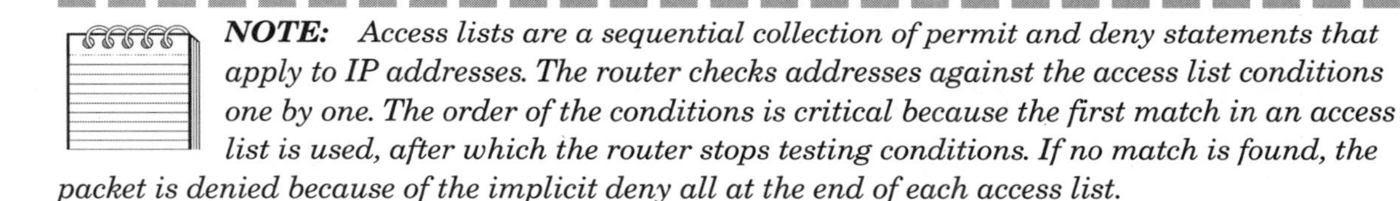

NOTE: *Access lists are a sequential collection of permit and deny statements that apply to IP addresses. The router checks addresses against the access list conditions one by one. The order of the conditions is critical because the first match in an access list is used, after which the router stops testing conditions. If no match is found, the packet is denied because of the implicit deny all at the end of each access list.*

b. Apply the distribution list to the OSPF interface that you are receiving the routes from:

```
R4(config)#router ospf 64
R4(config-router)#distribute-list 1 in ethernet 0
```

In order for the routes to be filtered, you must clear the routing table. To do this, use the command **clear ip route ***, which removes all the routes from the table. The distribute list will then be applied, preventing the four routes from being added into the table:

```
R4#clear ip route *
```

To verify that the distribute list is working properly, show the routing table on R4. The following is the output. Note that the four external routes are no longer present.

```
R4#show ip route
Codes: C - connected, S - static, I - IGRP, R - RIP, M - mobile, B - BGP
       D - EIGRP, EX - EIGRP external, O - OSPF, IA - OSPF inter area
       N1 - OSPF NSSA external type 1, N2 - OSPF NSSA external type 2
       E1 - OSPF external type 1, E2 - OSPF external type 2, E - EGP
       i - IS-IS, L1 - IS-IS level-1, L2 - IS-IS level-2, * - candidate default
       U - per-user static route, o - ODR

Gateway of last resort is not set

     152.1.0.0/16  is variably subnetted, 8 subnets, 4 masks
C       152.1.1.128/25  is directly connected, Ethernet0
O IA    152.1.2.128/25  [110/710] via 152.1.2.2, 00:00:02, Serial0
O IA    152.1.11.0/24   [110/656] via 152.1.2.2, 00:00:02, Serial0
O IA    152.1.1.0/25    [110/154] via 152.1.1.129, 00:00:02, Ethernet0
```

```
C       152.1.0.1/32     is directly connected, Loopback1
C       152.1.2.0/30     is directly connected, Serial0
C       152.1.2.4/32     is directly connected, Loopback0
O IA    152.1.1.76/30    [110/2208] via 152.1.2.2, 00:00:02, Serial0
```

Because the distribute list only filters routes from appearing in the routing table and not OSPF LSAs, the routes still exist in the OSPF database. Therefore, they will still be flooded to all downstream neighbors. The following is output from the **show ip route** command on R5. Note that the external routes are still in the routing table. In order to prevent these routes from being placed in the routing table, a distribute list must also be applied to R5 and R6.

```
R5#show ip route
Codes: C - connected, S - static, I - IGRP, R - RIP, M - mobile, B - BGP
       D - EIGRP, EX - EIGRP external, O - OSPF, IA - OSPF inter area
       E1 - OSPF external type 1, E2 - OSPF external type 2, E - EGP
       i - IS-IS, L1 - IS-IS level-1, L2 - IS-IS level-2, * - candidate default

Gateway of last resort is not set

     152.1.0.0 is variably subnetted, 11 subnets, 5 masks
O       152.1.1.128 255.255.255.128 [110/154] via 152.1.2.1, 00:57:08, Serial0
C       152.1.2.128 255.255.255.128 is directly connected, Serial1
O IA    152.1.11.0 255.255.255.0    [110/74] via 152.1.2.130, 00:57:08, Serial1
O IA    152.1.1.0 255.255.255.128   [110/218] via 152.1.2.1, 00:57:08, Serial0
C       152.1.2.0 255.255.255.252   is directly connected, Serial0
C       152.1.2.5 255.255.255.255   is directly connected, Loopback0
O E2    152.1.1.72 255.255.255.255  [110/20] via 152.1.2.1, 00:57:08, Serial0
O IA    152.1.1.76 255.255.255.252
           [110/1626] via 152.1.2.130, 00:57:08, Serial1
O E2    152.1.1.65 255.255.255.255  [110/20] via 152.1.2.1, 00:57:08, Serial0
O E2    152.1.1.69 255.255.255.255  [110/20] via 152.1.2.1, 00:57:09, Serial0
O E2    152.1.1.96 255.255.255.224  [110/20] via 152.1.2.1, 00:57:09, Serial0
```

In order to filter routes from the routing table, you need to identify network addresses that need to be filtered through the use of an access list and apply that list to an OSPF interface using a distribute list.

c. Define the access list on R5 and R6 to deny the loopback networks:

```
R5(config)#access-list 1 deny 152.1.1.72 0.0.0.0
R5(config)#access-list 1 deny 152.1.1.65 0.0.0.0
R5(config)#access-list 1 deny 152.1.1.69 0.0.0.0
R5(config)#access-list 1 deny 152.1.1.96 0.0.0.31
R5(config)#access-list 1 permit any

R6(config)#access-list 1 deny 152.1.1.72 0.0.0.0
R6(config)#access-list 1 deny 152.1.1.65 0.0.0.0
R6(config)#access-list 1 deny 152.1.1.69 0.0.0.0
R6(config)#access-list 1 deny 152.1.1.96 0.0.0.31
R6(config)#access-list 1 permit any
```

d. Apply the distribution list to the OSPF interface that you are receiving the routes from:

```
R5(config)#router ospf 64
R5(config-router)#distribute-list 1 in serial 0

R6(config)#router ospf 64
R6(config-router)#distribute-list 1 in serial 0/0
```

In order for the routes to be filtered, you must clear the routing table. To do this, use the command **clear ip route** *, which removes all the routes from the table. The distribute list is then applied, preventing the four routes from being added into the table:

```
R5#clear ip route *

R6#clear ip route *
```

To verify that the distribute list is working properly, show the routing table on R5 and R6. The following is the output from the **show ip route** command on R5. Note that the four external routes are no longer present:

```
R5#show ip route
Codes: C - connected, S - static, I - IGRP, R - RIP, M - mobile, B - BGP
       D - EIGRP, EX - EIGRP external, O - OSPF, IA - OSPF inter area
       E1 - OSPF external type 1, E2 - OSPF external type 2, E - EGP
       i - IS-IS, L1 - IS-IS level-1, L2 - IS-IS level-2, * - candidate default

Gateway of last resort is not set

     152.1.0.0 is variably subnetted, 7 subnets, 4 masks
O       152.1.1.128 255.255.255.128 [110/154] via 152.1.2.1, 00:03:14, Serial0
C       152.1.2.128 255.255.255.128 is directly connected, Serial1
O IA    152.1.11.0 255.255.255.0    [110/74] via 152.1.2.130, 00:03:14, Serial1
O IA    152.1.1.0 255.255.255.128   [110/218] via 152.1.2.1, 00:03:14, Serial0
C       152.1.2.0 255.255.255.252   is directly connected, Serial0
C       152.1.2.5 255.255.255.255   is directly connected, Loopback0
O IA    152.1.1.76 255.255.255.252
           [110/1626] via 152.1.2.130, 00:03:14, Serial1
```

30. *Configure the routers so that the source of each routing update is authenticated. Area 0 should use MD5 authentication, while area 3 and area 1 should use simple authentication. The key should be cisco plus the area number. For example, the key for area 0 is cisco0.*

Neighbor authentication enables the router to authenticate the source of each routing update that is received. An authentication key or password is exchanged between routers. If the keys do not match, the routing update is rejected.

Cisco uses two types of neighbor authentication: plain text and *Message Digest Algorithm Version 5* (MD5). Plain text authentication sends a key over the wire. Because the key is passed in plain text, it can be read during transit and therefore is not recommended.

MD5 authentication sends a message digest instead of the key. The MD5 algorithm is used to produce a "hash" of the key and this is what is sent. For additional information on the MD5 algorithm, reference RFC1321.

The question specifies that area 0 use MD5 authentication, enabling neighbor authentication on a router is done in two steps. First, authentication is enabled for a particular area under the routing process:

MD5 Neighbor Authentication

```
R3(config)#router ospf 64
R3(config-router)#area 0 authentication message-digest

R4(config)#router ospf 64
R4(config-router)#area 0 authentication message-digest

R5(config)#router ospf 64
R5(config-router)#area 0 authentication message-digest
```

The next step is to define the authentication key under the interface:

```
R3(config)#interface e0/0
R3(config-if)#ip ospf message-digest-key 1 md5 cisco0

R4(config)#interface e0
R4(config-if)#ip ospf message-digest-key 1 md5 cisco0
R4(config)#interface s0
R4(config-if)#ip ospf message-digest-key 1 md5 cisco0


R5(config)#interface s0
R5(config-if)#ip ospf message-digest-key 1 md5 cisco0
```

From R3, display the OSPF parameters with the command **show ip ospf interface e0/0.** Notice that MD5 is enabled on the interface and the key id is 1. This is very important because keys must match in each router.

```
R3#show ip ospf int e0/0
Ethernet0/0  is up, line protocol is up
  Internet Address 152.1.1.129/25, Area 0
  Process ID 64, Router ID 152.1.3.1, Network Type BROADCAST, Cost: 10
  Transmit Delay is 1 sec, State DR, Priority 1
  Designated Router (ID) 152.1.3.1, Interface address 152.1.1.129
  Backup Designated router (ID) 152.1.2.4, Interface address 152.1.1.254
  Timer intervals configured, Hello 10, Dead 40, Wait 40, Retransmit 5
    Hello due in 00:00:01
  Neighbor Count is 1, Adjacent neighbor count is 1
    Adjacent with neighbor 152.1.2.4  (Backup Designated Router)
  Message digest authentication enabled
    Youngest key id is 1
```

Notice that because you turned on authentication, the adjacency on the virtual link connecting area 2 to area 0 has not formed. The following is the output from the **show ip ospf virtual link** command on R5. Notice that the state of the adjacency is no longer full:

```
R5#show ip ospf virtual-links
Virtual Link to router 152.1.12.1 is up
  Transit area 1, via interface Serial1, Cost of using 64
  Transmit Delay is 1 sec, State POINT_TO_POINT,
  Timer intervals configured, Hello 10, Dead 40, Wait 40, Retransmit 5
    Hello due in 0:00:04
  Message digest authentication enabled
      No key configured, using default key id 0
```

Also note that R5 is no longer seeing the route 152.1.1.76 or 152.1.11.0 from R6:

```
R5#show ip route
Codes: C - connected, S - static, I - IGRP, R - RIP, M - mobile, B - BGP
       D - EIGRP, EX - EIGRP external, O - OSPF, IA - OSPF inter area
       E1 - OSPF external type 1, E2 - OSPF external type 2, E - EGP
       i - IS-IS, L1 - IS-IS level-1, L2 - IS-IS level-2, * - candidate default

Gateway of last resort is not set

     152.1.0.0 is variably subnetted, 5 subnets, 3 masks
O       152.1.1.128 255.255.255.128 [110/154] via 152.1.2.1, 00:04:34, Serial0
C       152.1.2.128 255.255.255.128 is directly connected, Serial1
O IA    152.1.1.0 255.255.255.128 [110/218] via 152.1.2.1, 00:04:35, Serial0
C       152.1.2.0 255.255.255.252 is directly connected, Serial0
C       152.1.2.5 255.255.255.255 is directly connected, Loopback0
```

Why is this? The answer can be found through some simple debugging. Turn **debug ip ospf events** on R5. The following is the output from the command. Notice that you are receiving a packet from 152.1.2.130 with a mismatch authentication type. The remote end uses type 0, which indicates no authentication and R5 is using type 2 authentication, which is MD5.

```
OSPF: Rcv pkt from 152.1.2.130, Serial1 : Mismatch Authentication type. Input packet
specified type 0, we use type 2
```

In order to fix this, you must enable authentication on the virtual link.

Similar to the previous steps, enabling neighbor authentication on virtual links is a two-step process. First, authentication is enabled for a particular area under the routing process. Authentication must be turned on for area 0 on R6 because the virtual link attaches to that area. Because this was already done in a previous step for R5, it does not need to be added again.

```
R6(config-router)#area 0 authentication message-digest
```

The next step is to define the authentication key under the interface. Since you are using a virtual link, this is where it is applied.

```
R6(config-router)#area 1 virtual-link 152.1.2.5 message-digest-key 1 md5 cisco0

R5(config-router)#area 1 virtual-link 152.1.12.1 message-digest-key 1 md5 cisco0
```

To verify that authentication is working properly across the virtual link, use the **show ip ospf virtual-links** command on R5. The following is the output. Notice that the virtual link is up, the state of the adjacency is full, and MD5 is enabled.

```
R5#show ip ospf virtual-links
Virtual Link to router 152.1.12.1 is up
  Transit area 1, via interface Serial1, Cost of using 64
  Transmit Delay is 1 sec, State POINT_TO_POINT,
  Timer intervals configured, Hello 10, Dead 40, Wait 40, Retransmit 5
    Hello due in 0:00:01
    Adjacency State FULL
  Message digest authentication enabled
    Youngest key id is 1
```

Also note that R5 is now seeing the two networks from R6:

```
R5#show ip route
Codes: C - connected, S - static, I - IGRP, R - RIP, M - mobile, B - BGP
       D - EIGRP, EX - EIGRP external, O - OSPF, IA - OSPF inter area
       E1 - OSPF external type 1, E2 - OSPF external type 2, E - EGP
       i - IS-IS, L1 - IS-IS level-1, L2 - IS-IS level-2, * - candidate default

Gateway of last resort is not set

     152.1.0.0 is variably subnetted, 7 subnets, 4 masks
O       152.1.1.128 255.255.255.128 [110/154] via 152.1.2.1, 00:01:13, Serial0
C       152.1.2.128 255.255.255.128 is directly connected, Serial1
O IA    152.1.11.0 255.255.255.0 [110/74] via 152.1.2.130, 00:01:04, Serial1
O IA    152.1.1.0 255.255.255.128 [110/218] via 152.1.2.1, 00:01:04, Serial0
C       152.1.2.0 255.255.255.252 is directly connected, Serial0
C       152.1.2.5 255.255.255.255 is directly connected, Loopback0
O IA    152.1.1.76 255.255.255.252[110/1626] via 152.1.2.130, 00:01:04, Serial1
```

The question specifies that all other areas use simple or plain text authentication. This leaves area 3 and area 1. Because area 2 uses a virtual link to attach to area 0, it must run MD5 over that link, as shown previously.

The second step in neighbor authentication is:

Plain Text Neighbor Authentication

```
R1(config)#router ospf 64
R1(config-router)#area 3 authentication


R2(config)#router ospf 64
R2(config-router)#area 3 authentication

R3(config)#router ospf 64
R3(config-router)#area 3 authentication

R5(config)#router ospf 64
R5(config-router)#area 1 authentication

R6(config)#router ospf 64
R6(config-router)#area 1 authentication
```

The next step is to define the authentication key under the interface.

```
R1(config)#interface e0/0
R1(config-if)#ip ospf authentication-key cisco3
R1(config)#interface s0.1
R1(config-if)#ip ospf authentication-key cisco3

R2(config)#interface e0
R2(config-if)#ip ospf authentication-key cisco3
R2(config)#interface s0.1
R2(config-if)#ip ospf authentication-key cisco3

R3(config)#interface s0/0
R3(config-if)#ip ospf authentication-key cisco3

R5(config)#interface s1
R5(config-if)#ip ospf authentication-key cisco1

R6(config)#interface s0/0
R6(config-if)#ip ospf authentication-key cisco1
```

31. *Configure* Dial on Demand routing *(DDR) between R1 and R6. R6 should call R1 when R6's serial connection is lost to R5. The backup should be invoked five seconds after the primary is lost, and once the primary is back, the router should wait 20 seconds before switching to the primary. Configure DDR so that R1 will accept R6's call, authenticate R6, and then securely call R6 back.*

This question requires the use of DDR on R1 and R6. In addition, you will need to configure both routers for PPP callback. PPP callback is an IOS feature that enables secure callback in a DDR environment. In this case, R1 will be the PPP callback client and R6 will be the PPP callback server. When R1 places a call to R6, the two routers will authenticate. If the authentication is successful, R6 then disconnects the call and calls back R1.

Start your DDR configuration by configuring R1 and R6 for ISDN. When configuring a BRI, the two main things that you need to know are the ISDN switch type and the SPID number if any are used. Some switches, such as the AT&T 5ESS, don't require any SPIDS when running 5ESS call control. For this case study, we used the ADTRAN Atlas 800 for an ISDN simulator. Depending on the switch you are using, this might change. The switch signaling is National ISDN 1 and the SPIDS are as follows:

R1	DN	SPID
BRI1	899-5101	5101
BRI2	899-5102	5102

R2	DN	SPID
BRI1	899-5201	5201
BRI2	899-5202	5202

The first step is to define the switch type under the global configuration. The following global command defines the switch type as National ISDN 1:

```
R6(config)#isdn  switch-type basic-ni1
R1(config)#isdn  switch-type basic-ni1
```

The next step is to configure the SPIDs under the BRI interface. The following interface command defines SPID 1 and 2:

```
R6(config-if)#isdn spid1 5201
R6(config-if)#isdn spid2 5202

R1(config-if)#isdn spid1 5201
R1(config-if)#isdn spid2 5202
```

Use the **show isdn status** command to verify that the ISDN circuit is active and the SPIDS are configured properly. The following is the output from the command. The first thing to verify is the layer 1 status. Notice that it is active on both routers, which means that the router senses the 2B1Q line coding of the BRI circuit.

The next thing to look for is the layer 2 status. Note that on both routers a TEI has been assigned for both channels. The last thing to check is the SPID status. On both routers, the SPIDs have been sent to the switch and are valid.

```
R6#show isdn status
The current ISDN Switchtype = basic-ni1
ISDN BRI1/0  interface
    Layer 1 Status:
        ACTIVE
```

```
    Layer 2 Status:
        TEI = 64, State = MULTIPLE_FRAME_ESTABLISHED
        TEI = 65, State = MULTIPLE_FRAME_ESTABLISHED
    Spid Status:
        TEI 64, ces = 1, state = 5(init)
            spid1 configured, no LDN, spid1 sent, spid1 valid
            Endpoint ID Info: epsf = 0, usid = 70, tid = 1
        TEI 65, ces = 2, state = 5(init)
            spid2 configured, no LDN, spid2 sent, spid2 valid
            Endpoint ID Info: epsf = 0, usid = 70, tid = 2
    Layer 3 Status:
        0 Active Layer 3 Call(s)
    Activated dsl 0 CCBs = 1
        CCB: callid=0x0, sapi=0, ces=1, B-chan=0
    Total Allocated ISDN CCBs = 1

R1#show isdn status
Global ISDN Switchtype = basic-ni
ISDN BRI0 interface
        dsl 0, interface ISDN Switchtype = basic-ni
    Layer 1 Status:
        ACTIVE
    Layer 2 Status:
        TEI = 64, Ces = 1, SAPI = 0, State = MULTIPLE_FRAME_ESTABLISHED
        TEI = 65, Ces = 2, SAPI = 0, State = MULTIPLE_FRAME_ESTABLISHED
    Spid Status:
        TEI 64, ces = 1, state = 8(established)
            spid1 configured, no LDN, spid1 sent, spid1 valid
            Endpoint ID Info: epsf = 0, usid = 70, tid = 1
        TEI 65, ces = 2, state = 5(init)
            spid2 configured, no LDN, spid2 sent, spid2 valid
            Endpoint ID Info: epsf = 0, usid = 70, tid = 2
    Layer 3 Status:
        0 Active Layer 3 Call(s)
    Activated dsl 0 CCBs = 0
    The Free Channel Mask:  0x80000003
    Total Allocated ISDN CCBs = 0
```

The question states that R6 should call R1 when the serial connection to R5 is lost. The backup should be invoked after five seconds and return to the primary once it has been up for 20 seconds. To accomplish this, you must use the backup interface command on R6 and set the backup delay accordingly.

The following command configures R6 so that the ISDN circuits back up the serial interface:

```
R6(config)#int s0/0
R6(config-if)#back interface bri 1/0
```

The question also specifies the time before switching to and from backup should be five and 20 seconds, respectively. The following command configures R6 so that it switches to backup after the serial interface has been down for five seconds and returns to the primary connection after it has been up for 20 seconds:

```
R6(config-if)#backup delay 5 20
```

Now that you have configured the BRI interface to back up the serial interface, you need to do some additional configurations to the ISDN interface.

a. Configure the encapsulation protocol for the interface to PPP. Enable multilink PPP on this interface so that both ISDN B channels can be used. Multilink PPP provides a method for spreading traffic across multiple links (in our case, B channels) while providing packet fragmentation and reassembly, proper sequencing, and load balancing on inbound and outbound traffic. Set the dialer load to 1 so that the second channel is brought up immediately. The dialer load threshold configures the interface to place a second call based on the traffic load on the primary B channel. The software monitors the traffic load and computes a moving average. If this average exceeds the value you set on the first B channel, the secondary line is activated.

```
R6(config-if)#encapsulation ppp
R6(config-if)#ppp multilink
R6(config-if)#dialer load-threshold 1

R1(config-if)#encapsulation ppp
R1(config-if)#ppp multilink
R1(config-if)#dialer load-threshold 1
```

b. Specify a dialer map on each router to map the destination name and directory number to an IP address:

```
R6(config-if)#dialer map ip 152.1.1.77 name R1 broadcast 8995101

R1(config-if)#dialer map ip 152.1.1.78 name R6 class dial1 broadcast 8995201
```

c. The third step is to specify interesting traffic that will cause the router to dial and keep the connection active. Any packet that matches the dialer group's specified trigger causes a connection to be made. The dialer group number is tied to a dialer list, which specifies the protocol and source destination addresses that are considered interesting:

```
R6(config-if)#dialer-group 1
R6(config)#dialer-list 1 protocol ip list 2
R6(config)#access-list 2 permit any

R1(config-if)#dialer-group 1
R1(config)#dialer-list 1 protocol ip list 1
R1(config)#access-list 1 permit any
```

d. Configure PPP callback on R6. This enables a router to request that a dial-up peer router call back. The callback feature can be used to control access and toll costs between the routers. When PPP callback is configured on the participating router, the calling router or client (in your case, R6) passes authentication information to R1 (the callback server), which uses the host name and dial string authentication information to determine whether to place a return call. If the authentication is successful, R1 disconnects and then places a return call to R6. The remote username of the return call is used to associate it with the initial call so that packets can be transmitted.

Both routers must be configured for PPP callback. R6 functions as a callback client and R1 is configured as a callback server. The callback client must be configured to initiate PPP callback requests, and the callback server must be configured to accept PPP callback requests and place return calls.

The **ppp callback request** command enables the interface to request ppp callbacks:

```
R6(config-if)#ppp callback request
```

The ppp authentication command enables chap or pap authentication on the interface. PPP supports two authentication protocols: the *Password Authentication Protocol* (PAP) and the *Challenge Handshake Authentication Protocol* (CHAP).

If PAP is used for authentication, the dialing router sends an authentication request to the server containing a username and password. This information is passed in the clear. The server compares this username and password to the list of defined users. If it matches, the authentication request is accepted andthe Cisco IOS software sends an authentication acknowledgment.

If CHAP is used, the server sends a request or "challenges" the remote device to respond. The challenge packet consists of an ID, a random number, and the host name of the local router. When the remote device receives the challenge packet, it concatenates the ID, the remote device's password, and the random number. It then encrypts it using the remote device's password. The remote device sends the results back to the server, along with the name associated with the password used in the encryption process. When the server receives the response, it uses the name it received to retrieve a password in its list of defined users. The retrieved password is the same password the remote device used in its encryption process. The access server then de-encrypts the concatenated information with the newly retrieved password. If the result matches the result sent in the response packet, authentication succeeds.

The benefit of using CHAP authentication is that the remote device's password is never transmitted in clear text. This prevents other devices from stealing the password and gaining illegal access to the network.

```
R6(config-if)#ppp authentication chap
R1(config-if)#ppp authentication chap
```

The following commands create a username-based authentication system on the router. The username and password are used during the authentication process:

```
R1(config)#username R6 password cisco
R6(config)#username R1 password cisco
```

The following command configures the interface to accept PPP callback requests:

```
R1(config-if)#ppp callback accept
```

This command enables callback security:

```
R1(config-if)#dialer callback-secure
```

You must configure a dialer map class for PPP callback, which identifies which router to return the call to by looking up the authenticated hostname in the dialer map statement:

```
R1(config)#map-class dialer dial1
R1(config-map-class)#dialer callback-server username
```

Once PPP callback is configured, use the **debug ppp authentication** and the **debug isdn q931** command on R6 to verify that it is working properly. The following is truncated output from the debugs after the cable was pulled from R6:

```
% UPDOWN: Interface Serial0/0, changed state to down ←cable is pulled
02:56:15: BRI1/0: Dialing cause ip (s=152.1.1.78, d=224.0.0.5) ←OPSF causes the
                                                                 dialing because it is
                                                                 interesting traffic
02:56:15: BRI1/0: Attempting to dial 8995101←R6 dials R1
%LINK-3-UPDOWN: Interface BRI1/0, changed state to up
%ISDN-6-LAYER2UP: Layer 2 for Interface BR1/0 , TEI 64 changed to up
02:56:15: ISDN BR1/0: TX ->  INFORMATION pd = 8  callref = (null)
        SPID Information i = '5201'
02:56:15: ISDN BR1/0: TX ->  SETUP pd = 8  callref = 0x05
02:56:15:         Bearer Capability i = 0x8890
02:56:15:         Channel ID i = 0x83
02:56:15:         Keypad Facility i = '8995101'
02:56:15: ISDN BR1/0: RX <-  INFORMATION pd = 8  callref = (null)
        ENDPOINT IDent i = 0xF081
02:56:15: ISDN BR1/0: Received EndPoint ID
02:56:15: ISDN BR1/0: RX <-  CALL_PROC pd = 8  callref = 0x85
02:56:15:         Channel ID i = 0x89
02:56:15: ISDN BR1/0: RX <-  CONNECT pd = 8  callref = 0x85
02:56:15:         Channel ID i = 0x89
%LINK-3-UPDOWN: Interface BRI1/0:1, changed state to up
02:56:15: BR1/0 :1 PPP: Treating connection as a callout
02:56:15: ISDN BR1/0: TX ->  CONNECT_ACK pd = 8  callref = 0x05
%ISDN-6-LAYER2UP: Layer 2 for Interface BR1/0, TEI 65 changed to up
02:56:15: ISDN BR1/0: TX ->  INFORMATION pd = 8  callref = (null)
        SPID Information i = '5202'
02:56:15: BR1/0:1 PPP: Phase is AUTHENTICATING, by both ←Two way chap authentication
02:56:15: BR1/0:1 CHAP: O CHALLENGE id 8 len 23 from "R6"
02:56:15: BR1/0:1 CHAP: I CHALLENGE id 8 len 23 from "R1"
02:56:15: BR1/0:1 CHAP: O RESPONSE id 8 len 23 from "R6"
02:56:15: BR1/0:1 CHAP: I SUCCESS id 8 len 4
02:56:15: BR1/0:1 CHAP: I RESPONSE id 8 len 23 from "R1"
```

```
02:56:15: BR1/0:1 CHAP: O SUCCESS id 8 len 4
02:56:15: BRI1/0:1: Callback negotiated - waiting for server disconnect ←PPP callback
                                                                        is negotiated
02:56:15: ISDN BR1/0: RX <-  INFORMATION pd = 8  callref = (null)
        ENDPOINT IDent i = 0xF082
02:56:15: ISDN BR1/0: Received EndPoint ID
%LINK-3-UPDOWN: Interface Virtual-Access1, changed state to up
02:56:15: Vi1 PPP: Treating connection as a callout
02:56:15: ISDN BR1/0: RX <-  DISCONNECT pd = 8  callref = 0x85
02:56:15:        Cause i = 0x8290 - Normal call clearing ←Call is disconnected by R1
%LINK-3-UPDOWN: Interface BRI1/0:1, changed state to down
02:56:15: Callback client for R1 8995101 created
02:56:15: BRI1/0:1: disconnecting call
02:56:15: ISDN BR1/0: TX ->  RELEASE pd = 8  callref = 0x05
%LINK-3-UPDOWN: Interface Virtual-Access1, changed state to down
02:56:15: ISDN BR1/0: RX <-  RELEASE_COMP pd = 8  callref = 0x85
02:56:30: ISDN BR1/0: RX <-  SETUP pd = 8  callref = 0x08 ←Incoming call from R1 to R6
02:56:30:        Bearer Capability i = 0x8890
02:56:30:        Channel ID i = 0x89
02:56:30:        Signal i = 0x40 - Alerting on - pattern 0
02:56:30:        Called Party Number i = 0xC1, '8995201'←R1 Directory Number
02:56:32: BR1/0:1 CHAP: I SUCCESS id 9 len 4
02:56:34: BR1/0:2 CHAP: I SUCCESS id 4 len 4 ←Two way chap authentication
02:56:34: BRI1/0:2: No callback negotiated
%LINEPROTO-5-UPDOWN: Line protocol on Interface BRI1/0:2, changed state to up
02:56:34: Callback received on Virtual-Access1 from R1 8995101
02:56:34: Freeing callback to R1 8995101
%ISDN-6-CONNECT: Interface BRI1/0:2 is now connected to 8995101 R1
```

32. *Run OSPF on the ISDN circuit between R1 and R6, and make sure that all the routes are learned via R6 during backup.*

The following commands place the BRI interface of R6 and R1 in area 2:

```
R6(config-router)#network 152.1.1.76 0.0.0.3 area 2
R1(config-router)#network 152.1.1.76 0.0.0.3 area 2
```

Notice, however, that R1 is not forming an adjacency after the ISDN connection is established. In order to troubleshoot this, use the **debug ip ospf** events command on R1. Shown below is the output from the debug. Notice that R1 is receiving a Hello packet with the NSSA bit set, indicating that R6 is a NSSA router. Remember from the earlier discussion if the options in the Hello packet do not match, the routers will not form an adjacency.

```
03:07:28: OSPF: Hello from 152.1.1.77 with mismatched NSSA option bit
03:07:35: OSPF: service_maxage: Trying to delete MAXAGE LSA
```

In order to correct this, area 2 on R1 needs to be configured as an NSSA area. The following command configures area 2 on R1 as an NSSA:

```
R1(config-router)#area 2 nssa
```

The second part of the question specifies that R6 should have visibility to all networks while in dial backup. To verify this, the routing table is displayed on R6; the following is the output. Notice that only the loopback interfaces on R1 are seen. The reason for this is area 2 is now discontinuous from area 0, so a virtual link will need to be added.

```
R6#show ip route
Codes: C - connected, S - static, I - IGRP, R - RIP, M - mobile, B - BGP
       D - EIGRP, EX - EIGRP external, O - OSPF, IA - OSPF inter area
       N1 - OSPF NSSA external type 1, N2 - OSPF NSSA external type 2
       E1 - OSPF external type 1, E2 - OSPF external type 2, E - EGP
       i - IS-IS, L1 - IS-IS level-1, L2 - IS-IS level-2, * - candidate default
       U - per-user static route, o - ODR

Gateway of last resort is not set


     152.1.0.0/16  is variably subnetted, 6 subnets, 4 masks
C       152.1.11.0/24  is directly connected, Ethernet0/0
C       152.1.12.0/24  is directly connected, Loopback0
C       152.1.1.76/30  is directly connected, BRI1/0
C       152.1.1.77/32  is directly connected, BRI1/0
O E2    152.1.1.65/32  [110/20] via 152.1.1.77, 00:00:19, BRI1/0
O E2    152.1.1.96/27  [110/20] via 152.1.1.77, 00:00:19, BRI1/0
```

The following commands add a virtual link connecting area 2 to area 0. Area 0 is also configured for authentication, so the virtual link must be configured for authentication as well.

```
R1(config-router)#area 3 virtual-link 152.1.3.1
R1(config-router)#area 0 authentication message-digest
R1(config-router)#area 3 virtual-link 152.1.3.1 message-digest-key 1 md5 cisco0

R3(config-router)#area 3 virtual-link 152.1.1.97 message-digest-key 1 md5 cisco0
```

Now display the routing table on R6, which is shown here. Notice all of the routes are now visible.

```
R6#show ip route
Codes: C - connected, S - static, I - IGRP, R - RIP, M - mobile, B - BGP
       D - EIGRP, EX - EIGRP external, O - OSPF, IA - OSPF inter area
       N1 - OSPF NSSA external type 1, N2 - OSPF NSSA external type 2
       E1 - OSPF external type 1, E2 - OSPF external type 2, E - EGP
       i - IS-IS, L1 - IS-IS level-1, L2 - IS-IS level-2, * - candidate default
       U - per-user static route, o - ODR

Gateway of last resort is not set

     152.1.0.0/16  is variably subnetted, 10 subnets, 7 masks
O IA    152.1.1.128/25  [110/2154] via 152.1.1.77, 00:04:09, BRI1/0
C       152.1.11.0/24   is directly connected, Ethernet0/0
C       152.1.12.0/24   is directly connected, Loopback0
O IA    152.1.1.0/25    [110/2208] via 152.1.1.77, 00:04:09, BRI1/0
O IA    152.1.1.0/26    [110/1571] via 152.1.1.77, 00:06:04, BRI1/0
O IA    152.1.2.0/30    [110/2736] via 152.1.1.77, 00:04:09, BRI1/0
```

```
O IA     152.1.1.80/29    [110/2144] via 152.1.1.77, 00:08:56, BRI1/0
C        152.1.1.76/30    is directly connected, BRI1/0
O E2     152.1.1.65/32    [110/20] via 152.1.1.77, 00:08:57, BRI1/0
O E2     152.1.1.96/27    [110/20] via 152.1.1.77, 00:08:57, BRI1/0
```

33. *Make sure that periodic OSPF Hellos and LSAs are suppressed and the link is only brought up by OSPF when a topology change occurs.*

To prevent periodic Hellos and LSAs from causing the ISDN link from coming up, the OSPF demand circuit is used. The OSPF on demand circuit is an enhancement to the OSPF protocol that enables efficient operation over demand circuits like ISDN, X.25, SVCs, and dial-up lines. This feature supports RFC 1793, Extending OSPF to Support Demand Circuits.

If demand circuits are not used, OSPF periodic Hello and LSA updates would be exchanged between routers that connect the on-demand link, even when no changes occurred.

The OSPF demand circuit suppresses periodic Hellos, and periodic refreshes of LSAs are not flooded over the demand circuit. Therefore, the link is only brought up when LSAs are exchanged for the first time, or when a change occurs in the information they contain. This enables the underlying data link layer to be closed when the network topology is stable.

This feature is useful when you want to connect telecommuters or branch offices to an OSPF backbone at a central site. In this case, OSPF for on-demand circuits enables the benefits of OSPF over the entire domain without excess connection costs. Periodic refreshes of Hello updates, LSA updates, and other protocol overhead are prevented from enabling the on-demand circuit when no "real" data is transmitted.

Overhead protocols such as Hellos and LSAs are transferred over the on-demand circuit only upon initial setup and when they reflect a change in the topology. This means that critical changes to the topology that require new SPF calculations are transmitted in order to maintain network topology integrity. Periodic refreshes that do not include changes, however, are not transmitted across the link.

Because this is a point-to-point topology, only one end of the demand circuit must be configured with the **ospf demand circuit** command. The following command configures the ISDN link between R1 and R6 as an OSPF demand circuit:

```
R6(config)#int bri 1/0
R6(config-if)#ip ospf demand-circuit
```

To verify that the link is configured as an OSPF demand circuit, use the **show ip ospf interface** command. The following is the output from the command on R6. Notice that the BRI interface is a demand circuit and is configured not to age the LSAs. Under normal conditions, the OSPF LSAs will age after 30 minutes and will need to be refreshed. This feature prevents OSPF LSAs from aging out of the database.

```
R6#show ip ospf interface bri 1/0
BRI1/0  is up, line protocol is up (spoofing)
  Internet Address 152.1.1.78/30, Area 2
  Process ID 64, Router ID 152.1.12.1, Network Type POINT_TO_POINT, Cost: 1562
  Configured as demand circuit.
  Run as demand circuit.
  DoNotAge LSA allowed.
  Transmit Delay is 1 sec, State POINT_TO_POINT,
  Timer intervals configured, Hello 10, Dead 40, Wait 40, Retransmit 5
    Hello due in 00:00:00
  Neighbor Count is 1, Adjacent neighbor count is 1
    Adjacent with neighbor 152.1.1.97  (Hello suppressed)
  Suppress hello for 1 neighbor(s)
```

To verify that this is working properly, force R6 into backup by disconnecting the serial cable. R6 should dial R1 and then disconnect after the idle timer expires. The following is the routing table on R6 after the idle timer expires. Even though the ISDN link is not active, the routes are still in the table:

```
R6#show ip route
Codes: C - connected, S - static, I - IGRP, R - RIP, M - mobile, B - BGP
       D - EIGRP, EX - EIGRP external, O - OSPF, IA - OSPF inter area
       N1 - OSPF NSSA external type 1, N2 - OSPF NSSA external type 2
       E1 - OSPF external type 1, E2 - OSPF external type 2, E - EGP
       i - IS-IS, L1 - IS-IS level-1, L2 - IS-IS level-2, * - candidate default
       U - per-user static route, o - ODR

Gateway of last resort is not set

     152.1.0.0/16  is variably subnetted, 10 subnets, 7 masks
O IA    152.1.1.128/25  [110/2154] via 152.1.1.77, 00:05:39, BRI1/0
C       152.1.11.0/24   is directly connected, Ethernet0/0
C       152.1.12.0/24   is directly connected, Loopback0
O IA    152.1.1.0/26    [110/1571] via 152.1.1.77, 00:05:39, BRI1/0
O IA    152.1.1.0/25    [110/2208] via 152.1.1.77, 00:05:39, BRI1/0
O IA    152.1.2.0/30    [110/2736] via 152.1.1.77, 00:05:39, BRI1/0
O IA    152.1.1.80/29   [110/2144] via 152.1.1.77, 00:05:39, BRI1/0
C       152.1.1.76/30   is directly connected, BRI1/0
O E2    152.1.1.65/32   [110/20] via 152.1.1.77, 00:05:39, BRI1/0
O E2    152.1.1.96/27   [110/20] via 152.1.1.77, 00:05:39, BRI1/0
```

Display the OSPF neighbors on R6, and notice that an adjacency has been formed with R1, but the dead timer is not present. Under normal circumstances, if R6 did not receive a Hello packet from R1 within the dead interval, the neighbor would be declared down. Because OSPF demand circuits suppress Hellos, the dead interval is not used:

```
R6# show ip ospf neighbor

Neighbor ID     Pri   State           Dead Time   Address         Interface
152.1.1.97        1   FULL/  -            -        152.1.1.77      BRI1/0
```

34. *Because R6 is attached to the backbone, it should advertise a default route into the domain.*

By default, in normal areas, routers don't generate default routes. To have an OSPF router generate a default route, you must use the default-information originate command. This generates an external type-2 link with link-state ID 0.0.0.0 and network mask 0.0.0.0, which makes the router an ASBR.

A default route can be injected into a normal area in two ways. If the ASBR already has the default route, you can advertise 0.0.0.0 into the area. If the ASBR doesn't have the route, you can add the keyword **always** to the default-information originate command, and then advertise 0.0.0.0. Because R6 does not have a default route for itself, you must use the always keyword.

The following is the command to redistribute a default route into the domain from R6:

```
R6(config)#router ospf 64
R6(config-router)#default-information originate always
```

To verify that this is working properly, show the routing table on R1. The following is the output from the command. Note that that a default route has been added to R1's routing table:

```
R1#show ip route
Codes: C - connected, S - static, I - IGRP, R - RIP, M - mobile, B - BGP
       D - EIGRP, EX - EIGRP external, O - OSPF, IA - OSPF inter area
       N1 - OSPF NSSA external type 1, N2 - OSPF NSSA external type 2
       E1 - OSPF external type 1, E2 - OSPF external type 2, E - EGP
       i - IS-IS, L1 - IS-IS level-1, L2 - IS-IS level-2, * - candidate default
       U - per-user static route, o - ODR, P - periodic downloaded static route
       T - traffic engineered route

Gateway of last resort is 152.1.1.83 to network 0.0.0.0

     152.1.0.0/16  is variably subnetted, 12 subnets, 7 masks
O       152.1.1.128/25  [110/592] via 152.1.1.83, Serial0.1
O IA    152.1.2.128/25  [110/1302] via 152.1.1.83, Serial0.1
O IA    152.1.11.0/24   [110/1248] via 152.1.1.83, Serial0.1
O IA    152.1.1.0/25    [110/646] via 152.1.1.83, Serial0.1
C       152.1.1.0/26    is directly connected, FastEthernet0
O       152.1.2.0/30    [110/1174] via 152.1.1.83, Serial0.1
C       152.1.1.80/29   is directly connected, Serial0.1
O E2    152.1.1.72/32   [110/20] via 152.1.1.2, FastEthernet0
C       152.1.1.76/30   is directly connected, BRI0
C       152.1.1.65/32   is directly connected, Loopback0
O E2    152.1.1.69/32   [110/20] via 152.1.1.83, Serial0.1
C       152.1.1.96/27   is directly connected, Loopback1
O*E2 0.0.0.0/0 [110/1] via 152.1.1.83, Serial0.1
```

35. *Configure R6 to run IGRP to the backbone. Only send IGRP messages out interface Ethernet 1/0.*

Configuring IGRP is a two-step process. First, enable the routing process and then add interfaces under the process that will run the routing protocol.

a. The following command enables IGRP routing process 64 on R6:

```
R6(config)#router igrp 64
```

b. This router configuration command specifies a list of networks that the IGRP routing process will run on. This command sends IGRP updates to the interfaces that are specified. If an interface's network is not specified, it will not be advertised in any IGRP update:

```
R6(config-router)#network 152.1.0.0
```

Because IGRP is a classful protocol, when it is enabled, it is turned on for the classful network of 152.1.0.0. This encompasses all five subnets of 152.1.0.0 on R6. The passive interface command enables the user to turn off IGRP advertisements on a particular interface (subnet).

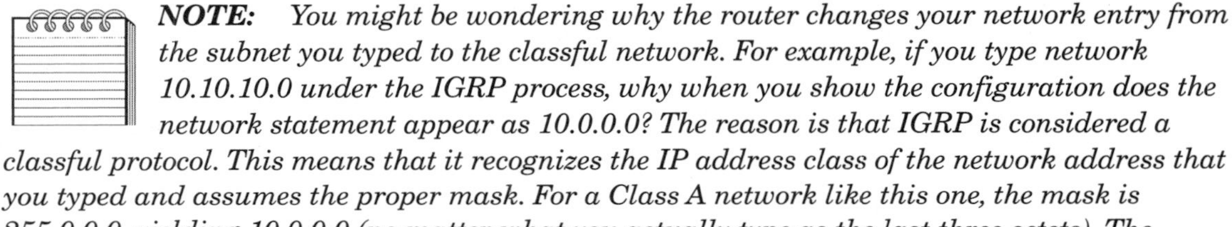

NOTE: *You might be wondering why the router changes your network entry from the subnet you typed to the classful network. For example, if you type network 10.10.10.0 under the IGRP process, why when you show the configuration does the network statement appear as 10.0.0.0? The reason is that IGRP is considered a classful protocol. This means that it recognizes the IP address class of the network address that you typed and assumes the proper mask. For a Class A network like this one, the mask is 255.0.0.0, yielding 10.0.0.0 (no matter what you actually type as the last three octets). The network statement tells the routing protocol to route on the interfaces where the network address matches the one specified in the network statement.*

The question also states that IGRP messages should only be sent out Ethernet 1/0. To do this, you must use the passive interface command under the routing process:

```
R6(config)#router igrp 64
R6(config-router)#passive-interface ethernet 0/0
R6(config-router)#passive-interface serial 0/0
R6(config-router)#passive-interface bri 1/0
R6(config-router)#passive-interface loopback 0
```

To verify that this is configured properly, use the **show ip protocols** command. The following is the truncated output. Notice that IGRP is enabled for network 152.1.0.0 and all interfaces other than Ethernet 1/0 are passive.

```
Routing Protocol is "igrp 64"
  Sending updates every 90 seconds, next due in 8 seconds
  Invalid after 270 seconds, hold down 280, flushed after 630
  Outgoing update filter list for all interfaces is not set
  Incoming update filter list for all interfaces is not set
  Default networks flagged in outgoing updates
  Default networks accepted from incoming updates
  IGRP metric weight K1=1, K2=0, K3=1, K4=0, K5=0
  IGRP maximum hopcount 100
  IGRP maximum metric variance 1
  Redistributing: igrp 64
```

```
  Routing for Networks:
    152.1.0.0
  Passive Interface(s):
    Ethernet0/0
    Serial0/0
    BRI1/0
    Loopback0
  Routing Information Sources:
    Gateway         Distance      Last Update
  Distance: (default is 100)
```

36. *R6 should only take in the following networks from the backbone:*

```
152.12.0.0
152.13.0.0
152.14.0.0
152.15.0.0
```

Verify the networks that R6 is learning via IGRP from the backbone with the command **show ip route igrp.** The following is the output from that command. R6 is receiving three additional networks from the backbone (152.16.0.0, 152.10.0.0 and 152.11.0.0):

```
R6#show ip route igrp
I     152.16.0.0/16 [100/1600] via 152.1.0.2, 00:00:20, Ethernet1/0
I     152.10.0.0/16 [100/1600] via 152.1.0.2, 00:00:20, Ethernet1/0
I     152.11.0.0/16 [100/1600] via 152.1.0.2, 00:00:20, Ethernet1/0
I     152.12.0.0/16 [100/1600] via 152.1.0.2, 00:00:20, Ethernet1/0
I     152.13.0.0/16 [100/1600] via 152.1.0.2, 00:00:20, Ethernet1/0
I     152.14.0.0/16 [100/1600] via 152.1.0.2, 00:00:20, Ethernet1/0
I     152.15.0.0/16 [100/1600] via 152.1.0.2, 00:00:20, Ethernet1/0
```

In order to filter incoming updates from the backbone, you must use a distribute list on R6. This a two-step process. First, an access list must be set up to identify and permit or deny a network based on the network address. Then the access list must be applied to routing updates through the use of a distribute list.

a. Define an access list on R6 denying networks 152.10.0.0, 152.11.0.0, and 152.16.0.0:

```
R6(config)#access-list 3 deny 152.10.0.0
R6(config)#access-list 3 deny 152.11.0.0
R6(config)#access-list 3 deny 152.16.0.0
R6(config)#access-list 3 permit any
```

b. Apply the access list to routing updates through the use of a distribute list under the IGRP routing process:

```
R6(config-router)#distribute-list 3 in ethernet 1/0
```

Verify that the routes are being filtered properly with the command **show ip route igrp.** The following is the output from that command. R6 is receiving only three networks; the rest are no longer in the table.

```
R6#show ip route igrp
I     152.12.0.0/16  [100/1600] via 152.1.0.2, 00:00:30, Ethernet1/0
I     152.13.0.0/16  [100/1600] via 152.1.0.2, 00:00:30, Ethernet1/0
I     152.14.0.0/16  [100/1600] via 152.1.0.2, 00:00:30, Ethernet1/0
I     152.15.0.0/16  [100/1600] via 152.1.0.2, 00:00:30, Ethernet1/0
```

37. *Advertise the routes learned from the backbone in the OSPF domain. Advertise as few of the networks as needed.*

In order to advertise the networks into the OSPF domain, you need to redistribute IGRP into OSPF. The following command advertises the IGRP learned routes into OSPF process 64 with a metric of 1000:

```
R6(config-router)#redistribute igrp 64 metric 1000
```

Because IGRP and OSPF use different metrics for calculating routes, it is necessary to tell R6 what it should set the metric to when it redistributes IGRP routes into OSPF. This can be done in several ways. The simplest is to add the metric after the redistribution command.

To verify that the routes are being redistributed properly, display the routing table on R1. The following is the output from the command. Note that R1 sees the four networks as OSPF External Type 2 routes.

```
R1#show ip route
Codes: C - connected, S - static, I - IGRP, R - RIP, M - mobile, B - BGP
       D - EIGRP, EX - EIGRP external, O - OSPF, IA - OSPF inter area
       N1 - OSPF NSSA external type 1, N2 - OSPF NSSA external type 2
       E1 - OSPF external type 1, E2 - OSPF external type 2, E - EGP
       i - IS-IS, L1 - IS-IS level-1, L2 - IS-IS level-2, * - candidate default
       U - per-user static route, o - ODR, P - periodic downloaded static route
       T - traffic engineered route

Gateway of last resort is 152.1.1.83 to network 0.0.0.0

     152.1.0.0/16 is variably subnetted, 12 subnets, 7 masks
O       152.1.1.128/25  [110/592] via 152.1.1.83, Serial0.1
O IA    152.1.2.128/25  [110/1302] via 152.1.1.83, Serial0.1
O IA    152.1.11.0/24   [110/1248] via 152.1.1.83, Serial0.1
O IA    152.1.1.0/25    [110/646] via 152.1.1.83, Serial0.1
C       152.1.1.0/26    is directly connected, FastEthernet0
O       152.1.2.0/30    [110/1174] via 152.1.1.83, Serial0.1
C       152.1.1.80/29   is directly connected, Serial0.1
O E2    152.1.1.72/32   [110/20] via 152.1.1.2, FastEthernet0
C       152.1.1.76/30   is directly connected, BRI0
C       152.1.1.65/32   is directly connected, Loopback0
O E2    152.1.1.69/32   [110/20] via 152.1.1.83, Serial0.1
C       152.1.1.96/27   is directly connected, Loopback1
O E2    152.12.0.0/16   [110/1000] via 152.1.1.83, Serial0.1
O E2    152.13.0.0/16   [110/1000] via 152.1.1.83, Serial0.1
O E2    152.14.0.0/16   [110/1000] via 152.1.1.83, Serial0.1
O E2    152.15.0.0/16   [110/1000] via 152.1.1.83, Serial0.1
O*E2    0.0.0.0/0  [110/1] via 152.1.1.83, Serial0.1
```

The second part of the question requires that you summarize the four networks into one advertisement. All the 152.x.0.0 prefixes can be advertised as an aggregate using the summary address command. Using this command for OSPF causes an OSPF ASBR to advertise one external route as an aggregate for all the redistributed routes that are covered by the address. For OSPF, this command summarizes only routes from other routing protocols that are being redistributed into OSPF.

Multiple groups of addresses can be summarized for a given level learned from other routing protocols. The metric used to advertise the summary is the smallest metric of all the more specific routes. This helps reduce the size of the routing table.

The first thing you must do is determine which address you can use to advertise all the networks. The easiest way to do this is to write the network numbers that you are trying to summarize in binary and then determine which bit positions match. These will become ones in the mask. For this example, the first 14-bit positions match, leaving us with 255.252.0.0 or a 14-bit mask.

```
152          12          ←IP address
10011000     00001100   ←Binary representation of IP address

152          13
10011000     00001101   ←Binary representation of IP address

152          14
10011000     00001110   ←Binary representation of IP address

152          15
10011000     00001111   ←Binary representation of IP address
--------------------------------------------------------------
11111111     11111100   ←Binary representation of mask 255.252.0.0
```

The following command summarizes the four networks into one advertisement:

```
R6(config-router)#summary-address 152.12.0.0 255.252.0.0
```

To verify that the summarization is working properly, display the routing table on R1. The following is the output from the **show ip route** command. Note: R1 no longer has the four networks in its table:

```
R1#show ip route
Gateway of last resort is 152.1.1.83 to network 0.0.0.0

     152.1.0.0/16  is variably subnetted, 12 subnets, 7 masks
O       152.1.1.128/25  [110/592] via 152.1.1.83, Serial0.1
O IA    152.1.2.128/25  [110/1302] via 152.1.1.83, Serial0.1
O IA    152.1.11.0/24   [110/1248] via 152.1.1.83, Serial0.1
O IA    152.1.1.0/25    [110/646] via 152.1.1.83, Serial0.1
C       152.1.1.0/26    is directly connected, FastEthernet0
O       152.1.2.0/30    [110/1174] via 152.1.1.83, Serial0.1
C       152.1.1.80/29   is directly connected, Serial0.1
```

```
O E2     152.1.1.72/32    [110/20] via 152.1.1.2, FastEthernet0
C        152.1.1.76/30    is directly connected, BRI0
C        152.1.1.65/32    is directly connected, Loopback0
O E2     152.1.1.69/32    [110/20] via 152.1.1.83, Serial0.1
C        152.1.1.96/27    is directly connected, Loopback1
O*E2 0.0.0.0/0  [110/1] via 152.1.1.83, Serial0.1
O E2     152.12.0.0/14    [110/1000] via 152.1.1.83, Serial0.1
```

38. *In addition to 152.1.11.0 and 152.1.12.0, which are already being advertised into the IGRP backbone, R6 should advertise the following networks:*

```
152.1.1.0
152.1.2.0
```

When redistributing from a classless protocol like OSPF into a classful protocol like IGRP, problems can arise if VLSM is used in the OSPF domain. OSPF only redistributes routes into the IGRP domain that have the same subnet mask as the IGRP domain. Since the link between R6 and the backbone is using a 24-bit mask, only routes with a 24-bit mask will be redistributed.

In this scenario, if the redistribute command is only used on R6, then the backbone would not see these networks because the mask of these routes does not match the mask of the network used on the IGRP domain between R6 and the backbone.

There are two ways to propagate networks 152.1.1.0 and 152.1.2.0 into the IGRP domain. The first is with a static route. A static route using a 24-bit mask can be created on R6 for the two networks. The static route would then be redistributed into IGRP. This method cannot be used because static routes are not permitted in this case study.

The second method is to summarize the networks to the same subnet mask length as the IGRP domain. To do this, use the OSPF summary address command on R6. The following is the command used to summarize the two networks to a 24-bit mask:

```
R6(config-router)#summary-address 152.1.1.0 255.255.255.0
R6(config-router)#summary-address 152.1.2.0 255.255.255.0
```

Because the question states that only the two specific networks should be redistributed from OSPF into IGRP, a route map is used in conjunction with the redistribute command.

The first step is to define an access list permitting the specified networks:

```
R6(config)#access-list 6 permit 152.1.1.0
R6(config)#access-list 6 permit 152.1.2.0
```

The second step is to define a route map on the router that permits the specified networks defined in the access list:

```
R6(config)#route-map summary permit 10
R6(config-route-map)#match ip address 6
```

The third step is to apply the route map to the redistribute command under the routing process:

```
R6(config-router)#redistribute ospf 64 route-map summary
```

Because theoretically you don't have access to the backbone router, use the **debug ip igrp transactions** command on R6 to verify which networks are being advertised. The following is the output from the debug. R6 is advertising four networks:

```
R6#debug ip igrp transactions
04:10:01: IGRP: sending update to 255.255.255.255 via Ethernet1/0  (152.1.0.1)
04:10:01:       subnet 152.1.11.0, metric=1100
04:10:01:       subnet 152.1.12.0, metric=501
04:10:01:       subnet 152.1.1.0, metric=6676
04:10:01:       subnet 152.1.2.0, metric=6676
```

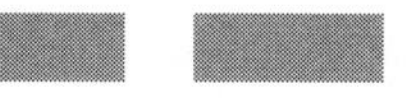

Complete Network Diagram

Figure 4-22 shows the complete network diagram with IP addressing. Table 4-7 shows the IP address work sheet.

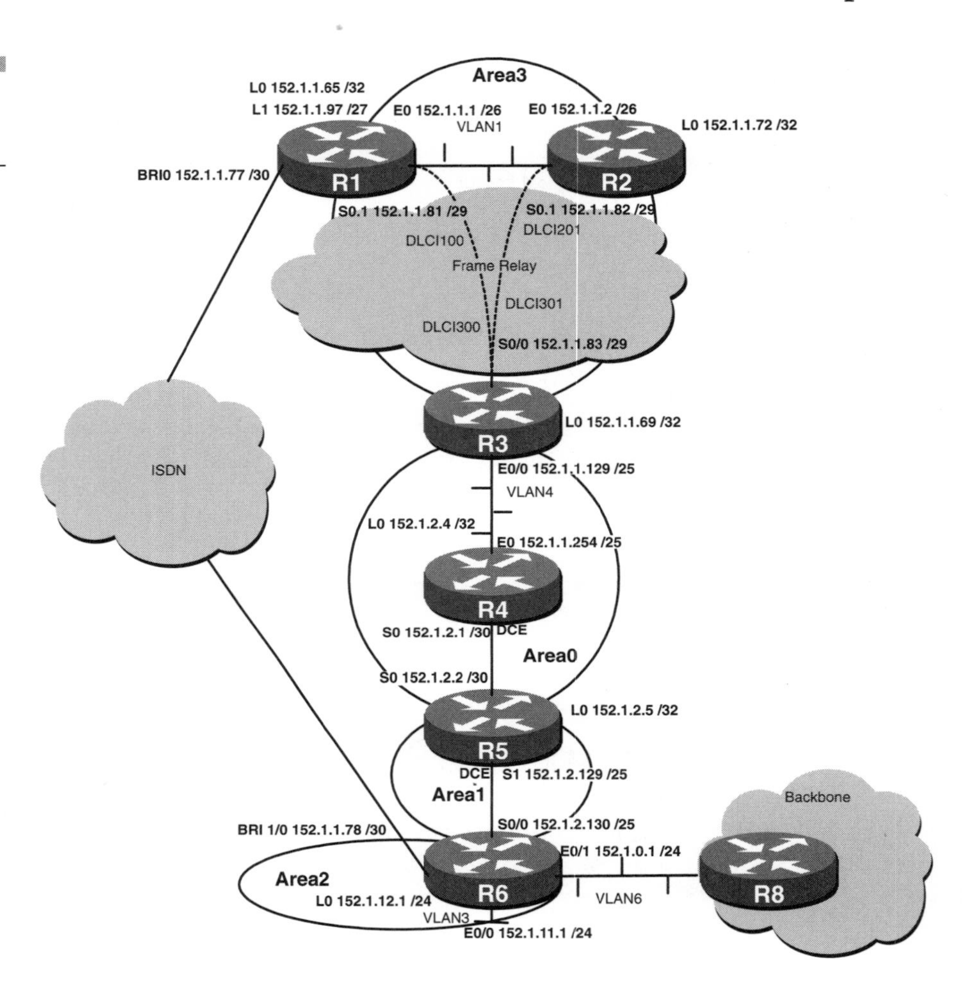

Figure 4-22 Network diagram with IP addressing

Table 4-7

IP Address Worksheet

IP Address	Mask	Subnet	Broadcast Address	Router	Interface
152.1.1.1	255.255.255.224	152.1.1.0	152.1.1.63	R1	E0
152.1.1.2	255.255.255.224	152.1.1.0	152.1.1.63	R2	E0
152.1.1.81	255.255.255.248	152.1.1.80	152.1.1.87	R1	S0.1
152.1.1.82	255.255.255.248	152.1.1.80	152.1.1.87	R2	S0.1
152.1.1.83	255.255.255.248	152.1.1.80	152.1.1.87	R3	S0
152.1.1.77	255.255.255.252	152.1.1.76	152.1.1.79	R1	BRI 1
152.1.1.78	255.255.255.252	152.1.1.76	152.1.1.79	R6	BRI 1
152.1.1.129	255.255.255.128	152.1.1.128	152.1.1.255	R3	E0
152.1.1.254	255.255.255.128	152.1.1.128	152.1.1.255	R4	E0
152.1.2.1	255.255.255.252	152.1.1.0	152.1.2.3	R4	S0
152.1.2.2	255.255.255.252	152.1.1.0	152.1.2.3	R5	S0
152.1.2.129	255.255.255.128	152.1.2.128	152.1.2.255	R5	S1
152.1.2.130	255.255.255.128	152.1.2.128	152.1.2.255	R6	S1
152.1.11.1	255.255.255.0	152.1.11.0	152.1.11.255	R6	E0/0
152.1.0.1	255.255.255.0	152.1.0.0	152.1.0.255	R6	E0/1
152.1.1.65	255.255.255.255			R1	L0
152.1.1.97	255.255.255.224	152.1.1.96	152.1.1.127	R1	L1
152.1.1.72	255.255.255.255			R2	L0
152.1.1.69	255.255.255.255			R3	L0
152.1.2.4	255.255.255.255			R4	L0
152.1.2.5	255.255.255.255			R5	L0
152.1.3.1	255.255.255.0	152.1.3.0	152.1.3.255	R3	L1
152.1.2.6	255.255.255.255			R4	L1

Router Configurations

R1

```
!
version 12.0
service timestamps debug uptime
service timestamps log uptime
no service password-encryption
no cdp run
!
hostname R1
!
!
username R6 password 0 cisco
memory-size iomem 25
ip subnet-zero
ip name-server 153.1.1.1
!
isdn switch-type basic-ni
!
!
!
interface Loopback0
 ip address 152.1.1.65 255.255.255.255
 no ip directed-broadcast
!
interface Loopback1
 ip address 152.1.1.97 255.255.255.224
 no ip directed-broadcast
!
interface Serial0
 no ip address
 no ip directed-broadcast
 encapsulation frame-relay IETF
 no fair-queue
 frame-relay lmi-type ansi
!
interface Serial0.1 point-to-point
 ip address 152.1.1.81 255.255.255.248
 no ip directed-broadcast
 ip ospf authentication-key cisco0
 ip ospf network non-broadcast
 ip ospf priority 0
 frame-relay interface-dlci 100
!
interface BRI0
 ip address 152.1.1.77 255.255.255.252
 no ip directed-broadcast
 encapsulation ppp
 dialer callback-secure
 dialer map ip 152.1.1.78 name R6 class dial1 broadcast 8995201
 dialer load-threshold 1 outbound
 dialer-group 1
 isdn switch-type basic-ni
```

```
 isdn spid1 5101
 isdn spid2 5102
 ppp callback accept
 ppp authentication chap
 ppp multilink
!
interface Ethernet0
 ip address 152.1.1.1 255.255.255.192
 no ip directed-broadcast
 ip ospf authentication-key cisco3
 ip ospf priority 0
 standby 1 ip 152.1.1.3
 standby 2 ip 152.1.1.4
 standby 1 priority 150
!
router ospf 64
 redistribute connected subnets route-map only_loopbacks
 network 152.1.1.0 0.0.0.63 area 3
 network 152.1.1.76 0.0.0.3 area 2
 network 152.1.1.80 0.0.0.7 area 3
 area 0 authentication message-digest
 area 2 nssa
 area 3 authentication
 area 3 virtual-link 152.1.3.1 message-digest-key 1 md5 cisco0
 auto-cost reference-bandwidth 900
!
ip classless
no ip http server
!
!
map-class dialer dial1
 dialer callback-server username
access-list 1 permit any
dialer-list 1 protocol ip list 1
route-map only_loopbacks permit 10
 match interface Loopback0 Loopback1
!
!
line con 0
 transport input none
line aux 0
line vty 0 4
 login
!
no scheduler allocate
end
```

R2

```
!
version 11.2
no service password-encryption
no service udp-small-servers
no service tcp-small-servers
no cdp run
```

```
!
hostname R2
!
!
!
interface Loopback0
 ip address 152.1.1.72 255.255.255.255
!
interface Ethernet0
 ip address 152.1.1.2 255.255.255.192
 ip ospf authentication-key cisco0
standby 1 ip 152.1.1.3
standby 2 ip 152.1.1.4
standby 2 priority 150
!
interface Serial0
 no ip address
 encapsulation frame-relay
 frame-relay lmi-type ansi
!
interface Serial0.1 point-to-point
 ip address 152.1.1.82 255.255.255.248
 ip ospf authentication-key cisco3
 ip ospf network non-broadcast
 ip ospf priority 0
 frame-relay interface-dlci 201
!
!
router ospf 64
 redistribute connected subnets route-map only_loopbacks
 network 152.1.1.80 0.0.0.7 area 3
 network 152.1.1.0 0.0.0.63 area 3
 area 3 authentication
 ospf auto-cost reference-bandwidth 900
!
no ip classless
route-map only_loopbacks permit 10
 match interface Loopback0
!
!
line con 0
line 1 16
line aux 0
line vty 0 4
 login
!
end
```

R3

```
!
version 11.1
service udp-small-servers
```

```
service tcp-small-servers
no cdp run
!
hostname R3
!
interface Loopback0
 ip address 152.1.1.69 255.255.255.255
!
interface Ethernet0/0
 ip address 152.1.1.129 255.255.255.128
 ip ospf message-digest-key 1 md5 cisco0
!
interface Serial0/0
 ip address 152.1.1.83 255.255.255.248
 encapsulation frame-relay IETF
 ip ospf authentication-key cisco3
 no frame-relay inverse-arp
 frame-relay lmi-type ansi
 frame-relay map ip 152.1.1.81 300 broadcast
 frame-relay map ip 152.1.1.82 301 broadcast
!
router ospf 64
 redistribute connected subnets route-map only_loopback0
 network 152.1.1.80 0.0.0.7 area 3
 network 152.1.1.128 0.0.0.127 area 0
 neighbor 152.1.1.82
 neighbor 152.1.1.81
 area 0 authentication message-digest
 area 3 authentication
 area 3 range 152.1.1.0 255.255.255.128
 area 3 virtual-link 152.1.1.97 message-digest-key 1 md5 cisco0
!
no ip classless
no cdp run
route-map only_loopback0 permit 10
 match interface Loopback0
!
!
line con 0
line aux 0
line vty 0 4
 login
!
end
```

R4

```
!
version 11.2
no service udp-small-servers
no service tcp-small-servers
no cdp run
!
```

```
hostname R4
!
!
ip host R1 2001 152.1.2.4
ip host R2 2002 152.1.2.4
ip host R3 2003 152.1.2.4
ip host R4 2004 152.1.2.4
ip host R5 2005 152.1.2.4
ip host R6 2006 152.1.2.4
ip host cat 2008 152.1.2.4
!
interface Loopback0
 ip address 152.1.2.4 255.255.255.255
!
interface Loopback1
 ip address 152.1.0.1 255.255.255.255
!
interface Ethernet0
 ip address 152.1.1.254 255.255.255.128
 ip ospf message-digest-key 1 md5 cisco0
!
interface Serial0
 ip address 152.1.2.1 255.255.255.252
 ip ospf message-digest-key 1 md5 cisco0
 clockrate 64000
!
router ospf 64
 network 152.1.1.128 0.0.0.127 area 0
 network 152.1.2.0 0.0.0.3 area 0
 distribute-list 1 in Ethernet0
 area 0 authentication message-digest
 ospf auto-cost reference-bandwidth 900
!
no ip classless
access-list 1 deny   152.1.1.72
access-list 1 deny   152.1.1.65
access-list 1 deny   152.1.1.69
access-list 1 deny   152.1.1.96 0.0.0.31
access-list 1 permit any
!
line con 0
line 1 16
 no exec
 transport input all
line aux 0
line vty 0 4
 login
!
end
```

R5

```
!
version 11.0
service udp-small-servers
```

```
service tcp-small-servers
no cdp run
!
hostname R5
!
!
!
interface Loopback0
 ip address 152.1.2.5 255.255.255.255
!

interface Serial0
 ip address 152.1.2.2 255.255.255.252
 ip ospf message-digest-key 1 md5 cisco0
!
interface Serial1
 ip address 152.1.2.129 255.255.255.128
 ip ospf authentication-key cisco1
 clockrate 64000
!
router ospf 64
 network 152.1.2.0 0.0.0.3 area 0
 network 152.1.2.128 0.0.0.127 area 1
 distribute-list 1 in Serial0
 area 1 authentication
 area 1 virtual-link 152.1.12.1 message-digest-key 1 md5 cisco0
 area 0 authentication message-digest
!
access-list 1 deny   152.1.1.72
access-list 1 deny   152.1.1.65
access-list 1 deny   152.1.1.69
access-list 1 deny   152.1.1.96 0.0.0.31
access-list 1 permit any
!
line con 0
line 1 16
 transport input all
line aux 0
 transport input all
line vty 0 4
 login
!
end
```

R6

```
!
version 11.2
service timestamps debug uptime
no service password-encryption
no service udp-small-servers
no service tcp-small-servers
no cdp run
!
hostname R6
```

```
!
!
username R1 password 0 cisco
ip subnet-zero
isdn switch-type basic-ni1
!
interface Loopback0
 ip address 152.1.12.1 255.255.255.0
!
interface Ethernet0/0
 ip address 152.1.11.1 255.255.255.0
!
interface Serial0/0
 backup delay 5 20
 backup interface BRI1/0
 ip address 152.1.2.130 255.255.255.128
 ip ospf authentication-key cisco1
!

interface Ethernet1/0
 ip address 152.1.0.1 255.255.255.0
!
!
interface BRI1/0
 ip address 152.1.1.78 255.255.255.252
 encapsulation ppp
 ip ospf demand-circuit
 isdn spid1 5201
 isdn spid2 5202
 dialer map ip 152.1.1.77 name R1 broadcast 8995101
 dialer load-threshold 1 outbound
 dialer-group 1
 no fair-queue
 ppp callback request
 ppp authentication chap
 ppp multilink
 hold-queue 75 in
!
router ospf 64
 summary-address 152.12.0.0 255.252.0.0
 summary-address 152.1.1.0 255.255.255.0
 summary-address 152.1.2.0 255.255.255.0
 redistribute igrp 64 metric 1000
 network 152.1.2.128 0.0.0.127 area 1
 network 152.1.1.76 0.0.0.3 area 2
 network 152.1.11.0 0.0.0.255 area 2
 default-information originate always
 distribute-list 1 in Serial0/0
 area 0 authentication message-digest
 area 1 authentication
 area 1 virtual-link 152.1.2.5 message-digest-key 1 md5 cisco0
 area 2 nssa
!
router igrp 64
 redistribute ospf 64 route-map summary
 passive-interface Ethernet0/0
 passive-interface Serial0/0
 passive-interface BRI1/0
 passive-interface Loopback0
```

```
 network 152.1.0.0
 default-metric 1500 10 255 255 1500
 distribute-list 3 in Ethernet1/0
!
no ip classless
access-list 1 deny   152.1.1.72
access-list 1 deny   152.1.1.65
access-list 1 deny   152.1.1.69
access-list 1 deny   152.1.1.96 0.0.0.31
access-list 1 permit any
access-list 2 permit any
access-list 3 deny   152.16.0.0
access-list 3 deny   152.10.0.0
access-list 3 deny   152.11.0.0
access-list 3 permit any
access-list 6 permit 152.1.1.0
access-list 6 permit 152.1.2.0
route-map summary permit 10
 match ip address 6
!
!
dialer-list 1 protocol ip list 2
!
line con 0
line aux 0
line vty 0 4
 login
!
end
```

R8

```
!
version 11.1
service slave-log
service udp-small-servers
service tcp-small-servers
!
hostname R8
!
!
ip subnet-zero
frame-relay switching
!
interface Loopback0
 ip address 152.10.10.10 255.255.255.0
!
interface Loopback1
 ip address 152.11.11.11 255.255.255.0
!
interface Loopback2
 ip address 152.12.12.12 255.255.255.0
!
interface Loopback3
 ip address 152.13.13.13 255.255.255.0
```

```
!
interface Loopback4
 ip address 152.14.14.14 255.255.255.0
!
interface Loopback5
 ip address 152.15.15.15 255.255.255.0
!
interface Loopback6
 ip address 152.16.16.16 255.255.255.0
!
interface Ethernet0
 ip address 152.1.0.2 255.255.255.0
 media-type 10BaseT
!
interface Ethernet1
 no ip address
 shutdown
!
interface Serial0
 no ip address
 encapsulation frame-relay IETF
 clockrate 64000
 frame-relay lmi-type ansi
 frame-relay intf-type dce
 frame-relay route 100 interface Serial3 300
!
interface Serial1
 no ip address
 encapsulation frame-relay IETF
 clockrate 64000
 frame-relay lmi-type ansi
 frame-relay intf-type dce
 frame-relay route 201 interface Serial3 301
!
interface Serial2
 no ip address
 encapsulation frame-relay IETF
 clockrate 64000
 frame-relay lmi-type ansi
 frame-relay intf-type dce
 frame-relay route 202 interface Serial4 302
!
interface Serial3
 no ip address
 encapsulation frame-relay IETF
 clockrate 64000
 frame-relay lmi-type ansi
 frame-relay intf-type dce
 frame-relay route 300 interface Serial0 100
 frame-relay route 301 interface Serial1 201
!
interface Serial4
 no ip address
 encapsulation frame-relay IETF
 clockrate 64000
```

```
 frame-relay lmi-type ansi
 frame-relay intf-type dce
 frame-relay route 302 interface Serial2 202
 frame-relay route 303 interface Serial5 703
!
interface Serial5
 no ip address
 encapsulation frame-relay IETF
 clockrate 64000
 frame-relay lmi-type ansi
 frame-relay intf-type dce
 frame-relay route 703 interface Serial4 303
!
interface Serial6
 no ip address
 shutdown
!
interface Serial7
 no ip address
 shutdown
!
router igrp 64
 network 152.1.0.0
 network 152.10.0.0
 network 152.11.0.0
 network 152.12.0.0
 network 152.13.0.0
 network 152.14.0.0
 network 152.15.0.0
 network 152.16.0.0
!
no ip classless
!
line con 0
line aux 0
line vty 0 4
 login
!
end
```

Case Study 4: IS-IS

IOS Requirements

Most of the routing protocols discussed became available in IOS 10.0, but all labs were performed using IOS 11.2.

Equipment Needed

The following equipment is needed to perform this case study. Table 4-8 lists the interface requirements for each router.

- One Catalyst 5500 Ethernet switch.
- One terminal server with one Ethernet and one serial port.
- One Cisco router with one serial port and one Ethernet port.
- One Cisco router with an Ethernet port and two serial ports.
- One Cisco router with an Ethernet port, one serial port, and one ISDN BRI interface.
- One Cisco router with one serial port and one ISDN BRI.
- Cisco IOS 11.2.
- A PC running a terminal emulation program for connecting to the console port of the terminal server
- Six Ethernet cables.
- Three Cisco DTE/DCE crossover cables.
- One Cisco Rolled cable.

Table 4-8

Interface requirements for each router

R1	1 Serial, 1 Ethernet
R2	1 Serial, 1 Ethernet (terminal server)
R3	2 Serials, 1 Ethernet
R4	1 Serial, 1 Ethernet, 1 BRI
R5	1 Serial, 1 BRI

Physical Connectivity Diagram

Figure 4-23 shows the physical connectivity for the routers in this case study.

Notes

- All IP addresses will be assigned from the class B network 152.1.0.0.
- No static or default routes should be used in this case study unless explicitly called for.
- Read through the entire case study before beginning.

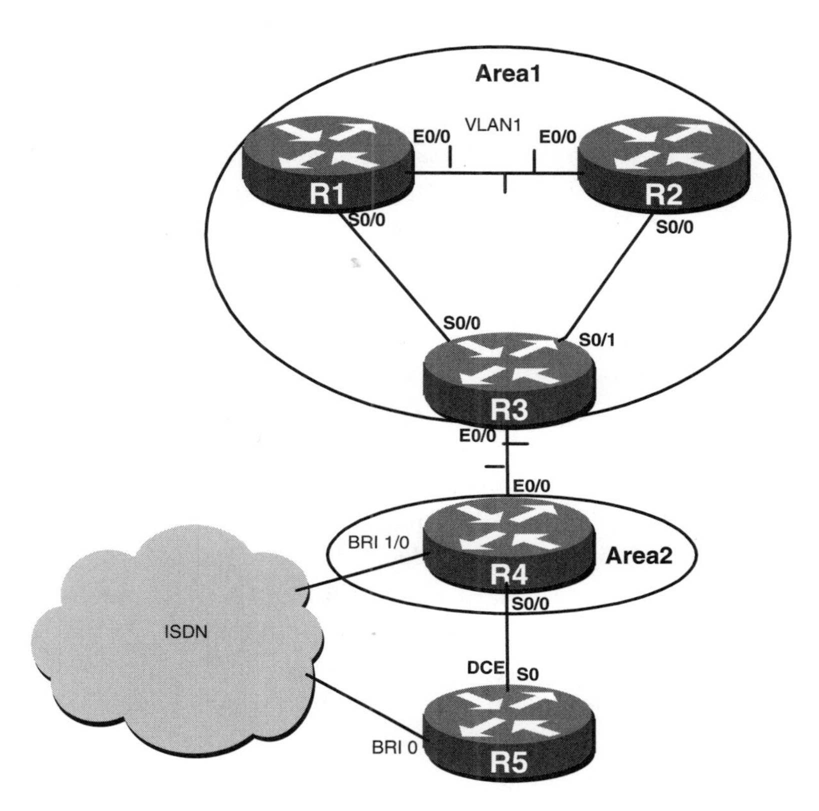

Figure 4-23
Network diagram

- Make sure you save your configurations regularly.
- All Level 1 passwords should be set to cisco; all Level 2 passwords should be set to ocsic.
- All area passwords should be set to cisco1 and all domain passwords should be set to ocsic1.
- The domain name on the Catalyst should be CCIE_Lab.

Questions

1. Cable the routers per Figure 4-23.
2. Erase any existing configuration from the routers and set each router's name. The name will be the corresponding router number; for example, router 1 will be R1.
3. Configure the terminal server (R2) so that all routers can be accessed by name through it. The port number of the terminal server attaches to the corresponding router. That is, port 1 of the terminal server goes to the console port of R1, while the Catalyst switch goes to port number 9. Use the Ethernet IP address as the IP address for reverse telneting.

4. Place Ethernet interface 0/0 of R1 and R2 in VLAN 1. The routers should be attached in ascending order, staring with port 1 on the Catalyst switch.
5. Assign IP addresses to Ethernet 0/0 of R1 and R2 from network 152.1.1.0. The subnet will contain 64 hosts. Assign the addresses in ascending order starting with R1.
6. Configure a loopback interface on R1, R2, and R3. The IP address should be 152.1.1.x, where x is the router name and has a 32-bit mask. For example, the loopback on R1 is 152.1.1.1.
7. Assign IP addresses to the interfaces connecting R1 and R3. Do not use more addresses than needed, and utilize the next available subnet. The addresses should be assigned in ascending order starting with R1.
8. Assign IP addresses to the interfaces connecting R2 and R3. Do not use more addresses than needed, and utilize the next available subnet. The addresses should be assigned in ascending order starting with R1.
9. Place Ethernet interface 0/0 of R3 and R4 in VLAN 2. The routers should be attached to the Catalyst switch in ascending order.
10. Assign an IP address to Ethernet 0/0 of R3 and R4. The subnet will contain 58 hosts. Assign the addresses in ascending order starting with R3.
11. Enable IS-IS on the Ethernet interfaces connecting R1 and R2. The interfaces will be in area 1, the Net will be 00.0001, and the system ID will be the MAC address of the Ethernet interface.
12. Enable IS-IS on the serial interfaces connecting R2 to R3 and R3 to R1. The interfaces will be in area 00.0001 and the system ID will be the MAC address of the Ethernet interface.
13. Enable IS-IS on the Ethernet interfaces connecting R3 and R4. The interfaces will be in area 2, the net will be 00.0002, and the system ID will be the MAC address of the Ethernet interface.
14. Make sure that R2 is elected the DR router for the broadcast network.
15. Configure R1 and R2 so that they only act as L1 routers.
16. R1 and R2 should be able to reach the Ethernet network connecting R3 to R4.
17. Advertise the loopback interfaces via IS-IS. The network should be advertised as internal by the router and no IS-IS packets should be sent over the loopback interfaces.
18. Configure R1 so that it authenticates its neighbor before forming an adjacency. The password used should be cisco.
19. Configure R3 so that it authenticates all L1 link state information within area 1. The password used should be cisco2.
20. Configure the IS-IS domain so that all L2 LSPs are authenticated.
21. Assign the following IP addresses to R4 and R5:

23. Configure RIP on R4 and R5. Do not send any RIP update packets out the interface connection R4 to R3.

R4 Interface S0/0	152.1.2.1 /24
R4 Interface Bri0	152.1.3.1 /24
R5 Interface Bri0	152.1.3.2 /24
R5 Interface S0	152.1.2.2 /24
R5 Interface Loopback 0	152.1.4.1 /24

24. Make sure that all routers can reach all the interfaces on all the other routers.

25. Configure DDR between R4 and R5. R4 should call R5 when the serial connection is lost. The backup should be invoked five seconds after the primary is lost, and once the primary is back, the router should switch back to the primary.

26. Make sure that RIP advertisements don't keep the link up if no other traffic is present.

Answer Guide

1. *Cable the routers per Figure 4-23.*

2. *Erase any existing configuration from the routers and set each router's name. The name will be the corresponding router number; for example, router 1 will be R1.*

The first step is to erase the configuration for all the routers as well as for the Catalyst switch. To erase the configuration on the router, perform the following command:

```
R1#write erase or R1#erase startup-config
```

This command erases the startup configuration stored in NVRAM on all platforms except the Cisco 7000, 7200, and 7500 series. In these series, the router erases or deletes the configuration pointed to by the CONFIG_FILE environment variable. If the CONFIG_FILE variable points to NVRAM, then the router erases NVRAM. If the CONFIG_FILE environment variable specifies a flash memory device and a configuration filename, then the command deletes the named file.

Erasing the configuration on the Catalyst 5500 is a bit different, so use this command:

```
CAT#clear config all
```

3. *Configure the terminal server (R2) so that all routers can be accessed by name through it. The port number of the terminal server attaches to the corresponding router. That is, port 1 of the terminal server goes to the console port of R1, and the Catalyst switch goes to port number 9. Use the Ethernet IP address as the IP address for reverse telneting.*

The terminal server provides access to all of your test routers via reverse telnet, which is the process of using telnet to make a connection out of an asynchronous port.

All the routers in the case study as well as the Catalyst switch will have their console port connected directly to one of the 16 asynchronous interfaces on the 2511RJ terminal server using a standard Cisco console rolled cable. The routers will be accessed using a reverse telnet connection. To make a reverse telnet connection, you must telnet to any active IP address on the box followed by 200x, where x is the port number that you want to access (**Telnet 1.1.1.1 2001**).

Use the following line commands to configure the terminal server:

```
R7(config)#line 1 16
R7(config-line)#transport input all
R7(config-line)#no exec
```

The command **transport input all** specifies the input transport protocol. By default, on IOS 11.1 and later, the transport input is set to none; prior to 11.1, the default was all. If the transport input is left to none, you will receive an error stating that the connection is refused by the remote host. The command no exec enables only outgoing connections for the line. This prevents the device that you are attached to from sending out unsolicited data. If the port receives unsolicited data, an EXEC process will start, which makes the line unavailable.

The next step is to configure hostnames on R4 so you can attach to any router simply by typing in the router name. The Cisco IOS software maintains a table of host names and their corresponding addresses. Similar to a DNS server, you can statically map host names to IP addresses. This is very useful and saves a lot of keystrokes when you have multiple devices connected to the terminal server.

The following global configuration commands are used to set the host names:

```
R2(config)#ip host r1 2001 152.1.2.4
R2(config)#ip host r2 2002 152.1.2.4
R2(config)#ip host r4 2004 152.1.2.4
R2(config)#ip host r5 2005 152.1.2.4
R2(config)#ip host r6 2006 152.1.2.4
R2(config)#ip host r7 2007 152.1.2.4
R2(config)#ip host cat 2009 152.1.2.4
```

4. *Place Ethernet interface 0 of R1 and R2 in VLAN 1. The routers should be attached in ascending order staring with port 1 on the Catalyst switch.*

   Because, by default, all ports on a Catalyst are in VLAN 1, no configuration is needed.

5. *Assign IP addresses to Ethernet 0 of R1 and R2 from network 152.1.1.0. The subnet will contain 64 hosts. Assign the addresses in ascending order starting with R1.*

   You will need 64 IP addresses, so you must use a 25-bit mask. This gives us 128 addresses, two of which can't be used, leaving 126. This is a bit more than you need, but if you used a

26-bit mask, you would be short. If you don't read the case study completely, this question can be tricky. Addresses 152.1.1.1 to 3 are used by the loopback interfaces of R1, R2, and R3 respectively. The next available subnet is 152.1.1.128. The following is the IP address assignment:

```
152.1.1.128  ←Network Address reserved
152.1.1.129  ←R1
152.1.1.130  ←R2
152.1.1.255  ←Broadcast Address reserved

R1(config)#int e0
R1(config-if)#ip add 152.1.1.129 255.255.255.128
R1(config-if)#no shut

R2(config-if)#int e0
R2(config-if)#ip address 152.1.1.130 255.255.255.128
R2(config-if)#no shut
```

6. *Configure a loopback interface on R1, R2, and R3. The IP address should be 152.1.1.x, where x is the router name and has a 32-bit mask. For example, the loopback on R1 is 152.1.1.1.*

```
R1(config)#int lo 0
R1(config-if)#ip add 152.1.1.1 255.255.255.255

R2(config)#int lo 0
R2(config-if)#ip add 152.1.1.2 255.255.255.255

R3(config)#int lo 0
R3(config-if)#ip add 152.1.1.3 255.255.255.255
```

7. *Assign IP addresses to the interfaces connecting R1 and R3. Do not use more addresses than needed, and utilize the next available subnet. The addresses should be assigned in ascending order starting with R1.*

   This is a simple subnetting example. For a serial interface, you only need two addresses, so use a 30-bit mask. The next available subnet is 152.1.1.4 /30. The following is the ip address assignment:

```
152.1.1.4  ←Network Address reserved
152.1.1.5  ←R1
152.1.1.6  ←R6
153.1.1.7  ←Broadcast Address reserved


R1(config)#int s0
R1(config-if)#ip add 152.1.1.5 255.255.255.252
R1(config-if)#no shut

R3(config)#int s0/0
R3(config-if)#ip add 152.1.1.6 255.255.255.252
R3(config-if)#no shut
```

8. *Assign IP addresses to the interfaces connecting R2 and R3. Do not use more addresses than needed, and utilize the next available subnet. The addresses should be assigned in ascending order starting with R1.*

   This is another simple subnetting example. For a serial interface, only two addresses are needed. So use a 30-bit mask that provides four addresses, two of which are reserved.

   The next available subnet is 152.1.1.48 /30. The following code shows the ip address assignments:

```
152.1.1.8  ←Network Address reserved
152.1.1.9  ←R2
152.1.1.10 ←R3
153.1.1.11 ←Broadcast Address reserved


R2(config)#int S0
R2(config-if)#ip add 152.1.1.9 255.255.255.252
R2(config-if)#no shut

R3(config)#int s0/1
R3(config-if)#ip add 152.1.1.10 255.255.255.252
R3(config-if)#no shut
```

9. *Place Ethernet interface 0 of R3 and R4 in VLAN 2. The routers should be attached to the Catalyst switch in ascending order.*

   Remember when the Catalyst first comes up, it is in a state where all ports are in VLAN 1. The Catalyst switch must have a domain name before it can use VLAN numbers other than 1. You can see in the **show vtp domain** output that the domain name has not been set on this switch.

```
Console> (enable) show vtp domain
Domain Name                      Domain Index VTP Version Local Mode  Password
-------------------------------- ------------ ----------- ----------- ----------
                                 1            2           server      -

Vlan-count Max-vlan-storage Config Revision Notifications
---------- ---------------- --------------- -------------
5          1023             0               disabled

Last Updater    V2 Mode  Pruning  PruneEligible on Vlans
--------------- -------- -------- -------------------------
0.0.0.0         disabled disabled 2-1000
```

   Set the vtp domain name with the command **set vtp domain CCIE_LAB:**

```
Console> (enable) set vtp domain CCIE_STUDY_GUIDE
VTP domain CCIE_LAB modified

Console> (enable) show vtp domain
Domain Name                      Domain Index VTP Version Local Mode  Password
```

```
-------------------------------- ------------ ----------- ----------- ----------
CCIE_LAB                          1            2           server      -

Vlan-count Max-vlan-storage Config Revision Notifications
---------- ---------------- --------------- -------------
5          1023             0               disabled

Last Updater    V2 Mode  Pruning  PruneEligible on Vlans
--------------- -------- -------- -------------------------
0.0.0.0         disabled disabled 2-1000
```

Use the **set vlan 2 5/3** command to move port 5/3 to VLAN 2. Notice that the switch automatically modifies VLAN 2 and removes port 5/3 from VLAN 1.

```
Console> (enable) set vlan 2 5/3
Vlan 2 configuration successful
VLAN 2 modified.
VLAN 1 modified.
VLAN  Mod/Ports
---- -----------------------
2     5/3
```

Use the **set vlan 2 5/4** command to move port 5/4 to VLAN 2:

```
Console> (enable) set vlan 2 5/4
VLAN 2 modified.
VLAN 1 modified.
VLAN  Mod/Ports
---- -----------------------
2     5/3 -4
```

Activate the VLAN with the command **set vlan 2**:

```
Console> (enable) set vlan 2
Vlan 2 configuration successful
```

10. *Assign an IP address to Ethernet 0 of R3 and R4. The subnet will contain 58 hosts. Assign the addresses in ascending order starting with R3.*

The subnet between R3 and R4 requires 58 hosts. For this, use a 26-bit mask. The following is the IP address assignment:

```
152.1.1.64  ←Network Address reserved
152.1.1.65  ←R3
152.1.1.66  ←R4
153.1.1.127 ←Broadcast Address reserved

R3(config)#int e0/0
R3(config-if)#ip add 152.1.1.65 255.255.255.192
R3(config-if)#no shut

R4(config-if)#int e0/0
R4(config-if)#ip add 152.1.1.66 255.255.255.192
R4(config-if)#no shut
```

11. *Enable IS-IS on Ethernet interfaces connecting R1 and R2. The interfaces will be in area 1, the Net will be 00.0100, and the system ID will be the MAC address of the Ethernet interface.*

Enabling IS-IS on a router is done in three steps:

First, an IS-IS process is defined using the global command **router isis**, which enables IS-IS on the router:

```
R1(config)#router isis

R2(config)#router isis
```

Next, a *network address* (NET) is defined under the IS-IS routing process. The NET specifies the area ID and the system ID for the router. The length of the NET can range from eight to 20 octets. The question specifies that the net should be 00.0001.xxxx.xxxx.xxxx where x is the MAC address of the Ethernet interface of the router. The following command defines the NET on each router:

```
R1(config)#router isis
R1(config-router)#net 00.0100.00e0.1e5b.2601

R2(config)#router isis
R2(config-router)#net 00.0100.00e0.1e81.81c1
```

The final step is to enable IS-IS on the interface of the router. This command must be added to any interface whose IP address should be advertised via IS-IS. The following commands enable IS-IS on the Ethernet interfaces connecting R1 and R2:

```
R1(config)#int e0/0
R1(config-if)#ip router isis

R2(config)#int e0/0
R2(config-if)#ip router isis
```

Use the **show clns interface e0/0** command to verify that IS-IS is configured on the interface. The following is the output from this command. Notice that on R1 interface e0/0 is configured for IS-IS and has formed one Level 1 adjacency. The command also shows the circuit type, timer intervals, the circuit ID, and the priority.

```
R1#show clns interface
Ethernet0/0  is up, line protocol is up
  Checksums enabled, MTU 1497, Encapsulation SAP
  ERPDUs enabled, min. interval 10 msec.
  RDPDUs enabled, min. interval 100 msec., Addr Mask enabled
  Congestion Experienced bit set at 4 packets
  CLNS fast switching enabled
  CLNS SSE switching disabled
  DEC compatibility mode OFF for this interface
  Next ESH/ISH in 26 seconds
```

```
  Routing Protocol: IS-IS
    Circuit Type: level-1-2
    Interface number 0x0, local circuit ID 0x1
    Level-1 Metric: 10, Priority: 64, Circuit ID: 0100.E01E.8181.01
    Number of active level-1 adjacencies: 1
    Level-2 Metric: 10, Priority: 64, Circuit ID: 0100.E01E.8181.01
    Number of active level-2 adjacencies: 1
    Next IS-IS LAN Level-1 Hello in 4 seconds
    Next IS-IS LAN Level-2 Hello in 4 seconds
```

To verify that R1 and R2 have formed an adjacency, use the **show clns is-neighbors.** The output from the command is shown below. The first four entries of the display show the system ID of R1's neighbor (which in this case is R2), the interface on which the neighbor is located, the state of the adjacency (Init or Up), and the adjacency type.

The next column is the priority of the neighbor. This is used for electing a DR on a broadcast network. The sixth column is the circuit ID, which is used by the router to uniquely identify the IS-IS interface. In this case because you are connected to a broadcast network, the circuit ID is concatenated with the system id of the network's DR, which is R2. The last column indicates the format of the adjacency. For integrated IS-IS, this will always be phase V.

```
R1#show clns is-neighbors

System Id      Interface   State  Type Priority  Circuit Id        Format
0100.E01E.8181 Et0/0        Up     L1L2 64/64      0100.E01E.8181.01  Phase V
```

12. *Enable IS-IS on the serial interfaces connecting R2 to R3 and R3 to R1. R3 will be in area 00.0100 and the system ID will be the MAC address of the Ethernet interface.*

The following commands enable IS-IS on the serial interfaces of R1 and R2:

```
R1(config)#int s0/0
R1(config-if)#ip router isis

R2(config)#int s0/0
R2(config-if)#ip router isis
```

On R3, you need to define an IS-IS process and a NET address for the router. The following commands enable the IS-IS process on R3 and assign a NET address:

```
R3(config)#router isis
R3(config-router)#net 00.0001.0010.7b1f.7941
```

The final step is to enable IS-IS on the serial interfaces:

```
R3(config-router)#int s0/0
R3(config-if)#ip router isis

R3(config-if)#int s0/1
R3(config-if)#ip router isis
```

Verify that R3 has formed an adjacency with R1 and R2 using the **show clns is-neighbors** command. The following is the output from the command. Notice that the circuit ID is only one octet long. The reason for this is that the networks are not broadcast. Therefore, the circuit ID is not concatenated with the DR system ID:

```
R3#show clns is-neighbors

System Id       Interface   State  Type Priority  Circuit Id         Format
0100.E01E.5B26 Se0/0        Up     L1L2 0 /0       00                 Phase V
0100.E01E.8181 Se0/1        Up     L1L2 0 /0       00                 Phase V
```

13. *Enable IS-IS on the Ethernet interfaces connecting R3 and R4. R4 will be in area 2, the net will be 00.0200, and the system ID will be the MAC address of the Ethernet interface.*

The following command enables IS-IS on the Ethernet interface of R3:

```
R3(config)#int e0/0
R3(config-if)#ip router isis
```

On R4, define an IS-IS process and a NET address for the router. The following commands enable the IS-IS process on R4 and assign a NET address:

```
R4(config)#router isis
R4(config-router)#net 00.0002.0010.7b0e.41e1
```

The final step is to enable IS-IS on the Ethernet interface:

```
R4(config)#int e0/0
R4(config-if)#ip router isis
```

14. *Make sure that R2 is elected as the DR router for the broadcast network.*

IS-IS elects a DR router on all broadcast networks. The reason is the same as for OSPF. Rather than have every router on the network advertise adjacency information with every other router, a DR distributes this information in a hierarchical fashion.

The DR election process is very similar to OSPF. Each IS-IS interface is assigned a priority in the range of 0 to 127, while the default for Cisco routers is 64. The priority is exchanged using the Hello packet.

The router with the highest priority becomes the DR. In the event of a tie, the router with the numerically highest system ID becomes the DR. If the priority of the interface is set to zero, then the router is ineligible to become the DR.

The question specifies that R2 should be elected the DR for the broadcast network. To accomplish this, you can set the priority of R1 lower, the priority of R2 higher, or set the priority of R1 to zero so that it is not eligible to participate in the DR election. The following command sets the IS-IS priority of the Ethernet interface of R1 to 0:

```
R1(config-if)#isis priority 0
```

To verify that R2 has become the DR for the network, use the **show clns is-neighbors** command. The following is the output from the command on R1. Notice that the circuit ID is R1's system ID plus 01. As stated earlier, on broadcast networks, the circuit ID is concatenated with the system ID of the network's DR.

```
R1#show clns is-neighbors

System Id      Interface   State  Type Priority  Circuit Id          Format
0100.107B.1F79 Se0/0        Up     L1L2 0 /0      00                   Phase V
0100.E01E.8181 Et0/0        Up     L1L2 64/64      0100.E01E.8181.01   Phase V
```

15. *Configure R1 and R2 so that they only act as L1 routers.*

By default, each IS-IS system is a L1/L2 router. As the network increases in size, using the default values can severely impact the performance of the network. Not only is the processing and maintenance of two LS databases a burden on the router's CPU and memory, but the L1 and L2 IS-IS PDU being originated by every router becomes a burden on the buffers and bandwidth. The following commands configure R1 and R2 as Level 1 routers only:

```
R1(config)#router isis
R1(config-router)#is-type level-1

R2(config)#router isis
R2(config-router)#is-type level-1
```

Use the **show clns protocol** command to verify that the routers are only acting as Level 1 routers. The following is the output from the command on R2. Notice that the IS type is now Level 1.

```
R2#show clns protocol
IS-IS Router: <Null Tag>
  System Id: 0100.E01E.8181.00  IS-Type: level-1
  Manual area address(es):
        00.02
  Routing for area address(es):
        00.00
  Interfaces supported by IS-IS:
        Serial0/0   - IP
        Ethernet0/0   - IP
  Redistributing:
    static
  Distance: 110
```

17. *Advertise the loopback interfaces via IS-IS. The network should be advertised as internal by the router and no IS-IS packets should be sent over the loopback interfaces.*

The first part of question specifies that the loopback interfaces should be advertised as internal by the router. In order to accomplish this, you must enable IS-IS on those interfaces:

```
R1(config)#int lo 0
R1(config-if)#ip router isis

R2(config)#int lo 0
R2(config-if)#ip router isis

R3(config)#int lo 0
R3(config-if)#ip router isis
```

Use the **show ip route** command on R4 to verify that the loopback interfaces are being advertised. The following is the output from the command. Notice that all three networks are L2 because they were learned from another area.

```
R4#show ip route
Codes: C - connected, S - static, I - IGRP, R - RIP, M - mobile, B - BGP
       D - EIGRP, EX - EIGRP external, O - OSPF, IA - OSPF inter area
       N1 - OSPF NSSA external type 1, N2 - OSPF NSSA external type 2
       E1 - OSPF external type 1, E2 - OSPF external type 2, E - EGP
       i - IS-IS, L1 - IS-IS level-1, L2 - IS-IS level-2, * - candidate default
       U - per-user static route, o - ODR

Gateway of last resort is not set

     152.1.0.0/16  is variably subnetted, 7 subnets, 4 masks
i L2     152.1.1.128/25  [115/30] via 152.1.1.65, Ethernet0/0
i L2     152.1.1.8/30    [115/20] via 152.1.1.65, Ethernet0/0
i L2     152.1.1.1/32    [115/30] via 152.1.1.65, Ethernet0/0
i L2     152.1.1.2/32    [115/30] via 152.1.1.65, Ethernet0/0
i L2     152.1.1.3/32    [115/20] via 152.1.1.65, Ethernet0/0
i L2     152.1.1.4/30    [115/20] via 152.1.1.65, Ethernet0/0
C        152.1.1.64/26  is directly connected, Ethernet0/0
```

The second part of the question states that no IS-IS packets should be sent over the loopback interfaces. In order for IS-IS to advertise the IP address, the protocol process is turned on for the interface.

When IS-IS is enabled on an interface protocol, packets are sent out that interface. To prevent this from happening, the **passive interface** command can be used. The following commands disable the sending of IS-IS protocol packets out of the loopback interfaces:

```
R1(config)#router isis
R1(config-router)#passive-interface loopback 0

R2(config)#router isis
R2(config-router)#passive-interface loopback 0

R3(config)#router isis
R3(config-router)#passive-interface loopback 0
```

To verify that the IS-IS updates are no longer being sent over the loopback interface, use the **show clns protocol** command. The following is the output from the command. Notice that interface loopback 0 is no longer processing IS-IS packets.

```
R1#show clns protocol
IS-IS Router: <Null Tag>
  System Id: 0000.E01E.5B26.00  IS-Type: level-1
  Manual area address(es):
        00.01
  Routing for area address(es):
        00.01
  Interfaces supported by IS-IS:
        Serial0/0 - IP
        Ethernet0/0 - IP
  Redistributing:
    static
  Distance: 110
```

18. *Configure R1 so that it authenticates its neighbor before forming an adjacency.*

Cisco IS-IS provides authentication at three levels: between neighbors, area-wide, and domain-wide. This question requires neighbor authentication between R1 and R2 and R1 and R3.

To authenticate between neighbors, the **isis password** command is used to configure a password on the appropriate interface. The command specifies the password and whether the password is for a Level 1 or Level 2 adjacency.

Because the adjacency between R1 and R3, and R1 and R2, is Level 1, only a Level 1 password is needed. The notes at the beginning of the case study state that the password cisco should be used for all level one adjacencies.

The following commands enable neighbor authentication between R1 and R3, and R1 and R2:

```
R1(config)#int e0/0
R1(config-if)#isis password cisco level-1

R1(config)#int s0/0
R1(config-if)#isis password cisco level-1
```

Display the IS-IS neighbors on R2 before enabling authentication. The following is the output from the command. Notice that the adjacency between R1 is no longer UP due to an authentication failure.

```
R2#show clns is-neighbors

System Id      Interface   State  Type Priority  Circuit Id          Format
0000.107B.1F79 Se0/0        Up     L1   0         01                  Phase V
0000.E01E.5B26 Et0/0        Init   L1   128       0000.E01E.8181.01   Phase V
```

Enable authentication on R2 and R3 with the following commands:

```
R2(config)#int e0/0
R2(config-if)#isis password cisco level-1

R3(config)#int S0/0
R2(config-if)#isis password cisco level-1
```

19. *Configure R3 so that it authenticates all L1 link state information within area 1.*

To authenticate LSPs within an area, the **area password** is used to specify the password under the IS-IS configuration. For the IS-IS password command used in the previous question, the password is carried in Hello packets and used to authenticate neighbors. The area password is carried in all L1 LSPs and is used to authenticate link state information.

This question can be a bit tricky if you do not understand how LSP authentication works. The question only states to configure R3, but when performing LSP area authentication, every router in the area must perform area authentication and must have the same password.

The notes in the beginning of the case study state that the password cisco1 should be used for area authentication.

The following commands enable area authentication for area 1:

```
R3(config)#router isis
R3(config-router)#area-password cisco1

R2(config)#router isis
R2(config-router)#area-password cisco1

R1(config)#router isis
R1(config-router)#area-password cisco1
```

20. *Configure the IS-IS domain so that all L2 LSPs are authenticated.*

To authenticate domain-wide, the **domain-password** command is used. The password is carried in L2 LSPs, regulating the exchange of Level 2 route information. Domain authentication does not authenticate neighbors but does authenticate L2 LSPs.

When performing domain-wide authentication, every router in the IS-IS domain must perform authentication and have the same password.

The notes at the beginning of the case study state that password **cisco2** should be used for domain authentication.

The following commands enable domain-wide authentication:

```
R3(config)#router isis
R3(config-router)#domain-password cisco2

R2(config)#router isis
R2(config-router)#domain-password cisco2

R1(config)#router isis
R1(config-router)#domain-password cisco2
```

Set the authentication password to cisco instead of cisco2 on R4:

```
R4(config)#router isis
R4(config-router)#domain-password cisco
```

Display the routing table. The following is the output from the command. Notice that the R4 does not see any of the networks in area 1:

```
R4#show ip route
Codes: C - connected, S - static, I - IGRP, R - RIP, M - mobile, B - BGP
       D - EIGRP, EX - EIGRP external, O - OSPF, IA - OSPF inter area
       N1 - OSPF NSSA external type 1, N2 - OSPF NSSA external type 2
       E1 - OSPF external type 1, E2 - OSPF external type 2, E - EGP
       i - IS-IS, L1 - IS-IS level-1, L2 - IS-IS level-2, * - candidate default
       U - per-user static route, o - ODR

Gateway of last resort is not set

     152.1.0.0/26  is subnetted, 1 subnets
C       152.1.1.64 is directly connected, Ethernet0/0
```

Enable the debugging of IS-IS update packets on R4 with the command **debug isis update-packets**. The following is the truncated output from the command. Notice that LSP authentication has failed.

```
R4#debug isis update-packets

ISIS-SNP: Rec L2 CSNP from 0000.107B.1F79 (Ethernet0/0 )
ISIS-Update: Sending L2 LSP 0000.107B.0E41.00-00, seq 2, ht 1000 on Ethernet0/0
ISIS-Update: Build L2 PSNP entry for 0000.107B.1F79.00-00, seq 0
ISIS-Update: Build L2 PSNP entry for 0000.107B.1F79.01-00, seq 0
ISIS-Update: Sending L2 PSNP on Ethernet0/0
ISIS-Update: Rec L2 LSP 0000.107B.1F79.00-00, seq E, ht 997,
ISIS-Update: from SNPA 0010.7b1f.7941 (Ethernet0/0 )
ISIS-Update: LSP authentication failed
ISIS-Update: Rec L2 LSP 0000.107B.1F79.01-00, seq 8, ht 997,
ISIS-Update: from SNPA 0010.7b1f.7941 (Ethernet0/0 )
ISIS-Update: LSP authentication failed
```

Display the IS-IS adjacencies on R4 with the command **show clns is-neighbors.** Notice that the adjacency is Up. The reason for this is no authentication is being used on the link; only LSP authentication is enabled. The neighbor-level password regulates the establishment of adjacencies and the domain authentication regulates the exchange of Level 2 route information. This is a common error; the neighbor relationship is Up, yet the router has no Level 2 routes in its database. The password must be consistent across the domain when using domain authentication.

```
R4#show clns is-neighbors

System Id      Interface   State  Type Priority  Circuit Id         Format
0000.107B.1F79 Et0/0        Up     L2   64        0000.107B.1F79.01  Phase V
```

Add the correct password to R4 with the following command:

```
R4(config)#router isis
R4(config-router)#domain-password cisco2
```

Now display the routing table on R4. Notice it is now receiving all of area 1's networks via R3:

```
R4#show ip route
Codes: C - connected, S - static, I - IGRP, R - RIP, M - mobile, B - BGP
       D - EIGRP, EX - EIGRP external, O - OSPF, IA - OSPF inter area
       N1 - OSPF NSSA external type 1, N2 - OSPF NSSA external type 2
       E1 - OSPF external type 1, E2 - OSPF external type 2, E - EGP
       i - IS-IS, L1 - IS-IS level-1, L2 - IS-IS level-2, * - candidate default
       U - per-user static route, o - ODR

Gateway of last resort is not set

     152.1.0.0/26  is subnetted, 1 subnets
C       152.1.1.64 is directly connected, Ethernet0/0

R4#show ip route
Codes: C - connected, S - static, I - IGRP, R - RIP, M - mobile, B - BGP
       D - EIGRP, EX - EIGRP external, O - OSPF, IA - OSPF inter area
       N1 - OSPF NSSA external type 1, N2 - OSPF NSSA external type 2
       E1 - OSPF external type 1, E2 - OSPF external type 2, E - EGP
       i - IS-IS, L1 - IS-IS level-1, L2 - IS-IS level-2, * - candidate default
       U - per-user static route, o - ODR

Gateway of last resort is not set

     152.1.0.0/16  is variably subnetted, 7 subnets, 4 masks
i L2    152.1.1.128/25 [115/30] via 152.1.1.65, Ethernet0/0
i L2    152.1.1.8/30 [115/20] via 152.1.1.65, Ethernet0/0
i L2    152.1.1.1/32 [115/20] via 152.1.1.65, Ethernet0/0
i L2    152.1.1.2/32 [115/20] via 152.1.1.65, Ethernet0/0
i L2    152.1.1.3/32 [115/10] via 152.1.1.65, Ethernet0/0
i L2    152.1.1.4/30 [115/20] via 152.1.1.65, Ethernet0/0
C       152.1.1.64/26  is directly connected, Ethernet0/0
```

21. *Assign the following IP addresses to R4 and R5:*

R4 Interface S0/0	152.1.2.1 /24
R4 Interface Bri0	152.1.3.1 /24
R5 Interface Bri0	152.1.3.2 /24
R5 Interface S0	152.1.2.2 /24
R5 Interface Loopback 0	152.1.4.1 /24

23. *Configure RIP on R4 and R5. Do not send any RIP update packets out the interface connection R4 to R3.*

RIP is a very simple protocol to configure. The RIP process is enabled on the router using the global command **router rip**. Then the network command is added under the routing process to specify which interfaces will receive and send RIP routing updates. It also specifies which networks will be advertised. The following commands enable RIP on R4 and R5:

```
R4(config)#router rip
R4(config-router)#network 152.1.0.0

R5(config)#router rip
R5(config-router)#network 152.1.0.0
```

Because RIP is a classful protocol, when it is enabled, it is turned on for the classful network of 152.1.0.0. This encompasses both subnets of 152.1.0.0 on R4. The passive interface command enables the user to turn off RIP advertisements on a particular interface (subnet).

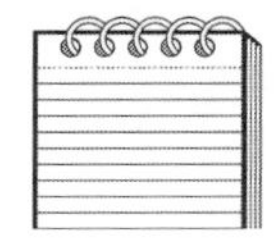

NOTE: *You might be wondering why the router changes your network entry from the subnet you typed into the classful network. For example, if you type network 152.1.2.0 under the RIP process, why when you show the configuration does the network statement appear as 152.1.0.0? The reason is that RIP is considered a classful protocol. This means that it recognizes the IP address class of the network address that you type and assumes the proper mask. For a Class B network like this one, the mask is 255.255.0.0, yielding 152.1.0.0 (no matter what you actually type as the last three octets). The network statement tells the routing protocol to route on the interfaces where the network address matches the one specified in the network statement.*

The second part of the question states that RIP messages should only be sent out Ethernet 0/0. To do this, you must use the passive interface command under the routing process.

You can display the interfaces that RIP is running on R4 with the **show ip protocol** command. The following is the output from the command. Notice that RIP is running on all three interfaces:

```
Routing Protocol is "rip"
  Sending updates every 30 seconds, next due in 16 seconds
  Invalid after 180 seconds, hold down 180, flushed after 240
  Outgoing update filter list for all interfaces is not set
  Incoming update filter list for all interfaces is not set
  Redistributing: rip
  Default version control: send version 1, receive any version
    Interface        Send  Recv   Key-chain
    Interface        Send  Recv   Key-chain
    Ethernet0/0      1     1 2
```

```
    Serial0/0        1      1 2
    BRI1/0           1      1 2
  Routing for Networks:
    152.1.0.0
  Routing Information Sources:
    Gateway          Distance      Last Update
    152.1.2.2             120      00:00:04
  Distance: (default is 120)
```

To prevent RIP updates from being sent out the Ethernet interface of R4, the passive interface command is used:

```
R4(config)#router rip
R4(config-router)#passive-interface e0/0
```

Now display the interfaces that RIP is running again on R4 with the **show ip protocol** command. The following is the output from the command. Notice that RIP is no longer running on Ethernet 0/0, which is now a passive interface:

```
Routing Protocol is "rip"
  Sending updates every 30 seconds, next due in 19 seconds
  Invalid after 180 seconds, hold down 180, flushed after 240
  Outgoing update filter list for all interfaces is not set
  Incoming update filter list for all interfaces is not set
  Redistributing: rip
  Default version control: send version 1, receive any version
    Interface        Send  Recv   Key-chain
    Interface        Send  Recv   Key-chain
    Serial0/0        1      1 2
    BRI1/0           1      1 2
  Routing for Networks:
    152.1.0.0
  Passive Interface(s):
    Ethernet0/0
  Routing Information Sources:
    Gateway          Distance      Last Update
    152.1.2.2             120      00:00:24
  Distance: (default is 120)
```

24. *Make sure that all routers can reach all the interfaces on all the other routers.*

In order to advertise the networks learned via RIP into the IS-IS and vice versa, mutual redistribution is needed.

The following command advertises the RIP-learned routes into IS-IS with a metric of 20:

```
R4(config-router)#redistribute rip metric 20
```

Because IS-IS and RIP use different metrics for calculating routes, it is necessary to tell R4 what it should set the metric to when it redistributes RIP routes into IS-IS. This can be done in several ways. The simplest is to add the metric after the redistribution command.

To verify that the routes are being redistributed properly, display the routing table on R3. The following is the output from the command. Note that R3 now sees three networks via R4.

```
R3#show ip route
Codes: C - connected, S - static, I - IGRP, R - RIP, M - mobile, B - BGP
       D - EIGRP, EX - EIGRP external, O - OSPF, IA - OSPF inter area
       N1 - OSPF NSSA external type 1, N2 - OSPF NSSA external type 2
       E1 - OSPF external type 1, E2 - OSPF external type 2, E - EGP
       i - IS-IS, L1 - IS-IS level-1, L2 - IS-IS level-2, * - candidate default
       U - per-user static route, o - ODR

Gateway of last resort is not set

     152.1.0.0/16  is variably subnetted, 10 subnets, 5 masks
i L1    152.1.1.128/25  [115/20] via 152.1.1.9, Serial0/1
                        [115/20] via 152.1.1.5, Serial0/0
C       152.1.1.8/30    is directly connected, Serial0/1
i L1    152.1.1.1/32    [115/10] via 152.1.1.5, Serial0/0
i L2    152.1.3.0/24    [115/30] via 152.1.1.66, Ethernet0/0
i L1    152.1.1.2/32    [115/10] via 152.1.1.9, Serial0/1
i L2    152.1.2.0/24    [115/30] via 152.1.1.66, Ethernet0/0
C       152.1.1.3/32    is directly connected, Loopback0
C       152.1.1.4/30    is directly connected, Serial0/0
i L2    152.1.4.0/24    [115/30] via 152.1.1.66, Ethernet0/0
C       152.1.1.64/26   is directly connected, Ethernet0/0
```

Now you must redistribute the routes learned via IS-IS into RIP. The following command redistributes IS-IS into RIP on R4:

```
R4(config)#router rip
R4(config-router)#redistribute isis metric 10
```

Display the routing table on R5 with the **show ip route** command. The following is the output. Notice that the only RIP routes that are being received are the host-specific loopback addresses.

```
R5#show ip route
Codes: C - connected, S - static, I - IGRP, R - RIP, M - mobile, B - BGP
       D - EIGRP, EX - EIGRP external, O - OSPF, IA - OSPF inter area
       E1 - OSPF external type 1, E2 - OSPF external type 2, E - EGP
       i - IS-IS, L1 - IS-IS level-1, L2 - IS-IS level-2, * - candidate default
       U - per-user static route

Gateway of last resort is not set

     152.1.0.0/16  is variably subnetted, 6 subnets, 2 masks
R       152.1.1.1/32 [120/10] via 152.1.2.1, 00:00:02, Serial0
R       152.1.1.2/32 [120/10] via 152.1.2.1, 00:00:03, Serial0
C       152.1.3.0/24 is directly connected, BRI0
R       152.1.1.3/32 [120/10] via 152.1.2.1, 00:00:03, Serial0
C       152.1.2.0/24 is directly connected, Serial0
C       152.1.4.0/24 is directly connected, Loopback0
```

When redistributing from a classless protocol like IS-IS into a classfull protocol like RIP, problems can arise if VLSM is used in the IS-IS domain. IS-IS only redistributes routes into the RIP domain that have the same subnet mask as the RIP domain or are host-specific. Because the link between R4 and R5 uses a 24-bit mask, only routes with a 24-bit mask and a 32-bit mask will be redistributed.

In this scenario, if the redistribute command is only used on R4, then R5 would not see these networks. This is because the mask of these routes does not match the mask of the network used on the RIP domain between R4 and R5.

Propagating IS-IS networks into the RIP domain can be done in two ways. The first is to create a static route using a 24-bit mask on R4 and for the network 152.1.1.0. The static route would then be redistributed into RIP. This method cannot be used, however, because static routes are not permitted in this case study.

The second method is to summarize the networks to the same subnet mask length as the RIP domain, 24 bits. To do this, use the IS-IS summary address command on R4. Any more-specific destinations addresses that fall within the summarization range are suppressed.

The following is the command used to summarize the two networks to a 24-bit mask:

```
R3(config)#router isis
R3(config-router)#summary-address 152.1.1.0 255.255.255.0
```

Display the routing table on R5 with the **show ip route** command. The following is the output. Notice that network 152.1.1.0 is now in the routing table. The reason that you no longer see the host-specific loopback addresses is that they were suppressed by R3 because they fall within the summary address.

```
R5#show ip route
Codes: C - connected, S - static, I - IGRP, R - RIP, M - mobile, B - BGP
       D - EIGRP, EX - EIGRP external, O - OSPF, IA - OSPF inter area
       E1 - OSPF external type 1, E2 - OSPF external type 2, E - EGP
       i - IS-IS, L1 - IS-IS level-1, L2 - IS-IS level-2, * - candidate default
       U - per-user static route

Gateway of last resort is not set

     152.1.0.0/24  is subnetted, 4 subnets
R       152.1.1.0 [120/10] via 152.1.2.1, 00:00:24, Serial0
C       152.1.3.0 is directly connected, BRI0
C       152.1.2.0 is directly connected, Serial0
C       152.1.4.0 is directly connected, Loopback0
```

25. *Configure DDR between R4 and R5. R4 should use the ISDN circuit for backup when the serial connection is lost. Do not use the backup interface command and do not use any static routes. The RIP update messages should not keep the ISND link active. Make sure that the ISDN link is only used if the primary is down.*

This question requires the use of DDR on R4 and R5. Start your DDR configuration by configuring R4 and R5 for ISDN. When configuring a BRI, the two main things that you need to know are the ISDN switch type and the SPID number, if any are used. Some switches like the AT&T 5ESS don't require any SPIDS when using 5ESS call control. For this case study, you will use the ADTRAN Atlas 800 for an ISDN simulator. Depending on the switch you are using, this might change. The switch signaling is National ISDN 1 and the SPIDS are as follows:

R5	DN	SPID
BRI1	899-5101	5101
BRI2	899-5102	5102

R4	DN	SPID
BRI1	899-5201	5201
BRI2	899-5202	5202

The first step is to define the switch type under the global configuration. The following global command defines the switch type as National ISDN 1:

```
R5(config)#isdn  switch-type basic-ni1

R4(config)#isdn  switch-type basic-ni1
```

The next step is to configure the SPIDs under the BRI interface. The following interface command defines SPID 1 and 2:

```
R5(config-if)#isdn spid1 5101
R5(config-if)#isdn spid2 5102

R4(config-if)#isdn spid1 5201
R4(config-if)#isdn spid2 5202
```

Use the **show isdn status** command to verify that the ISDN circuit is active and the SPIDS are configured properly.

The following is the output from the command. The first thing to verify is the layer 1 status. Notice that it is active on both routers, which means that the router senses the 2B1Q line coding of the BRI circuit:

```
R4#show isdn status
The current ISDN Switchtype = basic-ni1
ISDN BRI1/0  interface
    Layer 1 Status:
        ACTIVE
```

```
    Layer 2 Status:
        TEI = 64, State = MULTIPLE_FRAME_ESTABLISHED
        TEI = 65, State = MULTIPLE_FRAME_ESTABLISHED
    Spid Status:
        TEI 64, ces = 1, state = 5(init)
            spid1 configured, no LDN, spid1 sent, spid1 valid
            Endpoint ID Info: epsf = 0, usid = 70, tid = 1
        TEI 65, ces = 2, state = 5(init)
            spid2 configured, no LDN, spid2 sent, spid2 valid
            Endpoint ID Info: epsf = 0, usid = 70, tid = 2
    Layer 3 Status:
        0 Active Layer 3 Call(s)
    Activated dsl 0 CCBs = 1
        CCB: callid=0x0, sapi=0, ces=1, B-chan=0
    Total Allocated ISDN CCBs = 1


R5#show isdn status
The current ISDN Switchtype = basic-ni1
ISDN BRI1/0  interface
    Layer 1 Status:
        ACTIVE
    Layer 2 Status:
        TEI = 64, State = MULTIPLE_FRAME_ESTABLISHED
        TEI = 65, State = MULTIPLE_FRAME_ESTABLISHED
    Spid Status:
        TEI 64, ces = 1, state = 5(init)
            spid1 configured, no LDN, spid1 sent, spid1 valid
            Endpoint ID Info: epsf = 0, usid = 70, tid = 1
        TEI 65, ces = 2, state = 5(init)
            spid2 configured, no LDN, spid2 sent, spid2 valid
            Endpoint ID Info: epsf = 0, usid = 70, tid = 2
    Layer 3 Status:
        0 Active Layer 3 Call(s)
    Activated dsl 0 CCBs = 1
        CCB: callid=0x0, sapi=0, ces=1, B-chan=0
    Total Allocated ISDN CCBs = 1
```

The next thing to look for is the layer 2 status. Note that on both routers a TEI has been assigned for both channels. The last thing to check is the SPID status. On both routers, the SPIDs have been sent to the switch and are valid.

Some additional configuration to the ISDN interface is needed:

a. Configure the encapsulation protocol for the interface to PPP. Enable multilink PPP on this interface so that both ISDN B channels can be used. Multilink PPP provides a method for spreading traffic across multiple links (in our case, B channels) while providing packet fragmentation and reassembly, proper sequencing, and load balancing on inbound and outbound traffic. Set the dialer load to 1 so that the second channel is brought up immediately. The dialer load threshold configures the interface to place a second call based on the traffic load on the primary B channel. The software monitors the traffic load and computes a moving average. If this average exceeds the value you set on the first B channel, the secondary line is activated.

```
R5(config-if)#encapsulation ppp
R5(config-if)#ppp multilink
R5(config-if)#dialer load-threshold 1

R4(config-if)#encapsulation ppp
R4(config-if)#ppp multilink
R4(config-if)#dialer load-threshold 1
```

b. Now specify a dialer map on each router to map the destination name and directory number to an IP address.

```
R4(config-if)#dialer map ip 152.1.3.2 name R5 broadcast

R5(config-if)#dialer map ip 152.1.3.1 name R4 broadcast 8995201
```

c. The third step is to specify interesting traffic that will cause the router to dial and keep the connection active. Any packet that matches the dialer group's specified trigger causes a connection to be made. The dialer group number is tied to a dialer list, which specifies the protocol and source destination address that are considered interesting.

```
R5(config-if)#dialer-group 1
R5(config)#dialer-list 1 protocol ip list 2
R5(config)#access-list 2 permit any

R4(config-if)#dialer-group 1
R4(config)#dialer-list 1 protocol ip list 1
R4(config)#access-list 1 permit any
```

The **ppp authentication** command enables CHAP or PAP authentication on the interface:

```
R5(config-if)#ppp authentication chap

R4(config-if)#ppp authentication chap
```

Now create a username-based authentication system. The username and password are used during the authentication process:

```
R4(config)#username R5 password cisco

R5(config)#username R4 password cisco
```

To verify that the configuration is working properly, enable debug ppp authentication, debug q931, and debug dialer on R5. Disconnect the serial cable connecting R5 to R4. The following is the truncated output from the debug commands:

```
10:48:36: %LINK-3-UPDOWN: Interface Serial0, changed state to down ←cable is pulled
10:48:36: BRI0 DDR: Dialing cause ip (s=152.1.3.2, d=255.255.255.255) ←RIP causes the
                                                                        dialing because
                                                                        it is interesting
                                                                        traffic
```

```
10:48:36: BRI0 DDR: Attempting to dial 8995201 ←R5 dials R4
10:48:154618822656: %ISDN-6-LAYER2UP: Layer 2 for Interface BR0, TEI 65 changed
to up
10:48:156773458856: %LINK-3-UPDOWN: Interface BRI0:1, changed state to up
10:48:156774801043: isdn_call_connect: Calling lineaction of BRI0:1
10:48:156773458572: BR0:1 PPP: Treating connection as a callout
10:48:36: BR0:1 PPP: Phase is AUTHENTICATING, by both ←Two way chap authentication
10:48:36: BR0:1 CHAP: O CHALLENGE id 3 len 23 from "R5"
10:48:36: BR0:1 CHAP: I CHALLENGE id 3 len 23 from "R4"
10:48:36: BR0:1 CHAP: O RESPONSE id 3 len 23 from "R5"
10:48:36: BR0:1 CHAP: I SUCCESS id 3 len 4
10:48:36: BR0:1 CHAP: I RESPONSE id 3 len 23 from "R4"
10:48:36: BR0:1 CHAP: O SUCCESS id 3 len 4
10:48:36: %LINK-3-UPDOWN: Interface Virtual-Access1, changed state to upDialer s
tatechange to up Virtual-Access1Dialer call has been placed Virtual-Access1
10:48:36: Vi1 PPP: Treating connection as a callout
10:48:36: BRI0 DDR: Attempting to dial 8995201
10:48:161068426152: %LINK-3-UPDOWN: Interface BRI0:2, changed state to up
10:48:161069794467: isdn_call_connect: Calling lineaction of BRI0:2
10:48:161068425868: BR0:2 PPP: Treating connection as a callout
10:48:37: BR0:2 PPP: Phase is AUTHENTICATING, by both
10:48:37: BR0:2 CHAP: O CHALLENGE id 3 len 23 from "R5"
10:48:37: BR0:2 CHAP: I CHALLENGE id 3 len 23 from "R4"
10:48:37: BR0:2 CHAP: O RESPONSE id 3 len 23 from "R5"
10:48:37: BR0:2 CHAP: I SUCCESS id 3 len 4
10:48:37: BR0:2 CHAP: I RESPONSE id 3 len 23 from "R4"
10:48:37: BR0:2 CHAP: O SUCCESS id 3 len 4
10:48:37: %LINEPROTO-5-UPDOWN: Line protocol on Interface Serial0, changed state
 to down
10:48:37: %LINEPROTO-5-UPDOWN: Line protocol on Interface BRI0:1, changed state
to up
10:48:37: %LINEPROTO-5-UPDOWN: Line protocol on Interface Virtual-Access1, changed state
to up
10:48:38: %LINEPROTO-5-UPDOWN: Line protocol on Interface BRI0:2, changed state
to up
10:48:38: Virtual-Access1 DDR: dialer protocol up
R5#
10:48:43: %ISDN-6-CONNECT: Interface BRI0:1 is now connected to 8995201 ←ISDN dialback
                                                                         is successful,
                                                                         R5 is now
                                                                         connected to R4.
```

Because the question specifies that you cannot use floating static routes, you must run RIP on the ISDN interface. However, the question also states that RIP advertisements should not keep the link up if no other traffic is present.

Because any IP traffic is considered interesting, RIP protocol packets that are sent out every 30 seconds will keep the line up. The question states that the ISDN circuit should not stay active unless user traffic is present.

To accomplish this, you will use snapshot routing. This enables an ISDN hub and spoke network to be built without configuring and maintaining static routes. Snapshot routing is only supported on distance vector protocols, such as RIP and IGRP, for IP traffic. Without snapshot routing, running RIP on an ISDN circuit would mean that RIP packets would reset the ISDN idle timeout every 30 seconds, causing the ISDN interface to remain active, even if no user traffic is present. Snapshot defines an active and a quiet period. During the active period, a RIP-enabled snapshot router exchanges routing updates. If no active calls are made, the snapshot router initiates an ISDN call during the active period to send routing updates. During the quiet period , a snapshot router does not send routing updates and entries are frozen in the routing table. The active and quiet periods are user-defined. The minimum active period is five minutes and the minimum quiet period is eight minutes.

Any calls that bring up the ISDN interface also reset the snapshot routing process to the beginning of a new active period.

The following commands configure R5 as the snapshot client and R4 as the snapshot server.

This command defines the router as a snapshot client. The active time is set to five minutes and the quiet time is set to eight minutes.

```
R5(config-if)#snapshot client 5 8 dialer
```

This command defines the dialer string for snapshot updates:

```
R5(config-if)#dialer map snapshot 1 name R4 broadcast 8995201
```

This command defines R5 as the snapshot server:

```
R4(config-if)#snapshot server 5 dialer
R4(config-if)#dialer map snapshot 1 name R5 broadcast
```

To verify that the configuration is working properly, display the routing table on R5. The following is the output. Notice that R5 has two paths to network 152.1.1.0, one via the serial connection and one via the ISDN connection.

```
R5#show ip route
Codes: C - connected, S - static, I - IGRP, R - RIP, M - mobile, B - BGP
       D - EIGRP, EX - EIGRP external, O - OSPF, IA - OSPF inter area
       N1 - OSPF NSSA external type 1, N2 - OSPF NSSA external type 2
       E1 - OSPF external type 1, E2 - OSPF external type 2, E - EGP
       i - IS-IS, L1 - IS-IS level-1, L2 - IS-IS level-2, * - candidate default
       U - per-user static route, o - ODR

Gateway of last resort is not set

     152.1.0.0/16  is variably subnetted, 5 subnets, 2 masks
R       152.1.1.0/24 [120/10] via 152.1.2.1, 00:00:11, Serial0/0
                     [120/10] via 152.1.3.1, 00:00:11, BRI0
C       152.1.3.0/24  is directly connected, BRI0
C       152.1.3.1/32  is directly connected, BRI0
C       152.1.2.0/24  is directly connected, Serial0/0
C       152.1.4.0/24  is directly connected, Loopback0
```

You can see from the **show snapshot** command on R5 that R5 is the client and is currently in the active state. This will last for five minutes.

```
R5#show snapshot
BRI0 is up, line protocol is upSnapshot client
  Options: dialer support
```

```
  Length of active period:          5 minutes
  Length of quiet period:           8 minutes
  Length of retry period:           8 minutes
   For dialer address 1
    Current state: active, remaining/exchange time: 4/2  minutes
    Connected dialer interface:
       BRI0/0 :1
    Updates received this cycle: ip
```

Wait five minutes and display the snapshot status with the command **show snapshot.** In the following output, notice that the snapshot client is now in the quiet period. At this point, the router freezes the routing information received by the snapshot server and stops sending updates until the next data connection or the expiration of the quiet period.

```
R5#show snapshot
BRI0 is up, line protocol is upSnapshot client
  Options: dialer support
  Length of active period:          5 minutes
  Length of quiet period:           8 minutes
  Length of retry period:           8 minutes
   For dialer address 1
    Current state: quiet, remaining: 6 minutes
```

Now display the status of the ISDN interface with the command **show isdn status**. In the following output from the command, notice that no layer three calls are active.

```
R5#show isdn status
The current ISDN Switchtype = basic-ni1
ISDN BRI0/0  interface
    Layer 1 Status:
        ACTIVE
    Layer 2 Status:
        TEI = 64, Ces = 1, SAPI = 0, State = MULTIPLE_FRAME_ESTABLISHED
        TEI = 65, Ces = 2, SAPI = 0, State = MULTIPLE_FRAME_ESTABLISHED
    Spid Status:
        TEI 64, ces = 1, state = 5(init)
            spid1 configured, no LDN, spid1 sent, spid1 valid
            Endpoint ID Info: epsf = 0, usid = 70, tid = 1
        TEI 65, ces = 2, state = 5(init)
            spid2 configured, no LDN, spid2 sent, spid2 valid
            Endpoint ID Info: epsf = 0, usid = 70, tid = 2
    Layer 3 Status:
        0 Active Layer 3 Call(s)
    Activated dsl 0 CCBs = 0
    The Free Channel Mask:  0x80000003
    Total Allocated ISDN CCBs = 0
```

Now display the routing table on R5. Notice that the router still has a route to network 152.1.1.0 via the ISDN interface even though there is no active ISDN cal.

```
R5#show ip route
Codes: C - connected, S - static, I - IGRP, R - RIP, M - mobile, B - BGP
       D - EIGRP, EX - EIGRP external, O - OSPF, IA - OSPF inter area
       N1 - OSPF NSSA external type 1, N2 - OSPF NSSA external type 2
```

```
       E1 - OSPF external type 1, E2 - OSPF external type 2, E - EGP
       i - IS-IS, L1 - IS-IS level-1, L2 - IS-IS level-2, * - candidate default
       U - per-user static route, o - ODR

Gateway of last resort is not set

     152.1.0.0/24  is subnetted, 4 subnets
R       152.1.1.0 [120/10] via 152.1.2.1, 00:00:14, Serial0/0
                  [120/10] via 152.1.3.1, 00:10:14, BRI0
C       152.1.3.0 is directly connected, BRI0
C       152.1.2.0 is directly connected, Serial0/0
C       152.1.4.0 is directly connected, Loopback0
```

The last part of the question specifies that the ISDN circuit should not be used to route traffic unless the primary is down. Because RIP uses the hop count as its metric, it sees the path over the ISDN and the serial link as equal. The router will load balance across the two links. To prevent this from happening, you must increase the metric of the route learned over the ISDN interface.

To do this, you will use an offset to add one to the metric of network 152.1.1.0 when it is learned via the ISDN interface.

The first step is to define an access list to identify the network to which you want to add the offset:

```
R5(config)#access-list 1 permit 152.1.1.0
```

The next step is to apply an offset to the particular interface that the route will be learned on:

```
R5(config-router)#offset-list 2 in 1 bri 0
```

Now display the routing table on R5. Notice in the following output that the route to network 152.1.1.0 is no longer in the table. The reason, of course, is that the router only loads the best routes to a particular network into its forwarding table. Because the route learned via the ISDN interface has a metric of 11 versus the 10, it will only be loaded once the route via the serial interface is removed.

```
R5#show ip route
Codes: C - connected, S - static, I - IGRP, R - RIP, M - mobile, B - BGP
       D - EIGRP, EX - EIGRP external, O - OSPF, IA - OSPF inter area
       N1 - OSPF NSSA external type 1, N2 - OSPF NSSA external type 2
       E1 - OSPF external type 1, E2 - OSPF external type 2, E - EGP
       i - IS-IS, L1 - IS-IS level-1, L2 - IS-IS level-2, * - candidate default
       U - per-user static route, o - ODR

Gateway of last resort is not set

     152.1.0.0/16  is variably subnetted, 5 subnets, 2 masks
R       152.1.1.0/24  [120/10] via 152.1.2.1, 00:00:12, Serial0/0
C       152.1.3.0/24  is directly connected, BRI0
C       152.1.3.1/32  is directly connected, BRI0
```

```
C       152.1.2.0/24  is directly connected, Serial0/0
C       152.1.4.0/24  is directly connected, Loopback0
```

Now disconnect the serial cable between R5 and R4, and display the routing table on R5. The following is the output. Notice that the route via BRI0/0 is now loaded in the table.

```
R5#show ip route
Codes: C - connected, S - static, I - IGRP, R - RIP, M - mobile, B - BGP
       D - EIGRP, EX - EIGRP external, O - OSPF, IA - OSPF inter area
       N1 - OSPF NSSA external type 1, N2 - OSPF NSSA external type 2
       E1 - OSPF external type 1, E2 - OSPF external type 2, E - EGP
       i - IS-IS, L1 - IS-IS level-1, L2 - IS-IS level-2, * - candidate default
       U - per-user static route, o - ODR

Gateway of last resort is not set

     152.1.0.0/16  is variably subnetted, 4 subnets, 2 masks
R       152.1.1.0/24 [120/11] via 152.1.3.1, 00:00:03, BRI0
C       152.1.3.0/24  is directly connected, BRI0
C       152.1.3.1/32  is directly connected, BRI0
C       152.1.4.0/24  is directly connected, Loopback0
```

Complete Network Diagram

Figure 4-22 shows the complete network diagram with IP addressing. Table 4-7 shows the IP address work sheet.

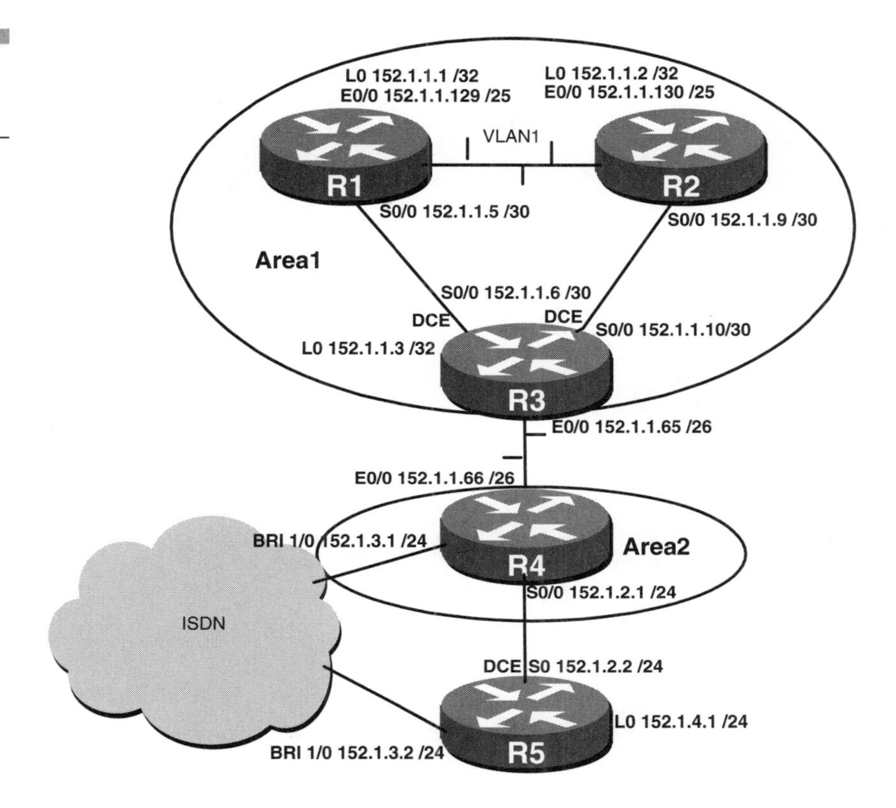

Figure 4-24 Network diagram with IP addressing

Router Configurations

R1

```
!
version 11.2
service timestamps debug uptime
service timestamps log uptime
no service password-encryption
no service udp-small-servers
no service tcp-small-servers
!
hostname R1
!
boot system flash c3620-js-mz_112-20_P.bin
!
clns routing
!
interface Loopback0
 ip address 152.1.1.1 255.255.255.255
!
interface Ethernet0/0
 ip address 152.1.1.129 255.255.255.128
 ip router isis
 isis password cisco level-1
 isis priority 0 level-1
!
interface Serial0/0
 ip address 152.1.1.5 255.255.255.252
 ip router isis
 isis password cisco level-1
!
router isis
 passive-interface Loopback0
 net 00.0100.00e0.1e5b.2601
 is-type level-1
 domain-password cisco2
 area-password cisco1
!
no ip classless
!
!
line con 0
line aux 0
line vty 0 4
 login
!
end
```

R2

```
!
version 11.2
no service password-encryption
```

```
no service udp-small-servers
no service tcp-small-servers
!
hostname R2
!
boot system flash slot0:c3620-js-mz_112-20_P.bin
!
ip host R1 2001 152.1.1.2
ip host R3 2003 152.1.1.2
ip host R4 2004 152.1.1.2
ip host R5 2005 152.1.1.2
ip host cat 2008 152.1.1.2
!
clns routing
!
interface Loopback0
 ip address 152.1.1.2 255.255.255.255
!
interface Ethernet0/0
 ip address 152.1.1.130 255.255.255.128
 ip router isis
 isis password cisco level-1
!
interface Serial0/0
 ip address 152.1.1.9 255.255.255.252
 ip router isis
!
router isis
 passive-interface Loopback0
 net 00.0100.00e0.1e81.81c1
 is-type level-1
 domain-password cisco2
 area-password cisco1
!
no ip classless
ip route 172.31.1.34 255.255.255.255 132.1.1.1
logging buffered 4096 debugging
!
!
line con 0
 exec-timeout 0 0
line 1 16
no exec
 transport input all
line aux 0
line vty 0 4
 login
!
end
```

R3

```
!
version 11.2
no service password-encryption
service udp-small-servers
```

```
service tcp-small-servers
!
hostname R3
!
boot system flash c3620-js-mz_112-14_P.exe
!
clns routing
!
interface Loopback0
 ip address 152.1.1.3 255.255.255.255
!
interface Ethernet0/0
 ip address 152.1.1.65 255.255.255.192
 ip router isis
!
interface Serial0/0
 ip address 152.1.1.6 255.255.255.252
 ip router isis
 clockrate 1000000
 isis password cisco level-1
!
interface Serial0/1
 ip address 152.1.1.10 255.255.255.252
 ip router isis
 clockrate 1000000
!
router isis
 summary-address 152.1.1.0 255.255.255.0
 passive-interface Loopback0
 net 00.0100.0010.7b1f.7941
 domain-password cisco2
 area-password cisco1
!
no ip classless
!
!
line con 0
line aux 0
line vty 0 4
 login
!
end
```

R4

```
!
version 11.2
no service password-encryption
no service udp-small-servers
no service tcp-small-servers
!
hostname R4
!
!
username R5 password 0 cisco
clns routing
```

```
isdn switch-type basic-ni1
!
interface Ethernet0/0
 ip address 152.1.1.66 255.255.255.192
 ip router isis
!
interface Serial0/0
 ip address 152.1.2.1 255.255.255.0
 no fair-queue
!
!
interface BRI1/0
 ip address 152.1.3.1 255.255.255.0
 encapsulation ppp
 isdn spid1 5201
 isdn spid2 5202
 dialer map snapshot 1 name R5 broadcast
 dialer map ip 152.1.3.2 name R5 broadcast
 dialer load-threshold 1 outbound
 dialer-group 1
 snapshot server 5 dialer
 no fair-queue
 ppp authentication chap
 hold-queue 75 in
!
router isis
 summary-address 152.1.1.0 255.255.255.0
 redistribute rip metric 20 metric-type internal level-2
 net 00.0002.0010.7b0e.41e1
 domain-password cisco2
!
router rip
 redistribute isis level-2 metric 10
 passive-interface Ethernet0/0
 network 152.1.0.0
!
no ip classless
access-list 1 permit any
!
dialer-list 1 protocol ip list 1
!
line con 0
line aux 0
line vty 0 4
 login
!
end
```

R5

```
!
version 12.0
service timestamps debug uptime
service timestamps log uptime
no service password-encryption
!
```

```
hostname R5
!
!
username R4 password 0 cisco
memory-size iomem 25
ip subnet-zero
!
isdn switch-type basic-ni
!
!
!
interface Loopback0
 ip address 152.1.4.1 255.255.255.0
 no ip directed-broadcast
!
interface Serial0
 ip address 152.1.2.2 255.255.255.0
 no ip directed-broadcast
 clockrate 1000000
!
interface BRI0
 ip address 152.1.3.2 255.255.255.0
 no ip directed-broadcast
 encapsulation ppp
 dialer map snapshot 1 name R4 broadcast 8995201
 dialer map ip 152.1.3.1 name R4 broadcast 8995201
 dialer-group 1
  isdn spid1 5101
 isdn spid2 5102
 snapshot client 5 8 dialer
 ppp authentication chap
!
router rip
offset-list 1 in 1 BRI0
network 152.1.0.0
!
no ip classless
no ip http server
!
access-list 1 permit any
dialer-list 1 protocol ip list 1
!
access-list 2 permit 152.1.1.0
!

line con 0
 transport input none
line aux 0
line vty 0 4
 login
!
no scheduler allocate
end
```

CHAPTER 5

A Case Study in DLSW and Bridging

Introduction

This chapter explores five key router technologies:

- Transparent bridging
- Dial backup for bridged circuits
- Data link switching (DLSW) with backup peers
- Integrated routing and bridging
- Multicast routing technology

Technology Overview

The following overview briefly describes these technologies.

Transparent Bridging

A bridge is a networking device that operates at the data link layer. The simplest bridge is a two-port transparent bridge. A bridge is capable of forwarding or blocking each frame that arrives on one of its ports. A bridge makes its forwarding/blocking decisions based on the destination MAC address of a data frame.

The advantage of a bridge is that it divides a network. A LAN is considered to be a single broadcast and collision domain. Each port of a bridge is its own collision domain; thus, the bridge can reduce network collisions. A bridge can also offer primitive network security by enabling frames to be filtered based on MAC addresses. Bridges are still necessary in certain situations because several commonly used protocols (such as NetBEUI) are non-routable.

Data Link Switching (DLSW)

DLSW provides a standard way of transporting NetBEUI and SNA traffic over an IP network. DLSW is similar to remote source route bridging, but it adds several enhanced features such as local termination, enhanced flow control, and better broadcast traffic handling.

In this case study, you will be using a feature of DLSW known as backup peers. It enables you to assign a secondary circuit path in case the primary path fails.

Integrated Routing and Bridging

With traditional bridging, a given protocol could either be bridged or routed on a given router. With the advent of concurrent routing and bridging, a given protocol can be bridged on some router interfaces and routed on others. Although it provides more flexibility, concurrent routing and bridging does not allow bridged and routed traffic to be mixed.

Integrated routing and bridging (IRB) enables a protocol to be routed on some interfaces and bridged on others. It enables bridged and routed traffic to be sent between the bridged and routed interfaces. IRB is configured by defining a *Bridge-Group Virtual Interface* (BVI). The BVI is a virtual interface within a router. The BVI only supports routing, not bridging. The BVI is used to forward bridged and routed traffic to the appropriate interfaces on the router.

Multicast Routing

Three primary types of transmission take place on a data network: unicast, broadcast, and multicast. A unicast transmission takes place when data is sent from one sender to one receiver. Broadcast transmissions occur when a sender sends data to all receivers. A multicast transmission occurs when a sender sends data to a subset of all the receivers in a network.

All receivers that receive the multicast transmission are said to be in the same multicast group. Membership in a multicast group is dynamic; a receiver can join and leave at any time.

All routers in a network that participate in multicast routing must run a multicast routing protocol. Routers also run the *Internet Group Management Protocol* (IGMP) to learn if any multicast group members are on their attached networks.

Multicast traffic is used on video-on-demand networks and on shout-down networks used in trading floors.

Case Study 5: DLSW and Bridging

IOS Requirements

All routers in this case study were running IOS 11.2 or higher.

Equipment Needed

The following equipment is needed to perform this case study:

- One Cisco router with two Ethernet ports and one serial port (R1)
- One Cisco router with two serial ports and an Ethernet port (R2)
- One Cisco router with two serial ports and a Token Ring interface (R3)
- One Cisco router with two serial ports, one Ethernet port, and one Fast Ethernet port (R4)
- One Cisco router with two serial ports, two Ethernet ports, and one ISDN port (R5)
- One Cisco router with one serial port, one Ethernet port, and one ISDN port (R6)
- One Cisco terminal server
- A Cisco Catalyst switch
- A Cisco IOS image on each router capable of supporting DLSW, multicast, and bridging
- A PC running a terminal emulation program for connecting to the console port of the terminal server
- Two Ethernet crossover cables
- An Ethernet hub
- Six Ethernet cables
- Five Cisco DTE/DCE crossover cables
- Two ISDN BRI circuits and two ISDN BRI cables
- Four workstations, three with Ethernet connections and one with a Token Ring connection

Obtaining ISDN Circuits

This case study requires two ISDN BRI circuits. You can either purchase the two ISDN circuits from your local telecommunications carrier or obtain an ISDN test switch, which you can use to provision your own BRI circuits. An Adtran Atlas 800 desktop ISDN switch is the option that was used for this case study. Details on configuring this ISDN switch for this case study can be found at the end of this chapter.

Physical Connectivity Diagram

Figure 5-1 shows the physical connectivity for the routers in your case study. Although it is not shown in the figure, an additional router, R7, is used as a terminal server.

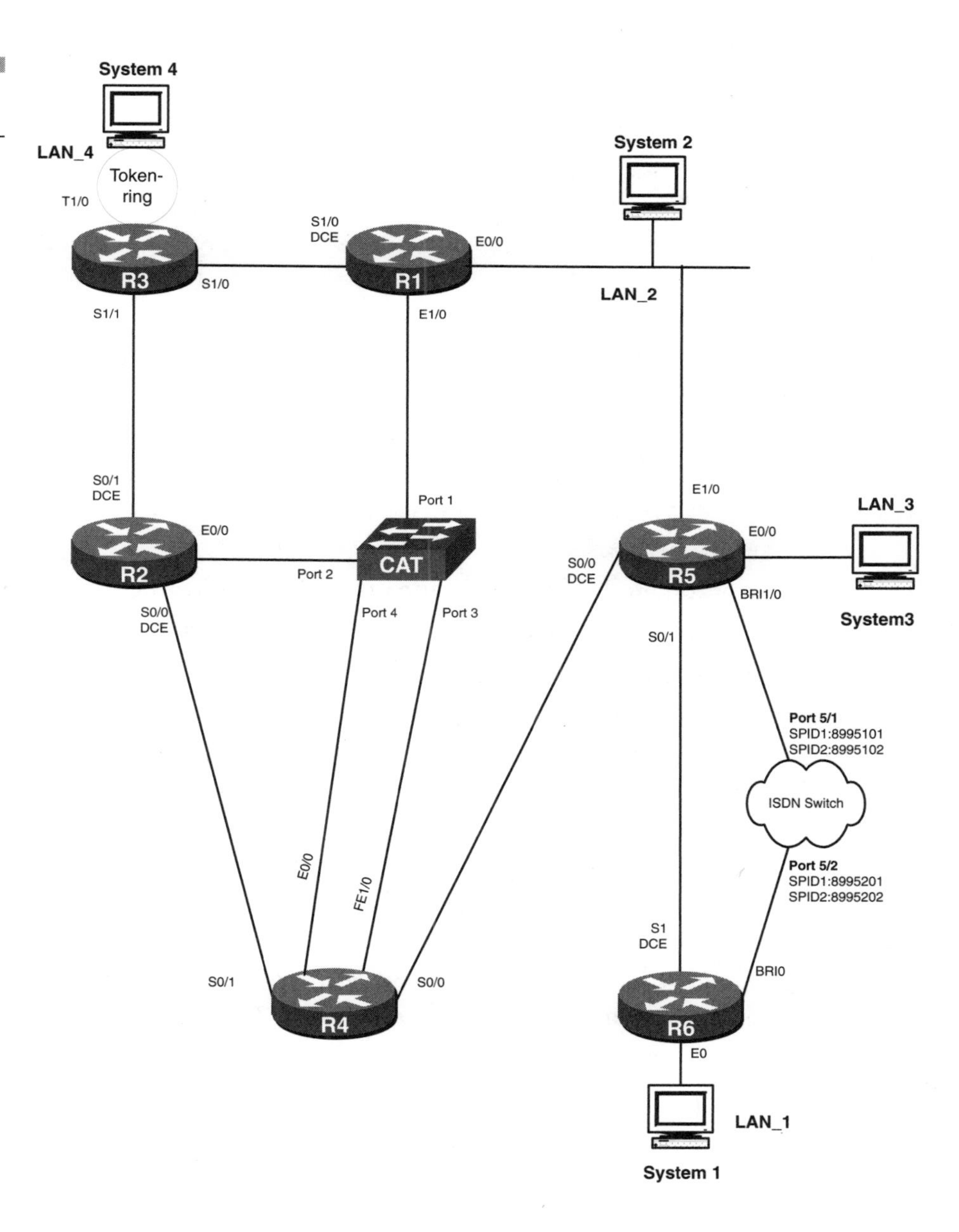

Figure 5-1 Physical connectivity

Questions

1. Cable the routers per Figure 5-1. Make sure that the DTE/DCE cables are connected per the figure.
2. Configure the terminal server so that all routers can be accessed by name through it. The port number of the terminal server attaches to the corresponding router; that is, port 1 of the terminal server goes to the console port of R1.
3. Set each router hostname per Figure 5-1.
4. Set the enable password on all routers to be **cisco**.
5. Configure the IP addressing on your network. All addresses should be assigned from the 152.3.0.0 network. Assign a loopback address to each router; use a 24-bit mask. The third octet should be the router number. For example, the loopback address on R1 should be 152.3.1.1 /24 .
6. LAN_3, LAN_2, and LAN_1 should be designed to handle 60 hosts each.
7. LAN_4 should be designed to handle 10 hosts.
8. The ISDN interfaces should be able to handle four hosts.
9. All serial interfaces, as well as the Ethernet core, should use as small a subnet as possible.
10. Configure R2's Ethernet interface to reside on VLAN 2 and configure R1's Ethernet interface to reside on VLAN 1.
11. Configure port 5 on the Catalyst switch to monitor all traffic from ports 1 and 2.
12. Verify and record the version of code running on the Catalyst switch.
13. Change the switching mode of the Catalyst to Store-and-Forward.
14. Change the baud rate of the console port of the Catalyst switch to 19,200 bits per second.
15. Configure the network to run OSPF. Use PPP encapsulation on all serial links.
16. Configure DLSW between LAN_4 (connected to R3) and LAN_2 (connected to R1 and R5). R1 should be the preferred peer to R3. If R1 fails, then R3 should automatically peer to R5.
17. Configure R3 so that its attached workstations will only be able to see two other workstations on the network. The two workstations that R3 should allow access to will have NetBIOS names: System3 and System9.
18. Configure transparent bridging between LAN_3 on R5 and LAN_1 on R6. Assume that stations on each of these LANs are running NetBEUI traffic.
19. Configure dial backup between R5 and R6. If the serial link fails between R5 and R6, the ISDN link should provide a backup path between the workstations on LAN_3 and LAN_1.

20. Configure R6 so that a workstation on LAN_4 with a source MAC address of 0000.8635.46e1 should have its packets rejected.
21. Enable IPX routing on R2 and R1.
22. Configure integrated routing and bridging between R2, R1, and R4. IP traffic should be routed and IPX traffic should be bridged.
23. Configure PIM Sparse mode on R1, R4, R5, and R6. Make the RP address 152.3.10.13 on R1.
24. Configure a static multicast group on R4. Make sure it is reachable and pingable from all routers.

Answer Guide

1. *Cable the routers per Figure 5-1. Make sure that the DTE/DCE cables are connected per the figure.*

   Cable the configuration per Figure 5-1 making sure to connect the DTE and DCE sides of the serial cables to the correct ports.

   The reader can use the Catalyst 1900 switch in place of a traditional Ethernet hub. On LAN_2, for example, you will connect E0/0 on R1, E1/0 on R5, and System2 into three ports of the Catalyst switch and place these three ports in their own VLAN.

   System3 and System1 can be connected to their respective routers with an Ethernet crossover cable. If you do not have an Ethernet crossover cable, you can create a new VLAN on the Catalyst switch and connect these workstations to the Catalyst switch.
2. *Configure the terminal server so that all routers can be accessed by name through it. The port number of the terminal server attaches to the corresponding router; that is, port 1 of the terminal server goes to the console port of R1.*

   The terminal server provides access to all your routers via reverse telnet. Reverse telnet is the process of using telnet to make a connection out of an asynchronous port. The terminal server should be configured as R7.

   All the routers in the case study have their console port connected directly to one of the 16 asynchronous interfaces on a Cisco 2511 terminal server router, using a standard Cisco rolled cable. The routers will be accessed using a reverse telnet connection. To make a reverse telnet connection, telnet to any active IP address on the box followed by 20xx, where xx is the terminal server port number that you want to access. If you have defined a loopback on the router at IP address 1.1.1.1, then the following command is used to reverse telnet to port 1 on the router: **Telnet 1.1.1.1 2001.**

Use the following line commands to configure the terminal server:

```
R7(config)#line 1 16
R7(config-line)#transport input all
R7(config-line)#no exec
```

The command **transport input all** specifies the input transport protocol. By default, on IOS 11.1 and higher, the transport input is set to none. Prior to 11.1, the default was all. If the transport input is left to none, you receive an error stating that the remote host refuses the connection. The command **no exec** enables only outgoing connections for the line. This prevents the device that you are attached to from sending out unsolicited data. If the port receives unsolicited data, an EXEC process starts, which makes the line unavailable.

The next step is to configure hostnames on the terminal server so you can attach to any router simply by typing in the router name. The Cisco IOS software maintains a table of host names and their corresponding addresses. Similar to a DNS server, you can statically map host names to an IP address. This is very useful and saves a lot of keystrokes when you have multiple devices connected to the terminal server.

The following global configuration commands are used to set the host names:

```
R7(config)#ip host R1 2001 1.1.1.1
R7(config)#ip host R2 2002 1.1.1.1
R7(config)#ip host R3 2002 1.1.1.1
R7(config)#ip host R4 2004 1.1.1.1
R7(config)#ip host R5 2005 1.1.1.1
R7(config)#ip host R6 2006 1.1.1.1
```

3. *Set each router hostname per Figure 5-1.*

   Connect to each of the routers and set their hostnames with the global command:

```
Router(config)#hostname R6
```

4. *Set the enable password on all routers to be* ***cisco****.*

   When a router is first configured, no default enable password is used. The enable password is used to prevent unauthorized users from changing the configuration. To set the enable password, use the following global configuration command:

```
R1(config)#enable password cisco
```

5. *Configure the IP addressing on your network. All addresses should be assigned from the 152.3.0.0 network. Assign a loopback address to each router; use a 24-bit mask.*

The question states that all addresses should be assigned from the class B network 152.3.0.0. A loopback address needs to be assigned to each of your six routers. The following addresses should be used for the loopback addresses of your six routers:

152.3.1.1/24	R1's Loopback
152.3.2.1/24	R2's Loopback
152.3.3.1/24	R3's Loopback
152.3.4.1/24	R4's Loopback
152.3.5.1/24	R5's Loopback
152.3.6.1/24	R6's Loopback

Notice that all six addresses are from the 152.3.0.0 address space.

6. *LAN_3, LAN_2, and LAN_1 should be designed to handle 60 hosts each.*

   Based on the question, the addresses assigned to LAN_3, LAN_2, and LAN_1 have two restrictions:

- The addresses must come from the 152.3.0.0 address space.
- The address space that is assigned to each LAN must be capable of supporting 60 hosts.

In order to support 60 hosts, a network must be assigned a 26-bit mask. This can be explained by looking at the mask bits. A 26-bit mask contains all 1s in the first three mask octets and two 1s in the last mask octet. The last mask octet can be written as:

```
1100 0000
```

Because six bits are dedicated to the host portion of the address, you can have $2^6 - 2 = 62$ useable addresses. The following addresses can be used for the three LANs:

152.3.7.1/26	LAN_2
152.3.7.129/26	LAN_3
152.3.7.193/26	LAN_1

7. *LAN_4 should be designed to handle 10 hosts.*

   In order to support 10 hosts, an address needs to have a 28-bit mask. A 28-bit mask contains all 1's in the first three octets (24 bits) and another four 1's in the last octet. The last octet of the mask is written as:

```
1111 0000
```

This gives you four bits that are used for the host address leaving you with $2^4 - 2 = 14$ useable addresses. The following address range can be used for LAN_4:

152.3.8.1/28	LAN_4

8. *The ISDN interfaces should be able to handle four hosts.*

Use a 29-bit mask for the ISDN interfaces. This gives you five network bits and three host bits in the last octet of the network mask. The last octet of the network mask can be written as:

```
1111 1000
```

You have three bits that are used for the host address, giving you $2^3 - 2 = 6$ host addresses. The following address range can be used for the ISDN interfaces:

152.3.9.1/29	ISDN Interfaces

9. *All serial interfaces, as well as the Ethernet core, should use as small a subnet as possible.*

A 30-bit subnet mask is the largest number of bits that can be used for a valid mask. A 30-bit mask is frequently used on point-to-point interfaces where only two addresses are needed. Such a mask contains six network bits in the fourth octet of the mask. The fourth octet of the mask can be written as:

```
1111 1100
```

A 30-bit mask leaves you with two bits, which are used for host addresses, giving you $2^2 - 2 = 2$ useable addresses per network.

Figure 5-2 shows your network configuration with IP addresses assigned to each router interface.

10. *Configure R2's Ethernet interface to reside on VLAN 2 and configure R1's Ethernet interface to reside on VLAN 1.*

A Catalyst 1920 Ethernet Switch is used for this lab. The Catalyst 1920 is a 24-port, 10Mbps Ethernet switch that also has two 100Mbps ports. It is capable of being configured into multiple VLANs. The two 100Mbps ports can also be configured as ISL trunks. The Catalyst 1920 can operate in two different modes:

- Command line mode
- Menu mode

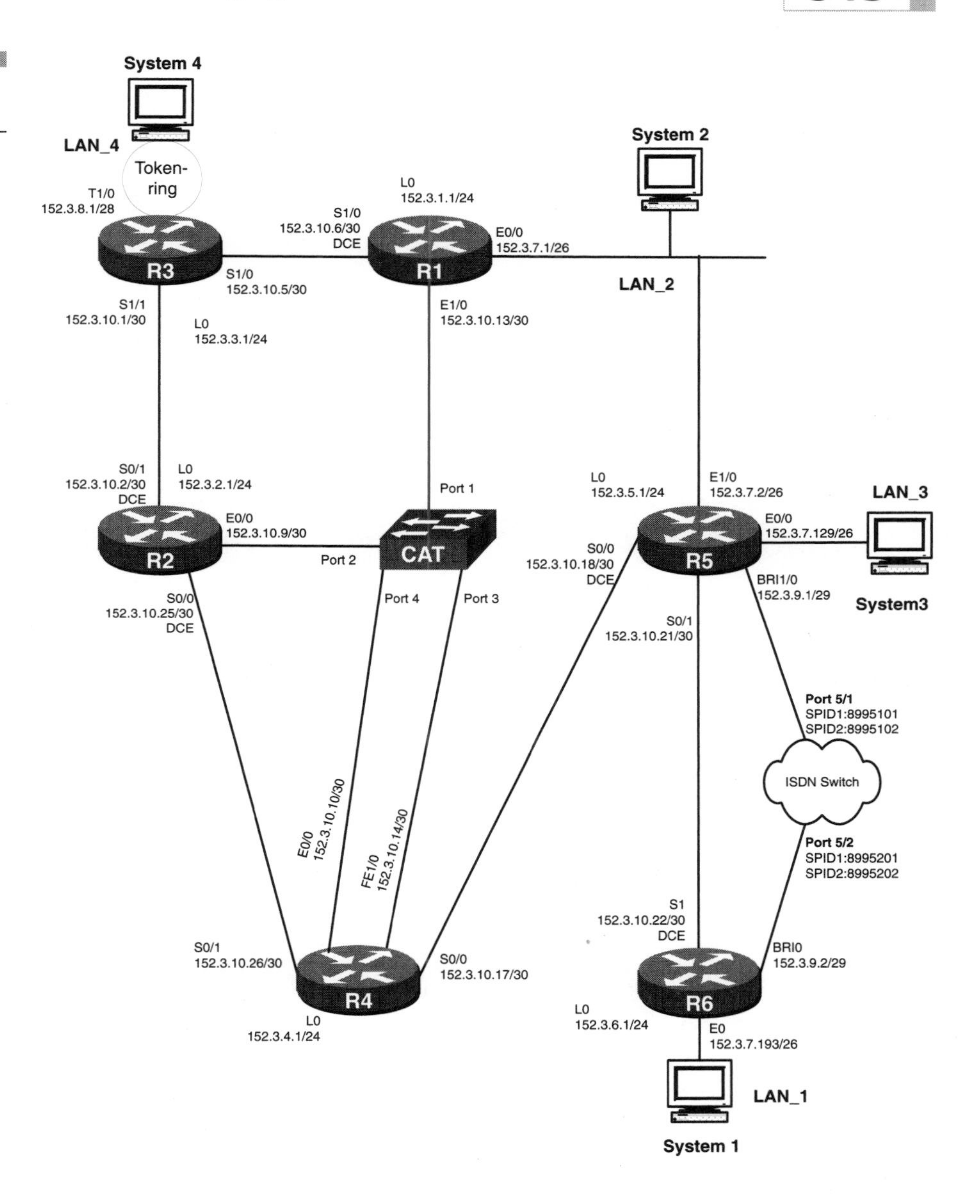

Figure 5-2
Network addressing

Use the Catalyst 1920 in Menu mode. The cable that is used to connect the terminal server to the Catalyst 1920 is the same rolled cable that is used to connect the terminal server and the routers. Once the connection is made between the terminal server and the Catalyst 1920, the main menu appears as shown here:

```
Catalyst 1900 - Main Menu

     [C] Console Settings
     [S] System
     [N] Network Management
     [P] Port Configuration
     [A] Port Addressing
     [D] Port Statistics Detail
     [M] Monitoring
     [V] Virtual LAN
     [R] Multicast Registration
     [F] Firmware
     [I] RS-232 Interface
     [U] Usage Summaries
     [H] Help
     [K] Command Line

     [X] Exit Management Console

Enter Selection:  V
```

Select option V to configure and add a VLAN. When a Catalyst switch is first powered on, all ports reside in VLAN 1. Catalyst ports 2 and 4 need to be configured in VLAN 2. In addition, R1, R5, and System 2 use ports 22, 23, and 24. Ports 22 through 24 reside in VLAN 3. This keeps you from having to use an additional Ethernet hub for connecting the stations on LAN_2.

The following shows the VLAN Configuration menu. VLAN 1 will already be configured when the switch is powered on. Before you assign a port to any VLAN other than VLAN 1, you must first define the VLAN. Select option A to add an additional VLAN.

```
Catalyst 1900 - Virtual LAN Configuration

    ----------------------- Information -----------------------------------
     VTP version: 1
     Configuration revision: 2
     Maximum VLANs supported locally: 1005
     Number of existing VLANs: 7
     Configuration last modified by: 0.0.0.0 at 00-00-0000 00:00:00

    ----------------------- Settings --------------------------------------
     [N] Domain name
     [V] VTP mode control              Server
     [F] VTP pruning mode              Disabled
     [O] VTP traps                     Enabled
```

```
    ---------------------- Actions -------------------------------------
     [L] List VLANs                   [A] Add VLAN
     [M] Modify VLAN                  [D] Delete VLAN
     [E] VLAN Membership              [S] VLAN Membership Servers
     [T] Trunk Configuration          [W] VTP password
     [P] VTP Statistics               [X] Exit to Main Menu

Enter Selection:  A
```

Then select the type of VLAN being added. In this case, you are adding an Ethernet VLAN:

```
This command selects the type of VLAN to be added.

The following VLAN types can be added:

[1]Ethernet, [2]FDDI, [3]Token-Ring, [4]FDDI-Net, or [5]Token-Ring-Net

Select a VLAN type [1-5]: Ethernet
```

The switch automatically selects the next available VLAN number for the new VLAN. You can see in this case that you have added VLAN 2 to your switch. Select option S to save your new VLAN configuration and exit to the previous menu.

```
Catalyst 1900 - Add Ethernet VLAN

    ---------------------- Settings ------------------------------------
     [N] VLAN Number                 2
     [V] VLAN Name                   VLAN0002
     [I] 802.10 SAID                 100002
     [M] MTU Size                    1500
     [L] Translational Bridge 1      0
     [J] Translational Bridge 2      0
     [T] VLAN State                  Enabled

     [S] Save and Exit                [X] Cancel and Exit

Enter Selection:  S
```

Next, add ports to your new VLAN. Select option E to go to the VLAN membership screen.

```
Catalyst 1900 - Virtual LAN Configuration

    ---------------------- Information ---------------------------------
     VTP version: 1
     Configuration revision: 2
     Maximum VLANs supported locally: 1005
     Number of existing VLANs: 7
     Configuration last modified by: 0.0.0.0 at 00-00-0000 00:00:00
```

```
---------------------- Settings -------------------------------------
  [N] Domain name
  [V] VTP mode control              Server
  [F] VTP pruning mode              Disabled
  [O] VTP traps                     Enabled

---------------------- Actions --------------------------------------
  [L] List VLANs                    [A] Add VLAN
  [M] Modify VLAN                   [D] Delete VLAN
  [E] VLAN Membership               [S] VLAN Membership Servers
  [T] Trunk Configuration           [W] VTP password
  [P] VTP Statistics                [X] Exit to Main Menu

Enter Selection:  E
```

The VLAN membership screen displays the VLAN assignment for every port on the Catalyst switch. Even though you have added VLAN 2 to your switch configuration, every port on the switch is still assigned to the default VLAN 1.

```
Catalyst 1900 - VLAN Membership Configuration

   Port  VLAN    Membership Type     Port  VLAN    Membership Type
   -----------------------------     -----------------------------
   1        1       Static           13       1       Static
   2        1       Static           14       1       Static
   3        1       Static           15       1       Static
   4        1       Static           16       1       Static
   5        1       Static           17       1       Static
   6        1       Static           18       1       Static
   7        1       Static           19       1       Static
   8        1       Static           20       1       Static
   9        1       Static           21       1       Static
   10       1       Static           22       1       Static
   11       1       Static           23       1       Static
   12       1       Static           24       1       Static
                                     AUI      1       Static
   A        1       Static
   B        1       Static

     [M] Membership type                 [V] VLAN assignment
     [R] Reconfirm dynamic membership    [X] Exit to previous menu
```

Select option V to assign a port to a specific VLAN. In this case, assign port 2 to VLAN 2.

```
Enter Selection:  V


This command assigns or reassigns ports to a VLAN.  If the port is configured
with dynamic VLAN membership, then assigning VLAN 0 causes the discovery of
VLAN membership to start all over again.

Port numbers should be separated by commas or spaces.  A port
number range may also be specified. The word ALL indicates all ports.
Example:  1, 7-11, AUI, 4, A

Enter port numbers:  2
```

```
Identify VLAN 0 - VLAN 1005:
Select [0-1005] 2
```

Once port 2 has been added to VLAN 2, the VLAN membership configuration appears, as shown in the following output. In addition, port 4 has to be added to VLAN 2 and ports 22 through 24 have to be added to VLAN 3. Remember that VLAN 3 needs to be defined before you can add ports to it:

```
Catalyst 1900 - VLAN Membership Configuration

   Port  VLAN    Membership Type      Port  VLAN    Membership Type
   -----------------------------      -----------------------------
   1        1       Static            13       1       Static
   2        2       Static            14       1       Static
   3        1       Static            15       1       Static
   4        1       Static            16       1       Static
   5        1       Static            17       1       Static
   6        1       Static            18       1       Static
   7        1       Static            19       1       Static
   8        1       Static            20       1       Static
   9        1       Static            21       1       Static
   10       1       Static            22       1       Static
   11       1       Static            23       1       Static
   12       1       Static            24       1       Static
                                      AUI      1       Static
   A        1       Static
   B        1       Static

     [M] Membership type               [V] VLAN assignment
     [R] Reconfirm dynamic membership  [X] Exit to previous menu
```

The following screen shows how the VLAN membership configuration should appear on the Catalyst 1920 after all the ports have been assigned to their proper VLANs:

```
Catalyst 1900 - VLAN Membership Configuration

   Port  VLAN    Membership Type      Port  VLAN    Membership Type
   -----------------------------      -----------------------------
   1        1       Static            13       1       Static
   2        2       Static            14       1       Static
   3        1       Static            15       1       Static
   4        2       Static            16       1       Static
   5        1       Static            17       1       Static
   6        1       Static            18       1       Static
   7        1       Static            19       1       Static
   8        1       Static            20       1       Static
   9        1       Static            21       1       Static
   10       1       Static            22       3       Static
   11       1       Static            23       3       Static
   12       1       Static            24       3       Static
                                      AUI      1       Static
   A        1       Static
   B        1       Static

     [M] Membership type               [V] VLAN assignment
     [R] Reconfirm dynamic membership  [X] Exit to previous menu
```

Now take a look at some of the other important Catalyst 1920 screens that can help when troubleshooting a configuration. As shown in the following output, each of the ports on the Catalyst switch maintain detailed statistics reports on the port status, frame counts, and errored frames:

```
Catalyst 1900 - Port 1 Statistics Report
    Receive Statistics                       Transmit Statistics
-------------------------------------- --------------------------------------
Total good frames                  1203 Total frames                      2111
Total octets                     103229 Total octets                    174473
Broadcast/multicast frames          982 Broadcast/multicast frames        2103
Broadcast/multicast octets        88875 Broadcast/multicast octets      173749
Good frames forwarded              1028 Deferrals                            1
Frames filtered                     175 Single collisions                    0
Runt frames                           0 Multiple collisions                  0
No buffer discards                    0 Excessive collisions                 0
                                        Queue full discards                  0
Errors:                                 Errors:
  FCS errors                          0   Late collisions                    0
  Alignment errors                    0   Excessive deferrals                0
  Giant frames                        0   Jabber errors                      0
  Address violations                  0   Other transmit errors              0

Select [A] Port addressing, [C] Configure port,
       [N] Next port, [P] Previous port, [G] Goto port,
       [R] Reset port statistics, or [X] Exit to Main Menu:
```

The Catalyst 1920 has several important status screens that can be used to verify that the attached VLANs and Ethernet ports are functioning properly. These reports can be accessed via the Usage Summaries menu, shown here:

```
Catalyst 1900 - Usage Summaries

     [P] Port Status Report
     [A] Port Addressing Report
     [E] Exception Statistics Report
     [U] Utilization Statistics Report
     [B] Bandwidth Usage Report

     [X] Exit to Main Menu

Enter Selection:  P
```

The port status report can be used to verify that specific ports are connected and functional. You can see from the following output that ports 1 through 4 and 22 through 24 have been enabled:

```
Catalyst 1900 - Port Status Report

 1   : Enabled                          13 : Suspended-no-linkbeat
 2   : Enabled                          14 : Suspended-no-linkbeat
 3   : Enabled                          15 : Suspended-no-linkbeat
 4   : Enabled                          16 : Suspended-no-linkbeat
 5   : Suspended-no-linkbeat            17 : Suspended-no-linkbeat
```

```
 6  : Suspended-no-linkbeat         18 : Suspended-no-linkbeat
 7  : Suspended-no-linkbeat         19 : Suspended-no-linkbeat
 8  : Suspended-no-linkbeat         20 : Suspended-no-linkbeat
 9  : Suspended-no-linkbeat         21 : Suspended-no-linkbeat
 10 : Suspended-no-linkbeat         22 : Enabled
 11 : Suspended-no-linkbeat         23 : Enabled
 12 : Suspended-no-linkbeat         24 : Enabled
                                    AUI: Enabled
 A  : Suspended-no-linkbeat
 B  : Suspended-no-linkbeat

 Monitor port: None, Network port: A, Trunk port: None
```

The port addressing report is used to view the MAC addresses of stations that are attached to specific ports of the Catalyst switch:

```
Catalyst 1900 - Port Addressing Report

Port                Addresses             Port                Addresses
------------------------------------- --------------------------------------
 1  :Dynamic        00-E0-1E-5B-26-11      13 :               Unaddressed
 2  :               Unaddressed            14 :               Unaddressed
 3  :Dynamic 2      Static 0               15 :               Unaddressed
 4  :Dynamic        00-E0-1E-5B-50-61      16 :               Unaddressed
 5  :               Unaddressed            17 :               Unaddressed
 6  :               Unaddressed            18 :               Unaddressed
 7  :               Unaddressed            19 :               Unaddressed
 8  :               Unaddressed            20 :               Unaddressed
 9  :               Unaddressed            21 :               Unaddressed
 10 :               Unaddressed            22 :Dynamic        00-E0-1E-5B-0D-71
 11 :               Unaddressed            23 :Dynamic        00-E0-1E-5B-26-01
 12 :               Unaddressed            24 :               Unaddressed
                                           AUI:               Unaddressed
 A  :               Unaddressed
 B  :               Unaddressed

Select [X] Exit to previous menu:  X
```

The Exception Statistics report is used to view port errors and security violations on the switch:

```
Catalyst 1900 - Exception Statistics Report (Frame counts)

        Receive    Transmit  Security             Receive    Transmit  Security
        Errors     Errors    Violations           Errors     Errors    Violations
     ------------------------------          ------------------------------
 1  :          0          0          0   13 :          0          0          0
 2  :          0          0          0   14 :          0          0          0
 3  :          0          0          0   15 :          0          0          0
 4  :          0          0          0   16 :          0          0          0
 5  :          0          0          0   17 :          0          0          0
 6  :          0          0          0   18 :          0          0          0
 7  :          0          0          0   19 :          0          0          0
 8  :          0          0          0   20 :          0          0          0
 9  :          0          0          0   21 :          0          0          0
 10 :          0          0          0   22 :          0          0          0
```

```
11 :          0           0           0  23 :          0           0           0
12 :          0           0           0  24 :          0           0           0
                                         AUI:          0           0           0
A  :          0           0           0
B  :          0           0           0
```

The Utilization Statistics report is used to monitor per-port transmission information on the switch:

```
Catalyst 1900 - Utilization Statistics Report (Frame counts)

          Receive     Forward    Transmit          Receive     Forward    Transmit
       --------------------------------         --------------------------------
1  :       1249        1072        2190  13 :          0           0           1
2  :        986         779        2015  14 :          0           0           1
3  :       1295        1070        2207  15 :          0           0           1
4  :       2244        2022         775  16 :          0           0           1
5  :          0           0           1  17 :          0           0           1
6  :          0           0           1  18 :          0           0           1
7  :          0           0           1  19 :          0           0           1
8  :          0           0           1  20 :          0           0           1
9  :          0           0           1  21 :          0           0           0
10 :          0           0           1  22 :        649         428        1659
11 :          0           0           1  23 :        785         565        1522
12 :          0           0           1  24 :          0           0           0
                                         AUI:          0           0        3137
A  :          0           0           0
B  :          0           0           0
```

The Bandwidth Usage report is used to monitor current and peak utilization on the switch. The switch records the switch's peak bandwidth as well as the time that the peak bandwidth occurred.

```
Catalyst 1900 - Bandwidth Usage Report

    ----------------------- Information -----------------------------------

     Current Bandwidth Usage                        0Mbps
     Peak Bandwidth Usage during this interval      0Mbps
     Peak Time recorded since start up              0d 00h 00m 45s

    ----------------------- Settings --------------------------------------

     [T] Capture time interval                      24 hour(s)
     [R] Reset capture
     [X] Exit to previous menu
```

The port configuration screen is used to verify and/or change port-specific parameters on the switch:

```
Catalyst 1900 - Port 1 Configuration

        Built-in 10Base-T
        802.1d STP State:  Forwarding     Forward Transitions:  1
```

```
---------------------- Settings ------------------------------------
 [D] Description/name of port
 [S] Status of port                            Enabled
 [F] Full duplex                               Disabled
 [I] Port priority (spanning tree)             128 (80 hex)
 [C] Path cost (spanning tree)                 100
 [H] Port fast mode (spanning tree)            Enabled

---------------------- Related Menus -------------------------------
 [A] Port addressing            [V] View port statistics
 [N] Next port                  [G] Goto port
 [P] Previous port              [X] Exit to Main Menu
```

The VLAN list screen is used to view every VLAN that is configured on the switch. Notice that your three VLANs (1, 2, and 3) have been configured on the switch:

```
This command displays a list of specified VLANs. You may specify a list
of VLAN numbers to display. A VLAN number ranges between 1-1005.
XB1 and XB2 in the display stands for Translational Bridge 1 and 2.

VLAN numbers should be separated by commas or spaces. A VLAN number range
may also be specified. The word ALL indicates all configured VLANs.
Example: 1, 2, 10-20, all

Enter VLAN numbers: all


VLAN  Name                State      Type             MTU   SAID      XB1  XB2
------------------------------------------------------------------------------
1     default             Enabled    Ethernet         1500  100001    1002 1003
2     VLAN0002            Enabled    Ethernet         1500  100002    0    0
3     VLAN0003            Enabled    Ethernet         1500  100003    0    0
1002  fddi-default        Suspended  FDDI             1500  101002    1    1003
1003  token-ring-defau    Suspended  Token-Ring       1500  101003    1    1002
1004  fddinet-default     Suspended  FDDI-Net         1500  101004    0    0
1005  trnet-default       Suspended  Token-Ring-Net   1500  101005    0    0
```

11. *Configure port 5 on the Catalyst switch to monitor all traffic from ports 1 and 2.*

Because the Catalyst is a switch, it behaves differently than a traditional hub. A given port only sees broadcast traffic for its VLAN, spanning tree BPDU packets, and any traffic for that specific destination port. In order to be able to put a LAN sniffer on the Catalyst switch and see traffic for multiple ports, a monitor port needs to be configured.

Three items need to be defined to configure a monitor port:

- You need to enable capturing monitor frames.
- You have to decide which port will act as the monitor port.
- You need to define which ports will have their traffic sent to the monitor port.

The monitor screen is accessed by selecting Monitoring (Option M) from the Main menu:

```
Catalyst 1900 - Main Menu

      [C] Console Settings
      [S] System
      [N] Network Management
      [P] Port Configuration
      [A] Port Addressing
      [D] Port Statistics Detail
      [M] Monitoring
      [V] Virtual LAN
      [R] Multicast Registration
      [F] Firmware
      [I] RS-232 Interface
      [U] Usage Summaries
      [H] Help
      [K] Command Line

      [X] Exit Management Console

Enter Selection:  M
```

Notice from the following monitoring configuration screen that monitoring is disabled by default, no monitor port has been assigned, and no ports have been added to the capture list:

```
Catalyst 1900 - Monitoring Configuration

     ----------------------- Settings ---------------------------------------
      [C] Capturing frames to the Monitor                 Disabled
      [M] Monitor port assignment                         None
      Current capture list:  No ports in list

     ----------------------- Actions ----------------------------------------
      [A] Add ports to capture list
      [D] Delete ports from capture list

      [X] Exit to Main Menu
```

Select option C to enable the capturing of frames to a monitor port:

```
Enter Selection:  C


This command enables or disables the monitoring (capturing) of frames from
ports that have been added to the capture list.
Actual monitoring takes places only if all of the following information
has been properly configured:  1) the capturing status, 2) the identity of
a port to which monitored frames are sent, and 3) a non-empty capture list.

Capturing frames to the Monitor may be [E]nabled or [D]isabled:

Current setting ===> Disabled

    New setting ===> Enabled
```

The capturing of frames has been enabled. Next, select option M to enable the monitor port.

```
Catalyst 1900 - Monitoring Configuration

     ---------------------- Settings -------------------------------------
      [C] Capturing frames to the Monitor             Enabled
      [M] Monitor port assignment                     None
      Current capture list:  No ports in list

     ---------------------- Actions --------------------------------------
      [A] Add ports to capture list
      [D] Delete ports from capture list

      [X] Exit to Main Menu

Enter Selection:  M


This command supplies the identity of the port to which monitored frames
are sent.

Actual monitoring takes places only if all of the following information
has been properly configured:  1) the capturing status, 2) the identity of
a port to which monitored frames are sent, and 3) a non-empty capture list.

Identify Port:  1 to 24[1-24], [AUI], [A], [B], or [N]one:
Select [1 - 24, AUI, A, B, N]:

Current setting ===> None

    New setting ===> 5  ← Port 5 will be the monitor port
```

The last step is to assign ports whose traffic will be captured and sent to the monitor port. Select option A to add ports to the capture list.

```
Catalyst 1900 - Monitoring Configuration

     ---------------------- Settings -------------------------------------
      [C] Capturing frames to the Monitor             Enabled
      [M] Monitor port assignment                     5
      Current capture list:  No ports in list

     ---------------------- Actions --------------------------------------
      [A] Add ports to capture list
      [D] Delete ports from capture list

      [X] Exit to Main Menu

Enter Selection:  A


This command adds ports to the capture list.
Actual monitoring takes places only if all of the following information
has been properly configured:  1) the capturing status, 2) the identity of
a port to which monitored frames are sent, and 3) a non-empty capture list.

Port numbers should be separated by commas or spaces.  A port
number range may also be specified. The word ALL indicates all ports.
Example:  1, 7-11, AUI, 4, A

Enter port numbers:  1-2  ← Ports 1 and 2 are added to the capture list
```

We see from the following output that the monitor port has been configured on the Catalyst:

```
Catalyst 1900 - Monitoring Configuration

    ----------------------- Settings ----------------------------------------
     [C] Capturing frames to the Monitor             Enabled
     [M] Monitor port assignment                     5
     Current capture list:  1-2

    ----------------------- Actions -----------------------------------------
     [A] Add ports to capture list
     [D] Delete ports from capture list

     [X] Exit to Main Menu

Enter Selection:
```

12. *Verify and record the version of code running on the Catalyst switch.*

The firmware configuration screen is used to verify the version of code that is running on the Catalyst switch. It also can be used to check on how much memory is in the Catalyst switch and to initiate a firmware upgrade.

The firmware configuration screen is accessed by selecting option F on the Main menu:

```
Catalyst 1900 - Main Menu

     [C] Console Settings
     [S] System
     [N] Network Management
     [P] Port Configuration
     [A] Port Addressing
     [D] Port Statistics Detail
     [M] Monitoring
     [V] Virtual LAN
     [R] Multicast Registration
     [F] Firmware
     [I] RS-232 Interface
     [U] Usage Summaries
     [H] Help
     [K] Command Line

     [X] Exit Management Console

Enter Selection:  F


 Catalyst 1900 - Firmware Configuration

    ----------------------- System Information ----------------------------
     FLASH:  1024K bytes
     V8.01.01     : Enterprise Edition
     Upgrade status:
     No upgrade currently in progress.
```

```
    ---------------------- Settings --------------------------------------
     [S] TFTP Server name or IP address
     [F] Filename for firmware upgrades
     [A] Accept upgrade transfer from other hosts    Enabled

    ---------------------- Actions ---------------------------------------
     [U] System XMODEM upgrade           [D] Download test subsystem (XMODEM)
     [T] System TFTP upgrade             [X] Exit to Main Menu
```

You can see that Version 8.01.01, Enterprise Edition is running on this switch.

13. *Change the switching mode of the Catalyst to Store-and-Forward.*

The Catalyst 1920 supports two switching modes: FragmentFree and Store-and-Forward. FragmentFree switching forwards frames after 64 bytes of the frame have been received. The Store-and-Forward switching mode waits until the entire frame has been received before forwarding it. To modify the switching mode of the Catalyst switch, select option S from the Main menu:

```
Catalyst 1900 - Main Menu

     [C] Console Settings
     [S] System
     [N] Network Management
     [P] Port Configuration
     [A] Port Addressing
     [D] Port Statistics Detail
     [M] Monitoring
     [V] Virtual LAN
     [R] Multicast Registration
     [F] Firmware
     [I] RS-232 Interface
     [U] Usage Summaries
     [H] Help
     [K] Command Line

     [X] Exit Management Console

Enter Selection:  S
```

Notice that the switch is currently set to FragmentFree switching mode.

```
Catalyst 1900 - System Configuration
        System Revision:  5   Address Capacity:  1024
        System UpTime:    0day(s) 00hour(s) 39minute(s) 06second(s)

    ---------------------- Settings --------------------------------------
     [N] Name of system
     [C] Contact name
     [L] Location
     [S] Switching mode                              FragmentFree
     [U] Use of store-and-forward for multicast      Disabled
     [A] Action upon address violation               Suspend
     [G] Generate alert on address violation         Enabled
     [I] Address aging time                          300 second(s)
     [P] Network port                                A
```

```
     [H] Half duplex back pressure   (10-mbps ports) Disabled
     [E] Enhanced congestion control (10-mbps ports) Disabled

    ----------------------- Actions -------------------------------------
     [R] Reset system                      [F] Reset to factory defaults
     [V] Reset VTP to factory defaults     [T] Reset to enable Bridge Groups
    ----------------------- Related Menus ---------------------------------
     [B] Broadcast storm control           [X] Exit to Main Menu

Enter Selection:  S


FragmentFree switching mode reduces bridge delay by making the
forwarding decision after 64 bytes have been received.
In contrast, Store-and-Forward switching mode
waits until the entire frame has been received before the forwarding
decision is made.  This command sets the switching mode.

Select Store-and-Forward[1], or FragmentFree[2]:

Current setting ===> FragmentFree

    New setting ===> Store-and-Forward  ← Select the new switching mode.
```

The switching mode has now been set to Store-and-Forward.

```
Catalyst 1900 - System Configuration
        System Revision:  5   Address Capacity:  1024
        System UpTime:    0day(s) 00hour(s) 39minute(s) 26second(s)

    ----------------------- Settings -------------------------------------
     [N] Name of system
     [C] Contact name
     [L] Location
     [S] Switching mode                              Store-and-Forward
     [U] Use of store-and-forward for multicast      Disabled
     [A] Action upon address violation               Suspend
     [G] Generate alert on address violation         Enabled
     [I] Address aging time                          300 second(s)
     [P] Network port                                A
     [H] Half duplex back pressure   (10-mbps ports) Disabled
     [E] Enhanced congestion control (10-mbps ports) Disabled

    ----------------------- Actions -------------------------------------
     [R] Reset system                      [F] Reset to factory defaults
     [V] Reset VTP to factory defaults     [T] Reset to enable Bridge Groups
    ----------------------- Related Menus ---------------------------------
     [B] Broadcast storm control           [X] Exit to Main Menu

Enter Selection:  X
```

14. *Change the baud rate of the console port of the Catalyst switch to 19200 bits per second.*

The console port settings are modified on the RS-232 interface screen. Select option I from the Main menu to access this screen.

```
Catalyst 1900 - Main Menu

     [C] Console Settings
```

```
     [S] System
     [N] Network Management
     [P] Port Configuration
     [A] Port Addressing
     [D] Port Statistics Detail
     [M] Monitoring
     [V] Virtual LAN
     [R] Multicast Registration
     [F] Firmware
     [I] RS-232 Interface
     [U] Usage Summaries
     [H] Help
     [K] Command Line

     [X] Exit Management Console

Enter Selection:  I
```

Select option B to change the baud rate.

```
Catalyst 1900 - RS-232 Interface Configuration

    ----------------------- Group Settings -------------------------------
     [B] Baud rate                                    9600 baud
     [D] Data bits                                    8 bit(s)
     [S] Stop bits                                    1 bit(s)
     [P] Parity setting                               None

    ----------------------- Settings -------------------------------------
     [M] Match remote baud rate (auto baud)           Enabled
     [A] Auto answer                                  Enabled
     [N] Number for dial-out connection
     [T] Time delay between dial attempts             300
     [I] Initialization string for modem

    ----------------------- Actions --------------------------------------
     [C] Cancel and restore previous group settings
     [G] Activate group settings

     [X] Exit to Main Menu

Enter Selection:  B


This command sets the speed in baud for the RS-232 interface.
Valid values are:  2400, 9600, 19200, 38400 and 57600.

This command parameter is part of a group of settings consisting
of the baud rate, the number of data and stop bits, and the
parity value.  These settings may be individually changed but new
values will only take effect when the entire group is activated.
Type G to activate group settings when done with selection.

Enter baud rate:

Current setting ===> 9600 baud

    New setting ===> 19200  ← New Baud Rate
```

```
 Catalyst 1900 - RS-232 Interface Configuration

 ---------------------- Group Settings -------------------------------
  [B] Baud rate                                     19200 baud
  [D] Data bits                                     8 bit(s)
  [S] Stop bits                                     1 bit(s)
  [P] Parity setting                                None

 ---------------------- Settings -------------------------------------
  [M] Match remote baud rate (auto baud)            Enabled
  [A] Auto answer                                   Enabled
  [N] Number for dial-out connection
  [T] Time delay between dial attempts              300
  [I] Initialization string for modem

 ---------------------- Actions --------------------------------------
  [C] Cancel and restore previous group settings
  [G] Activate group settings

  [X] Exit to Main Menu

Enter Selection:  X
```

15. *Configure the network to run OSPF. Use PPP encapsulation on all serial links.*

The basic addressing, encapsulation, clocking, and OSPF parameters need to be configured on each of your six routers. On R1, you need to configure four interfaces. Notice from the **show controller command** output shown below that S1/0 is a DCE interface and needs a clock rate statement under the interface.

```
R1#sh contr s 1/0
Interface Serial1/0
Hardware is Quicc 68360
DCE V.35, no clock  ← A DCE interface needs to have a clockrate defined
```

Because all your addresses start with 152, you can use a single network statement under your OSPF configuration to enable OSPF on all five router interfaces.

```
R1(config)# interface Loopback0
R1(config-if)# ip address 152.3.1.1 255.255.255.0
R1(config-if)# interface Ethernet0/0
R1(config-if)# ip address 152.3.7.1 255.255.255.192
R1(config-if)# interface Ethernet1/0
R1(config-if)# ip address 152.3.10.13 255.255.255.252
R1(config-if)# interface Serial1/0
R1(config-if)# ip address 152.3.10.6 255.255.255.252
R1(config-if)# encapsulation ppp
R1(config-if)# clockrate 800000
R1(config-if)# router ospf 64
R1(config-if)# network 152.0.0.0 0.255.255.255 area 0
```

R2 needs to have four interfaces configured. Note that S0/0 and S0/1 are both DCE interfaces.

```
R2(config)# interface Loopback0
R2(config-if)# ip address 152.3.2.1 255.255.255.0
R2(config-if)# interface Ethernet0/0
R2(config-if)# ip address 152.3.10.9 255.255.255.252
R2(config-if)# interface Serial0/0
R2(config-if)# ip address 152.3.10.25 255.255.255.252
R2(config-if)# encapsulation ppp
R2(config-if)# clockrate 800000
R2(config-if)# interface Serial0/1
R2(config-if)# ip address 152.3.10.2 255.255.255.252
R2(config-if)# encapsulation ppp
R2(config-if)# clockrate 800000
R2(config-if)# router ospf 64
R2(config-if)# network 152.0.0.0 0.255.255.255 area 0
```

R3 needs to have four interfaces configured. Notice that you have to set the LAN speed on Token Ring interface T1/0 .

```
R3(config)# interface Loopback0
R3(config-if)# ip address 152.3.3.1 255.255.255.0
R3(config-if)# interface Serial1/0
R3(config-if)# ip address 152.3.10.5 255.255.255.252
R3(config-if)# encapsulation ppp
R3(config-if)# interface TokenRing1/0
R3(config-if)# ip address 152.3.8.1 255.255.255.240
R3(config-if)# ring-speed 4
R3(config-if)# interface Serial1/1
R3(config-if)# ip address 152.3.10.1 255.255.255.252
R3(config-if)# encapsulation ppp
R3(config-if)# router ospf 64
R3(config-if)# network 152.0.0.0 0.255.255.255 area 0
```

R4 needs five interfaces configured. All serial interfaces on R4 are DTE interfaces.

```
R4(config)# interface Loopback0
R4(config-if)# ip address 152.3.4.1 255.255.255.0
R4(config-if)# interface Ethernet0/0
R4(config-if)# ip address 152.3.10.10 255.255.255.252
R4(config-if)# interface Serial0/0
R4(config-if)# ip address 152.3.10.17 255.255.255.252
R4(config-if)# encapsulation ppp
R4(config-if)# interface Serial0/1
R4(config-if)# ip address 152.3.10.26 255.255.255.252
R4(config-if)# encapsulation ppp
R4(config-if)# interface FastEthernet1/0
R4(config-if)# ip address 152.3.10.14 255.255.255.252
R4(config-if)# router ospf 64
R4(config-if)# network 152.0.0.0 0.255.255.255 area 0
```

R5 needs to have six interfaces configured. One of the serial interfaces is a DCE interface. R5 also has an ISDN BRI interface. For now, only configure an IP address and serial encapsulation (PPP) on this BRI interface.

```
R5(config)# interface Loopback0
R5(config-if)# ip address 152.3.5.1 255.255.255.0
R5(config-if)# interface Ethernet0/0
R5(config-if)# ip address 152.3.7.129 255.255.255.192
R5(config-if)# interface Serial0/0
R5(config-if)# ip address 152.3.10.18 255.255.255.252
R5(config-if)# encapsulation ppp
R5(config-if)# clockrate 800000
R5(config-if)# interface Serial0/1
R5(config-if)# ip address 152.3.10.21 255.255.255.252
R5(config-if)# encapsulation ppp
R5(config-if)# interface BRI1/0
R5(config-if)# ip address 152.3.9.1 255.255.255.248
R5(config-if)# encapsulation ppp
R5(config-if)# interface Ethernet1/0
R5(config-if)# ip address 152.3.7.2 255.255.255.192
R5(config-if)# router ospf 64
R5(config-if)# network 152.0.0.0 0.255.255.255 area 0
```

R6 has four interfaces that need to be configured. Interface S1 is a DCE interface, so you have to set the clock rate. For now, ISDN BRI interface BRI0 only has its IP address and serial encapsulation set.

```
R6(config)# interface Loopback0
R6(config-if)# ip address 152.3.6.1 255.255.255.0
R6(config-if)# interface Ethernet0
R6(config-if)# ip address 152.3.7.193 255.255.255.192
R6(config-if)# interface Serial1
R6(config-if)# ip address 152.3.10.22 255.255.255.252
R6(config-if)# encapsulation ppp
R6(config-if)# clockrate 800000
R6(config-if)# interface BRI0
R6(config-if)# ip address 152.3.9.2 255.255.255.248
R6(config-if)# encapsulation ppp
R6(config-if)# router ospf 64
R6(config-if)# network 152.0.0.0 0.255.255.255 area 0
```

Once the IP addresses and OSPF statements have been entered into each of the six routers, the routing tables should properly reflect the topology of the network. The routing tables of R3 and R5 are shown below. Notice that all the networks are being learned via OSPF.

R3's Routing Table

```
R3#show ip route
Codes: C - connected, S - static, I - IGRP, R - RIP, M - mobile, B - BGP
       D - EIGRP, EX - EIGRP external, O - OSPF, IA - OSPF inter area
       N1 - OSPF NSSA external type 1, N2 - OSPF NSSA external type 2
```

```
       E1 - OSPF external type 1, E2 - OSPF external type 2, E - EGP
       i - IS-IS, L1 - IS-IS level-1, L2 - IS-IS level-2, * - candidate default
       U - per-user static route, o - ODR

Gateway of last resort is not set

     152.3.0.0/16 is variably subnetted, 17 subnets, 6 masks
O       152.3.7.128/26 [110/84] via 152.3.10.6, 00:04:17, Serial1/0
O       152.3.10.16/30 [110/122] via 152.3.10.2, 00:04:17, Serial1/1
                       [110/122] via 152.3.10.6, 00:04:17, Serial1/0
O       152.3.10.24/30 [110/122] via 152.3.10.2, 00:04:17, Serial1/1
                       [110/122] via 152.3.10.6, 00:04:17, Serial1/0
C       152.3.10.0/30 is directly connected, Serial1/1
O       152.3.9.0/29 [110/1636] via 152.3.10.6, 00:04:17, Serial1/0
C       152.3.8.0/28 is directly connected, TokenRing1/0
C       152.3.10.2/32 is directly connected, Serial1/1
C       152.3.10.4/30 is directly connected, Serial1/0
C       152.3.10.6/32 is directly connected, Serial1/0
O       152.3.2.1/32 [110/65] via 152.3.10.2, 00:04:17, Serial1/1
C       152.3.3.0/24 is directly connected, Loopback0
O       152.3.10.8/30 [110/74] via 152.3.10.2, 00:04:17, Serial1/1
O       152.3.1.1/32 [110/65] via 152.3.10.6, 00:04:17, Serial1/0
O       152.3.7.0/26 [110/74] via 152.3.10.6, 00:04:17, Serial1/0
O       152.3.10.12/30 [110/74] via 152.3.10.6, 00:04:18, Serial1/0
O       152.3.4.1/32 [110/75] via 152.3.10.2, 00:04:18, Serial1/1
                     [110/75] via 152.3.10.6, 00:04:18, Serial1/0
152.3.5.1/32 [110/75] via 152.3.10.6, 00:04:18, Serial1/0
```

R5's Routing Table

```
R5#show ip route
Codes: C - connected, S - static, I - IGRP, R - RIP, M - mobile, B - BGP
       D - EIGRP, EX - EIGRP external, O - OSPF, IA - OSPF inter area
       N1 - OSPF NSSA external type 1, N2 - OSPF NSSA external type 2
       E1 - OSPF external type 1, E2 - OSPF external type 2, E - EGP
       i - IS-IS, L1 - IS-IS level-1, L2 - IS-IS level-2, * - candidate default
       U - per-user static route, o - ODR

Gateway of last resort is not set

     152.3.0.0/16 is variably subnetted, 16 subnets, 6 masks
C       152.3.7.128/26 is directly connected, Ethernet0/0
C       152.3.10.17/32 is directly connected, Serial0/0
C       152.3.10.16/30 is directly connected, Serial0/0
O       152.3.10.24/30 [110/68] via 152.3.7.1, 00:05:17, Ethernet1/0
O       152.3.10.0/30 [110/94] via 152.3.7.1, 00:05:17, Ethernet1/0
C       152.3.9.0/29 is directly connected, BRI1/0
O       152.3.8.0/28 [110/99] via 152.3.7.1, 00:05:17, Ethernet1/0
O       152.3.10.4/30 [110/74] via 152.3.7.1, 00:05:17, Ethernet1/0
O       152.3.2.1/32 [110/31] via 152.3.7.1, 00:05:17, Ethernet1/0
O       152.3.10.8/30 [110/30] via 152.3.7.1, 00:05:17, Ethernet1/0
O       152.3.3.1/32 [110/75] via 152.3.7.1, 00:05:17, Ethernet1/0
O       152.3.1.1/32 [110/11] via 152.3.7.1, 00:05:17, Ethernet1/0
C       152.3.7.0/26 is directly connected, Ethernet1/0
O       152.3.10.12/30 [110/20] via 152.3.7.1, 00:05:17, Ethernet1/0
O       152.3.4.1/32 [110/21] via 152.3.7.1, 00:05:17, Ethernet1/0
C       152.3.5.0/24 is directly connected, Loopback0
```

16. *Configure DLSW between LAN_4 (connected to R3) and LAN_2 (connected to R1 and R5). R1 should be the preferred peer to R3. If R1 fails, then R3 should automatically peer to R5.*

Figure 5-3 shows the required DLSW configuration. Notice that the primary DLSW circuit is between R3 and R1, while the backup DLSW circuit is between R3 and R5.

As shown in Figure 5-3, connect four Windows 95 workstations to your test bed. These four workstations will connect to routers R1, R3, R5, and R6. Use these workstations to verify that your bridging configurations are working properly.

Each of the four workstations is configured to only run NetBEUI, which is a non-routable protocol. The only way to transport NetBEUI across a network is via bridging or DLSW. Also make sure that you do not enable the IP protocol on your four workstations. This ensures that all the traffic between your workstations is bridged and not routed. The following figures demonstrate how to configure a workstation to run Windows networking using only the NetBEUI protocol.

Figure 5-4 shows the Windows Network Configuration screen. This screen is displayed when the Network icon is selected in the Windows control panel. Notice that the only protocol selected is NetBEUI. In addition, the Client for Microsoft Networks, the Microsoft Family Logon, and File and Printer Sharing for Microsoft Networks have been loaded. Also notice that your network card on the computer, shown in Figure 5-4, is a 3Com 10/100 card.

Clicking on the File and Print Sharing box brings you to the screen shown in Figure 5-5. You need to select file access in order for others to have access to your files.

Clicking on the network adapter card shown in Figure 5-4 takes you to the network card configuration screens. Figure 5-6 shows that the only protocol bound to the network adapter card is NetBEUI.

Clicking on the Identification tab shown in Figure 5-4 displays the identification screen shown in Figure 5-7. Notice that the name of this computer has been set to System3.

Multiple ways are available for determining the MAC address of a network card. 3COM installs a utility on the taskbar that can supply this information as well as provide network card diagnostics. Notice from Figure 5-8 that the MAC address of the network card on System3 is 00 00 86 35 46 E1.

Once a system has been configured with the previously shown steps, it will be ready to run Windows Networking. However, the system needs to be restarted before the changes take effect. Once the system reloads, supply a network login password to log into the network. The password is arbitrary and can be set to any value.

The question requires you to configure DLSW between R3 and R1. If R1 fails, then R3 should peer with R5. This can be accomplished with a DLSW backup peer. With DLSW backup peers, you configure a primary set and a backup set of peers. The DLSW peers

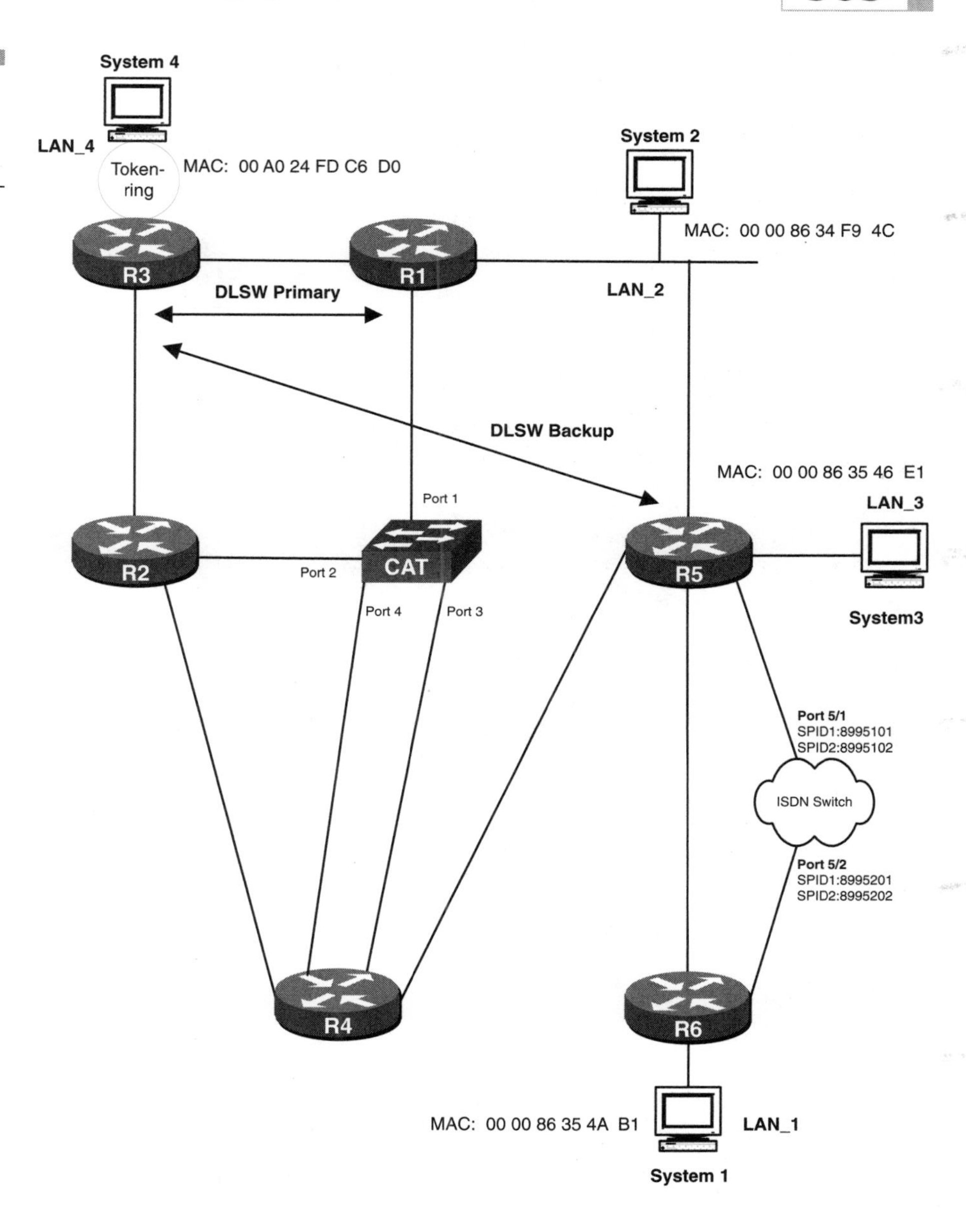

Figure 5-3
DLSW configuration and MAC address assignment

automatically switch to the backup peers once the connection between the primary peers is declared down.

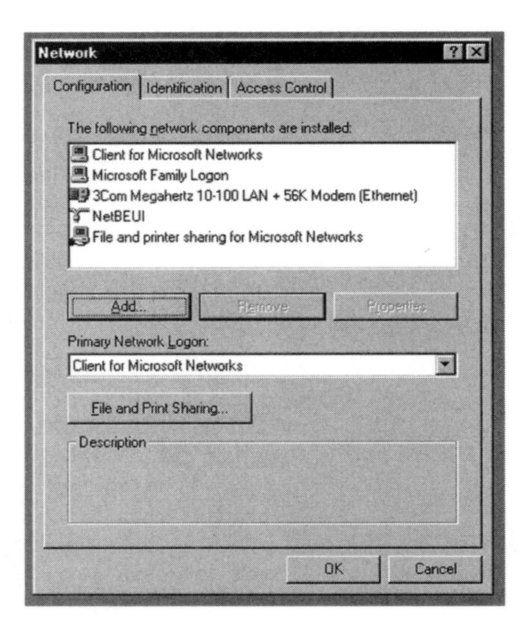

Figure 5-4 Windows networking configuration

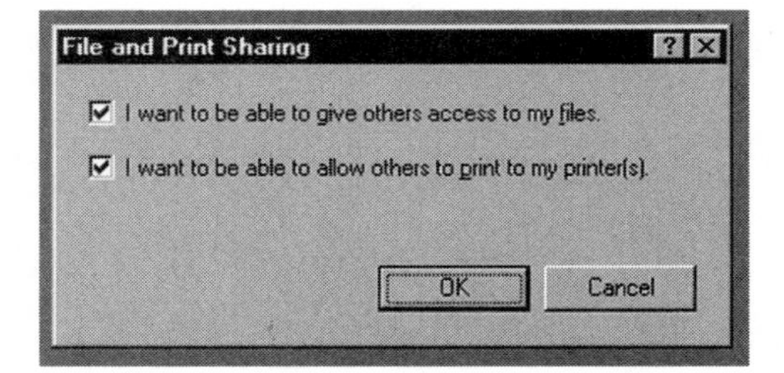

Figure 5-5 Windows file and print sharing configuration

Start by configuring R3, you must first define a virtual ring for your DLSW configuration. A virtual ring is a logically defined ring. It makes a network appear as a bridged network, when in reality, traffic is bridged at the edges of the network and transported in TCP.

```
R3(config)# source-bridge ring-group 169
```

Next, define your Token-Ring interface (T1/0) as your local DLSW peer id:

```
R3(config)# dlsw local-peer peer-id 152.3.8.1
```

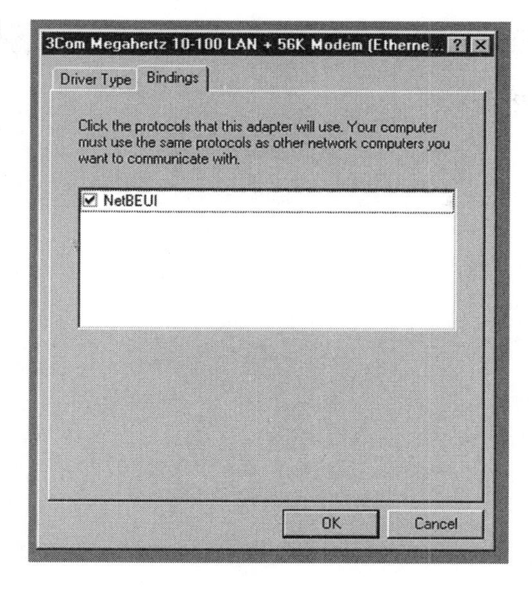

Figure 5-6 Network card bindings

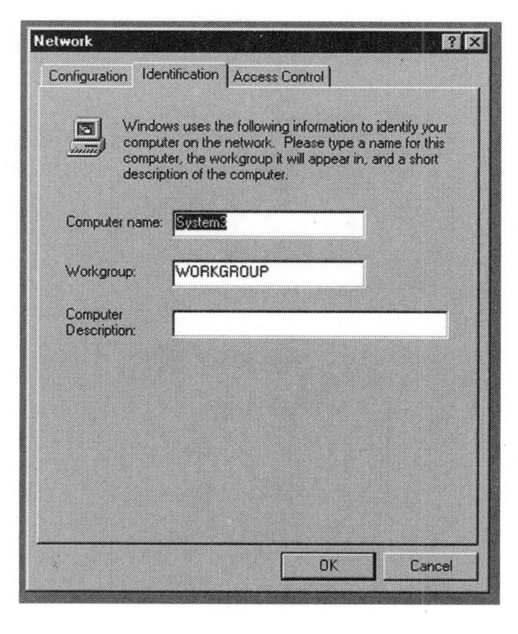

Figure 5-7 Network identification

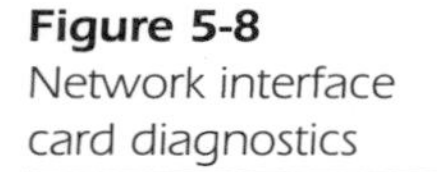

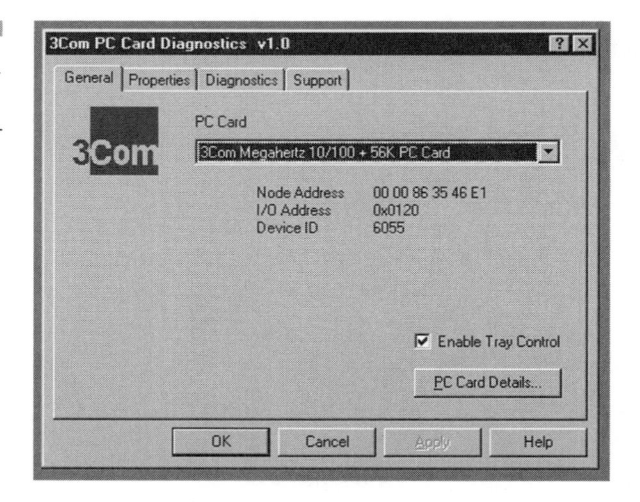

Figure 5-8 Network interface card diagnostics

Configuring a DLSW backup peer is a bit confusing the first time you do it. Initially, you need to configure your primary remote peer (152.3.7.1) with the **DLSW remote-peer** statement:

```
R3(config)# dlsw remote-peer 0 tcp 152.3.7.1
```

Next, configure your backup peer (152.3.7.2) as your remote peer, naming your primary peer (152.3.7.1) with the **backup-peer** command option. The syntax of the command may at first seem to be backwards, but it can be interpreted as follows: Configure 152.3.7.2 to be a remote DLSW backup peer. 152.3.7.2 should backup the primary peer at address 152.3.7.1:

```
R3(config)# dlsw remote-peer 0 tcp 152.3.7.2 backup-peer 152.3.7.1
```

Finally, configure the **source-bridge** ring-bridge-virtual ring statement under the Token Ring interface as well as the **source-bridge spanning** statement:

```
R3(config)# interface TokenRing1/0
R3(config-if)# source-bridge 1 2 169
R3(config-if)# source-bridge spanning
```

R1 is also configured with the same virtual ring number as R3:

```
R1(config)# source-bridge ring-group 169
```

You are only going to configure a dlsw local peer id for R1. Notice that the promiscuous statement has been added to the dlsw local peer statement. This enables R1 to accept a dlsw peer from any remote router requesting it:

```
R1(config)# dlsw local-peer peer-id 152.3.7.1 promiscuous
```

The **dlsw bridge-group 2** statement associates bridge-group 2 with the DLSW virtual ring 169:

```
R1(config)# dlsw bridge-group 2
```

Under interface E0/0 , define a bridge group and use bridge-group 2:

```
R1(config)# interface Ethernet0/0
R1(config-if)# bridge-group 2
```

Finally, define a spanning tree protocol to run on the router:

```
R1(config-if)# bridge 2 protocol ieee
```

R5 is configured similarly to R1:

a. Choose virtual ring 169.
b. Only configure a DLSW local peer id and use the promiscuous option.
c. Assign interface E1/0 to be in bridge-group 2.
d. Associate bridge-group 2 with the virtual ring.
e. Use the IEEE spanning tree protocol.

```
R5(config)# source-bridge ring-group 169
R5(config)# dlsw local-peer peer-id 152.3.7.2 promiscuous
R5(config)# dlsw bridge-group 2
R5(config)# interface Ethernet1/0
R5(config-if)# bridge-group 2
R5(config-if)# bridge 2 protocol ieee
```

Once routers R1, R3, and R5 have been configured, your primary DLSW peer should be active. Connect to R3 and type the **show dlsw peer** command. Notice that peer 152.3.7.1 is in a connected state. A DLSW-connected peer indicates that the peer is active and functional. Notice that transmit and receive traffic is coming into and out of your peer:

```
R3#show dlsw peer
Peers:                state     pkts_rx   pkts_tx  type  drops ckts TCP   uptime
 TCP 152.3.7.1       CONNECT      1017       714  conf      0    0   0 05:01:47
 TCP 152.3.7.2       DISCONN         0         0  conf      0    0   -        -
```

The **show dlsw capabilities** command displays information on the remote peer at address 152.3.7.1. Notice that the peer type of the remote peer is listed as being prom. This indicates that 152.3.7.1 is a promiscuous peer. Recall that your local peer on R1 was configured as a promiscuous peer.

```
R3#show dlsw capabilities
DLSw: Capabilities for peer 152.3.7.1(2065)
  vendor id (OUI)         : '00C' (cisco)
  version number          : 1
  release number          : 0
```

```
  init pacing window       : 20
  unsupported saps         : none
  num of tcp sessions      : 1
  loop prevent support     : no
  icanreach mac-exclusive  : no
  icanreach netbios-excl.  : no
  reachable mac addresses  : none
  reachable netbios names  : none
  cisco version number     : 1
  peer group number        : 0
  border peer capable      : no
  peer cost                : 3
  biu-segment configured   : no
  local-ack configured     : yes
  priority configured      : no
  peer type                : prom
  version string           :
Cisco Internetwork Operating System Software
IOS (tm) 3600 Software (C3620-JS-M), Version 11.2(20)P,  RELEASE SOFTWARE (fc1)
Copyright (c) 1986-1999 by cisco Systems, Inc.
Compiled Mon 11-Oct-99 21:14 by jaturner
```

The **show dlsw capabilities local** command is used to query key information about the local DLSW peer:

```
R3#show dlsw capabilities  local
DLSw: Capabilities for local peer
  vendor id (OUI)          : '00C' (cisco)
  version number           : 1
  release number           : 0
  init pacing window       : 20
  unsupported saps         : none
  num of tcp sessions      : 1
  loop prevent support     : no
  icanreach mac-exclusive  : no
  icanreach netbios-excl.  : no
  reachable mac addresses  : none
  reachable netbios names  : none
  cisco version number     : 1
  peer group number        : 0
  border peer capable      : no
  peer cost                : 3
  biu-segment configured   : no
  current border peer      : none
  version string           :
Cisco Internetwork Operating System Software
IOS (tm) 3600 Software (C3620-JS-M), Version 11.2(20)P,  RELEASE SOFTWARE (fc1)
Copyright (c) 1986-1999 by cisco Systems, Inc.
Compiled Mon 11-Oct-99 21:14 by jaturner
```

The **show dlsw reachability** command displays a list of cached NetBIOS names and MAC addresses. Notice that your local workstation's NetBIOS name (SYSTEM4) and remote workstation's NetBIOS name (SYSTEM2) have been learned and cached by R3. Recall that Figure 5-3 contains the MAC addresses of each workstation. You will notice from the following MAC addresses that the Token Ring workstation MAC address is unchanged, but that the Ethernet workstation MAC addresses have been converted to their connonical form:

```
R3#show dlsw reachability
DLSw Local MAC address reachability cache list
Mac Addr        status     Loc.  port                 rif
00a0.24fd.c6d0  FOUND      LOCAL   TokenRing1/0     06B0.0012.0A90

DLSw Remote MAC address reachability cache list
Mac Addr        status     Loc.  peer
0000.612c.9f32  FOUND      REMOTE  152.3.7.1(2065)
0007.78da.b08e  FOUND      REMOTE  152.3.7.1(2065)

DLSw Local NetBIOS Name reachability cache list
NetBIOS Name    status     Loc.  port                 rif
SYSTEM4         FOUND      LOCAL   TokenRing1/0     06B0.0012.0A90

DLSw Remote NetBIOS Name reachability cache list
NetBIOS Name    status     Loc.  peer
SYSTEM2         FOUND      REMOTE  152.3.7.1(2065) max-lf(17800)
```

Now switch to R1. The **show dlsw peer** command indicates that you have a connected peer at address 152.3.8.1. This is R3. Notice that the peer type is promiscuous. This is because a remote peer on R1 was not configured. You only configured R1 with a local peer id and the promiscuous option:

```
R1#show dlsw peer
Peers:                state     pkts_rx   pkts_tx  type  drops ckts TCP   uptime
 TCP 152.3.8.1       CONNECT        765      1071  prom      0    1   0 05:05:18
```

The **show dlsw reachability** command on R1 shows that R1 has learned and locally cached both of your workstation's NetBIOS names:

```
R1#show dlsw reachability
DLSw Local MAC address reachability cache list
Mac Addr        status     Loc.  port                 rif
0000.612c.9f32  FOUND      LOCAL   TBridge-002    --no rif--

DLSw Remote MAC address reachability cache list
Mac Addr        status     Loc.  peer
00a0.24fd.c6d0  FOUND      REMOTE  152.3.8.1(2065) max-lf(1500)

DLSw Local NetBIOS Name reachability cache list
NetBIOS Name    status     Loc.  port                 rif
SYSTEM2         FOUND      LOCAL   TBridge-002    --no rif--

DLSw Remote NetBIOS Name reachability cache list
NetBIOS Name    status     Loc.  peer
SYSTEM4         FOUND      REMOTE  152.3.8.1(2065) max-lf(1500)
```

The **show dlsw capabilities** command shows the peer capabilities of your remote peer at IP address 152.3.8.1. Notice that your remote peer is a configured peer and not a promiscuous peer because you configured this peer with a remote peer statement.

```
R1#show dlsw capabilities
DLSw: Capabilities for peer 152.3.8.1(2065)
  vendor id (OUI)         : '00C' (cisco)
  version number          : 1
  release number          : 0
  init pacing window      : 20
  unsupported saps        : none
  num of tcp sessions     : 1
  loop prevent support    : no
  icanreach mac-exclusive : no
  icanreach netbios-excl. : no
  reachable mac addresses : none
  reachable netbios names : none
  cisco version number    : 1
  peer group number       : 0
  border peer capable     : no
  peer cost               : 3
  biu-segment configured  : no
  local-ack configured    : yes
  priority configured     : no
  peer type               : conf
  version string          :
Cisco Internetwork Operating System Software
IOS (tm) 3600 Software (C3620-JS-M), Version 11.2(20)P,  RELEASE SOFTWARE (fc1)
Copyright (c) 1986-1999 by cisco Systems, Inc.
Compiled Mon 11-Oct-99 21:14 by jaturner
```

Now connect to R5. The **show dlsw peer** command indicates that no DLSW peers are active. This is due to the fact that R5 has been configured as a backup peer. Because the primary DLSW peer between R3 and R1 is still active, the backup peer has not been activated:

```
R5#show dlsw peer
Peers:                state     pkts_rx   pkts_tx  type  drops ckts TCP   uptime
```

The **show dlsw capabilities** command confirms that no remote peers are active:

```
R5#show dlsw capabilities
DLSw: No remote peer capabilities known at this time
```

The **show dlsw capabilities local** command indicates that you have configured your router as a DLSW local peer:

```
R5#show dlsw capabilities  local
DLSw: Capabilities for local peer
  vendor id (OUI)         : '00C' (cisco)
  version number          : 1
  release number          : 0
  init pacing window      : 20
```

```
  unsupported saps        : none
  num of tcp sessions     : 1
  loop prevent support    : no
  icanreach mac-exclusive : no
  icanreach netbios-excl. : no
  reachable mac addresses : none
  reachable netbios names : none
  cisco version number    : 1
  peer group number       : 0
  border peer capable     : no
  peer cost               : 3
  biu-segment configured  : no
  current border peer     : none
  version string          :
Cisco Internetwork Operating System Software
IOS (tm) 3600 Software (C3620-JS-M), Version 11.2(20)P,  RELEASE SOFTWARE (fc1)
Copyright (c) 1986-1999 by cisco Systems, Inc.
Compiled Mon 11-Oct-99 21:14 by jaturner
```

Now reconnect to R3. You are going to fail the peer between R1 and R3 and monitor how R3 automatically repeers to R5. Enable DLSW debugging with the **debug dlsw reach**, **debug dlsw peer**, and **debug dlsw local** commands:

```
R3#debug dlsw reach
DLSw reachability debugging is on at event level for all protocol traffic
R3#debug dlsw peer
DLSw peer debugging is on
R3#debug dlsw loc
DLSw local circuit debugging is on
```

The **show debug** command indicates that you have enabled debugging for both ends of your DLSW session (152.3.7.1 and 152.3.7.2):

```
R3#show debug
DLSw:
  DLSw Peer debugging is on
DLSw reachability debugging is on at event level for all protocol traffic
  DLSw basic debugging for peer 152.3.7.1(2065) is on
  DLSw basic debugging for peer 152.3.7.2(2065) is on
  DLSw Local Circuit debugging is on
```

The following screen print shows what the debug output looks like when the DLSW peer is up and active. Notice that the router sends periodic DLSW keepalive requests to your DLSW neighbor at IP address 152.3.7.1. Also notice that you receive a response for each keepalive request that is sent.

```
DLSw: Keepalive Request sent to peer 152.3.7.1(2065))  ← Keepalive Request
DLSw: Keepalive Response from peer 152.3.7.1(2065)  ← Keepalive Response
DLSw: Keepalive Request sent to peer 152.3.7.1(2065))  ← Keepalive Request
```

```
DLSw: Keepalive Response from peer 152.3.7.1(2065)  ← Keepalive Response
CSM: Received CLSI Msg : UDATA_STN.Ind   dlen: 215 from TokenRing1/0
CSM:   smac 80a0.24fd.c6d0, dmac c000.0000.0080, ssap F0, dsap F0
CSM: Received frame type NETBIOS DATAGRAM from 00a0.24fd.c6d0, To1/0
CSM: Received CLSI Msg : UDATA_STN.Ind   dlen: 84 from TokenRing1/0
CSM:   smac 80a0.24fd.c6d0, dmac c000.0000.0080, ssap F0, dsap F0
CSM: Received frame type NETBIOS NAME_QUERY from 00a0.24fd.c6d0, To1/0
CSM: Received CLSI Msg : UDATA_STN.Ind   dlen: 84 from TokenRing1/0
CSM:   smac 80a0.24fd.c6d0, dmac c000.0000.0080, ssap F0, dsap F0
CSM: Received frame type NETBIOS NAME_QUERY from 00a0.24fd.c6d0, To1/0
CSM: Received CLSI Msg : UDATA_STN.Ind   dlen: 84 from TokenRing1/0
CSM:   smac 80a0.24fd.c6d0, dmac c000.0000.0080, ssap F0, dsap F0
CSM: Received frame type NETBIOS NAME_QUERY from 00a0.24fd.c6d0, To1/0
```

Now disconnect the cable connected to interface E0/0 on R1. The following debug output is displayed. Notice that R3 sends repeated keepalive responses to the remote peer at IP address 152.3.7.1:

```
DLSw: Keepalive Response sent to peer 152.3.7.1(2065)) ← Repeated Keepalive Responses
DLSw: Keepalive Response sent to peer 152.3.7.1(2065))
DLSw: Keepalive Response sent to peer 152.3.7.1(2065))
DLSw: Keepalive Response sent to peer 152.3.7.1(2065))
After a period of time, R3 closes the peer connection to your remote peer at IP address
152.3.7.1:
DLSw: dlsw_tcpd_fini() for peer 152.3.7.1(2065)
DLSw: tcp fini closing connection for peer 152.3.7.1(2065) ← TCP connection to remote
                                                               peer is closed
DLSw: action_d(): for peer 152.3.7.1(2065)
DLSw: peer 152.3.7.1(2065), old state CONNECT, new state DISCONN
```

R3 then tries to establish a connection to its backup peer at IP address 152.3.7.2:

```
DLSw: action_a() attempting to connect peer 152.3.7.2(2065) ← Establishing a new
                                                                connection to backup peer
DLSw: action_a(): Write pipe opened for peer 152.3.7.2(2065)
DLSw: peer 152.3.7.2(2065), old state DISCONN, new state WAIT_RD
DLSw: passive open 152.3.7.2(11000) -> 2065
DLSw: action_c(): for peer 152.3.7.2(2065)
DLSw: peer 152.3.7.2(2065), old state WAIT_RD, new state CAP_EXG
DLSw: CapExId Msg sent to peer 152.3.7.2(2065)
DLSw: Recv CapExPosRsp Msg from peer 152.3.7.2(2065)
DLSw: action_e(): for peer 152.3.7.2(2065)
DLSw: Recv CapExId Msg from peer 152.3.7.2(2065)
DLSw: Pos CapExResp sent to peer 152.3.7.2(2065)
DLSw: action_e(): for peer 152.3.7.2(2065)
```

After peer negotiation has been completed, the new peer goes into a CONNECT state:

```
DLSw: peer 152.3.7.2(2065), old state CAP_EXG, new state CONNECT ← Backup peer is now
                                                                     connected
DLSw: peer_act_on_capabilities() for peer 152.3.7.2(2065)
DLSw: action_f(): for peer 152.3.7.2(2065)
DLSw: closing read pipe tcp connection for peer 152.3.7.2(2065)
```

Notice that R3 continues to try to establish a connection to its primary peer on R1 at IP address 152.3.7.1:

```
DLSw: action_a() attempting to connect peer 152.3.7.1(2065)
DLSw: Keepalive Response sent to peer 152.3.7.2(2065))
DLSw: CONN: peer 152.3.7.1 open failed, timed out [10]
DLSw: action_a(): CONN failed - retries 1
DLSw: peer 152.3.7.1(2065), old state DISCONN, new state DISCONN
DLSw: action_a() attempting to connect peer 152.3.7.1(2065)
DLSw: Keepalive Response sent to peer 152.3.7.2(2065))
DLSw: Keepalive Response sent to peer 152.3.7.2(2065))
DLSw: CONN: peer 152.3.7.1 open failed, timed out [10]
DLSw: action_a(): CONN failed - retries 2
DLSw: peer 152.3.7.1(2065), old state DISCONN, new state DISCONN
DLSw: Keepalive Response sent to peer 152.3.7.2(2065))
DLSw: action_a() attempting to connect peer 152.3.7.1(2065)
DLSw: Keepalive Response sent to peer 152.3.7.2(2065))
```

The **show dlsw peer** command on R3 shows that your backup peer is now connected and your primary peer is now disconnected:

```
R3#show dlsw peer
Peers:                  state     pkts_rx   pkts_tx  type  drops ckts TCP   uptime
 TCP 152.3.7.2         CONNECT         9         7  conf      0    0    0 00:02:43
 TCP 152.3.7.1         DISCONN      1139       815  conf      0    0    -        -
```

The **show dlsw reach** command indicates that your NetBIOS names are still being cached. You are now able to reach your remote workstation System2 via your new DLSW peer at IP address 152.3.7.2:

```
R3#show dlsw reach
DLSw Local MAC address reachability cache list
Mac Addr        status     Loc.  port                 rif
00a0.24fd.c6d0  FOUND      LOCAL   TokenRing1/0     06B0.0012.0A90

DLSw Remote MAC address reachability cache list
Mac Addr        status     Loc.  peer
0000.612c.9f32  FOUND      REMOTE  152.3.7.2(2065)

DLSw Local NetBIOS Name reachability cache list
NetBIOS Name    status     Loc.  port                 rif
SYSTEM4         FOUND      LOCAL   TokenRing1/0     06B0.0012.0A90

DLSw Remote NetBIOS Name reachability cache list
NetBIOS Name    status     Loc.  peer
SYSTEM2         FOUND      REMOTE  152.3.7.2(2065)  ← SYSTEM2 is now being learned via
                                                      our backup remote peer at IP
                                                      address 152.3.7.2
```

Now reconnect to R1. The **show dlsw peer** command indicates that you no longer have any active DLSW peers:

```
R1#show dlsw peer
Peers:                  state      pkts_rx    pkts_tx  type  drops ckts TCP   uptime
```

The **show dlsw reach** shows that you are not caching any remote NetBIOS peer names:

```
R1#show dlsw reach
DLSw Local MAC address reachability cache list
Mac Addr         status     Loc.  port              rif
0000.612c.9f32   FOUND      LOCAL   TBridge-002    --no rif--
0007.78da.b08e   FOUND      LOCAL   TBridge-002    --no rif--

DLSw Remote MAC address reachability cache list
Mac Addr         status     Loc.  peer

DLSw Local NetBIOS Name reachability cache list
NetBIOS Name     status     Loc.  port              rif
SYSTEM2          FOUND      LOCAL   TBridge-002    --no rif--

DLSw Remote NetBIOS Name reachability cache list
NetBIOS Name     status     Loc.  Peer
```

Now connect to R5. The **show dlsw peer** command indicates that you are connected to your remote peer on R3 at IP address 152.3.8.1. Notice that the connection type is promiscuous because you have not configured any remote DLSW peers on R5:

```
R5#show dlsw peer
Peers:                  state      pkts_rx    pkts_tx  type  drops ckts TCP   uptime
 TCP 152.3.8.1         CONNECT          9         11  prom      0    0    0 00:03:56
```

The **show dlsw reach** command indicates that R5 has learned and cached the NetBIOS names of the workstations on your two LANs:

```
R5#show dlsw reach
DLSw Local MAC address reachability cache list
Mac Addr         status     Loc.  port              rif
0000.612c.9f32   FOUND      LOCAL   TBridge-002    --no rif--
0007.78da.6480   FOUND      LOCAL   TBridge-002    --no rif--
00a0.24fd.c6d0   FOUND      LOCAL   TBridge-002    --no rif--

DLSw Remote MAC address reachability cache list
Mac Addr         status     Loc.  peer

DLSw Local NetBIOS Name reachability cache list
NetBIOS Name     status     Loc.  port              rif
SYSTEM2          FOUND      LOCAL   TBridge-002    --no rif--
SYSTEM4          FOUND      LOCAL   TBridge-002    --no rif--

DLSw Remote NetBIOS Name reachability cache list
NetBIOS Name     status     Loc.  Peer
```

The **show dlsw capabilities local** and **show dlsw capabilities remote** commands show that your peer is properly configured:

```
R5#show dlsw capabilities local
DLSw: Capabilities for local peer
  vendor id (OUI)         : '00C' (cisco)
  version number          : 1
  release number          : 0
  init pacing window      : 20
  unsupported saps        : none
  num of tcp sessions     : 1
  loop prevent support    : no
  icanreach mac-exclusive : no
  icanreach netbios-excl. : no
  reachable mac addresses : none
  reachable netbios names : none
  cisco version number    : 1
  peer group number       : 0
  border peer capable     : no
  peer cost               : 3
  biu-segment configured  : no
  current border peer     : none
  version string          :
Cisco Internetwork Operating System Software
IOS (tm) 3600 Software (C3620-JS-M), Version 11.2(20)P,  RELEASE SOFTWARE (fc1)
Copyright (c) 1986-1999 by cisco Systems, Inc.
Compiled Mon 11-Oct-99 21:14 by jaturner

R5#show dlsw capabilities remote
DLSw: Capabilities for peer 152.3.8.1(2065)
  vendor id (OUI)         : '00C' (cisco)
  version number          : 1
  release number          : 0
  init pacing window      : 20
  unsupported saps        : none
  num of tcp sessions     : 1
  loop prevent support    : no
  icanreach mac-exclusive : no
  icanreach netbios-excl. : no
  reachable mac addresses : none
  reachable netbios names : none
  cisco version number    : 1
  peer group number       : 0
  border peer capable     : no
  peer cost               : 3
  biu-segment configured  : no
  local-ack configured    : yes
  priority configured     : no
  peer type               : conf
  version string          :
Cisco Internetwork Operating System Software
IOS (tm) 3600 Software (C3620-JS-M), Version 11.2(20)P,  RELEASE SOFTWARE (fc1)
Copyright (c) 1986-1999 by cisco Systems, Inc.
Compiled Mon 11-Oct-99 21:14 by jaturner
```

Once the DLSW configuration has been completed, you should be able to verify that it is working. As shown in Figure 5-9, right-click on the Network Neighborhood icon on System2 and select the Find Computer option. When prompted for the name of a computer to find, enter System4.

System2 should be able to find System4 using Windows Networking as shown in Figure 5-10. Remember that System2 finds System4 via NetBEUI because the IP protocol on either System2 or System4 has not been enabled.

17. *Configure R3 so that its attached workstations can only see two other workstations on the network. The two workstations that R3 can allow access to will have NetBIOS names: System3 and System9.*

You can accomplish this task by configuring an access expression on R3. An access expression enables you to configure complex conditions under which bridged traffic can leave or enter an interface.

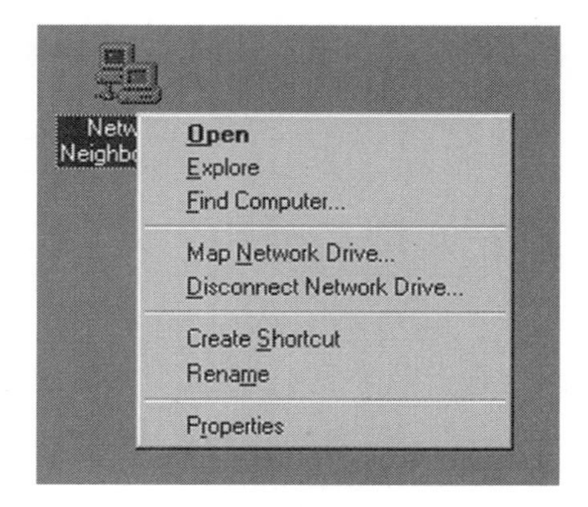

Figure 5-9 Network neighborhoods find computer

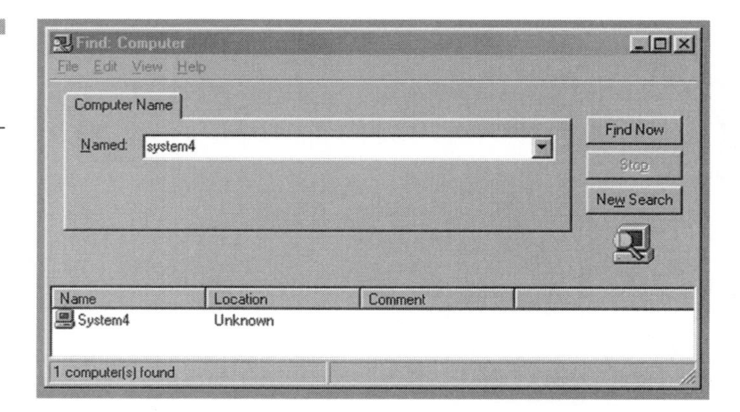

Figure 5-10 Network neighborhood

Workstations connected to R3 should only be able to see workstations with a NetBIOS name of System3 or System9. Start configuring your access expression by defining NetBIOS access lists on R3. First, define a NetBIOS access list that permits workstations with a NetBIOS name of SYSTEM3 and that denies all other NetBIOS names:

```
R3(config)# netbios access-list host test permit SYSTEM3
R3(config)# netbios access-list host test deny *
```

Next, define a NetBIOS access list that permits workstations with a NetBIOS name of SYSTEM9 and that denies all other NetBIOS names:

```
R3(config)# netbios access-list host rest permit SYSTEM9
R3(config)# netbios access-list host rest deny *
```

In order to activate the access list, it must be applied to the appropriate interface. The | character in the access expression is an "or" operator. It tells the router to apply the access expression if either the test or rest NetBIOS access list has a valid match.

```
R3(config)# interface Tok 1/0
R3(config-if)# access-expression input (netbios-host(test) | netbios-host(rest))
```

The access expression can be verified by using the **show access-expression** command:

```
R3#show access-expression
Interface TokenRing0/0:
   Input: (netbios-host(test) | netbios-host(rest))
```

18. *Configure transparent bridging between LAN_3 on R5 and LAN_1 on R6. Assume that stations on each of these LANS are running NetBEUI traffic.*

Configure transparent bridging by assigning each interface that is going to participate in a bridged domain to a specific bridge group. Choose bridge group 1 for this task. On R5, assign interface E0/0 and interface S0/1 to bridge group 1:

```
R5(config)# interface Ethernet0/0
R5(config-if)# bridge-group 1
R5(config-if)# interface Serial0/1
R5(config-if)# bridge-group 1
```

Next, define your spanning tree protocol. All bridges in a bridge group have to run the same spanning tree protocol. Use the IEEE version of spanning tree:

```
R5(config-if)# bridge 1 protocol ieee
```

Now you need to configure the same items on R6 as you did on R5. Assign interfaces E0 and S1 to bridge group 1:

```
R6(config)# interface Ethernet0
R6(config-if)# bridge-group 1
```

```
R6(config-if)# interface Serial1
R6(config-if)# bridge-group 1
```

Finally, enable the IEEE spanning tree protocol:

```
R6(config-if)# bridge 1 protocol ieee
```

After configuring routers R5 and R6, connect back to R5. The output of the **show bridge 1 verbose** command is shown below. Notice that ports E0/0 and S0/1 are in a forwarding state. Also notice that the MAC addresses of both System3 and System1 have been learned by the bridge:

```
R5#show bridge 1 verbose

Total of 300 station blocks, 297 free
Codes: P - permanent, S - self

BG Hash      Address      Action   Interface      VC    Age   RX count   TX count
 1 A7/0   0000.8635.46e1 forward  Ethernet0/0     -      4        123         57
 1 FB/0   0000.8635.4ab1 forward  Serial0/1       -      0        120         60

Flood ports          RX count    TX count
Ethernet0/0                63          63
Serial0/1                  63          63
BRI1/0                      0           0
```

Use the **show bridge group** command to display all the active bridge groups on R5. Notice that R5 has two bridge groups: group 1 and group 2. Bridge group 2 is the DLSW bridge group that you defined in a previous step, while Bridge group 1 is the bridge group that you just defined. Notice that bridge group 1 has two ports: interface E0/0 and interface S0/1 . Both of these ports are in a forwarding state:

```
R5#show bridge group
Bridge Group 1 is running the IEEE compatible Spanning Tree protocol

   Port 4 (Ethernet0/0) of bridge group 1 is forwarding
   Port 3 (Serial0/1) of bridge group 1 is forwarding

Bridge Group 2 is running the IEEE compatible Spanning Tree protocol

   Port 9 (Ethernet1/0) of bridge group 2 is forwarding
   Port 12 (DLSw Port0) of bridge group 2 is forwarding
```

Now connect to R6. The **show bridge 1 verbose** command indicates that both of the bridge group 1 interfaces on R6 (S1 and E0) are in a forwarding state:

```
R6#show bridge 1 verbose

Total of 300 station blocks, 298 free
Codes: P - permanent, S - self
```

```
BG Hash      Address        Action  Interface     VC    Age   RX count   TX count
 1 A7/0    0000.8635.46e1 forward  Serial1        -     2        57         57
 1 FB/0    0000.8635.4ab1 forward  Ethernet0      -     0       122         53

Flood ports         RX count     TX count
BRI0                       0           69
Ethernet0                 65            4
Serial1                    4           65
```

The **show bridge group** command verifies that you have three interfaces on R6 that are part of bridge group 1 (BRI0, E0, and S1):

```
R6#show bridge group

Bridge Group 1 is running the IEEE compatible Spanning Tree protocol

   Port 3 (BRI0) of bridge group 1 is forwarding
   Port 2 (Ethernet0) of bridge group 1 is forwarding
   Port 7 (Serial1) of bridge group 1 is forwarding
```

Once the configuration has been completed, verify that it is properly working by making sure that System3 on LAN_3 can see System1 on LAN_1 via Windows Networking. Figure 5-11 shows a user on System3 trying to locate System1 on the network. Remember that the selection screen shown in Figure 5-11 can be obtained by right-clicking on the Network Neighborhood icon on the Windows desktop.

System1 should be seen across the network, as shown in Figure 5-12.

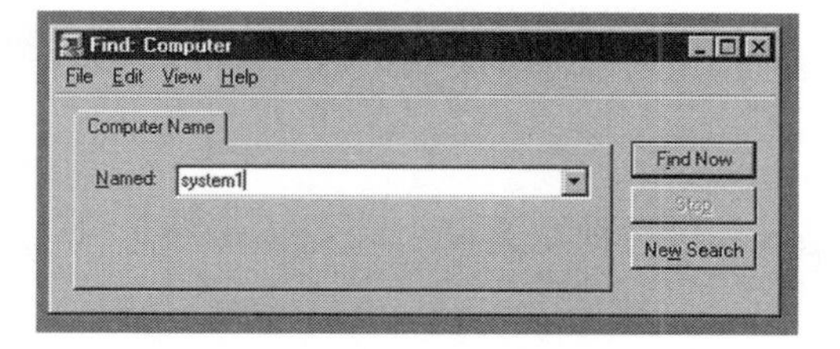

Figure 5-11 Network neighborhood find computer

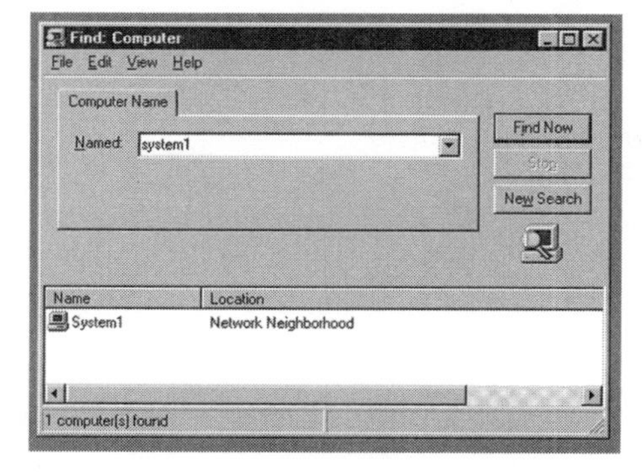

Figure 5-12 Network neighborhood

As an alternative to right-clicking on the Network Neighborhood icon to find a neighbor computer, you can double-click on the Network Neighborhood icon. This brings up a screen similar to the one shown in Figure 5-13. Notice that System1 has been found on the network.

19. *Configure dial backup between R5 and R6. If the serial link fails between R5 and R6, the ISDN link should provide a backup path between the workstations on LAN_3 and LAN_1.*

Start by configuring R5 for dial backup. First, define a username and password because you will be doing CHAP authentication. The username should be the hostname of the router on the far end of the ISDN call (in this case, you are calling R6). The password should be the same password on both R5 and R6 (make your shared password cisco):

```
R5(config)# username R6 password 0 cisco
```

Next, define your ISDN switch type. The Atlas 800 ISDN switch used in this case study is configured for National ISDN call control, so enter the **isdn switch-type basic-ni1** command:

```
R5(config)# isdn switch-type basic-ni1
```

When interface S0/1 on R5 goes down, configure R5 to dial R6. The **backup interface BRI1/0** command specifies that interface BRI1/0 will be activated when the S0/1 interface goes into a down state. The **backup delay 5 10** command specifies that your backup interface, BRI1/0 , will be activated five seconds after interface S0/1 goes into a down state and that interface BRI1/0 will be deactivated 10 seconds after interface S0/1 goes back up:

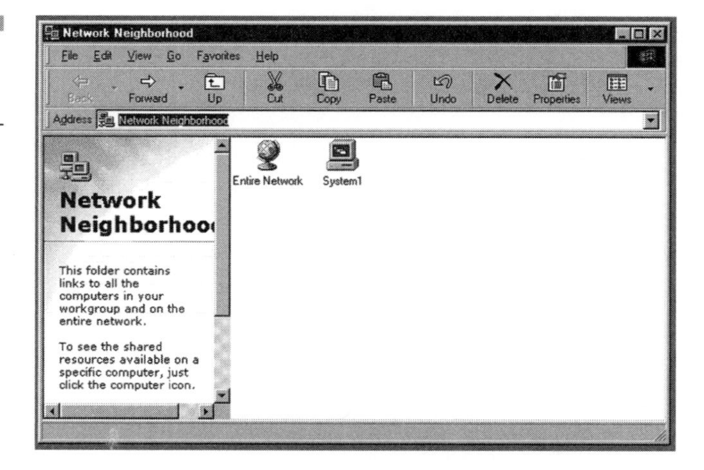

Figure 5-13 Network neighborhood

```
R5(config)# interface Serial0/1
R5(config-if)# backup interface BRI1/0
R5(config-if)# backup delay 5 10
```

Next, configure your ISDN interface, BRI 1/0 :

```
R5(config-if)# interface BRI1/0
```

First, assign your IP address (152.3.9.1) and serial encapsulation (PPP) to your interface:

```
R5(config-if)# ip address 152.3.9.1 255.255.255.248
R5(config-if)# encapsulation ppp
```

Next , define your SPID values. An ISDN BRI with National call control has a separate SPID value for each B channel. The SPID values are derived from the configuration of your Atlas 800 ISDN switch. This configuration can be found at the end of this chapter. Notice the format of the **isdn spid** command. Use your first SPID command as an example, **isdn spid1 5101 8995101**. The SPID is identified as being the SPID for the first or second B channel via use of the spid1 argument. Next, specify the actual SPID value (5101) followed by the directory number of the B channel (8995101):

```
R5(config-if)# isdn spid1 5101 8995101
R5(config-if)# isdn spid2 5102 8995102
```

Define a dialer map statement that associates reaching your far end router (R6) via dialing 8995201. Specify the bridge option to indicate that bridged traffic applies to this specific dialer map statement:

```
R5(config-if)# dialer map bridge name R6 broadcast 8995201
```

The **dialer-group 1** command is used in defining interesting traffic to the interface and defines the dialer access group that the interface belongs to:

```
R5(config-if)# dialer-group 1
```

Select CHAP as your authentication method with the **ppp authentication chap** command:

```
R5(config-if)# ppp authentication chap
```

Your last interface command is **bridge-group 1.** This command assigns Interface BRI1/0 to bridge group 1. This is the same bridge group that your primary interface (S0/1) is assigned to.

```
R5(config-if)# bridge-group 1
R5(config-if)#exit
```

Next, you need to go back to Global Interface Configuration mode and define your interesting traffic. The **dialer-list 1 protocol bridge list 201** command associates dialer group 1 with this dialer list command. It tells you that the interesting bridged traffic for interface BRI1/0 is defined by access list number 201.

```
R5(config)# dialer-list 1 protocol bridge list 201
```

Finally, define your access list. Permit all bridged traffic to act as interesting traffic.

```
R5(config)# access-list 201 permit 0x0000 0xFFFF
```

The configuration for R6 is similar to the configuration for R5. The one minor difference is the dialer map command. Notice that you do not specify a dial string. This means that R6 does not dial R5; R5 dials R6. The ISDN specific configuration commands for R6 are shown here:

```
R6(config)# username R5 password 0 cisco
R6(config)# isdn switch-type basic-ni1
R6(config)# interface BRI0
R6(config-if)# ip address 152.3.9.2 255.255.255.248
R6(config-if)# encapsulation ppp
R6(config-if)# isdn spid1 5201 8995201
R6(config-if)# isdn spid2 5202 8995202
R6(config-if)# dialer map bridge name R5 broadcast
R6(config-if)# dialer-group 1
R6(config-if)# ppp authentication chap
R6(config-if)# bridge-group 1
R6(config-if)#exit
R6(config)# access-list 201 permit 0x0000 0xFFFF
R6(config)# dialer-list 1 protocol bridge list 201
```

Now reconnect to R5 and type the **show isdn status** command. Notice that the router has not yet activated the ISDN circuit. This is due to the fact that you are using a backup interface. The router does not activate the ISDN circuit until it needs to activate the backup interface (BRI1/0). When using a backup interface, it is important to make sure that the SPIDs entered for the BRI circuit are correct. If not, you will not find out that the SPIDs are incorrect until it is time to activate the circuit.

```
R5#show isdn status
The current ISDN Switchtype = basic-ni1
ISDN BRI1/0 interface
    Layer 1 Status:
        DEACTIVATED
    Layer 2 Status:
        Layer 2 NOT Activated
    Spid Status:
```

```
        TEI Not Assigned, ces = 1, state = 1(terminal down)
            spid1 configured, spid1 NOT sent, spid1 NOT valid
        TEI Not Assigned, ces = 2, state = 1(terminal down)
            spid2 configured, spid2 NOT sent, spid2 NOT valid
            Layer 3 Status:
        0 Active Layer 3 Call(s)
            Activated dsl 8 CCBs = 0
            Total Allocated ISDN CCBs = 0
```

The **show interface bri 1/0** command indicates that the interface is in a standby mode:

```
R5#show interface bri 1/0
BRI1/0 is standby mode, line protocol is down
  Hardware is QUICC BRI with U interface
  Internet address is 152.3.9.1/29
  MTU 1500 bytes, BW 64 Kbit, DLY 20000 usec, rely 255/255, load 1/255
  Encapsulation PPP, loopback not set
  Last input never, output never, output hang never
  Last clearing of "show interface" counters never
  Input queue: 0/75/0 (size/max/drops); Total output drops: 0
  Queueing strategy: weighted fair
  Output queue: 0/1000/64/0 (size/max total/threshold/drops)
     Conversations  0/0/256 (active/max active/max total)
     Reserved Conversations 0/0 (allocated/max allocated)
  5 minute input rate 0 bits/sec, 0 packets/sec
  5 minute output rate 0 bits/sec, 0 packets/sec
     0 packets input, 0 bytes, 0 no buffer
     Received 0 broadcasts, 0 runts, 0 giants, 0 throttles
     0 input errors, 0 CRC, 0 frame, 0 overrun, 0 ignored, 0 abort
     0 packets output, 0 bytes, 0 underruns
     0 output errors, 0 collisions, 1 interface resets
     0 output buffer failures, 0 output buffers swapped out
     0 carrier transitions
```

The **show interface S 0/1** command shows that the interface is in an up/up state. You can see that interface BRI 1/0 is listed as the backup interface for S0/1 . Notice that your **backup delay 5 10** command has resulted in the failure delay being set to five seconds and the secondary disable delay being set to 10 seconds.

```
R5#show interface s 0/1
Serial0/1 is up, line protocol is up
  Hardware is QUICC Serial
  Internet address is 152.3.10.21/30
  Backup interface BRI1/0 , kickin load not set, kickout load not set
      failure delay 5 sec, secondary disable delay 10 sec
  MTU 1500 bytes, BW 1544 Kbit, DLY 20000 usec, rely 255/255, load 1/255
  Encapsulation PPP, loopback not set, keepalive set (10 sec)
  LCP Open
  Open: IPCP, CDPCP, BRIDGECP
  Last input 00:00:01, output 00:00:04, output hang never
  Last clearing of "show interface" counters never
  Input queue: 0/75/0 (size/max/drops); Total output drops: 0
  Queueing strategy: weighted fair
```

```
  Output queue: 0/1000/64/0 (size/max total/threshold/drops)
     Conversations  0/1/256 (active/max active/max total)
     Reserved Conversations 0/0 (allocated/max allocated)
  5 minute input rate 0 bits/sec, 0 packets/sec
  5 minute output rate 0 bits/sec, 0 packets/sec
     8764 packets input, 386615 bytes, 0 no buffer
     Received 8707 broadcasts, 0 runts, 0 giants, 0 throttles
     0 input errors, 0 CRC, 0 frame, 0 overrun, 0 ignored, 0 abort
     3780 packets output, 189359 bytes, 0 underruns
     0 output errors, 0 collisions, 24 interface resets
     0 output buffer failures, 0 output buffers swapped out
     11 carrier transitions
     DCD=up  DSR=up  DTR=up  RTS=up  CTS=up
```

Because you are still running over your primary circuit, the **show bridge 1 verbose** command indicates that interfaces E0/0 and S0/1 are still forwarding traffic. You can see that the receive and transmit counts on interface BRI 1/0 are still 0.

```
R5#show bridge 1 verbose

Total of 300 station blocks, 298 free
Codes: P - permanent, S - self

BG Hash      Address      Action  Interface      VC    Age   RX count   TX count
 1 A7/0   0000.8635.46e1 forward  Ethernet0/0     -     3        123         57
 1 FB/0   0000.8635.4ab1 forward  Serial0/1       -     3         29         24

Flood ports          RX count    TX count
Ethernet0/0                89         110
Serial0/1                 110          89
BRI1/0                      0           0
```

The **show bridge group** command tells you that interface BRI1/0 is part of bridge group 1, but that it is currently in a down state:

```
R5#show bridge group

Bridge Group 1 is running the IEEE compatible Spanning Tree protocol

   Port 4 (Ethernet0/0) of bridge group 1 is forwarding
   Port 3 (Serial0/1) of bridge group 1 is forwarding
   Port 5 (BRI1/0 ) of bridge group 1 is down

Bridge Group 2 is running the IEEE compatible Spanning Tree protocol

   Port 9 (Ethernet1/0) of bridge group 2 is forwarding
   Port 12 (DLSw Port0) of bridge group 2 is forwarding
```

Now fail the primary link and see how the ISDN backup circuit gets established. First, enable some key debug prompts so that you can see the backup process occur. Enable PPP authentication, PPP negotiation, ISDN layer 3 (q931) debugging, and ISDN layer 2 (q921) debugging:

```
R5#debug ppp authentication
PPP authentication debugging is on
R5#debug ppp negotiation
PPP protocol negotiation debugging is on
R5#debug isdn q931
ISDN Q931 packets debugging is on
R5#debug isdn q921
ISDN Q921 packets debugging is on
```

With the previous debug statements enabled, pull the S0/1 cable on R5. The following debug output appears:

```
%LINEPROTO-5-UPDOWN: Line protocol on Interface Serial0/1, changed state to down
%LINK-3-UPDOWN: Interface Serial0/1, changed state to down  ← Interface S0/1  will go
                                                              down when the interface
                                                              cable is pulled

Se0/1 IPCP: State is Closed  ← Once the interface goes down, all Layer 2 and Layer 3
                               protocols that were opened will be closed
Se0/1 CDPCP: State is Closed
Se0/1 BNCP: State is Closed
Se0/1 PPP: Phase is TERMINATING
Se0/1 LCP: State is Closed
Se0/1 PPP: Phase is DOWN
Se0/1 PPP: Unsupported or un-negotiated protocol. Link cdp


Se0/1 IPCP: Remove route to 152.3.10.22  ← R5 will remove its route to its directly
                                           connected neighbor
%LINK-3-UPDOWN: Interface BRI1/0:1, changed state to down
BR1/0:1 LCP: State is Closed
BR1/0:1 PPP: Phase is DOWN
%LINK-3-UPDOWN: Interface BRI1/0:2, changed state to down
BR1/0:2 LCP: State is Closed
BR1/0:2 PPP: Phase is DOWN
```

Notice that once the S0/1 interface is down, the BRI 1/0 interface begins to activate itself.

```
ISDN BR1/0: TX ->  IDREQ  ri = 86  ai = 127  ← The ISDN interface will send an ID
                                               request to the ISDN switch for the first
                                               B channel

ISDN BR1/0: RX <-  IDASSN  ri = 86  ai = 64  ← The ISDN switch will reply with an ID
                                               assignment for the first B channel
ISDN BR1/0: TX ->  SABMEp sapi = 0  tei = 64
%LINK-3-UPDOWN: Interface BRI1/0, changed state to up  ← The BRI 1/0 interface will be
                                                         declared up

ISDN BR1/0: RX <-  UAf sapi = 0  tei = 64
%ISDN-6-LAYER2UP: Layer 2 for Interface BR1/0, TEI 64 changed to up
ISDN BR1/0: TX ->  INFOc sapi = 0  tei = 64  ns = 0  nr = 0  i = 0x08007B3A0435313031
    INFORMATION pd = 8  callref = (null)
        SPID Information i = '5101'  ← The router will send the SPID1 value to the ISDN
                                       switch
ISDN BR1/0: RX <-  RRr sapi = 0  tei = 64  nr = 1
ISDN BR1/0: RX <-  INFOc sapi = 0  tei = 64  ns = 0  nr = 1  i = 0x08007B3B02F081
```

```
    INFORMATION pd = 8  callref = (null)
        ENDPOINT IDent i = 0xF081
ISDN BR1/0: TX ->  RRr sapi = 0  tei = 64  nr = 1
ISDN BR1/0: Received EndPoint ID
ISDN BR1/0: TX ->  IDREQ  ri = 1463  ai = 127  ← The ISDN interface will send an ID
                                                  request to the ISDN switch for the
                                                  second B channel

ISDN BR1/0: RX <-  IDASSN  ri = 1463  ai = 65  ← The ISDN switch will reply with an ID
                                                  assignment for the second B channel
ISDN BR1/0: TX ->  SABMEp sapi = 0  tei = 65
ISDN BR1/0: RX <-  UAf sapi = 0  tei = 65
%ISDN-6-LAYER2UP: Layer 2 for Interface BR1/0, TEI 65 changed to up
ISDN BR1/0: TX ->  INFOc sapi = 0  tei = 65  ns = 0  nr = 0  i = 0x08007B3A0435313032
    INFORMATION pd = 8  callref = (null)
        SPID Information i = '5102'  ← The router will send the SPID2 value to the ISDN
                                        switch

ISDN BR1/0: RX <-  RRr sapi = 0  tei = 65  nr = 1
ISDN BR1/0: RX <-  INFOc sapi = 0  tei = 65  ns = 0  nr = 1  i = 0x08007B3B02F082
    INFORMATION pd = 8  callref = (null)
        ENDPOINT IDent i = 0xF082
ISDN BR1/0: TX ->  RRr sapi = 0  tei = 65  nr = 1
ISDN BR1/0: Received EndPoint ID
ISDN BR1/0: RX <-  RRp sapi = 0  tei = 64 nr = 1
ISDN BR1/0: TX ->  RRf sapi = 0  tei = 64  nr = 1
```

Now view the status of the ISDN circuit with the **show isdn status** command. Notice that the ISDN circuit is now in an active state and that a *terminal endpoint identifier* (TEI) has been assigned for each of the two B channels on the ISDN circuit. In addition, the router has successfully sent both SPIDs to the switch. Notice that no ISDN calls are active.

```
R5#show isdn status
The current ISDN Switchtype = basic-ni1
ISDN BRI1/0 interface
    Layer 1 Status:
ACTIVE
    Layer 2 Status:
        TEI = 64, State = MULTIPLE_FRAME_ESTABLISHED
        TEI = 65, State = MULTIPLE_FRAME_ESTABLISHED
    Spid Status:
        TEI 64, ces = 1, state = 5(init)
            spid1 configured, spid1 sent, spid1 valid
            Endpoint ID Info: epsf = 0, usid = 70, tid = 1
        TEI 65, ces = 2, state = 5(init)
            spid2 configured, spid2 sent, spid2 valid
            Endpoint ID Info: epsf = 0, usid = 70, tid = 2
    Layer 3 Status:
        0 Active Layer 3 Call(s)
    Activated dsl 8 CCBs = 1
        CCB: callid=0x0, sapi=0, ces=1, B-chan=0
    Total Allocated ISDN CCBs = 1
```

You just saw from the output of the **show isdn status** command that R5 has still not placed a call to R6. You have also seen that the ISDN circuit on R5 is activated as soon as the BRI 1/0 interface goes into backup mode, but R5 does not actually place a call to R6 until some interesting traffic triggers the call. You can generate interesting traffic by going on to System3 and trying to view a file on System1 via Network Neighborhood. The following trace shows what happens when interesting traffic triggers an ISDN call from R5 to R6:

```
ISDN BR1/0: TX ->  INFOc sapi = 0  tei = 64  ns = 1  nr = 1  i =
0x08010105040288901801832C0738393935323031
    SETUP pd = 8  callref = 0x01
        Bearer Capability i = 0x8890
        Channel ID i = 0x83
        Keypad Facility i = '8995201'  ← R5 places a call to 899-5201.  This is the ISDN
                                          circuit connected to R6

ISDN BR1/0: RX <-  RRr sapi = 0  tei = 64  nr = 2
ISDN BR1/0: RX <-  INFOc sapi = 0  tei = 64  ns = 1  nr = 2  i = 0x08018102180189
    CALL_PROC pd = 8  callref = 0x81
        Channel ID i = 0x89
ISDN BR1/0: TX ->  RRr sapi = 0  tei = 64  nr = 2
ISDN BR1/0: RX <-  INFOc sapi = 0  tei = 64  ns = 2  nr = 2  i = 0x08018107180189
    CONNECT pd = 8  callref = 0x81  ← R5 receives a connect message from the ISDN switch
        Channel ID i = 0x89

ISDN BR1/0: TX ->  INFOc sapi = 0  tei = 64  ns = 2  nr = 3  i = 0x0801010F
    CONNECT_ACK pd = 8  callref = 0x01  ← R5 sends a connect acknowledge message to the
                                           ISDN switch


ISDN BR1/0: TX ->  RRr sapi = 0  tei = 64  nr = 3
%LINK-3-UPDOWN: Interface BRI1/0 :1, changed state to up  ← The ISDN Call on B-channel 1
                                                            is now active
BR1/0:1 PPP: Treating connection as a callout
BR1/0:1 PPP: Phase is ESTABLISHING, Active Open
BR1/0:1 LCP: O CONFREQ [Closed] id 1 len 15  ← R5 initiates an LCP configuration request

BR1/0:1 LCP:    AuthProto CHAP (0x0305C22305)  ← R5 is requesting CHAP authentication


BR1/0:1 LCP:    MagicNumber 0xE0B91E1E (0x0506E0B91E1E)
BR1/0:1 LCP: I CONFREQ [REQsent] id 1 len 15  ← R5 receives an LCP negotiation request
                                                from R6
BR1/0:1 LCP:    AuthProto CHAP (0x0305C22305)
BR1/0:1 LCP:    MagicNumber 0x00A6C01B (0x050600A6C01B)

BR1/0:1 LCP: O CONFACK [REQsent] id 1 len 15  ← R5 sends a configuration acknowledgement
                                                to R6

BR1/0:1 LCP:    AuthProto CHAP (0x0305C22305)
BR1/0:1 LCP:    MagicNumber 0x00A6C01B (0x050600A6C01B)
BR1/0:1 LCP: I CONFACK [ACKsent] id 1 len 15  ← R5 receives a configuration
                                                acknowledgement from R6
BR1/0:1 LCP:    AuthProto CHAP (0x0305C22305)
BR1/0:1 LCP:    MagicNumber 0xE0B91E1E (0x0506E0B91E1E)
BR1/0:1 LCP: State is Open


BR1/0:1 PPP: Phase is AUTHENTICATING, by both  ← Once the LCP negotiation is completed
                                                 the routers begin to authenticate to
                                                 each other
```

```
BR1/0:1 CHAP: O CHALLENGE id 1 len 23 from "R5"  ← R5 sends a CHAP challenge to R6

BR1/0:1 CHAP: I CHALLENGE id 1 len 23 from "R6"  ← R5 receives a CHAP challenge from R6

BR1/0:1 CHAP: O RESPONSE id 1 len 23 from "R5"
BR1/0:1 CHAP: I SUCCESS id 1 len 4  ← R6 sends R5 a CHAP success message
BR1/0:1 CHAP: I RESPONSE id 1 len 23 from "R6"
BR1/0:1 CHAP: O SUCCESS id 1 len 4  ← R5 sends R6 a CHAP success message

BR1/0:1 PPP: Phase is UP
BR1/0:1 IPCP: O CONFREQ [Closed] id 1 len 10  ← Once authentication has completed
                                                 network layer attributes are negotiated
BR1/0:1 IPCP:    Address 152.3.9.1 (0x030698030901)
BR1/0:1 CDPCP: O CONFREQ [Closed] id 1 len 4
BR1/0:1 BNCP: O CONFREQ [Closed] id 1 len 4
BR1/0:1 IPCP: I CONFREQ [REQsent] id 1 len 10
BR1/0:1 IPCP:    Address 152.3.9.2 (0x030698030902)
BR1/0:1 IPCP: O CONFACK [REQsent] id 1 len 10
BR1/0:1 IPCP:    Address 152.3.9.2 (0x030698030902)
BR1/0:1 CDPCP: I CONFREQ [REQsent] id 1 len 4
BR1/0:1 CDPCP: O CONFACK [REQsent] id 1 len 4
BR1/0:1 BNCP: I CONFREQ [REQsent] id 1 len 4
BR1/0:1 BNCP: O CONFACK [REQsent] id 1 len 4
BR1/0:1 IPCP: I CONFACK [ACKsent] id 1 len 10
BR1/0:1 IPCP:    Address 152.3.9.1 (0x030698030901)
BR1/0:1 IPCP: State is Open
BR1/0:1 CDPCP: I CONFACK [ACKsent] id 1 len 4
BR1/0:1 CDPCP: State is Open
BR1/0:1 BNCP: I CONFACK [ACKsent] id 1 len 4
BR1/0:1 BNCP: State is Open
BR1/0 IPCP: Install route to 152.3.9.2  ← R5 installs a route to its directly connected
                                           neighbor interface on R6

%LINEPROTO-5-UPDOWN: Line protocol on Interface BRI1/0:1, changed state to up
%ISDN-6-CONNECT: Interface BRI1/0 :1 is now connected to 8995201 R6  ← The call has been
                                                                        connected
```

The following output shows that the **show isdn status** command now indicates that you have an active ISDN call:

```
R5#show isdn status
The current ISDN Switchtype = basic-ni1
ISDN BRI1/0 interface
    Layer 1 Status:
        ACTIVE
    Layer 2 Status:
        TEI = 64, State = MULTIPLE_FRAME_ESTABLISHED
        TEI = 65, State = MULTIPLE_FRAME_ESTABLISHED
    Spid Status:
        TEI 64, ces = 1, state = 5(init)
            spid1 configured, spid1 sent, spid1 valid
            Endpoint ID Info: epsf = 0, usid = 70, tid = 1
        TEI 65, ces = 2, state = 5(init)
            spid2 configured, spid2 sent, spid2 valid
            Endpoint ID Info: epsf = 0, usid = 70, tid = 2
```

```
Layer 3 Status:
    1 Active Layer 3 Call(s)
Activated dsl 8 CCBs = 2
    CCB: callid=0x0, sapi=0, ces=1, B-chan=0
    CCB: callid=0x8001, sapi=0, ces=1, B-chan=1
Total Allocated ISDN CCBs = 2
```

The **show dialer interface bri 1/0** command shows dial-specific information for the ISDN interface on R5. Notice that the reason the call was placed was that bridged traffic qualified as interesting traffic (Dial reason: bridge (0xF0F0)). The current call remains up for an additional 82 seconds. The link resets the interesting traffic counter for any interesting traffic that is sent.

```
R5#show dialer int bri 1/0

BRI1/0 - dialer type = ISDN

Dial String      Successes   Failures    Last called   Last status
8995201                  1          0    00:00:56       successful
0 incoming call(s) have been screened.

BRI1/0:1 - dialer type = ISDN
Idle timer (120 secs), Fast idle timer (20 secs)
Wait for carrier (30 secs), Re-enable (15 secs)
Dialer state is data link layer up
Dial reason: bridge (0xF0F0)
Time until disconnect 82 secs
Connected to 8995201 (R6)

BRI1/0:2 - dialer type = ISDN
Idle timer (120 secs), Fast idle timer (20 secs)
Wait for carrier (30 secs), Re-enable (15 secs)
Dialer state is idle
```

The **show bridge group** command indicates that interface S0/1 is still part of bridge group 1, but that the interface is down. Interface BRI 1/0 is now in a forwarding state.

```
R5#show bridge group

Bridge Group 1 is running the IEEE compatible Spanning Tree protocol

   Port 4 (Ethernet0/0) of bridge group 1 is forwarding
   Port 3 (Serial0/1) of bridge group 1 is down  ← S0/1 is still part of Bridge Group 1
                                                     but it is down

   Port 5 (BRI1/0) of bridge group 1 is forwarding  ← BRI 1/0 is now in a forwarding
                                                       state

Bridge Group 2 is running the IEEE compatible Spanning Tree protocol

   Port 9 (Ethernet1/0) of bridge group 2 is forwarding  ← Notice that Bridge Group 2 is
                                                            not affected by the fact that
                                                            R5 is in dial backup
   Port 12 (DLSw Port0) of bridge group 2 is forwarding
```

The **show bridge 1 verbose** command shows that the BRI interface is now transmitting bridged traffic over the ISDN link:

```
R5#show bridge 1 verbose

Total of 300 station blocks, 297 free
Codes: P - permanent, S - self

BG Hash      Address        Action   Interface       VC     Age    RX count    TX count
 1 A7/0    0000.8635.46e1 forward  Ethernet0/0      -      0          12           0
 1 FB/0    0000.8635.4ab1 forward  BRI1/0           -      0           6           6

Flood ports           RX count    TX count
Ethernet0/0                 96         116
Serial0/1                  110          90
BRI1/0                       6           6  ← Bridge traffic is now flowing over the ISDN
                                              circuit
```

After the interesting traffic timeout expires, the ISDN call comes down. Notice that you have not reconnected the S0/1 cable on R5. The serial connection between R5 and R6 is still down. The ISDN call is being dropped because the interesting traffic timer has expired. Any additional interesting traffic causes the link to reactivate.

```
%ISDN-6-DISCONNECT: Interface BRI1/0:1  disconnected from 8995201 R6, call lasted 139
seconds  ← The call will be disconnected after the interesting traffic timer expires

%LINK-3-UPDOWN: Interface BRI1/0:1, changed state to down
%LINEPROTO-5-UPDOWN: Line protocol on Interface BRI1/0:1, changed state to down
```

The **show isdn status** command indicates that R6 has already negotiated the SPIDs for the ISDN circuit. Remember that R5 did not negotiate the SPIDs until the backup interface (BRI1/0) became active. The ISDN interface on R6 is not configured as a backup interface. Thus, as soon as the router boots up, it attempts to activate the ISDN circuit.

```
R6#show isdn status
The current ISDN Switchtype = basic-ni1
ISDN BRI0 interface
    Layer 1 Status:
ACTIVE  ← The ISDN circuit has been activated
    Layer 2 Status:
        TEI = 64, State = MULTIPLE_FRAME_ESTABLISHED
        TEI = 65, State = MULTIPLE_FRAME_ESTABLISHED
    Spid Status:
        TEI 64, ces = 1, state = 5(init)
            spid1 configured, spid1 sent, spid1 valid  ← SPID1 has been sent to
                                                          the ISDN switch and validated

            Endpoint ID Info: epsf = 0, usid = 70, tid = 1
        TEI 65, ces = 2, state = 5(init)
            spid2 configured, spid2 sent, spid2 valid  ← SPID2 has been sent to
                                                          the ISDN switch and validated
```

```
          Endpoint ID Info: epsf = 0, usid = 70, tid = 2
    Layer 3 Status:
        0 Active Layer 3 Call(s)
    Activated dsl 0 CCBs = 1
CCB: callid=0x0, sapi=0, ces=1, B-chan=0
    Total Allocated ISDN CCBs = 1
```

The **show bridge 1 verbose** command indicates that interface BRI0 is now sending and receiving bridged traffic:

```
R6#show bridge 1 verbose

Total of 300 station blocks, 300 free
Codes: P - permanent, S - self


Flood ports          RX count    TX count
BRI0                       12         125
Ethernet0                 119           6
Serial1                     6         110
```

The **show bridge group** command shows that interface BRI 0 is now a part of the bridge group, while interface S1 is still a part of the bridge group but is down:

```
R6#show bridge group
Bridge Group 1 is running the IEEE compatible Spanning Tree protocol

   Port 3 (BRI0) of bridge group 1 is forwarding
   Port 2 (Ethernet0) of bridge group 1 is forwarding
   Port 7 (Serial1) of bridge group 1 is down
```

Enable PPP authentication, PPP negotiation, and ISDN q931 debugging with the following debug commands:

```
R6#debug ppp authentication
PPP authentication debugging is on
R6#debug ppp negotiation
PPP protocol negotiation debugging is on
R6#debug isdn q931
ISDN Q931 packets debugging is on
```

Take a look at what happens when a call is received by R6. The first event that occurs is that the serial interface S1 goes into a down state:

```
%LINEPROTO-5-UPDOWN: Line protocol on Interface Serial1, changed state to down
%LINK-3-UPDOWN: Interface Serial1, changed state to down  ← Primary line failure
PPP encapsulation is closed on interface S1:
```

```
Se1 IPCP: State is Closed  ← IP is being closed
Se1 CDPCP: State is Closed  ← CDP is being closed
Se1 BNCP: State is Closed  ← Bridging is being closed
Se1 PPP: Phase is TERMINATING
Se1 LCP: State is Closed
Se1 PPP: Phase is DOWN
Se1 PPP: Unsupported or un-negotiated protocol. Link cdp
Se1 IPCP: Remove route to 152.3.10.21  ← The route to our directly connected neighbor is
                                          being deleted

ISDN BR0: RX <-  SETUP pd = 8  callref = 0x02  ← R6 receives an incoming call from R5

        Bearer Capability i = 0x8890
        Channel ID i = 0x89
        Calling Party Number i = '!', 0x80, '0008995101'
        Called Party Number i = 0xC1, '8995201'  ← The incoming call comes from 8995201

%LINK-3-UPDOWN: Interface BRI0:1, changed state to up
BR0:1 PPP: Treating connection as a callin
BR0:1 PPP: Phase is ESTABLISHING, Passive Open
BR0:1 LCP: State is Listen
ISDN BR0: TX ->  CONNECT pd = 8  callref = 0x82
        Channel ID i = 0x89
ISDN BR0: RX <-  CONNECT_ACK pd = 8  callref = 0x02
PPP negotiation begins over the ISDN backup line:
BR0:1 LCP: I CONFREQ [Listen] id 2 len 15  ← R5 sends R6 a PPP configuration request

BR0:1 LCP:    AuthProto CHAP (0x0305C22305)
BR0:1 LCP:    MagicNumber 0xE0D76365 (0x0506E0D76365)
BR0:1 LCP: O CONFREQ [Listen] id 2 len 15
BR0:1 LCP:    AuthProto CHAP (0x0305C22305)
BR0:1 LCP:    MagicNumber 0x00C50567 (0x050600C50567)
BR0:1 LCP: O CONFACK [Listen] id 2 len 15  ← R6 sends R5 a configuration acknowledgement

BR0:1 LCP:    AuthProto CHAP (0x0305C22305)
BR0:1 LCP:    MagicNumber 0xE0D76365 (0x0506E0D76365)
BR0:1 LCP: I CONFACK [ACKsent] id 2 len 15  ← R5 sends R6 a configuration
                                               acknowledgement

BR0:1 LCP:    AuthProto CHAP (0x0305C22305)
BR0:1 LCP:    MagicNumber 0x00C50567 (0x050600C50567)
BR0:1 LCP: State is Open  ← LCP negotiation has been successfully completed
```

After LCP negotiation has been completed, the authentication phase begins:

```
BR0:1 PPP: Phase is AUTHENTICATING, by both
BR0:1 CHAP: O CHALLENGE id 2 len 23 from "R6"
BR0:1 CHAP: I CHALLENGE id 2 len 23 from "R5"
BR0:1 CHAP: Waiting for peer to authenticate first
BR0:1 CHAP: I RESPONSE id 2 len 23 from "R5"
BR0:1 CHAP: O SUCCESS id 2 len 4
BR0:1 CHAP: Processing saved Challenge, id 2
BR0:1 CHAP: O RESPONSE id 2 len 23 from "R6"
BR0:1 CHAP: I SUCCESS id 2 len 4
BR0:1 PPP: Phase is UP  ← Authentication has successfully completed
```

After authentication has completed, the NCP negotiation begins:

```
BR0:1 IPCP: O CONFREQ [Closed] id 2 len 10
BR0:1 IPCP:    Address 152.3.9.2 (0x030698030902)
BR0:1 CDPCP: O CONFREQ [Closed] id 2 len 4
BR0:1 BNCP: O CONFREQ [Closed] id 2 len 4
BR0:1 IPCP: I CONFREQ [REQsent] id 2 len 10
BR0:1 IPCP:    Address 152.3.9.1 (0x030698030901)
BR0:1 IPCP: O CONFACK [REQsent] id 2 len 10
BR0:1 IPCP:    Address 152.3.9.1 (0x030698030901)
BR0:1 CDPCP: I CONFREQ [REQsent] id 2 len 4
BR0:1 CDPCP: O CONFACK [REQsent] id 2 len 4
BR0:1 BNCP: I CONFREQ [REQsent] id 2 len 4
BR0:1 BNCP: O CONFACK [REQsent] id 2 len 4
BR0:1 IPCP: I CONFACK [ACKsent] id 2 len 10
BR0:1 IPCP:    Address 152.3.9.2 (0x030698030902)
BR0:1 IPCP: State is Open  ← IP NCP has opened

BR0:1 CDPCP: I CONFACK [ACKsent] id 2 len 4
BR0:1 CDPCP: State is Open  ← Cisco Discovery Protocol NCP has opened

BR0:1 BNCP: I CONFACK [ACKsent] id 2 len 4
BR0:1 BNCP: State is Open  ← Bridging NCP has opened
```

After NCP is completed, R6 installs a route to its directly connected neighbor:

```
BR0 IPCP: Install route to 152.3.9.1
%LINEPROTO-5-UPDOWN: Line protocol on Interface BRI0:1, changed state to up
%ISDN-6-CONNECT: Interface BRI0:1 is now connected to 0008995101 R5
```

The **show dialer interface** bri 0 command shows that a single incoming call has been accepted from 8995101. Notice that the call will be disconnected in another 64 seconds.

```
R6#show dialer interface bri 0

BRI0 - dialer type = ISDN

Dial String      Successes   Failures    Last called   Last status
0 incoming call(s) have been screened.

BRI0:1 - dialer type = ISDN
Idle timer (120 secs), Fast idle timer (20 secs)
Wait for carrier (30 secs), Re-enable (15 secs)
Dialer state is data link layer up
Time until disconnect 64 secs
Connected to 0008995101 (R5)

BRI0:2 - dialer type = ISDN
Idle timer (120 secs), Fast idle timer (20 secs)
Wait for carrier (30 secs), Re-enable (15 secs)
Dialer state is idle
```

With the **show bridge 1 verbose** command, interface BRI0 is now forwarding bridged traffic:

```
R6#show bridge 1 verbose

Total of 300 station blocks, 298 free
Codes: P - permanent, S - self

BG Hash      Address      Action  Interface      VC    Age   RX count   TX count
 1 A7/0    0000.8635.46e1 forward BRI0            -     1          28         29
 1 FB/0    0000.8635.4ab1 forward Ethernet0       -     0          32         27

Flood ports         RX count    TX count
BRI0                       2         140
Ethernet0                132          10
Serial1                    8         114
```

The **show isdn status** command shows that you now have an active call on the router:

```
R6#show isdn status
The current ISDN Switchtype = basic-ni1
ISDN BRI0 interface
    Layer 1 Status:
        ACTIVE
    Layer 2 Status:
        TEI = 64, State = MULTIPLE_FRAME_ESTABLISHED
        TEI = 65, State = MULTIPLE_FRAME_ESTABLISHED
    Spid Status:
        TEI 64, ces = 1, state = 5(init)
            spid1 configured, spid1 sent, spid1 valid
            Endpoint ID Info: epsf = 0, usid = 70, tid = 1
        TEI 65, ces = 2, state = 5(init)
            spid2 configured, spid2 sent, spid2 valid
            Endpoint ID Info: epsf = 0, usid = 70, tid = 2
    Layer 3 Status:
        1 Active Layer 3 Call(s)
    Activated dsl 0 CCBs = 2
        CCB: callid=0x0, sapi=0, ces=1, B-chan=0
        CCB: callid=0x2, sapi=0, ces=1, B-chan=1
    Total Allocated ISDN CCBs = 2
```

The following router output is shown when the call is disconnected, after the idle timer expires:

```
%ISDN-6-DISCONNECT: Interface BRI0:1  disconnected from 0008995101 R5, call lasted 120
seconds
%LINK-3-UPDOWN: Interface BRI0:1, changed state to down
%LINEPROTO-5-UPDOWN: Line protocol on Interface BRI0:1, changed state to down
```

20. *Configure R6 so that a workstation on LAN_3 with a source MAC address of 0000.8635.46e1 should have its packets rejected.*

First, connect to R6 and verify that your bridged traffic is taking the serial link between R5 and R6. Use the **show bridge 1 verbose** command to verify this.

```
R6#show bridge 1 verbose

Total of 300 station blocks, 298 free
Codes: P - permanent, S - self

BG Hash      Address      Action   Interface      VC    Age   RX count   TX count
 1 A7/0   0000.8635.46e1 forward  Serial1         -     1        58         57
 1 FB/0   0000.8635.4ab1 forward  Ethernet0       -     0       125         53

Flood ports          RX count    TX count
BRI0                        0          73
Ethernet0                  68           5
Serial1                     5          68
```

Now enter router configuration mode using the **config term** command. Enter the following commands:

```
R6(config)#bridge 1 address 0000.8635.46e1 discard
R6(config)#exit
```

Now verify that you are blocking any traffic that has a destination MAC address equal to 0000.8635.46e1 using the **show bridge 1 verbose** command:

```
R6#show bridge 1 verbose

Total of 300 station blocks, 298 free
Codes: P - permanent, S - self

BG Hash      Address      Action   Interface      VC    Age   RX count   TX count
 1 A7/0   0000.8635.46e1 discard      -           -     P        59         57
                        ↑
                    Traffic from MAC address 0000.8635.46e1 is being discarded

 1 FB/0   0000.8635.4ab1 forward  Ethernet0       -     1       128         53

Flood ports          RX count    TX count
BRI0                        0          77
Ethernet0                  71           6
Serial1                     6          71
```

Once the MAC filter has been added to R6, verify that it is working by doing the following: Connect to System3 and try to find System1 on the network via Windows Networking. The screen shown in Figure 5-14 indicates the results of the search for System1. Notice that System1 cannot be found on the network.

21. *Enable IPX routing on R2 and R1.*

Start your configuration by connecting to R1. You will be routing IP traffic and bridging IPX traffic. You have already configured your IP addresses, so you now need to configure your IPX addresses. Because you will be bridging IPX, use the same IPX network number on R1's E1/0 interface and R2's E0/0 interface. Figure 5-15 shows the IPX addressing scheme for this configuration.

In order to run IPX on a router, you must first enable IPX routing. The **ipx routing 0001.0001.0001** command enables IPX routing and uses 1.1.1 as the IPX node number for any interfaces (such as serial interfaces) that do not have MAC addresses:

```
R1(config)#ipx routing 0001.0001.0001
```

Next, configure a loopback interface, which is assigned IPX network number 1. Use this loopback to make sure that you are learning IPX routes dynamically through the IPX EIGRP routing protocol:

```
R1(config)#interface Loopback0
R1(config-if)#ipx network 1
```

Interface E1/0 is assigned IPX network address 7:

```
R1(config-if)#interface Ethernet1/0
R1(config-if)#ipx network 7
```

The IPX RIP routing protocol is automatically enabled on the router. You will be running IPX EIGRP, so specifically enable it on the router. Also enable IPX EIGRP on IPX network 7 and IPX network 1:

```
R1(config-if)#ipx router eigrp 100
R1(config-ipx-router)#network 7
R1(config-ipx-router)#network 1
```

IPX RIP should be disabled on any networks that are running EIGRP. Therefore, disable IPX RIP on networks 7 and 1:

```
R1(config-router)#ipx rip
R1(config-ipx-router)#no network 7
R1(config-ipx-router)#no network 1
```

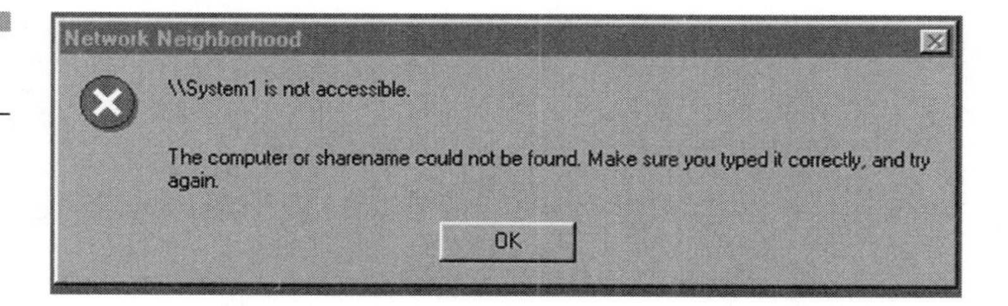

Figure 5-14 MAC address filter

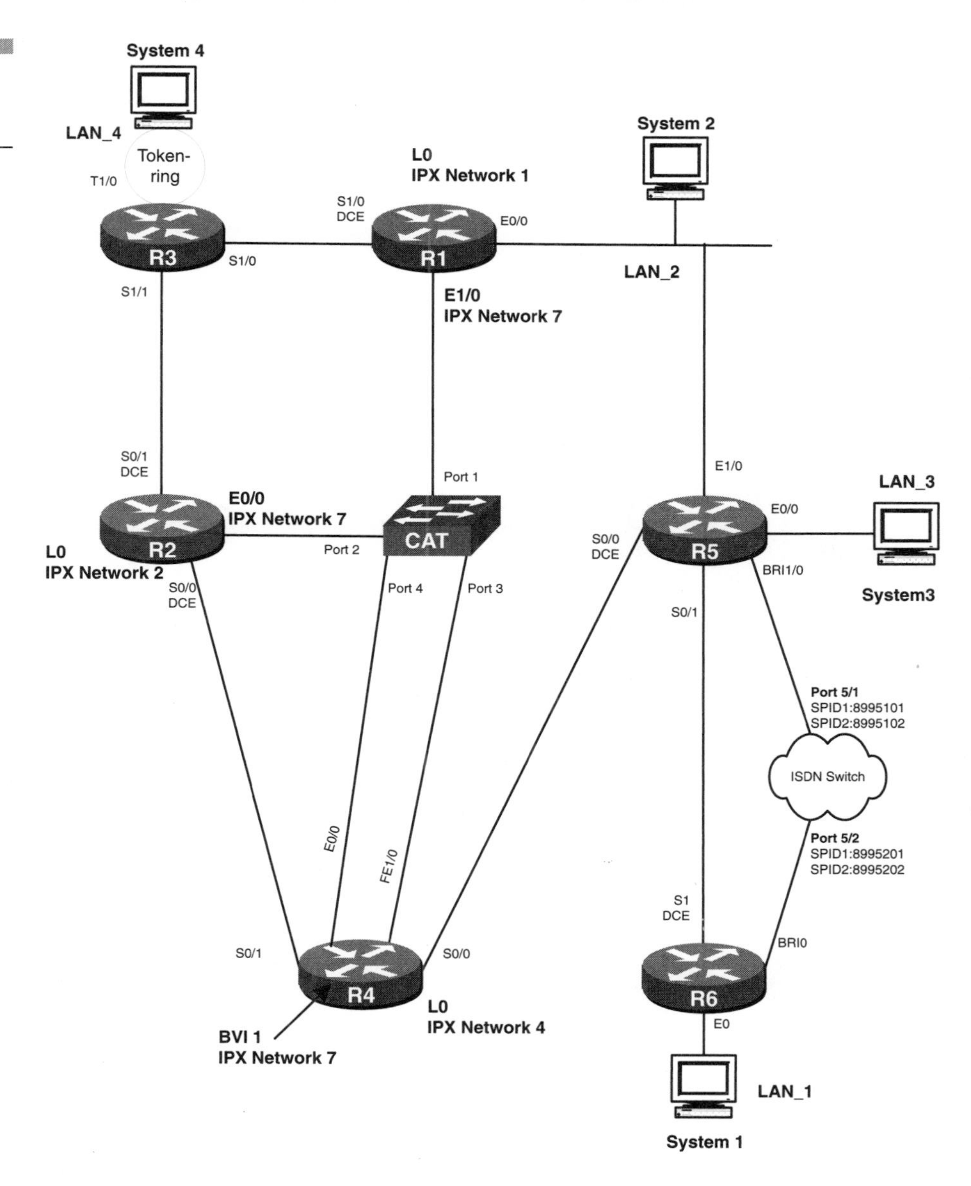

Figure 5-15 IPX network addresses

Now connect to R2. First, enable IPX on the router with the **ipx routing 0002.0002.0002** command.

```
R2(config)#ipx routing 0002.0002.0002
```

Add an IPX loopback address, which you will later use as a test point:

```
R2(config)#interface Loopback0
R2(config-if)#ipx network 2
```

Interface E0/0 is assigned to IPX network 7. Notice that this is the same IPX network that interface E1/0 of R1 has been assigned. Remember that you are bridging IPX between routers R1, R2, and R4. Thus, R1, R2, and R4 are assigned the same IPX network addresses.

```
R2(config-if)#interface Ethernet0/0
R2(config-if)#ipx network 7
```

You will be running IPX EIGRP on IPX network 7 and IPX network 2:

```
R2(config-if)#ipx router eigrp 100
R2(config-ipx-router)#network 7
R2(config-ipx-router)#network 2
```

Because you are running IPX EIGRP on networks 7 and 2, disable IPX RIP on both of these networks:

```
R2(config-router)#ipx rip
R2(config-ipx-router)#no network 7
R2(config-ipx-router)#no network 2
```

22. *Configure integrated routing and bridging between R2, R1, and R4. IP traffic should be routed and IPX traffic should be bridged.*

Connect to R4 and enable IPX routing on R4 with the **ipx routing 0004.0004.0004** command:

```
R4(config)#ipx routing 0004.0004.0004
```

Integrated routing and bridging is enabled globally with the **bridge irb** command:

```
R4(config)#bridge irb
```

IPX network 4 is assigned to a loopback interface. It will be used as a test point:

```
R4(config)#interface Loopback0
R4(config-if)#ipx network 4
```

Interfaces E0/0 and FastE1/0 will both be assigned to bridge group 1. Notice that an IPX address is not assigned to either of these interfaces. This is how you would normally configure the bridging of a protocol. Recall that R2 is connected to interface E0/0 via the Catalyst switch and R1 is connected to interface FastE1/0 via the Catalyst switch. Both R1 and R2 have been assigned to the same IPX network (IPX network 7). By putting interfaces E0/0 and FastE1/0 of R4 in the same bridge group, you can bridge IPX traffic between routers R1 and R2:

```
R4(config-if)#interface Ethernet0/0
R4(config-if)#bridge-group 1

R4(config-if)#interface FastEthernet1/0
R4(config-if)#bridge-group 1
```

Next, define a *Bridge-Group Virtual Interface* (BVI) on R4. The BVI number (in this case, BVI 1 has been created) is associated with the bridge group 1 that is defined on interfaces E0/0 and FastE1/0 . A BVI acts like a normal routed interface that does not support bridging. The BVI represents the corresponding bridge group (in this case, bridge group 1) to routed interfaces within the router. When you enable routing for a given protocol on the BVI, packets coming from a routed interface but destined for a host in a bridged domain are routed to the BVI and are forwarded to the corresponding bridged interface. All traffic routed to the BVI is forwarded to the corresponding bridge group as bridged traffic. All routable traffic received on a bridged interface is routed to other routed interfaces as if it is coming directly from the BVI.

```
R4(config-if)#interface BVI1
R4(config-if)#no ip address
R4(config-if)#no ip directed-broadcast
R4(config-if)#ipx network 7
```

Run IPX EIGRP on networks 4 and 7:

```
R4(config-if)#ipx router eigrp 100
R4(config-ipx-router)#network 4
R4(config-ipx-router)#network 7
```

Because you are running IPX EIGRP on your router interfaces, you can disable IPX RIP:

```
R4(config-router)#ipx rip
R4(config-ipx-router)#no network 4
R4(config-ipx-router)#no network 7
R4(config-ipx-router)#exit
```

You must configure the spanning tree protocol on the router, so use the IEEE version of spanning tree:

```
R4(config)#bridge 1 protocol ieee
```

You also need to configure which protocols will be bridged and which protocols will be routed on the router. The following statements do several things:

- Configure the router to only route the IP protocol.
- Configure the router to both bridge and route the IPX protocol. IPX will be bridged within bridge group 1 and will be routed outside bridge group 1:

```
R4(config)#bridge 1 route ip
R4(config)#bridge 1 route ipx
R4(config)#no bridge 1 bridge ip
```

After configuring the three routers, connect to R1 and type the **show ipx route** command on R1. Notice that R1 is learning two networks via IPX EIGRP: network 2 and network 4. IPX network 4 is a loopback interface on R4. This network resides outside of the BVI that you have configured.

```
R1#show ipx route
Codes: C - Connected primary network,    c - Connected secondary network
       S - Static, F - Floating static, L - Local (internal), W - IPXWAN
       R - RIP, E - EIGRP, N - NLSP, X - External, A - Aggregate
       s - seconds, u - uses

4 Total IPX routes. Up to 1 parallel paths and 16 hops allowed.

No default route known.

C          1 (UNKNOWN),        Lo0
C          7 (NOVELL-ETHER),   Et1/0
E          2 [409600/0] via        7.00e0.1e5b.0e21, age 00:21:41,
                         1u, Et1/0
E          4 [409600/0] via        7.00e0.1e5b.5061, age 00:14:05,
                         11u, Et1/0
```

Now connect to R2 and find out the full IPX address of interface E0/0 by typing the **show ipx interface E0/0** command. Notice that the IPX address of interface E0/0 is 7.00e0.1e5b.0e21.

```
R2#show ipx interface e 0/0
Ethernet0/0 is up, line protocol is up
  IPX address is 7.00e0.1e5b.0e21, NOVELL-ETHER [up]
  Delay of this IPX network, in ticks is 1 throughput 0 link delay 0
  IPXWAN processing not enabled on this interface.
  IPX SAP update interval is 1 minute(s)
  IPX type 20 propagation packet forwarding is disabled
  Incoming access list is not set
  Outgoing access list is not set
  IPX helper access list is not set
```

```
  SAP GNS processing enabled, delay 0 ms, output filter list is not set
  SAP Input filter list is not set
  SAP Output filter list is not set
  SAP Router filter list is not set
  Input filter list is not set
  Output filter list is not set
  Router filter list is not set
  Netbios Input host access list is not set
  Netbios Input bytes access list is not set
  Netbios Output host access list is not set
  Netbios Output bytes access list is not set
  Updates each 60 seconds, aging multiples RIP: 3 SAP: 3
  SAP interpacket delay is 55 ms, maximum size is 480 bytes
  RIP interpacket delay is 55 ms, maximum size is 432 bytes
  IPX accounting is disabled
  IPX fast switching is configured (enabled)
  RIP packets received 3, RIP packets sent 1
  SAP packets received 0, SAP packets sent 1
```

Reconnect to R1 and ping the E0/0 interface of R2 with the **ping ipx 7.00e0.1e5b.0e21** command, which will be successful. The E1/0 interface of R1 is also assigned to IPX network 7. This successful ping verifies that you are bridging IPX traffic between routers R1, R4, and R2:

```
R1#ping ipx 7.00e0.1e5b.0e21

Type escape sequence to abort.
Sending 5, 100-byte IPX cisco Echoes to 7.00e0.1e5b.0e21, timeout is 2 seconds:
!!!!!
Success rate is 100 percent (5/5), round-trip min/avg/max = 4/8/12 ms
```

Now connect to R4 and find the full IPX address of the BVI by typing the **show ipx int bvi 1** command. Notice that the BVI's IPX address is 7.00e0.1e5b.5061.

```
R4#show ipx int bvi 1
BVI1 is up, line protocol is up
  IPX address is 7.00e0.1e5b.5061, NOVELL-ETHER [up]
      ↑
      This is the IPX address of the BVI interface

  Delay of this IPX network, in ticks is 2 throughput 0 link delay 0
  IPXWAN processing not enabled on this interface.
  IPX SAP update interval is 60 seconds
  IPX type 20 propagation packet forwarding is disabled
  Incoming access list is not set
  Outgoing access list is not set
  IPX helper access list is not set
  SAP GNS processing enabled, delay 0 ms, output filter list is not set
  SAP Input filter list is not set
  SAP Output filter list is not set
```

```
  SAP Router filter list is not set
  Input filter list is not set
  Output filter list is not set
  Router filter list is not set
  Netbios Input host access list is not set
  Netbios Input bytes access list is not set
  Netbios Output host access list is not set
  Netbios Output bytes access list is not set
  Updates each 60 seconds aging multiples RIP: 3 SAP: 3
  SAP interpacket delay is 55 ms, maximum size is 480 bytes
  RIP interpacket delay is 55 ms, maximum size is 432 bytes
  RIP response delay is not set
  IPX accounting is disabled
  IPX fast switching is configured (enabled)
  RIP packets received 0, RIP packets sent 12
  SAP packets received 0, SAP packets sent 1
```

Now try to ping the BVI on R4 with the **ping ipx 7.00e0.1e5b.5061** command. The ping is successful and verifies that the BVI is capable of being pinged:

```
R1#ping ipx 7.00e0.1e5b.5061

Type escape sequence to abort.
Sending 5, 100-byte IPX cisco Echoes to 7.00e0.1e5b.5061, timeout is 2 seconds:
!!!!!
Success rate is 100 percent (5/5), round-trip min/avg/max = 4/8/12 ms
```

Now connect to R2 and display the IPX routing table with the **show ipx route** command. Notice that R2 has learned two routes dynamically via IPX EIGRP.

```
R2#show ipx route
Codes: C - Connected primary network,    c - Connected secondary network
       S - Static, F - Floating static, L - Local (internal), W - IPXWAN
       R - RIP, E - EIGRP, N - NLSP, X - External, A - Aggregate
       s - seconds, u - uses

4 Total IPX routes. Up to 1 parallel paths and 16 hops allowed.

No default route known.

C          2 (UNKNOWN),        Lo0
C          7 (NOVELL-ETHER),   Et0/0
E          1 [409600/0 ] via        7.00e0.1e5b.2611, age 00:23:06,
                         11u, Et0/0
E          4 [409600/0 ] via        7.00e0.1e5b.5061, age 00:15:30,
                         1u, Et0/0
```

Reconnect to R1 and use the **show ipx interface e 1/0** command to determine the full IPX address of the E1/0 interface of R1. Notice that the IPX address is 7.00e0.1e5b.2611.

```
R1#show ipx int e 1/0
Ethernet1/0 is up, line protocol is up
  IPX address is 7.00e0.1e5b.2611, NOVELL-ETHER [up]
  Delay of this IPX network, in ticks is 1 throughput 0 link delay 0
```

```
  IPXWAN processing not enabled on this interface.
  IPX SAP update interval is 1 minute(s)
  IPX type 20 propagation packet forwarding is disabled
  Incoming access list is not set
  Outgoing access list is not set
  IPX helper access list is not set
  SAP GNS processing enabled, delay 0 ms, output filter list is not set
  SAP Input filter list is not set
  SAP Output filter list is not set
  SAP Router filter list is not set
  Input filter list is not set
  Output filter list is not set
  Router filter list is not set
  Netbios Input host access list is not set
  Netbios Input bytes access list is not set
  Netbios Output host access list is not set
  Netbios Output bytes access list is not set
  Updates each 60 seconds, aging multiples RIP: 3 SAP: 3
  SAP interpacket delay is 55 ms, maximum size is 480 bytes
  RIP interpacket delay is 55 ms, maximum size is 432 bytes
  IPX accounting is disabled
  IPX fast switching is configured (enabled)
  RIP packets received 35, RIP packets sent 1
  SAP packets received 0, SAP packets sent 1
```

Now go back to R2. Make sure that you can ping the E1/0 interface of R1. You can see below that the ping is successful. Once again, this verifies that you are bridging IPX traffic between R1, R2, and R4:

```
R2#ping ipx 7.00e0.1e5b.2611

Type escape sequence to abort.
Sending 5, 100-byte IPX cisco Echoes to 7.00e0.1e5b.2611, timeout is 2 seconds:
!!!!!
Success rate is 100 percent (5/5), round-trip min/avg/max = 4/7/8 ms
```

Verify that you can ping the BVI of R4:

```
R2#ping ipx 7.00e0.1e5b.5061

Type escape sequence to abort.
Sending 5, 100-byte IPX cisco Echoes to 7.00e0.1e5b.5061, timeout is 2 seconds:
!!!!!
Success rate is 100 percent (5/5), round-trip min/avg/max = 4/7/8 ms
```

Now connect to R4. Display the status of all IPX interfaces on the router with the **show ipx interface brief** command. Notice that the BVI has been assigned to IPX network number 7 and is in an up state.

```
R4#show ipx interface brief
Interface            IPX Network Encapsulation Status                 IPX State
Ethernet0/0          unassigned  not config'd  up                     n/a
Serial0/0            unassigned  not config'd  up                     n/a
Serial0/1            unassigned  not config'd  up                     n/a
```

```
FastEthernet1/0       unassigned  not config'd  up               n/a
BVI1                  7           NOVELL-ETHER  up               [up]
Loopback0             4           UNKNOWN       up               [up]
```

Display the routing table on R4 with the **show ipx route** command. Notice that R4 has learned about the loopbacks on Routers R2 and R1 dynamically via IPX EIGRP. Also notice that the next hop for both dynamic routes is via the BVI.

```
R4#show ipx route
Codes: C - Connected primary network,    c - Connected secondary network
       S - Static, F - Floating static, L - Local (internal), W - IPXWAN
       R - RIP, E - EIGRP, N - NLSP, X - External, A - Aggregate
       s - seconds, u - uses, U - Per-user static

4 Total IPX routes. Up to 2 parallel paths and 16 hops allowed.

No default route known.

C          4 (UNKNOWN),        Lo0
C          7 (NOVELL-ETHER),   BV1
E          1 [512000/0] via        7.00e0.1e5b.2611, age 00:19:24,
                          1u, BV1
E          2 [512000/0] via        7.00e0.1e5b.0e21, age 00:19:28,
                          1u, BV1
```

Verify that you can ping Routers R1 and R2 from R4:

```
R4#ping ipx 7.00e0.1e5b.0e21

Type escape sequence to abort.
Sending 5, 100-byte IPXcisco Echoes to 7.00e0.1e5b.0e21, timeout is 2 seconds:
!!!!!
Success rate is 100 percent (5/5), round-trip min/avg/max = 4/4/4 ms


R4#ping ipx 7.00e0.1e5b.2611

Type escape sequence to abort.
Sending 5, 100-byte IPXcisco Echoes to 7.00e0.1e5b.2611, timeout is 2 seconds:
!!!!!
Success rate is 100 percent (5/5), round-trip min/avg/max = 1/4/8 ms
```

You can display which interfaces are being bridged and which interfaces are being routed on an interface-by-interface basis with the **show interfaces irb** command. Notice that you are allowing IPX to be both routed and bridged on the Ethernet interfaces of R4. IPX is also being routed on the BVI.

```
R4#show interfaces irb

Ethernet0/0

 Routed protocols on Ethernet0/0:
```

```
  ip         ipx

 Bridged protocols on Ethernet0/0:
  appletalk  decnet     ipx

 Software MAC address filter on Ethernet0/0
  Hash Len    Address      Matches  Act      Type
  0x00:  0 ffff.ffff.ffff       158 RCV Physical broadcast
  0x2A:  0 0900.2b01.0001         0 RCV DEC spanning tree
  0x58:  0 0100.5e00.0006         0 RCV IP multicast
  0x5B:  0 0100.5e00.0005        81 RCV IP multicast
  0x7F:  0 00e0.1e5b.5061        11 RCV Interface MAC address
  0x7F:  1 00e0.1e5b.5061         0 RCV Bridge-group Virtual Interface
  0xC0:  0 0100.0ccc.cccc        12 RCV CDP
  0xC2:  0 0180.c200.0000         2 RCV IEEE spanning tree
  0xC2:  1 0180.c200.0000         0 RCV IBM spanning tree

Serial0/0

 Routed protocols on Serial0/0:
  ip

Serial0/1

 Routed protocols on Serial0/1:
  ip

FastEthernet1/0

 Routed protocols on FastEthernet1/0:
  ip         ipx

 Bridged protocols on FastEthernet1/0:
  appletalk  decnet     ipx

 Software MAC address filter on FastEthernet1/0
  Hash Len    Address      Matches  Act      Type
  0x00:  0 ffff.ffff.ffff       169 RCV Physical broadcast
  0x2A:  0 0900.2b01.0001         0 RCV DEC spanning tree
  0x6E:  0 00e0.1e5b.5070         7 RCV Interface MAC address
  0x7F:  0 00e0.1e5b.5061         5 RCV Bridge-group Virtual Interface
  0xC0:  0 0100.0ccc.cccc        12 RCV CDP
  0xC2:  0 0180.c200.0000       359 RCV IEEE spanning tree
  0xC2:  1 0180.c200.0000         0 RCV IBM spanning tree

BVI1

 Routed protocols on BVI1:
  ipx

Loopback0

 Routed protocols on Loopback0:
  ip         ipx
```

23. *Configure PIM Sparse mode on R1, R4, R5, and R6. Make the* rendezvous point *(RP) address 152.3.10.13 on R1.*

Figure 5-16 shows your required multicast configuration.

Figure 5-16 Multicast configuration

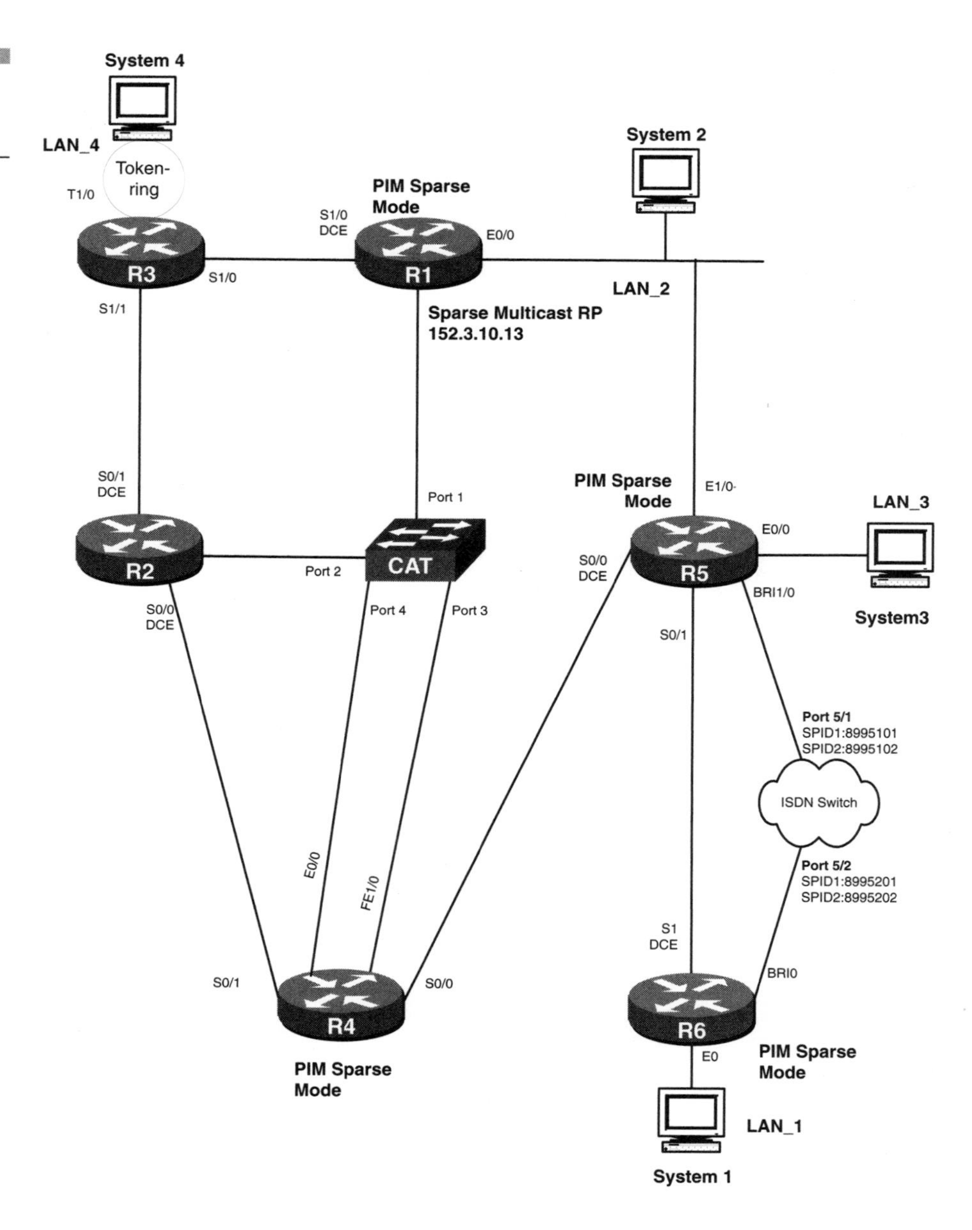

Start your multicast configuration by connecting to R1. IP multicast routing must be globally enabled with the **ip multicast-routing** command:

```
R1(config)#ip multicast-routing
```

Next, individually configure each interface that will be participating in multicast routing. First, configure the multicast type that you will be using on the network. You will be using PIM Sparse mode, so use the **ip pim sparse-mode** command:

```
R1(config-if)#interface Ethernet0/0
R1(config-if)#ip pim sparse-mode
```

Cisco Group Management Protocol (CGMP) is a Cisco proprietary protocol that runs between a Cisco router and a Catalyst switch. It is used only to send multicast traffic to switch ports that have multicast group members on them:

```
R1(config-if)#ip cgmp
```

Interface E1/0 on R1 is configured similarly to interface E0/0 :

```
R1(config-if)#interface Ethernet1/0
R1(config-if)#ip pim sparse-mode
R1(config-if)#ip cgmp
```

Now connect to R4. As with R1, you must globally enable multicast routing:

```
R4(config)#ip multicast-routing
```

Notice that you do not enable the CGMP protocol on interface S0/0 . CGMP can only be enabled on a LAN interface:

```
R4(config)#interface Serial0/0
R4(config-if)#ip pim sparse-mode
```

Next, enable PIM sparse mode and CGMP on interface FE1/0 :

```
R4(config-if)#interface FastEthernet1/0
R4(config-if)#ip pim sparse-mode
R4(config-if)#ip cgmp
R4(config-if)#exit
```

In a PIM Sparse mode network, each multicast sender forwards multicast traffic to a central point, called a *rendezvous point* (RP). The RP then forwards multicast traffic to all multicast receivers in a particular multicast group. R1 has been chosen as the RP. The RP in a router configuration can be specified in several ways. Explicitly configure the RP with the **ip pim rp-address 152.3.10.13** command. Notice that you do not have to define the RP address on the RP router (R1).

```
R4(config)#ip pim rp-address 152.3.10.13
```

R5 is also configured for multicast routing. Notice that you also have to configure the RP address on R5.

```
R5(config)#ip multicast-routing

R5(config)#interface Ethernet0/0
R5(config-if)#ip pim sparse-mode
R5(config-if)#ip cgmp

R5(config-if)#interface Serial0/0
R5(config-if)#ip pim sparse-mode

R5(config-if)#interface Serial0/1
R5(config-if)#ip pim sparse-mode

R5(config-if)#interface Ethernet1/0
R5(config-if)#ip pim sparse-mode
R5(config-if)#ip cgmp
R5(config-if)#exit

R5(config)#ip pim rp-address 152.3.10.13
```

R6 also needs to be configured for multicast routing:

```
R6(config)#ip multicast-routing

R6(config)#interface Ethernet0
R6(config-if)#ip pim sparse-mode
R6(config-if)#ip cgmp

R6(config-if)#interface Serial1
R6(config-if)#ip pim sparse-mode
R6(config-if)#exit

R6(config)#ip pim rp-address 152.3.10.13
```

24. *Configure a static multicast group on R4. Make sure it is reachable and pingable from all routers.*

A static multicast group can be defined with the **ip igmp join-group** command. The **ip igmp join-group 225.69.69.69** command is used to configure the router interface to act as though receivers are available for the 225.69.69.69 multicast group on the interface.

Connect to interface FE1/0 on R4 and add the static multicast group:

```
R4(config)#interface FastEthernet1/0
R4(config-if)#ip igmp join-group 225.69.69.69
```

Now connect to R1 and display the multicast routing table with the **show ip mroute** command. Notice that your static multicast group configured on R4, 225.69.69.69, has an entry in the multicast routing table.

```
R1#show ip mroute
IP Multicast Routing Table
Flags: D - Dense, S - Sparse, C - Connected, L - Local, P - Pruned
       R - RP-bit set, F - Register flag, T - SPT-bit set, J - Join SPT
Timers: Uptime/Expires
Interface state: Interface, Next-Hop, State/Mode

(*, 225.69.69.69), 01:00:18/00:02:41, RP 152.3.10.13, flags: SJC
  Incoming interface: Null, RPF nbr 0.0.0.0
  Outgoing interface list:
    Ethernet1/0 , Forward/Sparse, 00:59:59/00:02:04

(*, 224.0.1.40), 01:01:22/00:00:00, RP 152.3.10.13, flags: SJCL
  Incoming interface: Null, RPF nbr 0.0.0.0
  Outgoing interface list:
    Ethernet0/0, Forward/Sparse, 01:00:01/00:02:33
    Ethernet1/0, Forward/Sparse, 01:01:20/00:02:05
```

The **show ip pim interface** command displays router interfaces that are enabled for multicast routing. Recall that you enabled multicast routing on two interfaces on R1: E0/0 and E1/0 .

```
R1#show ip pim interface

Address          Interface          Mode        Nbr    Query     DR
                                                Count  Intvl
152.3.7.1        Ethernet0/0         Sparse      1      30      152.3.7.2
152.3.10.13      Ethernet1/0         Sparse      1      30      152.3.10.14
```

The **show ip pim rp** command displays information on the multicast RP. Notice that R1 realizes that it is the RP, even though you have not explicitly configured R1 as the RP.

```
R1#show ip pim rp
Group: 225.69.69.69, RP: 152.3.10.13, next RP-reachable in 00:01:17
Group: 224.0.1.40, RP: 152.3.10.13, next RP-reachable in 00:01:15
```

The **show ip igmp interface** command displays the multicast status of each interface on the router:

```
R1#show ip igmp interface
Ethernet0/0 is up, line protocol is up
  Internet address is 152.3.7.1, subnet mask is 255.255.255.192
  IGMP is enabled on interface
  Current IGMP version is 2
  CGMP is enabled on interface
        ↑
        IGMP and CGMP have been enabled on interface E0/0

  IGMP query interval is 60 seconds
  IGMP querier timeout is 120 seconds
  IGMP max query response time is 10 seconds
  Inbound IGMP access group is not set
  Multicast routing is enabled on interface
  Multicast TTL threshold is 0
  Multicast designated router (DR) is 152.3.7.2
  IGMP querying router is 152.3.7.1 (this system)
  No multicast groups joined
Ethernet1/0 is up, line protocol is up
  Internet address is 152.3.10.13, subnet mask is 255.255.255.252
  IGMP is enabled on interface
  Current IGMP version is 2
  CGMP is enabled on interface
  IGMP query interval is 60 seconds
  IGMP querier timeout is 120 seconds
  IGMP max query response time is 10 seconds
  Inbound IGMP access group is not set
  Multicast routing is enabled on interface
  Multicast TTL threshold is 0
  Multicast designated router (DR) is 152.3.10.14
  IGMP querying router is 152.3.10.13 (this system)
  Multicast groups joined: 224.0.1.40
Serial1/0 is up, line protocol is up
  Internet address is 152.3.10.6, subnet mask is 255.255.255.252
  IGMP is disabled on interface
  Multicast routing is disabled on interface
        ↑
        Interface S1/0  has not been configured for multicast routing


  Multicast TTL threshold is 0
  No multicast groups joined
Loopback0 is up, line protocol is up
  Internet address is 152.3.1.1, subnet mask is 255.255.255.0
  IGMP is disabled on interface
  Multicast routing is disabled on interface
  Multicast TTL threshold is 0
  No multicast groups joined
```

Ping the static multicast group, which you defined on R4 at address 225.69.69.69.69 and your ping will be successful:

```
R1#ping 225.69.69.69

Type escape sequence to abort.
Sending 1, 100-byte ICMP Echos to 225.69.69.69, timeout is 2 seconds:

Reply to request 0 from 152.3.10.14, 4 ms
Reply to request 0 from 152.3.10.14, 12 ms
Now connect to R4.  Display the multicast routing table:
R4#show ip mroute
IP Multicast Routing Table
Flags: D - Dense, S - Sparse, C - Connected, L - Local, P - Pruned
       R - RP-bit set, F - Register flag, T - SPT-bit set, J - Join SPT
       X - Proxy Join Timer Running
Timers: Uptime/Expires
Interface state: Interface, Next-Hop or VCD, State/Mode

(*, 225.69.69.69), 00:32:26/00:00:00, RP 152.3.10.13, flags: SJPCL
  Incoming interface: FastEthernet1/0, RPF nbr 152.3.10.13
  Outgoing interface list: Null

(*, 224.0.1.40), 00:32:26/00:00:00, RP 152.3.10.13, flags: SJPCL
  Incoming interface: FastEthernet1/0, RPF nbr 152.3.10.13
  Outgoing interface list: Null
```

The **show ip pim rp** command displays RP information for R4:

```
R4#show ip pim rp
Group: 225.69.69.69, RP: 152.3.10.13, v1, uptime 00:32:33, expires never
Group: 224.0.1.40, RP: 152.3.10.13, v1, uptime 00:32:33, expires 00:03:40
```

The **show ip pim neighbor** command confirms that you have PIM multicast neighbors on interface S0/0 (R5) and interface FE1/0 (R1):

```
R4#show ip pim neighbor
PIM Neighbor Table
Neighbor Address  Interface           Uptime    Expires   Ver  Mode
152.3.10.18       Serial0/0            00:32:47  00:01:13  v1   Sparse
152.3.10.13       FastEthernet1/0      00:32:26  00:01:04  v1   Sparse
```

The **show ip igmp groups** command indicates that R4 is a member of two multicast groups:

```
R4#show ip igmp groups
IGMP Connected Group Membership
Group Address    Interface          Uptime    Expires   Last Reporter
225.69.69.69     FastEthernet1/0    00:33:10  never     152.3.10.14
224.0.1.40       FastEthernet1/0    00:33:11  never     152.3.10.14
```

Complete Network Diagram

Figure 5-17 shows the final configuration for your network.

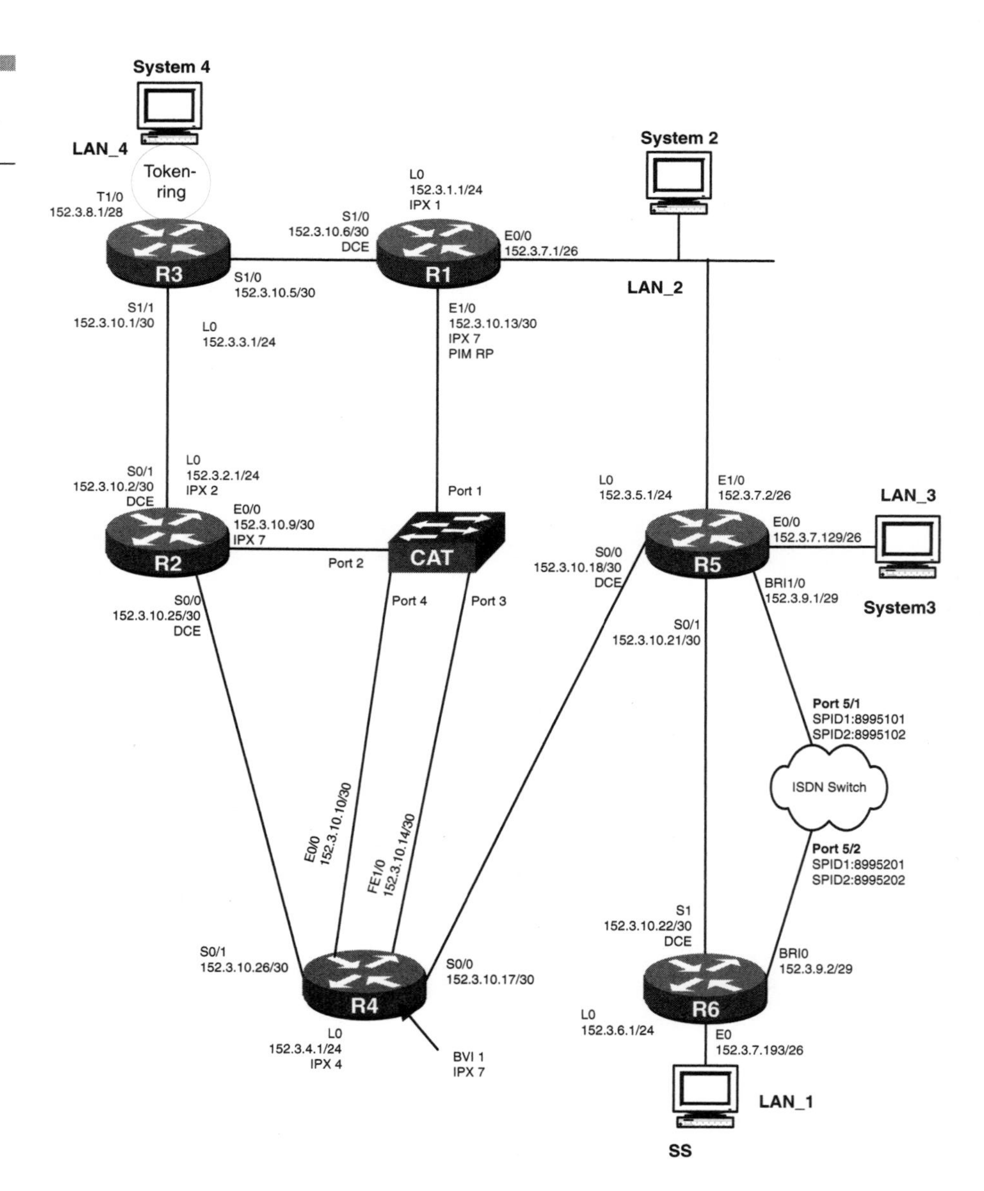

Figure 5-17 Completed network diagram

Router Configurations

The following section contains the complete router configurations for all routers used in this case study, including the router used as our terminal server.

R1

```
!
version 11.2
service timestamps debug uptime
service timestamps log uptime
no service password-encryption
no service udp-small-servers
no service tcp-small-servers
!
hostname R1
!
!
ip multicast-routing
ip dvmrp route-limit 20000
ipx routing 0001.0001.0001
source-bridge ring-group 169
dlsw local-peer peer-id 152.3.7.1 promiscuous
dlsw bridge-group 2
!
interface Loopback0
 ip address 152.3.1.1 255.255.255.0
 ipx network 1
!
interface Ethernet0/0
 ip address 152.3.7.1 255.255.255.192
 ip pim sparse-mode
 ip cgmp
 bridge-group 2
!
interface Ethernet1/0
 ip address 152.3.10.13 255.255.255.252
 ip pim sparse-mode
 ip cgmp
 ipx network 7
!
interface Serial1/0
 ip address 152.3.10.6 255.255.255.252
 encapsulation ppp
 clockrate 800000
!
router ospf 64
 network 152.0.0.0 0.255.255.255 area 0
!
ip classless
!
!
```

```
!
ipx router eigrp 100
 network 7
 network 1
!
!
ipx rip
 no network 7
!
!
!
bridge 2 protocol ieee
!
line con 0
line aux 0
line vty 0 4
 login
!
end
```

R2

```
!
version 11.2
no service password-encryption
no service udp-small-servers
no service tcp-small-servers
!
hostname R2
!
!
ipx routing 0002.0002.0002
!
interface Loopback0
 ip address 152.3.2.1 255.255.255.0
 ipx network 2
!
interface Ethernet0/0
 ip address 152.3.10.9 255.255.255.252
 ipx network 7
!
interface Serial0/0
 ip address 152.3.10.25 255.255.255.252
 encapsulation ppp
 no fair-queue
 clockrate 800000
!
interface Serial0/1
 ip address 152.3.10.2 255.255.255.252
 encapsulation ppp
 clockrate 800000
!
router ospf 64
 network 152.0.0.0 0.255.255.255 area 0
!
```

```
no ip classless
!
!
!
ipx router eigrp 100
 network 7
 network 2
!
!
ipx rip
 no network 7
!
!
!
!
line con 0
line aux 0
line vty 0 4
 login
!
end
```

R3

```
!
version 11.2
no service password-encryption
no service udp-small-servers
no service tcp-small-servers
!
hostname R3
!
netbios access-list host test permit SYSTEM3
netbios access-list host test deny *
netbios access-list host rest permit SYSTEM9
netbios access-list host rest deny *
!
source-bridge ring-group 169
dlsw local-peer peer-id 152.3.8.1
dlsw remote-peer 0 tcp 152.3.7.1
dlsw remote-peer 0 tcp 152.3.7.2 backup-peer 152.3.7.1
!
interface Loopback0
 ip address 152.3.3.1 255.255.255.0
!
interface Serial1/0
 ip address 152.3.10.5 255.255.255.252
 encapsulation ppp
 no fair-queue
!
interface TokenRing1/0
 ip address 152.3.8.1 255.255.255.240
 ring-speed 4
 access-expression input (netbios-host(test) | netbios-host(rest))
 source-bridge 1 2 169
```

```
 source-bridge spanning
!
interface Serial1/1
 ip address 152.3.10.1 255.255.255.252
 encapsulation ppp
!
router ospf 64
 network 152.0.0.0 0.255.255.255 area 0
!
no ip classless
!
!
line con 0
line aux 0
line vty 0 4
 login
!
end
```

R4

```
!
version 12.0
service timestamps debug uptime
service timestamps log uptime
no service password-encryption
!
hostname R4
!
!
ip subnet-zero
ip multicast-routing
ip dvmrp route-limit 20000
ipx routing 0004.0004.0004
!
bridge irb
!
!
!
!
interface Loopback0
 ip address 152.3.4.1 255.255.255.0
 no ip directed-broadcast
 ip ospf interface-retry 0
 ipx network 4
!
interface Ethernet0/0
 ip address 152.3.10.10 255.255.255.252
 no ip directed-broadcast
 ip ospf interface-retry 0
```

```
 bridge-group 1
!
interface Serial0/0
 ip address 152.3.10.17 255.255.255.252
 no ip directed-broadcast
 ip pim sparse-mode
 encapsulation ppp
 ip ospf interface-retry 0
!
interface Serial0/1
 ip address 152.3.10.26 255.255.255.252
 no ip directed-broadcast
 encapsulation ppp
 ip ospf interface-retry 0
!
interface FastEthernet1/0
 ip address 152.3.10.14 255.255.255.252
 no ip directed-broadcast
 ip pim sparse-mode
 ip ospf interface-retry 0
 ip igmp join-group 225.69.69.69
 ip cgmp
 bridge-group 1
!
interface BVI1
 no ip address
 no ip directed-broadcast
 ipx network 7
!
router ospf 64
 network 152.0.0.0 0.255.255.255 area 0
!
no ip classless
ip pim rp-address 152.3.10.13
!
ipx router eigrp 100
 network 4
!
ipx rip
 no network 4
!
bridge 1 protocol ieee
 bridge 1 route ip
 bridge 1 route ipx
 no bridge 1 bridge ip
!
line con 0
 transport input none
line aux 0
line vty 0 4
 login
!
end
```

R5

```
!
version 11.2
no service password-encryption
no service udp-small-servers
no service tcp-small-servers
!
hostname R5
!
!
username R6 password 0 cisco
ip multicast-routing
ip dvmrp route-limit 20000
isdn switch-type basic-ni1
source-bridge ring-group 169
dlsw local-peer peer-id 152.3.7.2 promiscuous
dlsw bridge-group 2
!
interface Loopback0
 ip address 152.3.5.1 255.255.255.0
!
interface Ethernet0/0
 ip address 152.3.7.129 255.255.255.192
 ip pim sparse-mode
 ip cgmp
 no keepalive
 bridge-group 1
!
interface Serial0/0
 ip address 152.3.10.18 255.255.255.252
 ip pim sparse-mode
 encapsulation ppp
 no fair-queue
 clockrate 800000
!
interface Serial0/1
 backup delay 5 10
 backup interface BRI1/0
 ip address 152.3.10.21 255.255.255.252
 ip pim sparse-mode
 encapsulation ppp
 bridge-group 1
!
interface BRI1/0
 ip address 152.3.9.1 255.255.255.248
 encapsulation ppp
 isdn spid1 5101 8995101
 isdn spid2 5102 8995102
 dialer map bridge name R6 broadcast 8995201
 dialer-group 1
 ppp authentication chap
 bridge-group 1
!
```

```
interface Ethernet1/0
 ip address 152.3.7.2 255.255.255.192
 ip pim sparse-mode
 ip cgmp
 bridge-group 2
!
router ospf 64
 network 152.0.0.0 0.255.255.255 area 0
!
no ip classless
ip pim rp-address 152.3.10.13
access-list 201 permit 0x0000 0xFFFF
!
dialer-list 1 protocol bridge list 201
bridge 1 protocol ieee
bridge 2 protocol ieee
!
line con 0
line aux 0
line vty 0 4
 login
!
end
```

R6

```
!
version 11.2
no service password-encryption
no service udp-small-servers
no service tcp-small-servers
!
hostname R6
!
!
username R5 password 0 cisco
ip multicast-routing
ip dvmrp route-limit 20000
isdn switch-type basic-ni1
!
interface Loopback0
 ip address 152.3.6.1 255.255.255.0
!
interface Ethernet0
 ip address 152.3.7.193 255.255.255.192
 ip pim sparse-mode
 ip cgmp
 no keepalive
 bridge-group 1
!
interface Serial1
 ip address 152.3.10.22 255.255.255.252
```

```
 ip pim sparse-mode
 encapsulation ppp
 clockrate 800000
 bridge-group 1
!
interface BRI0
 ip address 152.3.9.2 255.255.255.248
 encapsulation ppp
 isdn spid1 5201 8995201
 isdn spid2 5202 8995202
 dialer map bridge name R5 broadcast
 dialer-group 1
 ppp authentication chap
 bridge-group 1
!
router ospf 64
 network 152.0.0.0 0.255.255.255 area 0
!
no ip classless
ip pim rp-address 152.3.10.13
access-list 201 permit 0x0000 0xFFFF
dialer-list 1 protocol bridge list 201
bridge 1 protocol ieee
bridge 1 address 0000.8635.46e1 discard
!
line con 0
line aux 0
line vty 0 4
 login
!
end
```

Catalyst Switch

```
Catalyst 1900 - Virtual LAN Configuration

    ----------------------- Information ------------------------------------
     VTP version: 1
     Configuration revision: 2
     Maximum VLANs supported locally: 1005
     Number of existing VLANs: 7
     Configuration last modified by: 0.0.0.0 at 00-00-0000 00:00:00

    ----------------------- Settings ---------------------------------------
     [N] Domain name
     [V] VTP mode control                Server
     [F] VTP pruning mode                Disabled
     [O] VTP traps                       Enabled

    ----------------------- Actions ----------------------------------------
     [L] List VLANs                      [A] Add VLAN
     [M] Modify VLAN                     [D] Delete VLAN
     [E] VLAN Membership                 [S] VLAN Membership Servers
     [T] Trunk Configuration             [W] VTP password
     [P] VTP Statistics                  [X] Exit to Main Menu
```

```
      Catalyst 1900 - VLAN Membership Configuration

   Port  VLAN    Membership Type      Port  VLAN    Membership Type
   ------------------------------      ------------------------------
   1        1       Static            13       1       Static
   2        2       Static            14       1       Static
   3        1       Static            15       1       Static
   4        2       Static            16       1       Static
   5        1       Static            17       1       Static
   6        1       Static            18       1       Static
   7        1       Static            19       1       Static
   8        1       Static            20       1       Static
   9        1       Static            21       3       Static
   10       1       Static            22       3       Static
   11       1       Static            23       3       Static
   12       1       Static            24       3       Static
                                      AUI      1       Static
   A        1       Static
   B        1       Static

     [M] Membership type               [V] VLAN assignment
     [R] Reconfirm dynamic membership  [X] Exit to previous menu
```

```
VLAN  Name                State      Type             MTU   SAID      XB1  XB2
------------------------------------------------------------------------------
1     default             Enabled    Ethernet         1500  100001    1002 1003
2     VLAN0002            Enabled    Ethernet         1500  100002    0    0
3     VLAN0003            Enabled    Ethernet         1500  100003    0    0
1002  fddi-default        Suspended  FDDI             1500  101002    1    1003
1003  token-ring-defau    Suspended  Token-Ring       1500  101003    1    1002
1004  fddinet-default     Suspended  FDDI-Net         1500  101004    0    0
1005  trnet-default       Suspended  Token-Ring-Net   1500  101005    0    0
```

Terminal Server (R7)

```
!
version 11.2
no service password-encryption
no service udp-small-servers
no service tcp-small-servers
!
hostname R7
!
enable password cisco
!
no ip domain-lookup
ip host R2 2002 1.1.1.1
ip host R3 2003 1.1.1.1
ip host R4 2004 1.1.1.1
```

```
ip host R5 2005 1.1.1.1
ip host R6 2006 1.1.1.1
ip host R8 2008 1.1.1.1
ip host R1 2001 1.1.1.1
ip host cat 2007 1.1.1.1
!
interface Loopback0
 ip address 1.1.1.1 255.255.255.255
!
line con 0
line 1 16
 no exec
 transport input all
line aux 0
line vty 0 4
 login
!
end
```

Adtran Atlas 800 ISDN Switch Configuration

The Adtran Atlas 800 is being used in this chapter as your ISDN switch. The Atlas 800 can be populated with both ISDN BRI and ISDN PRI cards. The advantage of using the Atlas 800 is its ease of use and adherence to ISDN standards. In addition, the Atlas' small form factor makes it a portable unit that can be set up on a desktop. The base unit of the Atlas 800 comes with two ISDN PRI circuits. Additional BRI cards and PRI cards also can be added. The Atlas 800 chassis can hold up to eight additional cards. Each BRI card contains eight ISDN BRI U interface circuits, while each PRI card contains four ISDN PRI circuits.

The Atlas 800 chassis has a 10baseT Ethernet connection, which is used for system management. The Atlas 800 also has a control port, which can be used for local terminal access. The Atlas 800 is managed via a simple menu system. The management screens are the same whether you telnet to the Atlas via the Ethernet port or directly connect to the unit via the rear control port. Additional information on the Atlas 800 can be found on the Adtran Web page at `www.Adtran.com`.

The Atlas 800 is simple to configure. This section outlines the key configuration screens for setting up these labs.

The System Info screen is shown below. Key system information such as firmware levels and system uptime can be found on this screen.

```
CCIE LAB STUDY GUIDE/System Info
System Info   | System Name          CCIE LAB STUDY GUIDE
System Status | System Location      Adtran ATLAS 800
```

```
System Config | System Contact          Adtran ATLAS 800
System Utility| Firmware Revision       ATLAS 800  Rev. H 09/18/98 09:11:41
Modules       | System Uptime           0 days  1 hours  36 min  55 secs
Dedicated Maps| Startup Mode            Power cycle
Dial Plan     | Current Time/Date (24h) Saturday March  6  16:55:12  1999
              | Installed Memory        Flash:1048576 bytes  DRAM:8388608 bytes
              | Serial Number           847B8304
              | Boot ROM Rev            C 11/18/97
```

The Modules screen displays all the active modules that are installed in the Atlas 800. Notice in the following screen print that your system has four occupied slots. Slot 0 is the system controller, which is part of the base chassis. Slot 1 contains a four-port ISDN PRI card. Slot 5 contains an eight-port ISDN BRI card. Slot 8 contains a four-port V.35 card. For this case study, only use the BRI card.

```
CCIE LAB STUDY GUIDE/Modules
System Info   | Slt   Type    Menu    Alarm    Test   State    Status      Rev
System Status |  0  Sys Ctrl  [+]     [OK]     [OFF]  ONLINE  Online        T
System Config |  1  T1/PRI-4  [+]     [OK]     [OFF]  ONLINE  Online        A
System Utility|  2  EMPTY                             ONLINE  Empty         -
Modules       |  3  EMPTY                             ONLINE  Empty         -
Dedicated Maps|  4  EMPTY                             ONLINE  Empty         -
Dial Plan     |  5  UBRI-8    [+]     [OK]     [OFF]  ONLINE  Online        C
              |  6  EMPTY                             ONLINE  Empty         -
              |  7  EMPTY                             ONLINE  Empty         -
              |  8  V35Nx-4   [+]     [OK]     [OFF]  ONLINE  Online        M
```

The Dial Plan screen is used to configure *directory numbers* (DNs) for each of the BRI circuits. Notice from the following screen print that BRI #1 in slot #5 has been assigned 8995101 as its number for channel B1. BRI #2 in slot 5 has been assigned 8995201 as its number for channel B1.

```
CCIE LAB STUDY GUIDE/Dial Plan/User Term
Network Term|  #   Slot/Svc   Port/Link  Sig  In#Accept  Out#Rej   Ifce Config
User Term   |  1   5)UBRI-8    1)BRI 5/1       [8995101]    [+]      [8995101]
Global Param|  2   5)UBRI-8    2)BRI 5/2       [8995201]    [+]      [8995201]
```

The Interface Configuration screen is used to set the switch type for each circuit. Notice from the following screen that the National ISDN switch type has been selected for this circuit.

```
CCIE LAB STUDY GUIDE/Dial Plan/User Term[1]/Interface Configuration
Incoming Number Accept List| Switch Type          National ISDN
Outgoing Number Reject List| SPID list            [8995101]
Interface Configuration    | Strip MSD            None
                           | Source ID            0
                           | Outgoing Caller ID   Send as provided
```

The SPID list screen is used to configure the SPID and call capability for each B channel. Notice from the following screen print that the SPID for channel B1 has been set to 5101, while the SPID for channel B2 has been set to 5102. Each of these two B channels is enabled for 64K, 56K, audio, and speech capabilities.

```
CCIE LAB STUDY GUIDE/Dial Plan/User Term[1]/Interface Configuration/SPID list
SPID list| #     Phone #          SPID #      Calls  D64    D56    Audio  Speech
         | 1     8995101           5101          2   Enable Enable Enable Enable
         | 2     8995102           5102          2   Enable Enable Enable Enable
```

CHAPTER 6

A Case Study in IPSec, VoIP, and ATM LANE

Introduction

This case study explores the following topics:

- IP Security (IPSec)
- LAN Emulation (LANE)
- Network Address Translation (NAT)
- Voice over IP (VoIP)
- Service Assurance Agent (SAA)
- Committed Access Rate (CAR)
- Queuing

Technology Overview

IPSEC

When asked what IPSec is, most people will say encryption for IP packets. However, IPSec is not just encryption; it is a framework of open standards for ensuring secure private communications over IP networks. IPSec ensures data integrity and authenticity as well as confidentiality (encryption) across IP networks.

IPSec protocols offer security services at the IP layer through *Authentication Header* (AH) and the *Encapsulating Security Payload* (ESP) protocols. These security services include access control, connectionless integrity, data authentication, protection against replay, and encryption.

Authentication Header (AH)

AH mode provides authentication for the IP header and the payload contained in the IP packet. It does this by using a keyed hash through a shared secret value. The AH service protects the external IP header and payload. It protects all of the fields in the IP header that do not change during transit. Fields such as the *Time to Live* (TTL), *Type of Service* (TOS), and Header checksum are zeroed before the integerity check value is calculated. This is necessary because the values may change during transit and the sender has no way of predicting what they might be when they reach the reciever. The AH header is placed in the packet beween the original IP header and the payload, as shown below.

```
--------------------------------
|orig IP hdr   |    |     |      |
|(any options)| AH | TCP | Data |
--------------------------------
|<------- authenticated ------->|
```

Encapsulating Security Payload (ESP)

ESP performs payload encryption at the IP packet layer. It can also provide authentication through the use of an optional *Integrity Check Value* (ICV). If authentication is used, the value is calculated after the encryption is done.

The following is what an IP packet would look like if ESP with authentication were used in Tunnel mode. The two modes of IPSec will be discussed later in the chapter.

```
----------------------------------------------------------
| new IP hdr  |     | orig IP hdr   |   |    | ESP    | ESP|
|             | ESP |               |TCP|Data|Trailer|Auth|
----------------------------------------------------------
                    |<--------- encrypted ---------->|
              |<----------- authenticated ---------->|
```

IPSEC Modes of Operation

IPSec can be implemented using two options. The first is for the end stations to provide security directly (Transport mode). The advantage of this is it has no impact on the network. The network design, topology, and routing have no impact on the security services. The disadvantage is that all endstations must run IPSec software and users must be "IPSec-aware," meaning that they are responsible for determining what traffic is to be encrypted and or authenticated. This requires a certain understanding by the end user because in most cases configuration changes are needed.

The second option is totally transparent to the end user. In Tunnel mode, the network provides the security. The advantage, of course, is the end users are not directly involved and do not need to be IPSec-aware. The disadvantage is that the network design, topology, and routing must all be considered. Also, additional processing power is required in the routers performing the IPSec service.

Transport Mode

In Transport mode, only the IP payload is encrypted, while the original headers are left intact. The ESP header is inserted after the IP header and before the upper layer protocol header. The upper layer protocols are encrypted and authenticated along with the ESP header. ESP does not authenticate the IP header itself. Authentication protects the external IP header along with the data payload, protecting all the fields in the header that do not change in transit.

```
-------------------------------------------------
|orig IP hdr   | ESP |     |      |   ESP   | ESP|
|              | Hdr | TCP | Data | Trailer |Auth|
-------------------------------------------------
                     |<----- encrypted ---->|
               |<------ authenticated ----->|
```

Table 6-1 shows the services provided by the different IPSec protocols in **transport mode**.

Tunnel Mode

In Tunnel mode, the entire IP packet is encrypted and becomes the payload of a new IP packet. The new IP header has the destination address of its IPSec peer. All the information from the packet, including the header, is protected. Tunnel mode protects against traffic analysis because the true endpoints of the packet cannot be determined, only the endpoints of the IPSec tunnel. If AH is used, both the original header and the new header along with the payload are

Table 6-1

Transport Mode Services Provided

Security Service	AH	ESP	AH & ESP
Access control	Yes	Yes	Yes
Connectionless integrity data origin authentication	Authenticates parts of the original IP header along with upper layer protocols	Encrypts and optionally authenticates upper layer protocols and does not authenticate the original IP header	Encrypts and authenticates upper layer protocols and authenticates parts of the original IP header
Encryption	No	Yes	Yes
Hiding the original source and destination	No	No	No

authenticated. If ESP is used with the authentication option and not AH, only the original IP datagram and the ESP header are protected. The new IP header is not authenticated. Table 6-2 shows the services provided by the different IPSec protocols in Tunnel mode.

How IPSec Works

Because the scenarios in this case study use Tunnel mode with Pre-share authentication and IKE, this section will describe how this works.

With IPSec, you define which data should be secured through the use of access lists. If the traffic matches the access list, it is protected using the defined security protocols. At this point, IKE negotiation begins. IKE provides authentication of the IPSec peer, negotiates the IPSec *security association* (SA), and establishes the IPSec key.

The peer that initiates the session will send its configured ISAKMP policy to its remote peer. The policy contains the defined encryption algorithm, the hash algorithm, the authentication method, the Diffe-Hellman group, and the lifetime.

The remote peer then will try to find a matching policy. A match is made when the receiving peer finds a configured policy that has the same encryption algorithm, hash algorithm, authentication type, and Diffe-Hellman parameters as the ones being sent by its peer. The lifetime of the remote peer must also be less than or equal to the one configured on the router. If no matching policy is found, IKE will refuse negotiation and IPSEC will not be established. If a policy is agreed upon, IKE will complete the negotiation process and an SA will be created.

Table 6-2 Tunnel Mode Services Provided

Security Service	AH	ESP	AH & ESP
Access control	Yes	Yes	Yes
Connectionless integrity data origin authentication	Authenticates all of the original IP datagram and parts of the new IP header	Encrypts and optionally authenticates the original IP datagram and does not authenticate the new IP header	Encrypts and authenticates the original IP datagram and authenticates parts of new IP header
Encryption	No	Yes	Yes
Hiding the original source and destination	Yes	Yes	Yes

Once established, the SA is then applied to all traffic that matches that particular access list that caused the initial negotiation. The corresponding inbound SA is used when processing the incoming traffic from that peer. The SA will have a lifetime equal to the negotiated value. Once the SA expires, a new SA will need to be established.

LANE

LANE is defined by the ATM forum as a technology to interconnect legacy LANs using ATM-attached devices. An *Emulated LAN* (ELAN) appears to be one bridged segment.

When talking about LANE, it is important to understand the terminology. LANE defines a *LAN emulation client* (LEC), a *LAN emulation configuration server* (LECS), and a *broadcast and unknown server* (BUS). The (LEC) is any system that supports LANE. The client emulates a LAN interface to the higher level protocols.

The client performs data forwarding, address resolution, and registration of MAC addresses with the LANE server and communicates with other LECs via the ATM VCCs.

The LECS is responsible for maintaining the database of ELANs and the LES that controls the ELAN. It accepts queries from the LECs and responds with the ATM address of the LES.

The LES is a central control to which all LECS attach. Each LEC maintains a Control Direct VCC to the LES. This VCC is used to forward registration and control information. The LES maintains a point-to-multipoint VCC, known as the Control Distribute VCC, to all LECs. The Control Distribute VCC is used only to forward control information.

The BUS acts as a central point for distributing broadcasts and multicasts. Since ATM is a point-to-point technology, it lacks the capability to support any-to-any or broadcast-type traffic. LANE solves this problem by centralizing the broadcast support in the BUS. Each LEC must set up a Multicast Send VCC to the BUS. The BUS then adds the LEC as a leaf to its Point-to-Multipoint VCC (known as the Multicast Forward VCC). The BUS also acts as a multicast server. LANE is defined on *ATM Adaptation Layer 5* (AAL5), which specifies a simple trailer to be appended to a frame before it is broken into ATM cells. The problem is that no means exists for differentiating ATM cells from different senders when multiplexed on a virtual channel. It is assumed that the cells received are in sequence, and when the *End of Message* (EOM) cell arrives, you should just have to reassemble all the cells that have already arrived.

The BUS takes the sequence of cells on each Multicast Send VCC and reassembles them into frames. When a full frame is received, it is queued for sending to all of the LECs on the Multicast Forward VCC. This way, all the cells from a particular data frame are guaranteed to be sent in order and are not interleaved with cells from any other data frames on the Point-to-Multipoint VCC.

An ELAN provides layer 2 communication between all users on the ELAN, which can be though of as a *virtual LAN* (VLAN) or a single broadcast domain. When a LEC comes up, it must first find the LECS to discover which ELAN to join. The LEC queries the ATM switch via the *Interim Local Management Interface* (ILMI). The switch contains the ATM address of the LECS. The LEC then can use UNI signaling to contact the LECS.

Once the LEC has determined the ATM address of the LECs, it creates a signaling packet using that address. It signals the LECs to determine which ELAN it is a member of and to determine the ATM address of the LES that serves that ELAN.

After the LEC has discovered the ATM address of the desired LES, it drops the connection to the LECS and creates a signaling packet with the ATM address of the LES. Once it is connected to the LES, it registers its MAC and ATM addresses for the ELAN. This information is maintained so that no two LECs can register the same MAC or ATM addresses. The LES adds the LEC as a leaf of its Point-to-Multipoint Control Distribute VCC. Finally, the LES issues the LEC a successful LE_JOIN_RESPONSE that contains a *LANE Client ID* (LECID), which is an identifier that is unique to the new client. This ID is used by the LEC to filter its own broadcasts from the BUS.

After the LEC has successfully joined the LES, its first task is to find the ATM address of the BUS and join the broadcast group. To do this, it sends a special LE_ARP packet on the Control Direct VCC to the LES. The LES recognizes that the LEC is looking for the BUS, responds with the ATM address of the BUS, and forwards that response on the Control Distribute VCC.

When the LEC has the ATM address of the BUS, its next action is to create a signaling packet with that address and signal a Multicast Send VCC. Upon receipt of the signaling request, the BUS adds the LEC as a leaf on its Point-to-Multipoint Multicast Forward VCC. At this time, the LEC has become a member of the ELAN.

When a LEC has a data packet to send to an unknown destination, it issues an ARP request to the LES using the Control Direct VCC. The LES forwards the request on the Control Distribute VCC to all LEC stations. In parallel, the unicast data packets are sent to the BUS to be forwarded to all endpoints. Unicast packets continue using the BUS until the ARP request has been resolved.

The ARP response contains the ATM address of the LEC that is associated with the MAC address being sought. The LEC now signals the other LEC directly and sets up a Data Direct VCC that will be used for unicast data between the LECs.

NAT

NAT is a router function that provides the translation from one IP address to another. Address translation is required for customers who have private (or unregistered addresses) and want to access a public service (where publicly registered addresses are used).

One of the greatest problems facing the Internet today is the issue of address depletion. NAT promises to relieve some of this pressure by allowing organizations to reuse globally unique IP addresses in other parts of their network.

NAT also enables organizations to use unregistered IP addresses within the network as long as the addresses are translated to globally unique Internet-registered addresses before they leave that domain.

NAT Terminology

When dealing with NAT on a Cisco router, it is important to understand the terminology used (see Figure 6-1):

- **Inside local address**: The IP address that is assigned to a host on the inside network. This address is probably not an IP address assigned by the *Network Information Center* (NIC) or service provider.
- **Inside global address**: A NIC-registered IP address that is used to represent one or more inside local IP addresses to the outside world.
- **Outside local address:** The IP address of an outside host as it appears to the inside network. Not necessarily a legitimate address, it is allocated from address space routable on the inside.
- **Outside global address**: The IP address assigned to a host on the outside network by the host's owner. The address is allocated from a globally routable address or network space.

Voice Technology

Voice over IP (VoIP) is a technology that enables analog traffic (telephone calls and faxes) to be carried over an IP network. As shown in Figure 6-2, with VoIP technology, a *digital signal processor* (DSP) that resides in a Cisco Voice Network Module takes analog traffic (voice and fax), converts the analog traffic to digital signals, compresses the digital signals, and then sends the analog traffic as IP packets. These voice packets are then sent as encapsulated IP traffic using the ITU H.323 specification, which defines the transmission of voice, data, and video across a network.

Voice traffic is very sensitive to delays. This is because voice traffic has the characteristics of constant bit rate traffic. That is, a constant stream of information is being sent. Unlike data packets, if voice packets are lost or received in error, it does no good to retransmit the original

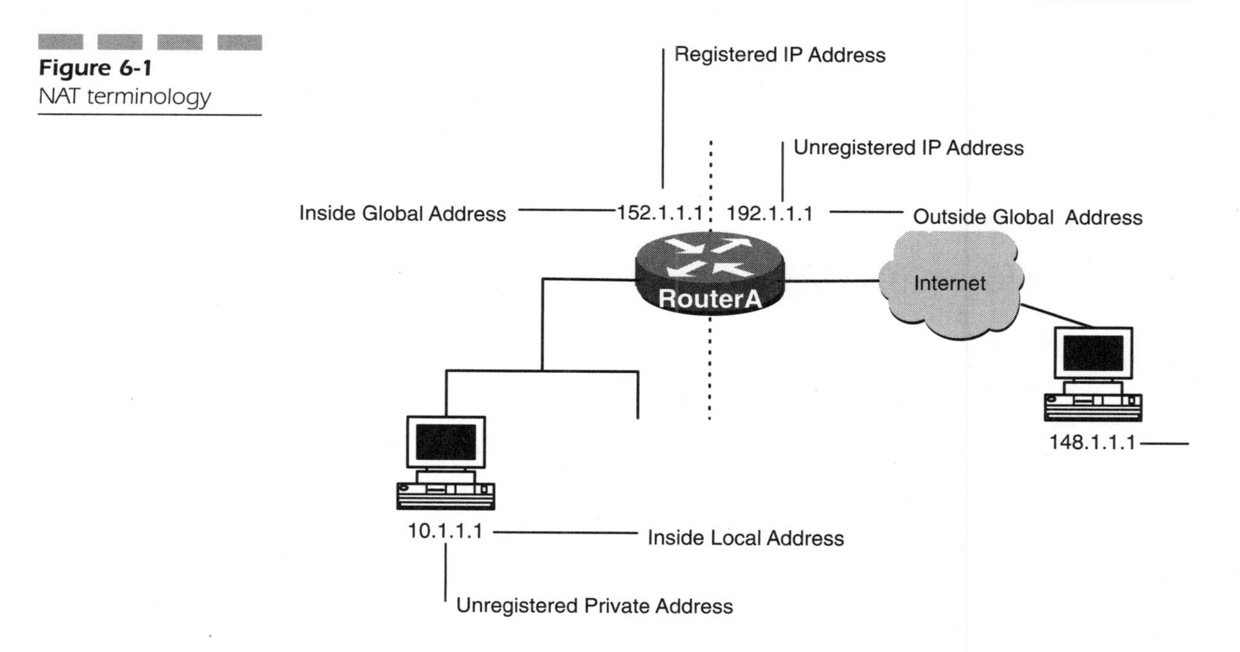

Figure 6-1
NAT terminology

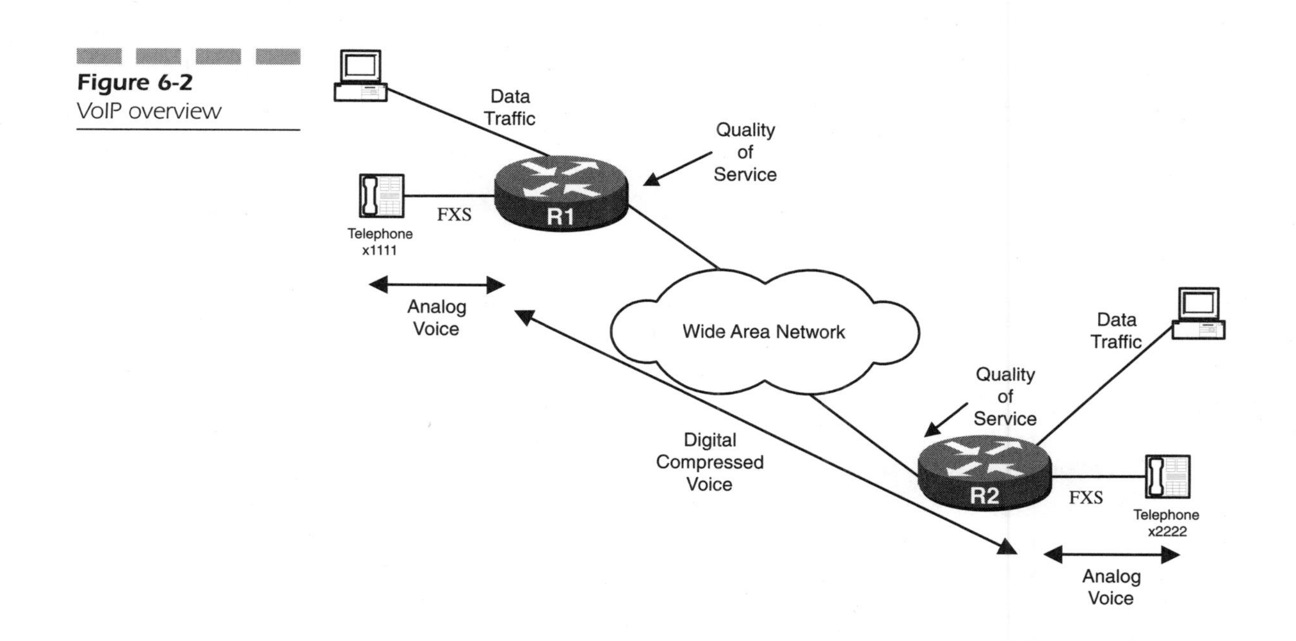

Figure 6-2
VoIP overview

traffic because a constant stream of traffic is being sent in real time. Because of the delay sensitivity of voice traffic, it is important to properly design your network. The Cisco IOS supports several protocols and enhanced features, which can improve the *quality of service* (QoS) of voice traffic. In order to bring voice and fax traffic into the router, you must install special *voice interface cards* (VICs) in the router.

VoIP Technology

The current public phone network takes analog traffic from a customer's home or office and brings it into a voice switch, such as a Lucent 5ESS or a Nortel DMS-250. Once at the voice switch, the analog signal is converted into a digital signal using a technology known as *pulse code modulation* (PCM). The analog signal is converted into a digital signal by sampling the analog signal 8,000 times per second. Each of the 8,000 samples is given an eight-bit value. This gives a sample rate of 64,000 bits/second.

Voice Compression

Voice compression techniques have been standardized by the ITU and are supported by the major vendors of VoIP equipment. The most popular voice compression technique is the G.729 algorithm. This G.729 specification describes a compression scheme where voice is coded into 8kb/s streams. Encapsulation overhead, which is added to the 8kb/s voice stream, actually creates a VoIP call, which is in the range of 22kb/s. This number can be brought down to the 11kb/s range when using header compression.

Compressed Voice Quality

A common test that is used to determine the quality of sound produced by different types of voice compression is the *mean opinion score* (MOS). MOS works by having several people listen to a voice recording and judge the quality of the voice. Samples are rated on a scale of 1 (poor) to 5 (excellent). Although very subjective, the MOS test is the best measurement of voice quality. Uncompressed voice, sometimes referred to as Toll quality voice, is given a MOS score of 5. The G.729 algorithm, which is the most widely used voice compression method, is usually given a MOS rating around 4.0.

It should be noted that voice compression is a type of compression that is known as lossey compression. Lossey compression is any type of compression where the original signal cannot be recreated. Voice compression and graphic compression, such as JPEG, are examples of lossey compression. Care must be taken in designing networks that use lossey compression. Multiple compression hops can greatly reduce the quality of the voice, and therefore the MOS score.

Voice Packet Delay and Jitter

An important consideration to take into account when implementing a voice network is to minimize the end-to-end delay. Because voice traffic is real-time traffic, it is very susceptible to delay. It is desirable to keep the end-to-end delay in a voice network to less than 200 milliseconds. Several types of delay exist:

- *Propagation delay*, which is caused by the electrical voice signals traveling via a fiber or copper-based medium.
- *Serialization delay*, which is caused by the devices that handle voice packets. Several types of serialization delays exist, such as the following:
  - Compression algorithm-induced delays are considered a type of serialization delay. The serialization delay introduced by the G.729 algorithm is in the range of 15 ms. In contrast, the delay introduced by the standard uncompressed PCM algorithm is in the range of five ms.
  - A second type of serialization delay is the time it takes to generate a voice packet. A VoIP data frame is generated every 10 milliseconds. Two data frames are encapsulated within a single voice packet. This creates a packet delay in the range of 20 milliseconds.
  - A third type of serialization delay is the time it takes for voice packets to be moved to the output queue. The Cisco IOS has several QOS features that are designed to minimize this type of serialization delay.

Jitter is defined as a variation between when a voice packet is supposed to be received and when it actually is received. Jitter can be thought of as variable latency for different packets or cells. It causes a disruption to the real-time voice stream. The Cisco IOS employs buffering techniques, which help adjust for packet jitter.

Echo

Echo occurs when a speaker hears his or her own voice in the telephone receiver while speaking. When properly timed, echo is reassuring to the speaker. On the other hand, if the echo is greater than 25 milliseconds, it can be an annoyance. Echo is caused by a mismatch in impedance when the voice signal is converted at the central office from a four-wire signal at the switch to a two-wire signal at a local loop. The Cisco IOS provides an echo cancellation capability.

Voice Interface Cards

Cisco supports a variety of WIC cards and network modules that enable voice traffic to be brought into a router. These interfaces include FXS, FXO, E&M, and T1 or E1 trunk interfaces.

Service Assurance Agent (SAA)

Service Assurance Agent (SAA), formerly called *Response Time Reporter* (RTR), is a Cisco IOS feature that enables the measurement of various end-to-end network parameters. Today's customers are demanding *Service Level Agreements* (SLAs) that guarantee minimum acceptable levels of service. SAA provides for the measurement of key network metrics, which enable an SLA to be guaranteed. As shown in Figure 6-3, SAA works by sending test packets from an SAA enabled router to other devices in the network. In the figure, R1 is sending a test echo packet to R2 every 30 seconds and a test echo packet to R3 every 40 seconds. In the case of the test echo packets, SAA does not have to be enabled on R2 and R3. In fact, Routers 2 and 3 do not even have to be router devices; they could be any device that is capable of responding to an echo ping.

SAA provides for the following measurements:

- *IP/ICMP echo*: The *Internet Control Message Protocol* (ICMP) echo operation measures end-to-end response time between a Cisco router and devices using IP. SAA also enables a user to measure QoS between endpoints by setting the TOS bits on an IP packet.

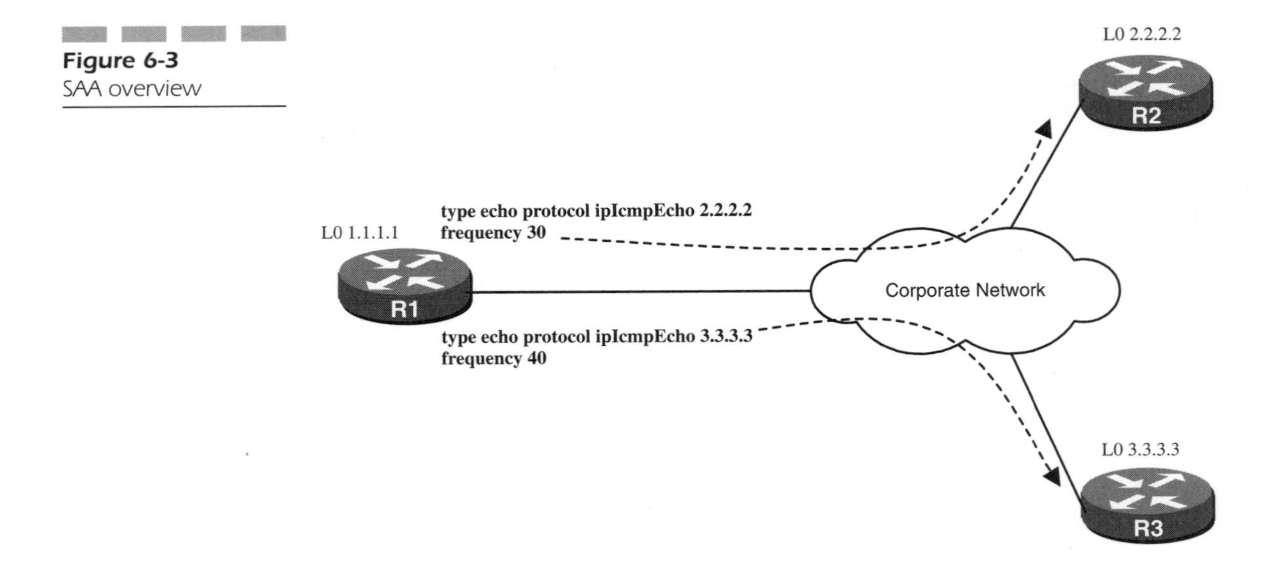

Figure 6-3 SAA overview

- *SNA echo*: The SNA echo operation measures end-to-end response times between a Cisco router and devices using SNA.
- *IP/ICMP path echo*: Path echo operations record statistics for each hop along the path that the operation takes to reach its destination. The path is discovered using trace route.
- *TCP connection*: The TCP connection operation is used to discover the time it takes to connect a target device. When the target is a Cisco router, SAA makes a TCP connection to any port number specified by the user. This operation is useful in simulating TCP (telnet or HTTP) connection times.
- *UDP echo*: The UDP echo operation calculates UDP response times between a Cisco router and any IP-enabled device. SAA sends a UDP datagram to any port number specified by the user.
- *Jitter and UDP Plus*: The UDP Plus operation measures additional items beyond the UDP operation. The UDP Plus operation measures per direction packet-loss and jitter.
- *HTTP*: The HTTP operation measures the *Round Trip Time* (RTT) taken to connect and access data from an HTTP server. Three HTTP server response time measurements exist:
  - *DNS lookup*: Round-trip time measured to perform a domain name system lookup.
  - *TCP connect*: Round-trip time measured to perform a TCP connection to the HTTP server.
  - *HTTP transaction time*: Round-trip time measured to send a request and get a response back from the HTTP server.
- *DHCP*: The SAA DHCP probe measures the round-trip time taken to discover a DHCP server and obtain a lease from it. After obtaining an IP address from the DHCP server, the IP address is released.
- DLSw+: The SAA *Data Link Switching* (DLSw+) probe measures the DLSw+ protocol stack and network response times between DLSw peers.
- DNS: The DNS response time is computed by taking the difference between the time taken to send a DNS request and to receive a reply.

Committed Access Rate (CAR)

Committed Access Rate (CAR), shown in Figure 6-4, is a Cisco IOS feature that provides several important capabilities:

- *Traffic matching*: The identification of interesting traffic for rate limiting and/or precedence setting. The Cisco IOS supports a variety of traffic matching criteria including:

- IP precedence
- QoS group
- MAC address
- IP access list

- *Traffic Measurement*: CAR uses a token bucket algorithm to accurately measure network traffic. Although a bit complex, the following is an overview of the CAR token bucket algorithm.

  As each packet has the CAR limit applied, tokens are removed from the bucket in accordance with the byte size of the packet. Tokens are replenished at regular intervals in accordance with the configured committed rate. The maximum number of tokens that can ever be in the bucket is determined by the normal burst size. If a packet arrives and the available tokens are less than the byte size of the packet, the extended burst comes into play. If no extended burst capability is available, which can be achieved by setting the extended burst value equal to the normal burst value, then the operation is as in a standard token bucket (the packet will be dropped if tokens are unavailable). However, if an extended burst capability is configured (the extended burst is greater than the normal burst), then the stream is allowed to borrow more tokens.

- *Configurable action policies*: The IOS can be configured to determine how conforming traffic (traffic that falls within the measured rate limits) and non-conforming traffic (traffic that falls outside the rate limits) are handled. The options for conforming and non-conforming traffic are as follows:

  - The traffic can be transmitted.
  - A specific IP precedence value can be set in the packet header and the packet can then be transmitted.

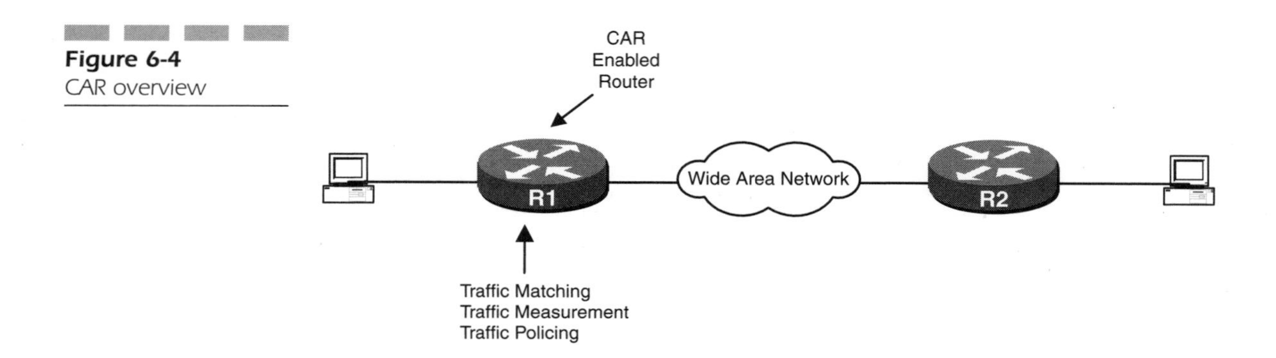

Figure 6-4
CAR overview

- The packet can be discarded.
- CAR can continue and evaluate the next rate policy.
- CAR can set a specific IP precedence and then continue evaluating the next rate policy.
- The packet can be assigned to a QoS group and then be transmitted.
- The packet can be assigned to a QoS group and then the next rate policy can be evaluated.

CAR Packet Classification

CAR provides for packet classification, which is the capability to assign traffic up to eight *classes of service* (COS). This is done by using the three precedence bits in the Type of Service field in the IP header. Packets can be classified based on various policies such as the following:

- Physical port
- Source or destination IP or MAC address
- Application port
- IP protocol type
- Access list or extended access list criteria

Queuing

Queuing enables congestion control by determining the order in which packets are sent out an interface based on priorities assigned to those packets. The IOS contains four types of queuing protocols. When no congestion exists, packets are sent out an interface as soon as they arrive. When traffic arrives at an outgoing interface faster than it can be transmitted, the IOS applies the configured queuing mechanism to determine which packets will be sent.

The four types of queuing mechanisms are as follows:

- *First In, First Out (FIFO):* With FIFO, queuing packets are sent out the interface in the same order that they arrive.
- *Weighted Fair Queuing (WFQ):* WFQ provides queuing based on traffic application flows. For serial interfaces running at speeds of two Mbps and below, WFQ is the default queuing mechanism.
- *Priority Queuing:* With priority queuing, four queues are defined. Traffic from higher queues is always sent before traffic from lower queues. The disadvantage of priority queuing is that lower priority traffic does not get sent if higher priority traffic is being queued.

- *Custom Queuing*: Custom queuing defines up to 16 different queues that are each serviced in a round robin fashion. The configuration of custom queuing requires that the number of bytes to be sent from each queue before servicing the next queue be set. This enables the user to configure the percentage of the circuit's bandwidth that each queue will receive.

Case Study 6: IPSEC, VoIP, and ATM LANE

IOS Requirements

This lab requires IOS 12.0 with support for IPSec, VoIP, and ATM LANE.

Equipment Needed

The following equipment is needed to perform this case study.

- One Cisco router with one Ethernet interface and two FXS voice ports
- One Cisco router with one Ethernet interface and two serial interfaces
- One Cisco router with four serial interfaces and one FXS voice port
- One Cisco router with two Ethernet interfaces and one serial interface
- One Cisco terminal server with one serial interface and one Ethernet interface
- Two Cisco routers with one Ethernet interface and one ATM OC-3 interface
- One Catalyst 5500 switch
- One LS1010 Lightstream ATM switch with two OC-3 ports
- A PC running a terminal emulation program for connecting to the console port of the terminal server
- Seven Ethernet cables
- Four Cisco DTE/DCE crossover cables
- Two single mode or multimode fiber cables
- Three analog phones

4. Configure the terminal server (R5) so that all routers can be accessed by name through it. The port number of the terminal server attaches to the corresponding router; that is, port 1 of the terminal server goes to the console port of R1, and the Catalyst switch goes to port number 9. Use the loopback address as the IP address for reverse telneting.
5. Configure loopback 0 on all the routers. The IP address should be 135.25.x.1 where x is the router number. Use a 32-bit mask.
6. Assign IP addresses to the serial interface between R3 and R4, and R3 and R5. Use the subnet 135.25.11.0 and do not allocate more addresses than needed. Use the first available subnet and assign the addresses in ascending order, starting with the connection between R3 and R4.
7. Assign IP addresses to the serial interface between R1 and R3. Use the subnet 135.25.9.0. Do not allocate more addresses than needed. Use the first available subnet on the S0/0 interfaces and the second available subnet on the S0/1 interfaces.
8. Place the following interfaces in the assigned VLANs (ports should be attached to the Catalyst in ascending order, starting with R1):
   - E1/0 of R2 and E0/0 of R1 in VLAN 1
   - E0/0 of R5 and E0 of R6 in VLAN 2
   - E0 of R7 in VLAN 3
   - E0/0 of R4 in VLAN 4
   - E1/0 of R4 in VLAN 5
9. Assign private IP addresses of 10.1.1.1 /24 to E0/0 of R4.
10. Assign an IP address to the LAN between R2 and R1 that enables 30 useable host addresses. Assign the IP addresses in an ascending order starting with R2.
11. R4 contains a FTP server that needs to be accessed via networks off R3 and R5. The subnet contains only the FTP server, so assign the next available subnet from 135.25.11.0. Do not allocate more addresses than needed and use the first available subnet.
12. Place all networks, except for VLAN 4, VLAN 2, VLAN 3, R6, and R7, in OSPF Area 0.
13. Configure R4, R3, and R5 so that all telnet traffic originating from the router uses the source IP address of loopback 0.
14. Configure R4, R5, and R3 to accept telnet access. Only permit telnet traffic that is sourced from the loopback interface of any of the three routers.
15. Configure R3 and R4 so that when a telnet session is established from one router to the other, the session is encrypted. IKE should be used to create the SA using peer-share. The loopback interfaces also should be used as the peer addresses.
16. Configure R4 so that any packets originating from the 10.1.1.0 address have their source IP address changed to the IP address of Serial 0/0.

17. Assign private IP addresses of 135.25.12.1 /24 to E0/0 of R5 and place the interface in area 0.
18. Configure R3 and R4 so that traffic between E0/0 of R4 to E0/0 of R3 is encrypted and authenticated. IKE should be used to create the SA using peer-share. Triple Des should be used for encryption and MD5 should be used for authentication. The IP address of the serial interface should be used as the peer and the key should be cisco1.
19. Configure R6 and R7 for LANE. R6 should be the LECS, LES, and the BUS, while R7 should be the LEC. Use mkt as the emulated ELAN name. Configure the LECS so that it participates in SSRP. The IP address of ATM0 R6 is 172.16.0.1 /16, and the IP address of R7 ATM0 R7 is 172.16.0.2 /16.
20. Assign IP address 2.2.2.2 /24 to Ethernet 0 of R7, 1.1.1.1 /24 to Ethernet 0 of R6, and 1.1.1.2 to E0/0 of R5. Make sure that 2.2.2.2 can reach 1.1.1.1. Do not enable routing on R6 or R7 and do not use static routes.
21. Configure R2 and R3 to handle voice traffic. Also configure the routers with the directory numbers shown.
22. Configure R2 so that when the handset at extension 2222 is picked up, it places a call to extension 3333.
23. Configure R3 so that extension 3333 can place a call to either extension 2222 or extension 2223.
24. Configure R3 so that you can only dial 611 to reach extension 2222. You should still have to dial 2223 to reach extension 2223.
25. Configure custom queuing on R2 so that voice is given 75 percent of the bandwidth and data is given 25 percent of the bandwidth.
26. Configure priority queuing on R3 to make voice high priority.
27. Configure RSVP between R1 and R3 S0/0 interfaces so that the maximum single flow allowed is equal to 400kb/sec.
28. Configure RSVP between R1 and R3 S0/1 interfaces so that the maximum single flow allowed is equal to 100kb/sec.
29. Configure IP precedence on R3. Calls placed to extension 2222 should have an IP Prec bit set to critical (5), while calls placed to extension 2223 should have an IP Prec bit set to priority (1).
30. You want to monitor the round-trip delay between R3 and R2. Configure the network so that a monitor packet is sent every 10 seconds for 30 minutes.
31. Configure R2 so that all other routers in the network can load R2's IOS image.
32. Configure the network so that traffic coming from R2 into R1 meets the following requirements:

- The first 496,000 bits/sec of traffic with a source IP address of 135.25.2.1 should be transmitted with an IP precedence of 6. Additional traffic should be sent with an IP precedence of 3.
- The first 96,000 bits/sec of Web traffic should be transmitted. Additional Web traffic should be dropped.
- Up to 200,000 bits/sec of other types of traffic should be transmitted with an IP precedence of 1. Additional traffic should be dropped.

Answer Guide

1. *Cable the routers per Figure 6-5. Make sure that the correct ports are configured for DTE/DCE (the DCE clock should be set to 800K on all routers).*

   This question is straightforward. The only thing that needs to be done is to set the clock rate on the DCE side of each serial link.

To set the clock rate, use the following interface command:

```
R1(config-if)#clock rate 800000
```

Use the **show controller interface** command to verify that the DCE cable is connected to the correct serial port. The following is truncated output from the command. Note that the cable attached is a V35 DCE cable and the clockrate is set to 800K.

```
R1#show controllers s 0
HD unit 0, idb = 0x970D0, driver structure at 0x9AE28
buffer size 1524  HD unit 0, V.35 DCE cable, clockrate 800000
cpb = 0xE1, eda = 0x4940, cda = 0x4800
RX ring with 16 entries at 0xE14800
```

2. *Erase any existing configuration from the routers and set each router's name. The name will be the corresponding router number; for example, router 1 will be R1.*

   The first step is to erase the configuration for all the routers as well as the Catalyst switch. To erase the configuration on the router, perform the following command:

```
R1#write erase or R1#erase startup-config
```

This command erases the startup configuration stored in NVRAM on all platforms except the Cisco 7000, 7200, and 7500 series. On the Cisco 7000, 7200, and 7500 series, the router erases or deletes the configuration pointed to by the CONFIG_FILE environment variable. If the CONFIG_FILE variable points to NVRAM, then the router erases NVRAM. If the CONFIG_FILE environment variable specifies a flash memory device and a configuration filename, then the command deletes the named file.

Erasing the configuration on the Catalyst 5500 is a bit different, so use the command:

```
CAT#clear config all
```

3. *Set the enable secret on each router to be cisco.*

When a router is first configured, there is no default enable password or enable secret. The enable secret is used to keep unauthorized users from changing the configuration. Both the enable password and the enable secret accomplish the same thing; that is, they allow you to establish an encrypted password that users must enter to access enable mode (the default) or any privilege level you specify. The enable secret uses a one-way hash algorithm to encrypt the password; the enable password uses a much weaker algorithm that can be easily broken. Tools exist on the Internet such as GetPass V1.1, which enables you to enter the encrypted password and provides the clear text password.

Cisco strongly recommends that you use the enable secret command because of the improved encryption algorithm.

To set the enable secret, use the following global configuration command:

```
R5(config)#enable secret cisco
```

4. *Configure the terminal server (R5) so that all routers can be accessed by name through it. The port number of the terminal server attaches to the corresponding router; that is, port 1 of the terminal server goes to the console port of R1, and the Catalyst switch goes to port number 9. Use the loopback address as the IP address for reverse telneting.*

All the routers in the case study, as well as the Catalyst switch, will have their console ports connected directly to one of the 16 asynchronous interfaces on the 2511RJ terminal server using a standard Cisco console rolled cable. The routers will be accessed using a reverse telnet connection. To make a reverse telnet connection, telnet to any active IP address on the box followed by 200x, where x is the port number that you want to access (Telnet 135.25.5.1 2001).

Use the following line commands to configure the terminal server:

```
R5(config)#line 1 16
R5(config-line)#transport input all
R5(config-line)#no exec
```

The command **transport input all** specifies the input transport protocol. By default, on IOS 11.1 and later, the transport input is set to none; prior to 11.1, the default was all. If the transport input is left to none, you will receive an error stating that the connection is refused by the remote host. The command no exec allows only outgoing connections for the line. This prevents the device that you are attached to from sending out unsolicited data. If the port receives unsolicited data, an EXEC process starts, making the line unavailable.

The next step is to configure hostnames on R5 so you can attach to any router simply by typing in the router name. The Cisco IOS software maintains a table of host names and their corresponding addresses. Similar to a DNS server, you can statically map host names to IP addresses. This is very useful and saves a lot of keystrokes when you have multiple devices connected to the terminal server.

The following global configuration commands are used to set the host names:

```
R5(config)#ip host R1 2001 135.25.5.1
R5(config)#ip host R2 2002 135.25.5.1
R5(config)#ip host R4 2004 135.25.5.1
R5(config)#ip host R5 2005 135.25.5.1
R5(config)#ip host R6 2006 135.25.5.1
R5(config)#ip host R7 2007 135.25.5.1
R5(config)#ip host R8 2008 135.25.5.1
R5(config)#ip host cat 2009 135.25.5.1
```

5. *Configure loopback 0 on all the routers. The IP address should be 135.25.x.1 where x is the router number. Use a 32-bit mask.*

```
R1(config)#interface loopback 0
R1(config-if)#ip add 135.25.1.1 255.255.255.255

R2(config)#interface loopback 0
R2(config-if)#ip add 135.25.2.1 255.255.255.255

R3(config)#interface loopback 0
R3(config-if)#ip add 135.25.3.1 255.255.255.255

R4(config)#interface loopback 0
R4(config-if)#ip add 135.25.4.1 255.255.255.255

R5(config)#interface loopback 0
R5(config-if)#ip add 135.25.5.1 255.255.255.255

R6(config)#interface loopback 0
R6(config-if)#ip add 135.25.6.1 255.255.255.255

R7(config)#interface loopback 0
R7(config-if)#ip add 135.25.7.1 255.255.255.255
```

6. *Assign IP addresses to the serial interface between R3 and R4, and R3 and R5. Use the subnet 135.25.11.0 and do not allocate more addresses than needed. Use the first available subnet and assign the addresses in ascending order, starting with the connection between R3 and R4.*

   This question involves simple subnetting. You need to pick a subnet that contains two useable addresses. For this, use a 30-bit mask that provides four addresses, of which two are usable. The following is the IP address assignment:

```
135.25.11.0 /30← Network Address
135.25.11.1 /30← R3
135.25.11.2 /30← R4
135.25.11.3 /30← Broadcast Address

135.25.11.4 /30← Network Address
135.25.11.5 /30← R3
135.25.11.6 /30← R5
135.25.11.7 /30← Broadcast Address
```

   The following commands define IP addresses on R3 and R4:

```
R3(config)#interface s2/0
R3(config-if)#ip address 135.25.11.1 255.255.255.252

R4(config)#interface s0/0
R4(config-if)#ip address 135.25.11.2 255.255.255.252

R3(config-if)#interface s2/1
R3(config-if)#ip address 135.25.11.5 255.255.255.252

R5(config)#interface s0/0
R5(config-if)#ip address 135.25.11.6 255.255.255.252
```

7. *Assign IP addresses to the serial interface between R1 and R3. Use the subnet 135.25.9.0 and do not allocate more addresses than needed.Use the first available subnet on the S0/0 interfaces and the second available subnet on the S0/1 interfaces.*

   You need to pick a subnet that contains two useable addresses. For this, use a 30-bit mask that provides four addresses, of which two are usable. The following is the IP address assignment:

```
135.25.9.0 /30← Network Address
135.25.9.1 /30← R1 S0/0
135.25.9.2 /30← R3 S0/0
135.25.9.3 /30← Broadcast Address

135.25.9.4 /30← Network Address
135.25.9.5 /30← R1 S0/1
135.25.9.6 /30← R3 S0/1
135.25.9.7 /30← Broadcast Address
```

```
2     enet   100002     1500    -       -       -       -    -        0       0

VLAN AREHops STEHops Backup CRF
---- ------- ------- ----------
```

9. *Assign the private IP address of 10.1.1.1 /24 to E0/0 of R4.*

   The following command assigns an IP address on R4:

```
R4(config)#interface e0/0
R4(config-if)#ip address 10.1.1.1 255.255.255.0
```

10. *Assign an IP address to the LAN between R2 and R1 that enables 30 useable host addresses. Assign the IP addresses in an ascending order, starting with R2.*

    You need to pick a subnet that contains 30 useable addresses. For this, you must use a 27-bit mask that provides 32 addresses, of which 30 are usable. The following is the IP address assignment:

```
135.25.7.0 /27 ← Network Address
135.25.7.1 /27 ← R2
135.25.7.2 /27 ← R1
135.25.7.31 /27← Broadcast Address
```

    The following commands define IP addresses on R2 and R1:

```
R1(config)#interface E0/0
R1(config-if)#ip address 135.25.7.2 255.255.255.224

R2(config)#interface E1/0
R2(config-if)#ip address 135.25.7.1 255.255.255.224
```

11. *R4 contains a FTP server that needs to be accessed via networks off of R3 and R5. The subnet only contains the FTP server, so assign the next available subnet from 135.25.11.0. Do not allocate more addresses than needed and use the first available subnet.*

    This subnet requires two host addresses. For this, use a 30-bit mask. The next available subnet is 135.25.11.8; the following is the IP address assignment:

```
135.25.11.8 /30 ← Network Address
135.25.11.9 /30 ← R4
135.25.11.10 /30← FTP Server
135.25.11.11 /30← Broadcast Address
```

The following command defines IP addresses on R4:

```
R4(config)#interface e1/0
R4(config-if)#ip add 135.25.11.9 255.255.255.252
```

12. *Place all networks, except for VLAN 4, VLAN 2,VLAN 3, R6, and R7, in OSPF area 0.*

Enabling OSPF on a router is accomplished in two steps. First, an OSPF process is defined and then an interface is added to the process. The command to start an OSPF process is Router OSPF [Process #]. The process number is used internally by the router, and multiple OSPF routing processes can be configured on one router.

The following command enables OSPF process 64 on the router and assigns the Ethernet interface of R3, R4, and R5 to area 0:

```
R1(config-router)#network 135.25.0.0 0.0.255.255 area 0

R2(config-router)#network 135.25.0.0 0.0.255.255 area 0

R3(config-router)#network 135.25.11.0 0.0.0.255 area 0

R4(config-router)#network 135.25.11.0 0.0.0.255 area 0

R5(config-router)#network 135.25.11.0 0.0.0.255 area 0
```

The following command assigns the loopback interfaces R3, R4, and R5 to area 0:

```
R4(config-router)#network 135.25.4.1 0.0.0.0 area 0

R3(config-router)#network 135.25.3.1 0.0.0.0 area 0

R5(config-router)#network 135.25.5.1 0.0.0.0 area 0
```

13. *Configure R4, R3, and R5 so that all telnet traffic originating from the router uses the source IP address of loopback 0.*

Normally, the Cisco IOS uses the IP address of the closest interface to the destination as the source of a telnet packet, which originates from the router. The Cisco IOS enables you to set which source IP address the router uses for any telnet session, which originates from the router.

This feature is especially useful in cases when the router has many interfaces and you want to ensure that all telnet packets from a particular router have the same source IP address.

The following command configures the router to use the loopback IP address as the source for all telnet packets:

```
R4(config)#ip telnet source-interface loopback 0

R3(config)#ip telnet source-interface loopback 0

R5(config)#ip telnet source-interface loopback 0
```

14. *Configure R4, R5, and R3 to accept telnet access. Only permit telnet traffic that originates from the loopback interface of any of the other two routers.*

Each telnet<$vty ports;ATM LANE> port on the router is known as a virtual terminal. In most cases, a maximum of five *virtual terminal* (VTY) ports are on the router, allowing five concurrent telnet sessions.

In order to telnet to a router, a password must first be configured under the VTY. If a password is not configured, you will get the following error when you attempt to connect:

```
R3#135.25.4.1
Trying 135.25.4.1 ... Open


Password required, but none set

[Connection to 135.25.4.1 closed by foreign host]
```

The following commands configure a password on all VTY lines of the router:

```
R3(config)#line vty 0 4
R3(config-line)#password cisco

R4(config)#line vty 0 4
R4(config-line)#password cisco

R5(config)#line vty 0 4
R5(config-line)#password cisco
```

Now verify that the password is set correctly by telneting from R3 to R4. Notice that you are now prompted for a password.

```
R3#telnet 135.25.4.1
Trying 135.25.4.1 ... Open


User Access Verification

Password:
R4>
```

The question specifies that the router should only accept a telnet connection with a source IP address of the loopback interface of one of the other two routers. To control telnet access into the router, an access class is used. The following commands define the access class on the three routers:

```
R3(config)#access-list 1 permit 135.25.4.1
R3(config)#access-list 1 permit 135.25.5.1

R3(config)#line vty 0 4
R3(config-line)#access-class 1 in

R4(config)#access-list 1 permit 135.25.3.1
R4(config)#access-list 1 permit 135.25.5.1

R4(config)#line vty 0 4
R4(config-line)#access-class 1 in

R5(config)#access-list 1 permit 135.25.3.1
R5(config)#access-list 1 permit 135.25.4.1

R5(config)#line vty 0 4
R5(config-line)#access-class 1 in
```

Now verify that the commands are working correctly by telneting from R3 to the loopback interface of R4. Use the extended **telnet** command to source the packet from the serial interface of R3. Below is the output from the command, notice that the telnet connection is refused.

```
R3#telnet 135.25.4.1 /source-interface s0/0
Trying 135.25.4.1 ...
% Connection refused by remote host
```

Now try telneting from R3 to R4, and source the packet from the loopback interface. Below is the output from the command, notice that the telnet command is now accepted.

```
R3#telnet 135.25.4.1
Trying 135.25.4.1 ... Open


User Access Verification

Password:
```

15. *Configure R3 and R4 so that when a telnet session is established from one router to the other, the session is encrypted. IKE should be used to create the SA using pre-share. The loopback interfaces should be used as the peer addresses.*

The first step is to create an IKE crypto policy. IKE is used to create the SA between the routers. This negotiation process is protected, so the peers must first agree on a shared IKE policy. Under the IKE policy, the encryption algorithm, hash algorithm, authentication method, Diffie-Hellman group identifier, and SA lifetime are configured. With the exception of the SA lifetime, if these parameters do not match, the SA negotiation will fail. The following command configures the IKE policy on R3 and R4:

```
R3(config)#crypto isakmp policy 5

R4(config)#crypto isakmp policy 5
```

Up to 10,000 policies can be defined. The priority value (the number at the end of the command) is used to determine the priority of each policy. The lower the number, the higher the priority. During negotiation, the polices are evaluated in order of priority. If no policies are explicitly configured, then the default parameters are used to define the policy, as shown in Table 6-4.

The authentication type is then defined under the IKE policy. Three types of authentication can be used: RSA-SIG, RSA-ENCR, or pre-share. The question calls for pre-share keys. If RSA-Sig were used, a certificate authority would be needed. The following command enables the IKE policy to use pre-share keys:

```
R3(config)#crypto isakmp policy 5
R3(config-isakmp)#authentication pre-share

R4(config)#crypto isakmp policy 5
R4(config-isakmp)#authentication pre-share
```

Because pre-shared keys are being used, a shared key must be defined along with the peer's identity. The identity can be either the peer's IP address or name. The question

Table 6-4

Default Policy Parameters

Parameter	Default Value
Encryption Algorithm	DES
Hash Algorithm	SHA-1
Authentication Method	RSA signatures
Diffie-Hellman Identifier Group	Group 1 (768-bit DH)
SA's lifetime	86,400 seconds

specifies that the IP address of the loopback 0 should be used, and the shared secret should be cisco. The following command defines the key that will be used and the peer address:

```
R3(config)#crypto isakmp key cisco address 135.25.4.1

R4(config)#crypto isakmp key cisco address 135.25.3.1
```

View the crypto policy on R3 with the **show crypto isakmp policy** command. The following is the output from the command. Notice that two policies exist, the one you just created and a default policy. The only difference is the authentication method, the default policy, is using RSA-sig as the authentication method.

```
R3#show crypto isakmp policy
Protection suite of priority 5
        encryption algorithm:   DES - Data Encryption Standard (56 bit keys).
        hash algorithm:         Secure Hash Standard
        authentication method:  Pre-Shared Key
        Diffie-Hellman group:   #1 (768 bit)
        lifetime:               86400 seconds, no volume limit
Default protection suite
        encryption algorithm:   DES - Data Encryption Standard (56 bit keys).
        hash algorithm:         Secure Hash Standard
        authentication method:  Rivest-Shamir-Adleman Signature
        Diffie-Hellman group:   #1 (768 bit)
        lifetime:               86400 seconds, no volume limit
```

The next step is to define the transform set or sets that will be used. A transform is simply the algorithm or algorithms that the router is willing to use for the session. The various transform sets are offered to the receiver during IKE; the receiver selects the one that will be used.

Because the question calls for encryption only, just one transform needs to be defined. The following is a list of the transforms supported:

```
  h-md5-hmac     AH-HMAC-MD5 transform
  ah-sha-hmac    AH-HMAC-SHA transform
  comp-lzs       IP Compression using the LZS compression algorithm
  esp-3des       ESP transform using 3DES(EDE) cipher (168 bits)
  esp-des        ESP transform using DES cipher (56 bits)
  esp-md5-hmac   ESP transform using HMAC-MD5 auth
  esp-null       ESP transform w/o cipher
  esp-sha-hmac   ESP transform using HMAC-SHA auth
```

The following command defines the transform on R3 and R4:

```
R3(config)#crypto ipsec transform-set encr-only esp-3des

R4(config)#crypto ipsec transform-set encr-only esp-3des
```

The transform set that is going to be used can be viewed with the command, **show crypto ipsec transform-set.** The following is the output from the command on R3. Notice that the transform set encr-only is using esp-3des.

```
R3#show crypto ipsec transform-set

Transform set encr-only: { esp-3des  }
   will negotiate = { Tunnel,  },
```

The next step is to define the traffic that will be given security protection. This is done using an extended access list to identify the traffic. The question states that all traffic from Loopback 0 of R3 to Loopback 0 of R4 should be encrypted. Only the traffic that you want to protect needs to be defined. In access list terminology, permit means protect and deny means don't protect. Any traffic that is denied will pass in the clear.

The following commands define the access lists on R3 and R4:

```
R3(config)#access-list 101 permit ip host 135.25.3.1 host 135.25.4.1
R4(config)#access-list 101 permit ip host 135.25.4.1 host 135.25.3.1
```

The next step is to define a crypto map that combines the policy and traffic information. The crypto map contains the traffic that security must be applied to (defined by the access list), the actual algorithm to apply (defined by the transform), and the crypto end point (the remote peer). An IPSec crypto map is defined with a tag, a sequence number, and the encryption method.

The following commands define a crypto map on R3 and R4:

```
R3(config)#crypto map peer-r4 local-address loopback 0
R3(config)#crypto map peer-r4 10 ipsec-isakmp
R3(config-crypto-map)#set peer 135.25.4.1
R3(config-crypto-map)# set transform-set encr-only
R3(config-crypto-map)#match address 101

R4(config)#crypto map peer-r3 local-address loopback 0
R4(config)#crypto map peer-r3 10 ipsec-isakmp
R4(config-crypto-map)#set peer 135.25.3.1
R4(config-crypto-map)#set transform-set encr-only
R4(config-crypto-map)#match address 101
```

The last step is to apply the map to an interface:

```
R4(config)#interface s0/0
R3(config-if)#crypto map peer-r4 encr-only

R4(config)#interface s0/0
R4(config-if)#crypto map peer-r3 encr-only
```

Now display the active IPSec connections on R4 with the command **show crypto engine connections active.** The following is the output from the command. Notice that no active connections exist. This is because no traffic matching the crypto map has been sent.

```
R4#show crypto engine connections active

  ID Interface         IP-Address       State  Algorithm           Encrypt  Decrypt
   1 <none>            <none>           set    HMAC_SHA+DES_56_CB        0        0
Crypto adjacency count : Lock: 0, Unlock: 0
```

Enable the following debug commands on R4:

```
R4#debug crypto ipsec
R4#debug crypto isa
R4#debug crypto isakmp
```

Now telnet from R4 to the loopback address of R3. The following is the output from the debug commands:

```
R4# telnet 135.25.3.1 ← Telnet session to loopback on R3
Trying 135.25.3.1 ...
08:14:28: IPSEC(sa_request): , ← SA negotiation
  (key eng. msg.) src= 135.25.4.1, dest= 135.25.3.1, ← Using the loopback address as the
source (crypto map peer-r3 local-address loopback 0)
    src_proxy= 135.25.4.1/255.255.255.255/0/0 (type=1),
    dest_proxy= 135.25.3.1/255.255.255.255/0/0 (type=1),
    protocol= ESP, transform= esp-3des, ←Using 3DES
    lifedur= 3600s and 4608000kb,
    spi= 0x0(0), conn_id= 0, keysize= 0, flags= 0x4004
08:14:29: ISAKMP (0:1): beginning Quick Mode exchange, M-ID of -234684864
08:14:29: IPSEC(key_engine): got a queue event...
Open4:29: IPSEC(spi_response): getting spi 392368911 for SA

User Access Verification

Password: ← Password prompt on R3


        from 135.25.3.1      to 135.25.4.1      for prot 3
08:14:29: ISAKMP (1): sending packet to 135.25.3.1 (I) QM_IDLE
08:14:30: ISAKMP (1): received packet from 135.25.3.1 (I) QM_IDLE
08:14:30: ISAKMP (0:1): processing SA payload. message ID = -234684864
08:14:30: ISAKMP (0:1): Checking IPSec proposal 1
08:14:30: ISAKMP: transform 1, ESP_3DES
08:14:30: ISAKMP:   attributes in transform:
08:14:30: ISAKMP:      encaps is 1
08:14:30: ISAKMP:      SA life type in seconds
08:14:30: ISAKMP:      SA life duration (basic) of 3600
08:14:30: ISAKMP:      SA life type in kilobytes
```

Now display the active IPSec connections on R4 with the command **show crypto engine connections active.** The following is the output from the command. Notice that there are no active connections. This is because no traffic matching the crypto map has been sent.

```
R4#show crypto engine connections active

  ID Interface        IP-Address       State  Algorithm            Encrypt  Decrypt
   1 <none>           <none>           set    HMAC_SHA+DES_56_CB         0        0
Crypto adjacency count : Lock: 0, Unlock: 0
```

Turn on the following debug commands on R4:

```
R4#debug crypto ipsec
R4#debug crypto isakmp
```

Now ping from the Ethernet interface of R4 to the Ethernet interface of R3 using the extended ping command:

```
R4#ping
Protocol [ip]:
Target IP address: 135.25.12.1
Repeat count [5]:
Datagram size [100]:
Timeout in seconds [2]:
Extended commands [n]: y
Source address or interface: 10.1.1.1
Type of service [0]:
Set DF bit in IP header? [no]:
Validate reply data? [no]:
Data pattern [0xABCD]:
Loose, Strict, Record, Timestamp, Verbose[none]:
Sweep range of sizes [n]:
```

The following is an analysis from the output of the debugs:

```
The ping source and destination matched the access list 102, which is tied to crypo map
peer-r3 sequence number 20.
                                    ↓
02:33:34: IPSEC(sa_request): ,
  (key eng. msg.) src= 135.25.4.1, dest= 135.25.3.1,

The scr_proxy is the local tunnel end point and the dest_proxy is the remote crypto
endpoint.
                                    ↓

    src_proxy= 135.25.11.2/255.255.255.255/0/0 (type=1),
    dest_proxy= 135.25.12.0/255.255.255.0/0/0 (type=4),

The scr proxy is the source of the interesting traffic as defined by the access list.
Since NAT was being used, the src-proxy address is that of the serial interface.  The dest
proxy address is the destination address that the packet is going to.
                                   ↓

  protocol= ESP, transform= esp-3des esp-md5-hmac ,
```

```
    lifedur= 3600s and 4608000kb,
    spi= 0x0(0), conn_id= 0, keysize= 0, flags= 0x4004
```

The protocol and the transforms are specified by the crypto map peer-r3 20. SPI is zero, which means the main mode of negotiation is being started.

↓

```
02:33:34: ISAKMP (0:1): beginning Main Mode exchange
02:33:34: ISAKMP (1): sending packet to 135.25.3.1 (I) MM_NO_STATE
02:33:34: ISAKMP (1): received packet from 135.25.3.1 (I) MM_NO_STATE
02:33:34: ISAKMP (0:1): processing SA payload. message ID = 0
02:33:34: ISAKMP (0:1): Checking ISAKMP transform 1 against priority 5 policy
02:33:34: ISAKMP:      encryption DES-CBC
02:33:34: ISAKMP:      hash SHA
02:33:34: ISAKMP:      default group 1
02:33:34: ISAKMP:      auth pre-share
```

These are the attributes that are being offered by R3.

↓

```
02:33:34: ISAKMP (0:1): atts are acceptable. Next payload is 0
```

Policy 5 of this router matches the attributes that were sent by R3. Peer share authentication starts now.

↓

```
02:33:34: ISAKMP (0:1): SA is doing pre-shared key authentication

02:33:34: ISAKMP (1): SA is doing pre-shared key authentication

using id type ID_IPV4_ADDR
02:33:34: ISAKMP (1): sending packet to 135.25.3.1 (I) MM_SA_SETUP
02:33:34: ISAKMP (1): received packet from 135.25.3.1 (I) MM_SA_SETUP
02:33:34: ISAKMP (0:1): processing KE payload. message ID = 0
02:33:34: ISAKMP (0:1): processing NONCE payload. message ID = 0
```

Nonce from the far end is being processed

↓

```
02:33:34: ISAKMP (0:1): SKEYID state generated
02:33:34: ISAKMP (0:1): processing vendor id payload
02:33:34: ISAKMP (0:1): speaking to another IOS box!
02:33:34: ISAKMP (1): ID payload
        next-payload : 8
        type         : 1
        protocol     : 17
        port         : 500
        length       : 8
02:33:34: ISAKMP (1): Total payload length: 12
02:33:34: ISAKMP (1): sending packet to 135.25.3.1 (I) MM_KEY_EXCH
02:33:34: ISAKMP (1): received packet from 135.25.3.1 (I) MM_KEY_EXCH
02:33:34: ISAKMP (0:1): processing ID payload. message ID = 0
02:33:34: ISAKMP (0:1): processing HASH payload. message ID = 0
02:33:34: ISAKMP (0:1): SA has been authenticated with 135.25.3.1
```

```
At this point, preshare authentication has succeeded and ISAKMP has been negotiated.
                                                                  ↓
02:33:34: ISAKMP (0:1): beginning Quick Mode exchange, M-ID of -1594944679
Quick mode authentication starts.
              ↓
02:33:34: IPSEC(key_engine): got a queue event...
02:33:34: IPSEC(spi_response): getting spi 513345059 for SA
        from 135.25.3.1      to 135.25.4.1      for prot 3

ISAKMP gets the SPI from the IPSEC routine and sends it to the other side.
                     ↓
02:33:34: ISAKMP (1): sending packet to 135.25.3.1 (I) QM_IDLE
02:33:35: ISAKMP (1): received packet from 135.25.3.1 (I) QM_IDLE
02:33:35: ISAKMP (0:1): processing SA payload. message ID = -1594944679
02:33:35: ISAKMP (0:1): Checking IPSec proposal 1


R4 processes the IPSEC attributes offered by the remote end.  The following are the
attributes of the transform offered by the remote end.
                     ↓
02:33:35: ISAKMP: transform 1, ESP_3DES
02:33:35: ISAKMP:   attributes in transform:
02:33:35: ISAKMP:      encaps is 1
02:33:35: ISAKMP:      SA life type in seconds
02:33:35: ISAKMP:      SA life duration (basic) of 3600
02:33:35: ISAKMP:      SA life type in kilobytes
02:33:35: ISAKMP:      SA life duration (VPI) of  0x0 0x46 0x50 0x0
02:33:35: ISAKMP:      authenticator is HMAC-MD5
02:33:35: ISAKMP (0:1): atts are acceptable.

The IPSEC SA has been successfully negotiated.

Here R4 validates the proposal that it negotiated with the remote side.
                    ↓
02:33:35: IPSEC(validate_proposal_request): proposal part #1,
  (key eng. msg.) dest= 135.25.3.1, src= 135.25.4.1,
    dest_proxy= 135.25.12.0/255.255.255.0/0/0 (type=4),
    src_proxy= 135.25.11.2/255.255.255.255/0/0 (type=1),
    protocol= ESP, transform= esp-3des esp-md5-hmac ,
    lifedur= 0s and 0kb,
    spi= 0x0(0), conn_id= 0, keysize= 0, flags= 0x4
02:33:35: ISAKMP (0:1): processing NONCE payload. message ID = -1594944679
02:33:35: ISAKMP (0:1): processing ID payload. message ID = -1594944679
02:33:35: ISAKMP (0:1): processing ID payload. message ID = -1594944679
02:33:35: ISAKMP (0:1): Creating IPSec SAs
```

```
Crypto engine entries have been created. Inbound traffic has an spi 513345059 and conn_id
2000. Outbound traffic has a spi 71175277 and conn_id 2001.
                                                          ↓
02:33:35:         inbound SA from 135.25.3.1      to 135.25.4.1      (proxy 135.25.12.0
to 135.25.11.2    )
02:33:35:         has spi 513345059 and conn_id 2000 and flags 4
02:33:35:         lifetime of 3600 seconds
02:33:35:         lifetime of 4608000 kilobytes
02:33:35:         outbound SA from 135.25.4.1      to 135.25.3.1      (proxy 135
.25.11.2     to 135.25.12.0     )
02:33:35:         has spi 71175277 and conn_id 2001 and flags 4
02:33:35:         lifetime of 3600 seconds
02:33:35:         lifetime of 4608000 kilobytes
02:33:35: ISAKMP (1): sending packet to 135.25.3.1 (I) QM_IDLE
02:33:35: ISAKMP (0:1): deleting node -1594944679
02:33:35: IPSEC(key_engine): got a queue event...
02:33:35: IPSEC(initialize_sas): ,
  (key eng. msg.) dest= 135.25.4.1, src= 135.25.3.1,
    dest_proxy= 135.25.11.2/255.255.255.255/0/0 (type=1),
    src_proxy= 135.25.12.0/255.255.255.0/0/0 (type=4),
    protocol= ESP, transform= esp-3des esp-md5-hmac ,
    lifedur= 3600s and 4608000kb,
    spi= 0x1E990623(513345059), conn_id= 2000, keysize= 0, flags= 0x4
02:33:35: IPSEC(initialize_sas): ,
  (key eng. msg.) src= 135.25.4.1, dest= 135.25.3.1,
    src_proxy= 135.25.11.2/255.255.255.255/0/0 (type=1),
    dest_proxy= 135.25.12.0/255.255.255.0/0/0 (type=4),
    protocol= ESP, transform= esp-3des esp-md5-hmac ,
    lifedur= 3600s and 4608000kb,
    spi= 0x43E0C6D(71175277), conn_id= 2001, keysize= 0, flags= 0x4
02:33:35: IPSEC(create_sa): sa created,
  (sa) sa_dest= 135.25.4.1, sa_prot= 50,
    sa_spi= 0x1E990623(513345059),
    sa_trans= esp-3des esp-md5-hmac , sa_conn_id= 2000
02:33:35: IPSEC(create_sa): sa created,
  (sa) sa_dest= 135.25.3.1, sa_prot= 50,
    sa_spi= 0x43E0C6D(71175277),
    sa_trans= esp-3des esp-md5-hmac , sa_conn_id= 2001
```

Now display the active crypto connections with the command, **show crypto engine connection active.** The following is the output from the command. Notice there are two IPSEC SA's, one for incoming traffic **2000** and one for outgoing traffic **2001.**

```
R4#show crypto engine connections active
  ID Interface        IP-Address      State  Algorithm            Encrypt  Decrypt
   1 <none>           <none>          set    HMAC_SHA+DES_56_CB         0        0
2000 Serial0/0        135.25.11.2     set    HMAC_MD5+3DES_56_C         0        9
2001 Serial0/0        135.25.11.2     set    HMAC_MD5+3DES_56_C         9        0
Crypto adjacency count : Lock: 0, Unlock: 0
```

19. *Configure R6 and R7 for LANE. R6 should be the LECS, LES, and the BUS, while R7 should be the LEC. Use mkt as the emulated ELAN name. Configure the LECS so that it participates in SSRP. The IP address of ATM 0 R6 is 172.16.0.1 / 16 and the IP address of ATM1 R7 is 172.16.0.3 / 16.*

```
R5(config)#interface e0/0
R5(config-if)#ip address 1.1.1.2 255.255.255.0

R6(config)#interface e0
R6(config-if)#ip address 1.1.1.1 255.255.255.0

R7(config)#interface e0
R7(config-if)#ip address 2.2.2.2 255.255.255.0
```

The question specifies that 2.2.2.2 should be able to reach 1.1.1.1 without using routing. To accomplish this, you need to bridge using LANE. In order to do this, you first must disable routing on the router. The following command disables IP routing on R6 and R7:

```
R6(config)#no ip routing

R7(config)#no ip routing
```

To verify that this is working properly, ping 2.2.2.2 from R6. Use the extended ping command to source the packet from 1.1.1.1:

```
R6#ping
Protocol [ip]:
Target IP address: 2.2.2.2
Repeat count [5]:
Datagram size [100]:
Timeout in seconds [2]:
Extended commands [n]: y
Source address or interface: 1.1.1.1
Type of service [0]:
Set DF bit in IP header? [no]:
Validate reply data? [no]:
Data pattern [0xABCD]:
Loose, Strict, Record, Timestamp, Verbose[none]:
Sweep range of sizes [n]:
Sending 5, 100-byte ICMP Echos to 2.2.2.2, timeout is 2 seconds:
.!!!!
```

Now display the LANE information on R7. Notice that a new VCD has been created for the connection between R6 and R7.

```
R7#show lane
LE Client ATM0.2  ELAN name: mkt  Admin: up  State: operational
Client ID: 1                 LEC up for 8 minutes 29 seconds
ELAN ID: 0
Join Attempt: 1
Known LE Servers: 1
Last Fail Reason: Locally deactivate
HW Address: 00d0.bbfb.08c8   Type: ethernet              Max Frame Size: 1516
ATM Address: 47.00918100000000107BD56901.00D0BBFB08C8.02
```

```
VCD  rxFrames  txFrames  Type        ATM Address
  0         0         0  configure   47.00918100000000107BD56901.006070CD51ED.00
 70         1         4  direct      47.00918100000000107BD56901.006070CD51EB.01
 71         6         0  distribute  47.00918100000000107BD56901.006070CD51EB.01
 72         0        16  send        47.00918100000000107BD56901.006070CD51EC.01
 73        26         0  forward     47.00918100000000107BD56901.006070CD51EC.01
 74         5         3  data        47.00918100000000107BD56901.006070CD51EA.01
```

21. *Configure R2 and R3 to handle voice traffic. Also configure the routers with the directory numbers shown.*

Basic voice configuration is fairly simple. You need to configure the directory number of each handset using the dial-peer voice command.

First, define a voice pots peer with the dial-peer voice 1 pots command:

```
R2(config)#dial-peer voice 1 pots
```

Next, define the phone number of this port. This port should be assigned phone number 2222 with the **destination-pattern** 2222 command:

```
R2(config-dial-peer)#destination-pattern 2222
```

Then define which physical port will be associated with this dial peer:

```
R2(config-dial-peer)#port 0/0 /0
```

You then need to repeat the same commands for the second extension connected to R2:

```
R2(config)#dial-peer voice 3 pots
R2(config-dial-peer)#destination-pattern 2223
R2(config-dial-peer)#port 0/0 /1
```

Now connect to R3 and configure the directory number for port 1/0/0 with the following commands:

```
R3(config)#dial-peer voice 1 pots
R3(config-dial-peer)#destination-pattern 3333
R3(config-dial-peer)#port 1/0 /0
```

At this point, you should be able to pick up each handset and hear a dial tone. The **show voice port** command, shown in the following, can be used to verify the settings of a voice port on the router:

```
R2#show voice port

Foreign Exchange Station 0/0/0
 Type of VoicePort is FXS  ← Port type
 Operation State is DORMANT
 Administrative State is UP
 No Interface Down Failure
 Description is not set
```

```
R3#show dial-peer voice
VoiceOverIpPeer2
information type = voice,
tag = 2, destination-pattern = `2223',
answer-address = `', preference=0,
group = 2, Admin state is up, Operation state is up,
incoming called-number = `', connections/maximum = 0/unlimited,
application associated:
type = voip, session-target = `ipv4:135.25.2.1',
technology prefix:
ip precedence = 0, UDP checksum = disabled,
session-protocol = cisco, req-qos = best-effort,
acc-qos = best-effort,
fax-rate = voice,   payload size =  20 bytes
codec = g729r8,   payload size =  20 bytes,
Expect factor = 10, Icpif = 30,signaling-type = cas,
VAD = enabled, Poor QOV Trap = disabled,
Connect Time = 1727, Charged Units = 0,
Successful Calls = 6, Failed Calls = 0,
Accepted Calls = 6, Refused Calls = 0,
Last Disconnect Cause is "10  ",
Last Disconnect Text is "normal call clearing.",
Last Setup Time = 2274609.
VoiceOverIpPeer3
information type = voice,
tag = 3, destination-pattern = `2222',
answer-address = `', preference=0,
group = 3, Admin state is up, Operation state is up,
incoming called-number = `', connections/maximum = 0/unlimited,
application associated:
type = voip, session-target = `ipv4:135.25.2.1',
technology prefix:
ip precedence = 0, UDP checksum = disabled,
session-protocol = cisco, req-qos = best-effort,
acc-qos = best-effort,
fax-rate = voice,   payload size =  20 bytes
codec = g729r8,   payload size =  20 bytes,
Expect factor = 10, Icpif = 30,signaling-type = cas,
VAD = enabled, Poor QOV Trap = disabled,
Connect Time = 1406, Charged Units = 0,
Successful Calls = 15, Failed Calls = 0,
Accepted Calls = 15, Refused Calls = 0,
Last Disconnect Cause is "10  ",
Last Disconnect Text is "normal call clearing.",
Last Setup Time = 2277544.
```

The dial peers can also be confirmed with the **show dialplan** command:

```
R3#show dialplan number 2222
Macro Exp.: 2222

VoiceOverIpPeer3
information type = voice,
tag = 3, destination-pattern = `2222',
answer-address = `', preference=0,
```

```
group = 3, Admin state is up, Operation state is up,
incoming called-number = `', connections/maximum = 0/unlimited,
application associated:
type = voip, session-target = `ipv4:135.25.2.1',
technology prefix:
ip precedence = 5, UDP checksum = disabled,
session-protocol = cisco, req-qos = best-effort,
acc-qos = best-effort,
fax-rate = voice,   payload size =  20 bytes
codec = g729r8,   payload size =  20 bytes,
Expect factor = 10, Icpif = 30,signaling-type = cas,
VAD = enabled, Poor QOV Trap = disabled,
Connect Time = 1406, Charged Units = 0,
Successful Calls = 15, Failed Calls = 0,
Accepted Calls = 15, Refused Calls = 0,
Last Disconnect Cause is "10  ",
Last Disconnect Text is "normal call clearing.",
Last Setup Time = 2277544.
Matched: 2222   Digits: 4
Target: ipv4:135.25.2.1


R3#show dialplan number 2223
Macro Exp.: 2223

VoiceOverIpPeer2
information type = voice,
tag = 2, destination-pattern = `2223',
answer-address = `', preference=0,
group = 2, Admin state is up, Operation state is up,
incoming called-number = `', connections/maximum = 0/unlimited,
application associated:
type = voip, session-target = `ipv4:135.25.2.1',
technology prefix:
ip precedence = 1, UDP checksum = disabled,
session-protocol = cisco, req-qos = best-effort,
acc-qos = best-effort,
fax-rate = voice,   payload size =  20 bytes
codec = g729r8,   payload size =  20 bytes,
Expect factor = 10, Icpif = 30,signaling-type = cas,
VAD = enabled, Poor QOV Trap = disabled,
Connect Time = 1727, Charged Units = 0,
Successful Calls = 6, Failed Calls = 0,
Accepted Calls = 6, Refused Calls = 0,
Last Disconnect Cause is "10  ",
Last Disconnect Text is "normal call clearing.",
Last Setup Time = 2274609.
Matched: 2223   Digits: 4
Target: ipv4:135.25.2.1
```

24. *Configure R3 so that you can only dial 611 to reach extension 2222. You should still have to dial 2223 to reach extension 2223.*

This question requires that you configure a feature known as number expansion. Number expansion enables a partial or alternate number to be dialed, which will in turn connect

27. *Configure RSVP between R1 and R3 S0/0 interfaces so that the maximum single flow allowed is equal to 400kb/sec.*

RSVP is enabled on the physical interface of the router. It allows the maximum reservable and maximum single flow reservable bandwidth to be configured. By default, the maximum reservable bandwidth is set to 75 percent of the interface bandwidth. Start your RSVP configuration by connecting to R1 and defining the interface bandwidth to be 800 kb/sec. The 800 kb/sec number comes from the fact that you are clocking this interface at this rate.

```
R1(config)#interface Serial0/0
R1(config-if)#bandwidth 800
```

Next, set the RSVP-specific parameters on the interface. The **ip rsvp bandwidth 600 400** command sets the maximum reservable bandwidth on the interface to 600 kb/sec. The maximum single reservation is set to 400 kb/sec:

```
R1(config-if)#ip rsvp bandwidth 600 400
```

The same commands then need to be repeated on R3:

```
R3(config)#interface Serial0/0
R3(config-if)#bandwidth 800
R3(config-if)#ip rsvp bandwidth 600 400
```

The **show ip rsvp interface** command shown here confirms that your RSVP reservations have been made on R1 and R3:

```
R1#show ip rsvp interface
interface    allocated  i/f max  flow max pct UDP  IP   UDP_IP   UDP M/C
Se0/0        0M         600K     400K      0   0   0    0        0

R3#show ip rsvp interface
interface    allocated  i/f max  flow max pct UDP  IP   UDP_IP   UDP M/C
Se0/0        0M         600K     400K      0   0   0    0        0
```

28. *Configure RSVP between R1 and R3 S0/1 interfaces so that the maximum single flow allowed is equal to 100kb/sec.*

As in the previous question, begin by configuring the bandwidth of the interface. Because you are clocking the interface at 800 kb/sec, set the bandwidth to the same value:

```
R1(config)#interface Serial0/1
R1(config-if)#bandwidth 800
```

Next, define the maximum reservable bandwidth and the maximum single flow with the **ip rsvp bandwidth 600 100** command:

```
R1(config-if)#ip rsvp bandwidth 600 100
```

The same items need to be configured on R3:

```
R3(config)#interface Serial0/1
R3(config-if)#bandwidth 800
R3(config-if)#ip rsvp bandwidth 600 100
```

The **show ip rsvp interface** command can be used to verify that you have configured RSVP on both serial interfaces of your routers:

```
R1#show ip rsvp interface
interface     allocated  i/f max   flow max pct UDP   IP    UDP_IP    UDP M/C
Se0/0         0M         600K      400K      0    0     0     0         0
Se0/1         0M         600K      100K      0    0     0     0         0
```

29. *Configure IP Precedence on R3. Calls placed to extension 2222 should have an IP Prec bit set to critical (5). Calls placed to extension 2223 should have an IP Prec bit set to priority (1).*

Voice dial peer 2 points towards extension 2223. The ip precedence 1 command configures the dial peer to set the precedence bit to 1 on voice traffic:

```
R3(config)#dial-peer voice 2 voip
R3(config-dial-peer)#ip precedence 1
R3(config-dial-peer)#exit
```

Voice dial peer 3 will have its traffic precedence bit set to 5:

```
R3(config)#dial-peer voice 3 voip
R3(config-dial-peer)#ip precedence 5
```

Notice that the **show dial-peer voice** command confirms that both of the voice peers on R3 are using the proper IP precedence bits:

```
R3#show dial-peer voice
VoiceOverIpPeer2
information type = voice,
tag = 2, destination-pattern = `2223',
```

```
answer-address = `', preference=0,
group = 2, Admin state is up, Operation state is up,
incoming called-number = `', connections/maximum = 0/unlimited,
application associated:
type = voip, session-target = `ipv4:135.25.2.1',
technology prefix:
ip precedence = 1, UDP checksum = disabled,
session-protocol = cisco, req-qos = best-effort,
acc-qos = best-effort,
fax-rate = voice,   payload size =  20 bytes
codec = g729r8,   payload size =  20 bytes,
Expect factor = 10, Icpif = 30,signaling-type = cas,
VAD = enabled, Poor QOV Trap = disabled,
Connect Time = 1727, Charged Units = 0,
Successful Calls = 6, Failed Calls = 0,
Accepted Calls = 6, Refused Calls = 0,
Last Disconnect Cause is "10  ",
Last Disconnect Text is "normal call clearing.",
Last Setup Time = 2274609.
VoiceOverIpPeer3
information type = voice,
tag = 3, destination-pattern = `2222',
answer-address = `', preference=0,
group = 3, Admin state is up, Operation state is up,
incoming called-number = `', connections/maximum = 0/unlimited,
application associated:
type = voip, session-target = `ipv4:135.25.2.1',
technology prefix:
ip precedence = 5, UDP checksum = disabled,
session-protocol = cisco, req-qos = best-effort,
acc-qos = best-effort,
fax-rate = voice,   payload size =  20 bytes
codec = g729r8,   payload size =  20 bytes,
Expect factor = 10, Icpif = 30,signaling-type = cas,
VAD = enabled, Poor QOV Trap = disabled,
Connect Time = 1406, Charged Units = 0,
Successful Calls = 15, Failed Calls = 0,
Accepted Calls = 15, Refused Calls = 0,
Last Disconnect Cause is "10  ",
Last Disconnect Text is "normal call clearing.",
Last Setup Time = 2277544.
```

30. *You want to monitor the round-trip delay between R3 and R2. Configure the network so that a monitor packet is sent every 10 seconds for 30 minutes.*

Cisco's *Service Assurance Agent* (SAA), formerly known as *Round Trip Reporter* (RTR), enables a router to send out test packets that can measure a variety of network parameters such as round-trip delay, packet loss, and jitter. Multiple *response time reporters* (RTRs) can be set up on a router, and each RTR can measure a different network performance parameter.

SAA configuration is initiated by defining the RTR number:

```
R3(config)#rtr 1
```

The RTR test packet type is then defined. In this case, you will be running an echo packet and will use an ICMP echo packet as your test packet. Define the target for your round-trip test packet to be the loopback interface of R2:

```
R3(config-rtr)# type echo protocol ipIcmpEcho 135.25.2.1
```

Now configure the router to send an echo test packet every 10 seconds:

```
R3(config-rtr)# frequency 10
Setting frequency below 60 seconds may be undesirable.  ← Cisco warning message
```

The following commands configure the RTR probe to keep historical data on the round-trip data points:

```
R3(config-rtr)# lives-of-history-kept 2
R3(config-rtr)#buckets-of-history-kept 20
R3(config-rtr)#filter-for-history all
```

The final part of the RTR configuration is to define when the RTR probe starts to take measurements. The default for the RTR probe is to take measurements for one hour (3,600 seconds). You will configure the RTR probe to start taking measurements immediately and to continue taking measurements for 30 minutes (1,800 seconds):

```
R3(config)#rtr schedule 1 life 1800 start-time now
```

Once configured, several SAA commands can verify the configuration and view the round-trip echo measurements. The **show rtr application** command displays all the possible types of RTR probes that can be configured:

```
R3#show rtr application
        Response Time Reporter
Version: 2.1.1 Round Trip Time MIB
Max Packet Data Size (ARR and Data): 16384
Time of Last Change in Whole RTR: 09:09:08.000 UTC Thu Sep 28 2000
System Max Number of Entries: 500

Number of Entries configured: 1
    Number of active Entries: 1
   Number of pending Entries: 0
  Number of inactive Entries: 0

        Supported Operation Types
Type of Operation to Perform:  echo  ← Supported RTR probe types
Type of Operation to Perform:  pathEcho
```

```
Type of Operation to Perform:  udpEcho
Type of Operation to Perform:  tcpConnect
Type of Operation to Perform:  http
Type of Operation to Perform:  dns
Type of Operation to Perform:  jitter
Type of Operation to Perform:  dlsw
Type of Operation to Perform:  dhcp

        Supported Protocols
Protocol Type: ipIcmpEcho
Protocol Type: ipUdpEchoAppl
Protocol Type: snaRUEcho
Protocol Type: snaLU0EchoAppl
Protocol Type: snaLU2EchoAppl
Protocol Type: ipTcpConn
Protocol Type: httpAppl
Protocol Type: dnsAppl
Protocol Type: jitterAppl
Protocol Type: dlsw
Protocol Type: dhcp

Number of configurable probe is 499
```

The **show rtr collection-statistics** command displays information on RTR operation failures:

```
R3#show rtr collection-statistics
        Collected Statistics
Entry Number: 1
Start Time Index: 09:09:08.000 UTC Thu Sep 28 2000
Path Index: 1
Hop in Path Index: 1
Number of Failed Operations due to a Disconnect: 0  ← RTR failure statistics
Number of Failed Operations due to a Timeout: 0
Number of Failed Operations due to a Busy: 0
Number of Failed Operations due to a No Connection: 0
Number of Failed Operations due to an Internal Error: 0
Number of Failed Operations due to a Sequence Error: 0
Number of Failed Operations due to a Verify Error: 0
Target Address: 135.25.2.1
Corrupted packet count: 0
```

The RTR configuration can be verified with the **show rtr configuration** command:

```
R3#show rtr configuration
        Complete Configuration Table (includes defaults)
Entry Number: 1
Owner:
Tag:
Type of Operation to Perform: echo
Reaction and History Threshold (milliseconds): 5000
Operation Frequency (seconds): 10
Operation Timeout (milliseconds): 5000
```

```
Verify Data: FALSE
Status of Entry (SNMP RowStatus): active
Protocol Type: ipIcmpEcho  ← RTR test type
Target Address: 135.25.2.1  ← RTR target address
Source Address: 0.0.0.0
Target Port: 0
Source Port: 0
Request Size (ARR data portion): 28
Response Size (ARR data portion): 28
Control Packets: enabled
Loose Source Routing: disabled
LSR Path:
Type of Service Parameters: 0x0
Life (seconds): 1800
Next Scheduled Start Time: Start Time already passed  ← Test running
Entry Ageout: never
Connection Loss Reaction Enabled: FALSE
Timeout Reaction Enabled: FALSE
Threshold Reaction Type: never
Threshold Falling (milliseconds): 3000
Threshold Count: 5
Threshold Count2: 5
Reaction Type: none
Number of Statistic Hours kept: 2
Number of Statistic Paths kept: 1
Number of Statistic Hops kept: 1
Number of Statistic Distribution Buckets kept: 1
Statistic Distribution Interval (milliseconds): 20
Number of History Lives kept: 2
Number of History Buckets kept: 20
Number of History Samples kept: 1
History Filter Type: all
```

Previous samples can be viewed with the **show rtr history** command:

```
R3#show rtr history
        Point by point History
Entry    = Entry Number
LifeI    = Life Index
BucketI  = Bucket Index
SampleI  = Sample Index
SampleT  = Sample Start Time
CompT    = Completion Time (milliseconds)
Sense    = Response Return Code

Entry LifeI      BucketI    SampleI    SampleT    CompT      Sense      TargetAddr
1     1          1          1          120869     8          1          135.25.2.1
1     1          2          1          121868     4          1          135.25.2.1
1     1          3          1          122868     4          1          135.25.2.1
1     1          4          1          123868     4          1          135.25.2.1
1     1          5          1          124868     4          1          135.25.2.1
1     1          6          1          125868     4          1          135.25.2.1
1     1          7          1          126868     4          1          135.25.2.1
1     1          8          1          127868     4          1          135.25.2.1
1     1          9          1          128868     4          1          135.25.2.1
1     1          10         1          129868     4          1          135.25.2.1
1     1          11         1          130868     4          1          135.25.2.1
```

The **show rtr operational-state** command displays information on RTR memory usage and the remaining time for the current test:

```
R3#show rtr operational-state
        Current Operational State
Entry Number: 1
Modification Time: 09:09:08.000 UTC Thu Sep 28 2000
Diagnostics Text:
Last Time this Entry was Reset: Never
Number of Octets in use by this Entry: 4742
Connection Loss Occurred: FALSE
Timeout Occurred: FALSE
Over Thresholds Occurred: FALSE
Number of Operations Attempted: 13
Current Seconds Left in Life: 1678  ← Remaining time
Operational State of Entry: active
Latest Completion Time (milliseconds): 4
Latest Operation Start Time: 09:11:08.000 UTC Thu Sep 28 2000
Latest Operation Return Code: ok
Latest 135.25.2.1
```

31. *Configure R2 so that all other routers in the network can load R2's IOS image.*

The Cisco IOS contains a feature that enables the router to act as a TFTP server. This can be used to load IOS code throughout the network from a router acting as a TFTP server. Connect to R2 and confirm the name of the current IOS image that is loaded on the router with the **show flash** command:

```
R2#show flash

System flash directory:
File  Length   Name/status
  1   10036816  c3620-js-mz.121-2.bin
[10036880 bytes used, 6740336 available, 16777216 total]
16384K bytes of processor board System flash (Read/Write)
```

The router is configured as a TFTP server with the **tftp-server** command. Notice that the file that resides in the router flash is being served to the TFTP clients.

```
R2(config)#tftp-server flash:c3620-js-mz.121-2.bin
```

The configuration can be tested by connecting to R1. Issue the **copy tftp flash** command on R1:

```
R1#copy tftp flash

System flash directory:
File  Length   Name/status
```

```
  1   2259976  c3620-i-mz.112-7a.P
  2   2972356  c3620-d-mz_112-16_p.bin
  3   4214712  c3620-js-mz_112-20_P.bin
[9447236 bytes used, 7329980 available, 16777216 total]
Address or name of remote host [255.255.255.255]? 135.25.2.1
Source file name? c3620-js-mz.121-2.bin
Destination file name [c3620-js-mz.121-2.bin]?
Accessing file 'c3620-js-mz.121-2.bin' on 135.25.2.1...
Loading c3620-js-mz.121-2.bin from 135.25.2.1 (via Ethernet0/0): ! [OK]

Erase flash device before writing? [confirm]
Flash contains files. Are you sure you want to erase? [confirm]

Copy 'c3620-js-mz.121-2.bin' from server
  as 'c3620-js-mz.121-2.bin' into Flash WITH erase? [yes/no]y
Erasing device... eeeeeeeeeeeeeeeeeeeeeeeeeeeeeeeeeeeeeeeeeeeeeeeeeeeeeeeeeeeeeeee
...erased
Loading c3620-js-mz.121-2.bin from 135.25.2.1 (via Ethernet0/0):
!!!!!!!!!!!!!!!!!!!!!!!!!!!!!!!!!!!!!!!!!!!!!!!!!!!!!!!!!!!!!!!!!!!!!!!!!!!!!!!!!!!!!!!!!!
!!!!!!!!!!!!!!!!!!!!!!!!!!!!!!!!!!!!!!!!!!!!!!!!!!!!!!!!!!!!!!!!!!!!!!!!!!!!!!!!!!!!!!!!!!
!!!!!!!!!!!!!!!!!!!!!!!!!!!!!!!!!!!!!!!!!!!!!!!!!!!!!!!!!!!!!!!!!!!!!!!!!!!!!!!!!!!!!!!!!!
!!!!!!!!!!!!!!!!!!!!!!!!!!!!!!!!!!!!!!!!!!!!!!!!!!!!!!!!!!!!!!!!!!!!!!!!!!!!!!!!!!!!!!!!!!
!!!!!!!!!!!!!!!!!!!!!!!!!!!!!!!!!!!!!!!!!!!!!!!!!!!!!!!!!!!!!!!!!!!!!!!!!
[OK - 10036816/16777216 bytes]

Verifying checksum...  OK (0xDBFC)
Flash device copy took 00:02:29 [hh:mm:ss]
```

Once the flash has been updated, use the **show flash** command to confirm that the original files in the flash memory have been erased and that the flash now contains the proper IOS image:

```
R1#show flash

System flash directory:
File  Length   Name/status
  1   10036816  c3620-js-mz.121-2.bin
[10036880 bytes used, 6740336 available, 16777216 total]
16384K bytes of processor board System flash (Read/Write)
```

The router can now be reloaded with the **reload** command, which causes the router to be rebooted from the new IOS image:

```
R1#reload
Proceed with reload? [confirm]

%SYS-5-RELOAD: Reload requested
System Bootstrap, Version 11.1(7)AX [kuong (7)AX], EARLY DEPLOYMENT RELEASE SOFT
WARE (fc2)
Copyright (c) 1994-1996 by cisco Systems, Inc.
C3600 processor with 49152 Kbytes of main memory
Main memory is configured to 32 bit mode with parity disabled

program load complete, entry point: 0x80008000, size: 0x992534
```

```
Self decompressing the image : ################################################# .
Self decompressing the image :
#####################################################
```

32. *Configure the network so that traffic coming from R2 into R1 meets the following requirements:*

- The first 496,000 bits/sec of traffic with a source IP address of 135.25.2.1 should be transmitted with an IP precedence of 6. Additional traffic should be sent with an IP precedence of 3.
- The first 96,000 bits/sec of Web traffic should be transmitted. Additional Web traffic should be dropped.
- Up to 200,000 bits/sec of other types of traffic should be transmitted with an IP precedence of 1. Additional traffic should be dropped.

You need to configure three different CAR rate-limit commands in order to achieve the required objectives. For the first objective (setting the precedence of traffic originating from 135.25.2.1), create an input rate limit statement, because you want to limit the traffic as it comes from R2 into R1. Both the conform action and the exceed action will be able to transmit the traffic, but with different precedence bits set. The following command sets this up:

```
R1(config-if)# rate-limit input access-group 169 496000 24000 24000 conform-action set-
prec-transmit 6 exceed-action set-prec-transmit 1
```

You also need to configure an access list that will identify interesting traffic as traffic that has a source IP address of 135.25.2.1. The following access list will be associated with the previous **rate-limit** command. Notice that the access list number (169) is the same as the access group number specified in the **rate-limit** command.

```
R1(config)#access-list 169 permit ip 135.25.2.1 0.0.0.0 any
```

The next objective is to limit Web traffic to 96000 bits/sec. To achieve this, you need to configure a conform action that transmits the traffic and an exceed action that drops the traffic as shown in the following command:

```
R1(config-if)# rate-limit input access-group 170 96000 16000 16000 conform-action transmit
exceed-action drop
```

An access list is needed to only apply the rate limit to Web traffic. Notice that the access list number (170) is the same as the access group number specified in the **rate-limit** command.

```
R1(config)#access-list 170 permit tcp any any eq www
```

The final objective requires that you limit all other traffic to 200,000 bits/second and that you set the precedence to 1 on all transmitted traffic. To do this, you need to set a rate limit that transmits traffic up to the rate limit and then discards additional traffic. No access list is needed because all traffic applies to this statement:

```
R1(config-if)# rate-limit input 200000 12000 12000 conform-action set-prec-transmit 1
exceed-action drop
```

Once the CAR configuration has been completed, it can be verified by checking the interface that it was implemented on. CAR adds an extension to the normal **show interface** command, which becomes the **show interface rate-limit** command. You can see from the following command output that all three rate limits have been set on the interface:

```
R1#show int e 0/0  rate-limit
Ethernet0/0
  Input
    matches: access-group 169
      params:  496000 bps, 24000 limit, 24000 extended limit
      conformed 28 packets, 2296 bytes; action: set-prec-transmit 6
      exceeded 0 packets, 0 bytes; action: set-prec-transmit 1
      last packet: 60112ms ago, current burst: 0 bytes
      last cleared 00:05:30 ago, conformed 0 bps, exceeded 0 bps
    matches: access-group 170
      params:  96000 bps, 16000 limit, 16000 extended limit
      conformed 2 packets, 164 bytes; action: transmit
      exceeded 0 packets, 0 bytes; action: drop
      last packet: 40116ms ago, current burst: 0 bytes
      last cleared 00:04:22 ago, conformed 0 bps, exceeded 0 bps
    matches: all traffic
      params:  200000 bps, 12000 limit, 12000 extended limit
      conformed 4 packets, 328 bytes; action: set-prec-transmit 1
      exceeded 0 packets, 0 bytes; action: drop
      last packet: 116ms ago, current burst: 0 bytes
      last cleared 00:03:51 ago, conformed 0 bps, exceeded 0 bps
```

The **show access-lists** command can be used to verify the two CAR-related access lists that were configured:

```
R1#show access-lists

Extended IP access list 169
    permit ip host 135.25.2.1 any
Extended IP access list 170
    permit tcp any any eq www
```

Complete Network Diagram

Figure 6-6 shows the final network topology.

```
 destination-pattern 2223
 ip precedence 1
 session target ipv4:135.25.2.1
!
dial-peer voice 3 voip
 destination-pattern 2222
 ip precedence 5
 session target ipv4:135.25.2.1
!
num-exp 611 2222
!
interface Loopback0
 ip address 135.25.3.1 255.255.255.255
!
 interface Ethernet0/0
 ip address 135.25.12.1 255.255.255.0
 no ip directed-broadcast
!
interface Serial0/0
 bandwidth 800
 ip address 135.25.9.2 255.255.255.252
 priority-group 1
 ip rsvp bandwidth 600 400
!
interface Serial0/1
 bandwidth 800
 ip address 135.25.9.6 255.255.255.252
 priority-group 1
 clockrate 800000
 ip rsvp bandwidth 600 100
!
interface Serial2/0
 ip address 135.25.11.1 255.255.255.252
 no ip directed-broadcast
 no ip mroute-cache
 clockrate 800000
 crypto map peer-r4
!
interface Serial2/1
 ip address 135.25.11.5 255.255.255.252
 no ip directed-broadcast
 clockrate 800000
!
router ospf 64
 network 135.25.3.1 0.0.0.0 area 0
 network 135.25.11.0 0.0.0.255 area 0
 network 135.25.12.0 0.0.0.255 area 0
!
no ip classless
no ip http server
!
access-list 105 permit udp any any range 16384 16484
access-list 105 permit tcp any any eq 1720
access-list 1 permit 135.25.5.1
access-list 1 permit 135.25.4.1
access-list 101 permit ip host 135.25.3.1 host 135.25.4.1
access-list 102 permit ip 135.25.12.0 0.0.0.255 host 135.25.11.2
priority-list 1 protocol ip high list 105
priority-list 1 default medium
```

```
!
!
rtr 1
 type echo protocol ipIcmpEcho 135.25.2.1
 frequency 10
 lives-of-history-kept 2
 buckets-of-history-kept 20
 filter-for-history all
rtr schedule 1 life 1800 start-time now
!
line con 0
 transport input none
line aux 0
line vty 0 4
access-class 1 in
password cisco
login
!
end
```

R4

```
!
version 12.0
service timestamps debug uptime
service timestamps log uptime
no service password-encryption
!
hostname R4
!
ip telnet source-interface Loopback0
!
crypto isakmp policy 5
 authentication pre-share
crypto isakmp key cisco address 135.25.3.1
crypto isakmp key cisco1 address 135.25.11.2
!
!
crypto ipsec transform-set encr-only esp-3des
crypto ipsec transform-set encr&auth esp-3des esp-md5-hmac
!
!
crypto map peer-r3 local-address Loopback0
crypto map peer-r3 10 ipsec-isakmp
 set peer 135.25.3.1
 set transform-set encr-only
 match address 101
crypto map peer-r3 20 ipsec-isakmp
```

```
 set peer 135.25.3.1
 set transform-set encr&auth
 match address 102
!
!
interface Loopback0
 ip address 135.25.4.1 255.255.255.255
 no ip directed-broadcast
!
interface Ethernet0/0
 ip address 10.1.1.1 255.255.255.0
 no ip directed-broadcast
 ip nat inside
!
interface Serial0/0
 ip address 135.25.11.2 255.255.255.252
 no ip directed-broadcast
 ip nat outside
 no ip mroute-cache
 crypto map peer-r3
!
interface Ethernet1/0
 ip address 135.25.11.9 255.255.255.252
 no ip directed-broadcast
!
!
!
router ospf 64
 network 135.25.4.1 0.0.0.0 area 0
 network 135.25.11.0 0.0.0.255 area 0
!
ip nat inside source list 2 interface Serial0/0 overload
ip classless
no ip http server
!
access-list 1 permit 135.25.3.1
access-list 2 permit 10.1.1.0 0.0.0.255
access-list 101 permit ip host 135.25.4.1 host 135.25.3.1
access-list 101 permit ip host 135.25.3.1 host 135.25.4.1
access-list 102 permit ip host 135.25.11.2 135.25.12.0 0.0.0.255
!
line con 0
 transport input none
line aux 0
line vty 0 4
 access-class 1 in
 password cisco
 login
!
end
```

R5

```
!
version 12.0
no service udp-small-servers
no service tcp-small-servers
!
hostname R5
!
ip host R1 2001 135.25.5.1
ip host R2 2002 135.25.5.1
ip host R3 2003 135.25.5.1
ip host R4 2004 135.25.5.1
ip host R5 2005 135.25.5.1
ip host R6 2006 135.25.5.1
ip host R7 2007 135.25.5.1
ip host cat 2009 135.25.5.1
!
ip telnet source-interface Loopback0
!
interface Loopback0
 ip address 135.25.5.1 255.255.255.255
!
interface Ethernet0/0
 ip address 1.1.1.2 255.255.255.0
 no ip directed-broadcast
!
interface Serial0/0
 ip address 135.25.11.6 255.255.255.252
!
!
router ospf 64
 network 135.25.11.0 0.0.0.255 area 0
 network 135.25.5.1 0.0.0.0 area 0
!
no ip classless
access-list 1 permit 135.25.4.1
access-list 1 permit 135.25.3.1
!
line con 0
line 1 16
no exec
 transport input all
line aux 0
line vty 0 4
 access-class 1 in
 password cisco
 login
!
end
```

R6

```
version 12.1
service timestamps debug uptime
service timestamps log uptime
no service password-encryption
!
hostname R6
!
ip subnet-zero
no ip routing
!
!
lane database cisco
  name mkt server-atm-address 47.00918100000000107BD56901.006070CD51EB.01
  default-name mkt
no lane client flush
!
interface ATM0
 no ip address
 no ip route-cache
 no ip mroute-cache
 atm pvc 1 0 5 qsaal
 atm pvc 2 0 16 ilmi
 no atm ilmi-keepalive
 lane config auto-config-atm-address
 lane config database cisco
!
interface ATM0.1 multipoint
 ip address 172.16.0.1 255.255.0.0
 no ip route-cache
 no ip mroute-cache
 lane server-bus ethernet mkt
 lane client ethernet mkt
!
interface Ethernet0
 ip address 1.1.1.1 255.255.255.0
 no ip route-cache
 no ip mroute-cache
!
ip classless
no ip http server
!
!
line con 0
 transport input none
line aux 0
line vty 0 4
 login
!
end
```

R7

```
version 12.1
service timestamps debug uptime
service timestamps log uptime
no service password-encryption
!
hostname R6
!
ip subnet-zero
no ip routing
!
lane client flush
cns event-service server
!
interface Ethernet0
 ip address 2.2.2.2 255.255.255.0
 no ip route-cache
 no ip mroute-cache
 media-type auto-select
!
interface ATM1
 no ip address
 no ip route-cache
 no ip mroute-cache
 atm pvc 1 0 5 qsaal
 atm pvc 2 0 16 ilmi
 no atm ilmi-keepalive
!
interface ATM1.2 multipoint
 ip address 172.16.0.3 255.255.0.0
 no ip route-cache
 no ip mroute-cache
 lane client ethernet mkt
!
ip classless
no ip http server
!
line con 0
 transport input none
line aux 0
line vty 0 4
 login
!
end
```

LS1010

```
!
version 11.2
no service pad
no service password-encryption
no service udp-small-servers
no service tcp-small-servers
!
hostname LS1010
!
boot system flash slot0:ls1010-wp-mz.120-9.bin
!
!
atm lecs-address-default 47.0091.8100.0000.0010.7bd5.6901.0060.70cd.51ed.00 1
atm address 47.0091.8100.0000.0010.7bd5.6901.0010.7bd5.6901.00
atm router pnni
 node 1 level 56 lowest
  redistribute atm-static
!
!
interface ATM0/0/0
 atm maxvp-number 8
 atm maxvc-number 8192
!
interface ATM0/0/1
 mtu 8192
 atm maxvp-number 8
 atm maxvc-number 8192
!
interface ATM0/0/2
!
interface ATM0/0/3
!
interface ATM2/0/0
 no ip address
 atm maxvp-number 0
!
interface Ethernet2/0/0
 ip address 10.10.3.170 255.255.255.0
!
no ip classless
!
line con 0
line aux 0
line vty 0 4
 login
!
end
```

APPENDIX

Case Study Tearsheets

Case Study 1: BGP

IOS Requirements

BGP-4 first became available in IOS release 10.0, but this case study was performed using IOS 11.2.

Equipment Needed

The following equipment is needed to perform this case study:

- One Catalyst 1900 Ethernet switch
- One terminal server with one Ethernet and one serial port
- One Cisco router with one serial port and one Ethernet port
- Three Cisco routers with two serial ports and one Ethernet port
- One Cisco router with one Ethernet port and three serial ports
- One Cisco router with one Ethernet port and four serial ports
- One peripheral router with one Ethernet port and three serial ports. This router (R8) will be used as the Frame Relay switch as well as the BGP neighbor. (The student does not configure this router and should use the configuration provided on the CD-ROM.) Table 2-1 displays each router's interface requirements.
- Cisco IOS 11.2
- A PC running a terminal emulation program for connecting to the console port of the terminal server
- Six Ethernet cables
- Seven Cisco DTE/DCE crossover cables
- One Cisco Rolled cable

5. Place Ethernet interface 0 of R5, R6, R7, and R8 in Virtual Local Area Network (VLAN) 1. The routers should be attached in ascending order, starting with port 1 on the Catalyst switch.
6. Configure the router so that if an incorrect command is entered, the router does not try to translate it to an IP address.
7. The Mac addresses of the Ethernet interfaces attached to the VLAN should have the following format (routernumber. ASnumber .0000). For example, R7 would be set to

```
0007. 0400.0000.
```

8. Configure the Catalyst so that only the designated router can attach to the specific port.
9. Verify that step 8 works by swapping the cables between port 1 and port 2 of the Catalyst switch. The ports should go into blocking mode.
10. Normalize the port.
11. Assign IP addresses to Ethernet 0 of R5, R6, R7, and R8 from network 152.1.1.0. Do not allocate more addresses than needed; use the first available subnet and assign the addresses in ascending order starting with R5.
12. Place Ethernet 0 of R1 and R3 in VLAN 2; place Ethernet 1 of R1 and Ethernet 0 of R2 in VLAN 3.
13. Connect the rest of the routers serially as per Figure 2-6. Make sure that the correct ports are configured for DTE/DCE. The DCE clock will be set to 64K on all routers.
14. R3 and R2 are connected via Frame Relay to R1. Use physical interfaces on R3 and R2 and a logical interface on R1. Assign addresses to each interface using subnet 152.1.10.0. Do not allocate more addresses than needed; use the first subnet available and assign the addresses in ascending order, starting with R1. Configure the interfaces for Frame Relay. Make sure that the only DLCIs that are used are the ones provided. Make sure that all routers can reach each other.

    The following is the physical connectivity of R1, R2, and R3 to the Frame Relay Switch (R8).

Frame Switch (R8) interface S0	R1 Interface S0/0
Frame Switch (R8) interface S0	R2 Interface S0
Frame Switch (R8) interface S0	R3 Interface S0/1

15. Assign IP addresses to Ethernet 0 of R3 and R1 using the next available subnet in 152.1.8.0. The subnet should contain 48 host addresses, R1 should use the first address in the subnet, and R3 should use the last one.
16. Assign IP addresses to Ethernet 1 of R1 and Ethernet 0 of R2 using the first available subnet in 152.1.9.0. The subnet should contain 128 host addresses. R1 should use the first address in the subnet and R2 should use the last one.

17. Assign IP addresses to Serial 0 on R7 and serial S0/0 on R3 using the first available subnet in 152.1.20.0. The address should be assigned out in ascending order, starting with R3.
18. Assign IP address to the serial interfaces connecting R3 and R4, use the first available subnet in 152.1.12.0. The subnet should contain the minimal hosts needed and the IP addresses should be assigned in ascending order, staring with R3.
19. Assign IP addresses to the serial interfaces connecting R3 and R5, use the next available subnet in 152.1.12.0. The subnet should contain the minimal hosts needed and the IP addresses should be assigned in ascending order, staring with R3.
20. Assign an IP address to Serial 1 on R4 and R5 using the next available subnet in 152.1.12.0. The subnet should contain the minimal hosts needed and the IP addresses should be assigned in ascending order, starting with R4.
21. Assign an IP address to Serial 1 of R2 and Serial 0 of R6 using the first available subnet in 152.1.11.0. The subnet should contain the minimal hosts needed and the IP addresses should be assigned in ascending order, starting with R2.
22. Assign the following loopback addresses to the corresponding routers:

    R7 Loopback 0 152.1.20.7 /32

    R7 Loopback 1 152.1.20.17 /28

    R4 Loopback 0 152.1.12.4 /32

    R4 Loopback 0 152.1.13.1 /24

    R5 Loopback 0 152.1.12.5 /32

    R5 Loopback 1 152.1.14.1 /29

    R3 Loopback 0 152.1.10.3 /32

    R2 Loopback 0 152.1.10.2 /32

    R1 Loopback 0 152.1.10.1 /32

    R6 Loopback 0 152.1.11.6 /32

    R6 Loopback 1 152.1.11.129 /26

IGP Protocols

23. Place the serial connections between R3, R2, and R1 in OSPF Area 0.
24. Place VLAN 2 in OSPF area 1 and VLAN 3 in OSPF area 2. R1 should be the DR for Area 1 and R2 should be the DR in Area 2.
25. R1 should prefer the Frame Relay link over the Ethernet connection. All other connections should prefer the interface with the highest bandwidth.

26. Change the Hello interval on R3 to two minutes.
27. Configure EIGRP on the interface connecting R3 to R7. Do not send advertisements out any other interface. Enable EIGRP on all interfaces on R7, but do not send updates out the interface connecting to VLAN1.
28. R7 should be able to see all the networks behind R3. R3 only needs to see network 152.1.20.16/28 via R7.
29. Make sure R7 does not advertise any routes back to R3 that were learned from R3 and vice versa. Do not rely on split horizons!
30. Enable EIGRP using process 56 on R4 and R5. Don't run EIGRP on the interfaces that are connected to other ASs.
31. Make sure that the loopback interfaces on R2, R1, and R3 are being propagated via OSPF. OSPF should not be run on these interfaces, and the routes should not show up as OSPF external routes.

EBGP/IBGP

32. Place R3, R2, and R1 in AS 100. R1 will peer with R2 and R3. R1 is the only router in AS100 that should have two IBGP neighbors.
33. Place R4 and R5 in AS 200. R4 should peer with R5 and R3.
34. Place R6 in AS 300 and R7 in AS 400
35. R7 should BGP peer with R6, R5, and R8
36. R5 should BGP peer with R7, R8, R6, and R3.
37. R6 should BGP peer with R7, R8, R5, and R2.
38. R7, R5, and R6 should all see the following networks via BGP from R8:

    Network 148.1.0.0

    Network 148.2.0.0

    Network 148.3.0.0

    Network 148.4.0.0

39. AS400 should advertise local network 152.1.20.0 via BGP and AS300 should advertise local network 152.1.11.128 via BGP.
40. AS 200 should advertise local network 152.1.13.0 /24 and 152.1.14.0 / 29.
41. AS100 should only accept local routes from AS300. All other routes should be accepted via AS200.

42. AS100 should accept a default route from AS300 in case of a link failure between A200 and AS100. The default should not be advertised outside of AS100.
43. Advertise network 152.1.9.0 via IBGP on R1 and R2. Do not use the redistribute command.
44. AS 100 should advertise all local networks via BGP. To prevent AS 100 from being used as a transit network, AS300 should only accept local networks from AS100.
45. AS100 prefers that network 152.1.9.0 be reachable by the outside world via the link between AS100 and AS300.
46. Under no circumstances can AS100 be used as a transit network for AS200 to reach AS300. To accomplish this, no additional configurations should be added to any router other than R2.
47. AS200 and AS300 should not pass network 152.1.9.0. No additional configuration should be used to achieve this outside of AS100.
48. R3 should use R5 to reach network 152.1.14.0 and R4 to reach network 152.1.13.0. No additional configuration should be used to achieve this outside of AS200.
49. R7 should originate and advertise network 148.0.0.0, 148.5.0.0, 148.6.0.0, and 148.7.0.0. Add four loopback interfaces to R7: loopback 2 IP address 148.0.0.1/16, loopback 3 IP address 148.5.0.1 /16, loopback 4 IP address 148.6.0.1 /16, and loopback 5 IP address 148.7.0.1 /16.
50. R5 should advertise as few prefixes for 148.x.0.0 as needed.
51. R3 is running an IGP (EIGRP) on the private link between it and AS400. Make sure that the IGP route to network 152.1.20.16 is favored over the EBGP route.
52. The configuration on R3 is growing extremely large and approaching the limitation of the router's NVRAM. Reduce the size of the configuration on R3.
53. Configure R3 so that informational messages are logged to the internal buffer. Make sure the router can store up to 16K of information and the information is time stamped.
54. Configure R3 so that SNMP information can only be read by host 192.1.1.1.
55. Network 152.1.13.0 is suspected of being unstable, causing BGP UPDATE and WITHDRAWN messages to be repeatedly propagated on the network. Configure R3 so that if the network continues to FLAP, it will be withdrawn.
56. Enable R3 so that when its neighbors make a BGP change, the change will take effect without resetting the BGP TCP session. R3 should be able to initiate the change locally; no commands should have to be entered on the neighboring routers.

Case Study 2: Desktop Protocols

IOS Requirements

IOS 11.2 is used in this case study.

Equipment Needed

The following equipment is needed to perform this case study:

- One Cisco router with two serial ports
- One Cisco router with two serial ports and an ISDN BRI port
- Two Cisco routers with one serial port and one Ethernet port
- Two Cisco routers with one Ethernet port
- One Cisco router with one serial, one Ethernet, and one ISDN BRI port
- One Cisco router with three serial ports. This router will be used as the Frame Relay switch
- One Cisco terminal server. (One of the other Cisco routers could also double as a terminal server. In this case study, R7 will be your terminal server.)
- A Cisco IOS desktop image loaded on each router.
- A PC running a terminal emulation program for connecting to the console port of the terminal server
- Three Ethernet cables
- An Ethernet hub
- Five Cisco DTE/DCE crossover cables
- Two ISDN BRI circuits and two ISDN BRI cables

Obtaining ISDN Circuits

This case study requires two ISDN BRI circuits that will be used to configure IPX DDR on Routers R1 and R6. You have two options for obtaining BRI circuits for this case study:

- Purchase the two ISDN circuits from your local telecommunications carrier.
- Obtain an ISDN simulator you can use to provision your own BRI circuits. This is the option that is used in this case study and an Adtran Atlas 800 desktop ISDN switch is utilized here. Details on configuring this ISDN switch for this case study can be found at the end of this chapter.

Physical Connectivity Diagram

Figure 3-3 shows the physical connectivity for the routers in this case study.

Figure 3-3 Physical connectivity diagram

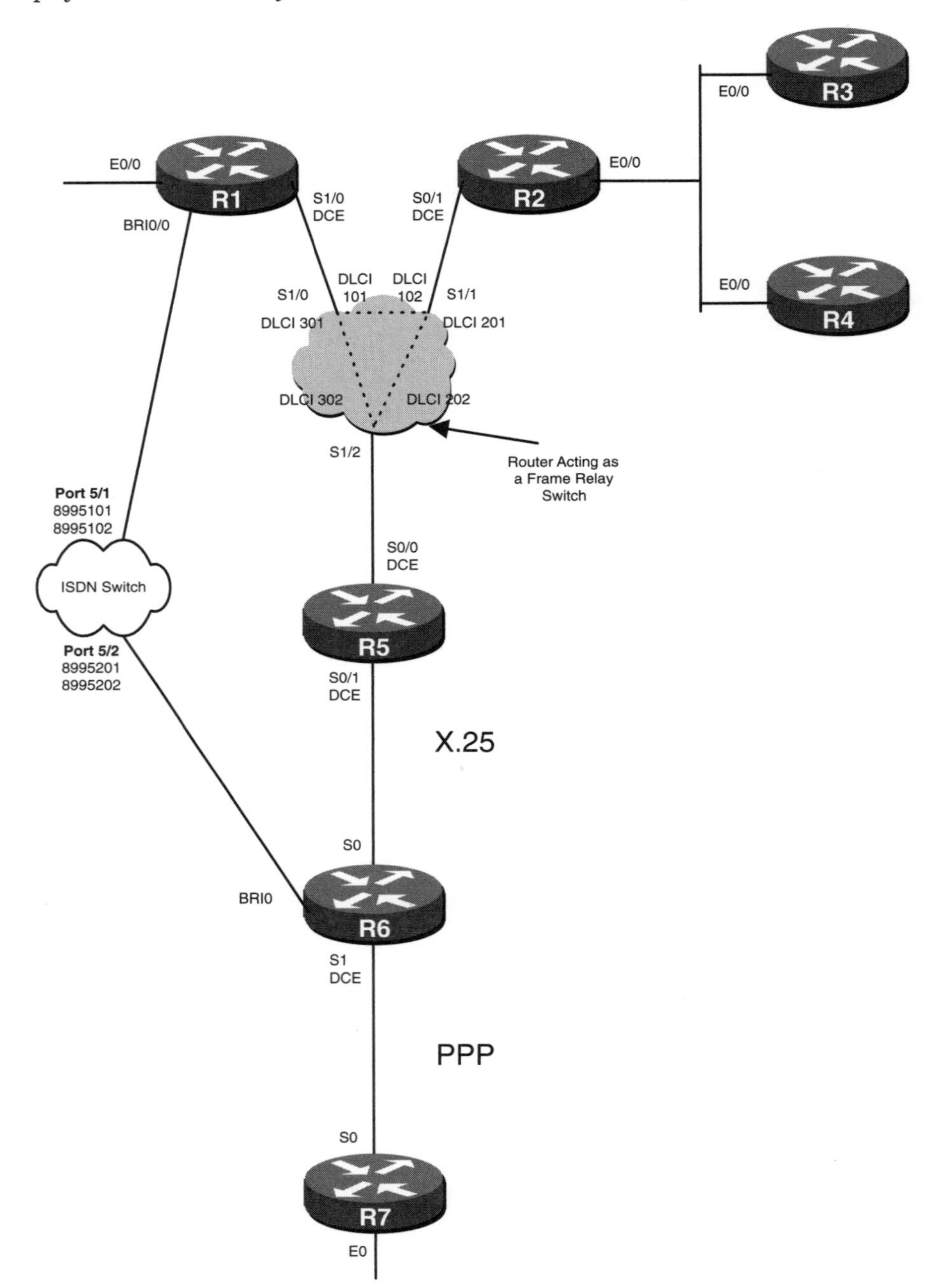

Questions

DECnet

1. Cable the routers and make sure that the DTE/DCE cables are connected as per Figure 3-3.
2. Configure the terminal server so that all the routers can be accessed by name through it. The port number of the terminal server attaches to the corresponding router. That is, port 1 of the terminal server goes to the console port of R1, and the router acting as a Frame Relay switch goes to port 8 of the terminal server.
3. Set each router hostname as per Figure 3-3.
4. Set the enable password on all routers to be **cisco**.
5. Configure the router acting as a Frame Relay switch so that a full mesh of PVCs is used between the three serial ports on the Frame Relay switch. Configure the PVCs as follows:

Frame Switch Port In	DLCI In	Frame Switch Port Out	DLCI Out
S1/0	101	S1/1	102
S1/0	301	S1/2	302
S1/1	102	S1/0	101
S1/1	201	S1/2	202
S1/2	202	S1/1	201
S1/2	302	S1/0	301

As an example, the first line of the table would be interpreted as follows: Any traffic that enters port S1/0 of the Frame Relay switch with a DLCI value of 101 will be sent out port S1/1 of the Frame Relay switch with a DLCI value of 102.

6. Configure R1 with a DECnet address of 2.1:
   a. Configure the router with an interface cost equal to the router number.
   b. R1 is connected to the Frame Relay network via interface S1/0. Configure R1 so that Frame Relay traffic only uses DLCI 101.
7. Configure R2 with a DECnet address of 2.2.
   a. Configure the router with an interface cost equal to the router number.
   b. R2 is connected to the Frame Relay network via interface S0/1. Configure R2 so that Frame Relay traffic to and from R1 uses DLCI 102, and Frame Relay traffic to and from R5 uses DLCI 201.

8. Configure R3 with a DECnet address of 2.3. Also configure it with an interface cost equal to the router number.
9. Configure R4 with a DECnet address of 2.4. Also configure it with an interface cost equal to the router number.
10. Configure R5 with a DECnet address of 2.5. Interface S0/1 of R5 should use X.25 encapsulation with an X.25 address of 5555. Interface S0/1 should act as an X.25 DCE.
    a. Configure the router with an interface cost equal to the router number.
    b. R5 is connected to the Frame Relay network via interface S0/0. Configure R5 so that Frame Relay traffic only uses DLCI 202.
11. Configure R6 with a DECnet address of 2.6. Interface S0 of R6 should use X.25 encapsulation with an X.25 address of 6666. Interface S0 should act as an X.25 DTE. Also configure the router with an interface cost equal to the router number.
12. Configure R7 with a DECnet address of 2.7. Also configure the router with an interface cost equal to the router number.
13. Configure R3 so that its routing table does not have an entry for R6.
14. Configure R7 for LAT translation. A LAT session request to a loopback interface on R7 should be translated into a telnet session into R7.
15. Verify that the configuration is working properly.

IPX

16. Configure R1 as follows:
    a. Interface E0/0 should be assigned IPX network number 12. It should use the default encapsulation for an IPX Ethernet interface.
    b. Create a loopback 0 interface with an IPX network number of 1.
    c. The IPX node number should be set to 0001.0001.0001.
    d. Interface S1/0 should be assigned IPX network number 8. Use a physical interface.
    e. Run IPX EIGRP on all interfaces except the loopback interface, which should run IPX RIP.
17. Configure R2 as follows:
    a. Interface E0/0 should be assigned two different IPX network numbers, 10 and 11.
    b. Create a loopback 0 interface with an IPX network number of two.
    c. The IPX node number should be set to 0002.0002.0002.
    d. Interface S0/1 should be assigned IPX network number 8. Use a physical interface.
    e. Run IPX EIGRP on all interfaces except the loopback interface, which should run IPX RIP.

18. Configure R3 as follows:
   - **a.** Interface E0/0 should be assigned IPX network number 11.
   - **b.** The IPX node number should be set to 0003.0003.0003.
   - **c.** Create a loopback 0 interface with an IPX network number of 3.
   - **d.** Run IPX EIGRP on all interfaces except the loopback interface, which should run IPX RIP.

19. Configure R4 as follows:
   - **a.** Interface E0/0 should be assigned IPX network number 10.
   - **b.** The IPX node number should be set to 0004.0004.0004.
   - **c.** Create a loopback 0 interface with an IPX network number of 4.
   - **d.** Run IPX EIGRP on all interfaces except the loopback interface that should run IPX RIP.

20. Configure R5 as follows:
   - **a.** Interface S0/0 should be assigned IPX network number 8. Use a physical interface.
   - **b.** The IPX node number should be set to 0005.0005.0005.
   - **c.** Create a loopback 0 interface with an IPX network number of 5.
   - **d.** Run IPX EIGRP on interface S0/0.
   - **e.** Run IPX RIP on the loopback interface.
   - **f.** Using internal network number 100, run IPX WAN with NLSP on interface S0/1.

21. Configure R6 as follows:
   - **a.** Interface S1 should be assigned IPX network number 9.
   - **b.** The IPX node number should be set to 0006.0006.0006.
   - **c.** Create a loopback 0 interface with an IPX network number of 6.
   - **d.** Run IPX EIGRP on interface S1.
   - **e.** Run IPX RIP on the loopback interface.
   - **f.** Using Internal network number 101, run IPX WAN with NLSP on interface S0.

22. Configure R7 as follows:
   - **a.** Interface S0 should be assigned IPX network number 9.
   - **b.** The IPX node number should be set to 0007.0007.0007.
   - **c.** Create a loopback 0 interface with an IPX network number of 7.
   - **d.** Run IPX EIGRP on all interfaces except the loopback interface that should run IPX RIP.
   - **e.** Interface E0 should be assigned IPX network number 13.

23. Configure two static SAP entries on R6 as follows:
   - **a.** The first static entry should be for a print server residing on IPX network 13 on R7.
   - **b.** The second static entry should be for a file server residing on IPX network 13 on R7.

24. Configure R1 so that it cannot see any print servers.

25. Configure R2 so that R3 and R4 do not have any entries for IPX network 7 in their routing tables.

26. Configure DDR between R1 and R6. Configure R1 so that it will call R6 when R1's Frame Relay connection is lost. Also configure DDR so that R6 will accept R1's call, authenticate R1, and then securely call R1 back. Use IPX network 14 for the ISDN BRI network.
27. Verify that the configuration is working properly.

AppleTalk

28. Configure R1 as follows:
    - a. Interface E0/0 should be assigned AppleTalk cable range 41-99, with an interface address of 69.1. The zone name for this interface should be LAN_1.
    - b. Interface S1/0 should be assigned AppleTalk cable range 100-199, with an interface address of 124.1. The zone name should be WAN_1_2_5.
    - c. Run AppleTalk EIGRP on interface S1/0 and RTMP on interface E0/0.
29. Configure R2 as follows:
    - a. Interface E0/0 should be assigned AppleTalk cable range 200-299, with an interface address of 234.2. The zone name should be set to LAN_2_3_4.
    - b. Interface S0/1 should be assigned AppleTalk cable range 100-199, with an interface address of 124.2. The zone name should be WAN_1_2_5.
    - c. Run AppleTalk EIGRP on interface S0/1 and RTMP on interface E0/0.
30. Configure R3 as follows:
    - a. Interface E0/0 should be assigned AppleTalk cable range 200-299, with an interface address of 234.3. The zone name should be set to LAN_2_3_4.
    - b. Run RTMP on this router.
31. Configure R4 as follows:
    - a. Interface E0/0 should be assigned AppleTalk cable range 200-299, with an interface address of 234.4. The zone name should be set to LAN_2_3_4.
    - b. Run RTMP on this router.
32. Configure R5 as follows:
    - a. Interface S0/0 should be assigned AppleTalk cable range 100-199, with an interface address of 124.5. The zone name should be set to WAN_1_2_5.
    - b. Run AppleTalk EIGRP on all AppleTalk interfaces on this router.
    - c. Interface S0/1 should be configured with IP address 142.2.1.17/29. Run IP EIGRP on this interface.

33. Configure R6 as follows:

  a. Do not run any AppleTalk routing on this router.
  b. Interface S0 should be assigned IP address 142.2.1.18/29.
  c. Interface S1 should be assigned IP address 142.2.1.9/29.
  d. Run IP EIGRP on this router.

34. Configure R7 as follows:

  a. Interface E0 should be assigned AppleTalk cable range 701-799, with an interface address of 720.14. The zone name should be set to LAN_7 with a secondary zone name of TopSecret. Run RTMP on this interface.
  b. Interface S0 should be configured with IP address 142.2.1.10/29. Run IP EIGRP on this interface.

35. Configure a GRE tunnel between R5 and R7.

  a. Run AppleTalk EIGRP over this tunnel.
  b. The tunnel cable range should be 500-700.
  c. The tunnel zone should be WAN_5_7.
  d. The tunnel endpoint on R5 should be AppleTalk address 545.5.
  e. The tunnel endpoint on R7 should be AppleTalk address 660.7.

36. Configure R2 so that R3 and R4 cannot reach any addresses in AppleTalk cable range 701-799.

37. Configure R4 so that it can no longer see any routes to the 41-99 cable range.

38. Verify that the configuration is working properly.

Case Study 3: OSPF

IOS Requirements

Most of the routing protocols discussed became available in IOS 10.0, however, all labs were performed using IOS 11.2.

Equipment Needed

The following equipment is needed to perform this case study. Table 4-6 outlines the interface requirements for each router.

- One Catalyst 5500 Ethernet switch.
- One terminal server with one Ethernet and one serial port.
- One Cisco router with two serial ports.
- Two Cisco routers with an Ethernet port, one serial port, and one ISDN BRI interface.
- One Cisco router with two serial ports and one Ethernet port.
- One peripheral router with one Ethernet port and three serial ports. This router (R8) will be used as the Frame Relay switch as well as the backbone router. **(The student does not configure this router and should use the configuration provided on the CD).**
- Cisco IOS 11.2.
- A PC running a terminal emulation program for connecting to the console port of the terminal server.
- Five Ethernet cables.
- Five Cisco DTE/DCE crossover cables.
- One Cisco Rolled cable.

Table 4-6 Interface requirements for each router

R1	1 Serial, 1 Ethernet, 1 BRI
R2	1 Serial, 1 Ethernet
R3	1 Serial, 1 Ethernet
R4	1 Serial, 1 Ethernet (terminal server)
R5	2 Serial
R6	1 Serial, 2 Ethernet, 1 BRI
R8	3 Serial, 1 Ethernet (Frame switch)

Physical Connectivity Diagram

Figure 4-21 shows the physical connecting network diagram for the routers in this case study.

Figure 4-21
Network diagram

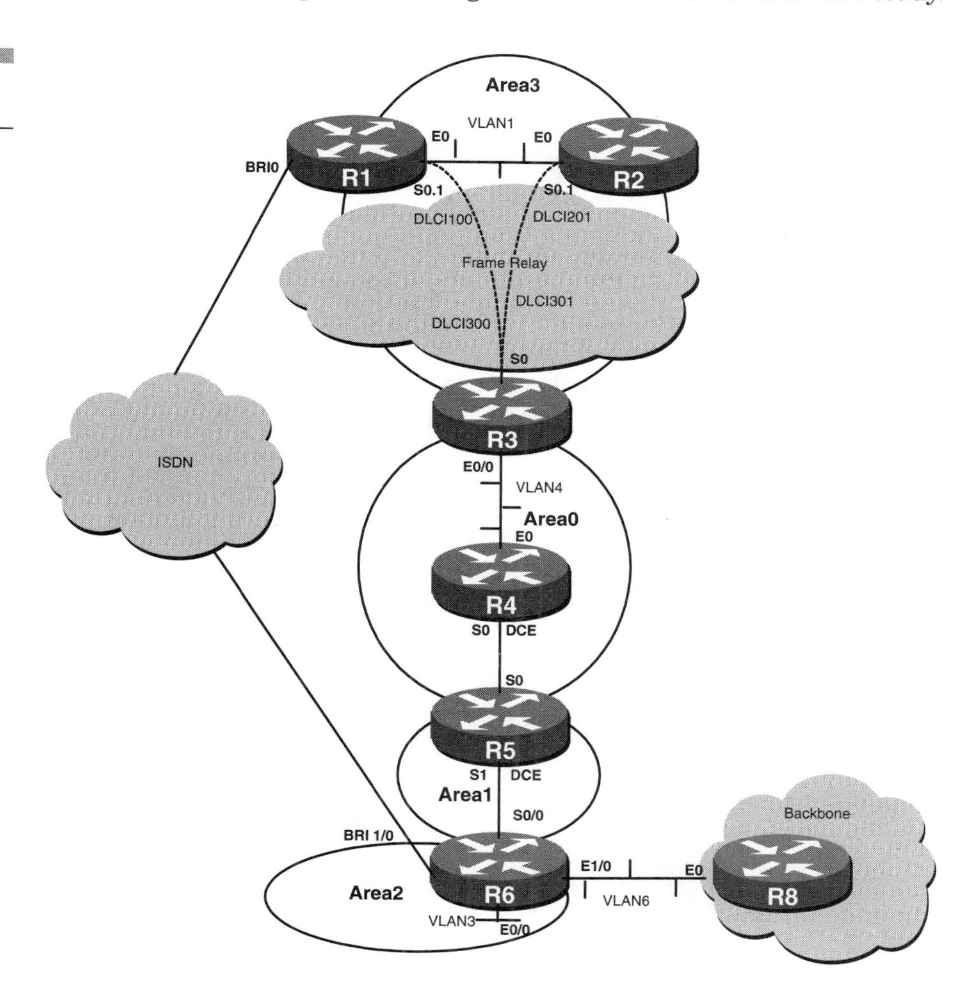

Notes

- All IP addresses will be assigned from the class B network 152.1.0.0.
- No static or default routes should be used in this case study unless explicitly called for.
- Use Process ID 64 for all OSPF configurations.
- Use AS 64 for all IGRP configurations.
- Do not use the network point-to-multipoint command under OSPF.

- No configuration is needed on R8; load the configuration that is provided off the CD.
- Read through the entire case study before beginning.
- Make sure you save your configurations regularly.
- The VTP domain name on the Catalyst should be CCIE_Lab.

Questions

1. Cable the routers per Figure 4-21 and load the configuration provided on the CD to R8. The configuration provided is from a 4500 with an Ethernet interface and two four-port serial network modules. If you plan on using a router other than a 4500 or one that has a different way of numbering its interfaces, such as a 3620, the interfaces provided in the configuration will need to be renamed.
2. Erase any existing configuration from the routers and set each router's name. The name will be the corresponding router number; for example, router 1 will be R1.
3. Set the enable secret on each router to be **cisco**.
4. Configure the terminal server (R4) so that all routers can be accessed by name through it. The port number of the terminal server attaches to the corresponding router; that is, port one of the terminal server goes to the console port of R1, while the Catalyst switch goes to port number 9. (Use the Ethernet IP address as the IP address for reverse telneting.)
5. Place Ethernet interface 0 of R1 and R2 in VLAN 1. The routers should be attached in ascending order staring with port 1 on the Catalyst switch.
6. A DNS server is located at 153.1.1.1. Configure R1 so that this server is used for address resolution.
7. Disable the CDP on all routers.
8. Assign IP addresses to Ethernet 0 of R1, and R2 from network 152.1.1.0. Do not allocate more addresses than needed. Use the first available subnet and assign the addresses in ascending order starting with R1.
9. VLAN 1 contains 28 servers in a server farm. Configure R1 and R2 so that in the event of a router failure the other router will take over routing for the subnet. Under normal conditions, the load should be split evenly between both routers. Fourteen of the servers are defined with a default gateway of 152.1.1.3 and the other fourteen are defined with a default gateway of 152.1.1.4.
10. R1 and R2 are connected via Frame Relay to R3. Use physical interfaces on R3 and R2 and a logical interface on R1. Assign addresses to each interface using the next available subnet of 152.1.1.0. Do not allocate more addresses than are needed and assign the addresses in ascending order starting with R1. Configure the interfaces for Frame Relay. Make sure

that the only DLCIs that are used are the ones provided and that all routers can reach each other.

The following is the physical connectivity of R1, R2, and R3 to the Frame Relay switch (R8).

Frame Switch (R8) interface S0	R1 Interface S0
Frame Switch (R8) interface S1	R2 Interface S0
Frame Switch (R8) interface S3	R3 Interface S0/0 (3600)

11. Assign IP addresses to BRI 0 of R1 and R6 from network 152.1.1.0. Do not allocate more addresses than needed. Use the next available subnet and assign the addresses in ascending order starting with R1.
12. Place Ethernet 0/0 of R3 and Ethernet 0 of R4. In VLAN 4, place Ethernet 0/0 of R6 in VLAN 3.
13. Assign IP addresses to Ethernet 0 of R3 and R4 from network 152.1.1.0. The subnet will contain 64 hosts. Assign the addresses in ascending order starting with R3.
14. Connect the rest of the routers serially per Figure 4-21. Make sure that the correct ports are configured for DTE/DCE (the DCE clock will be set to 64K on all routers unless otherwise specified).
15. Assign an IP address to S0 of R4 and R5 from network 152.1.2.0. Do not allocate more addresses than needed. Use the first available subnet and assign the addresses in ascending order starting with R4.
16. Assign IP addresses to S1 of R5 and S0/0 of R6 from network 152.1.10.0. Use an address that will support up to 80 hosts. Do not allocate more addresses than needed. Use the first available subnet and assign the addresses in ascending order starting with R5.
17. Set the maximum transmission unit on Serial 0 of R1 to 500.
18. Configure the Catalyst so that only hosts 152.1.1.1 and 152.1.1.2 have telnet access. Assign an address to the SCO interface of the catalyst switch, use the next available address on subnet 152.1.1.0/26.
19. Assign an IP address to Ethernet 0/0 of R6 from network 152.1.11.0. Use a subnet that supports up to 200 hosts.
20. Assign IP address 152.1.0.1 /24 to Ethernet 1/0 of R6, and place R6 E 1/0 in VLAN 6. The Ethernet interface from the backbone router R8 should also be placed in VLAN6.
21. Assign the following loopback addresses to the corresponding routers:

```
R1 Loopback 0   152.1.1.65 /32
R1 Loopback 1   152.1.1.97 /27
R2 Loopback 0   152.1.1.72 /32
R3 Loopback 0   152.1.1.69 /32
R4 Loopback 0   152.1.2.4  /32
R5 Loopback 0   152.1.2.5  /32
R3 Loopback 1   152.1.3.1  /24
R6 Loopback 0   152.1.12.1  /24
```

OSPF

22. Place VLAN 1 in OSPF area 3, and make sure that R1 is the DR for the area.
23. Place the Frame Relay cloud between R1, R2, and R3 in OSPF area 3.
24. Place VLAN4 and the serial interface connecting R4 and R5 in OSPF area 0.
25. Place the BRI interface on R6 and VLAN 3 in OSPF area 2.
26. Configure area 2 so that no external LSAs are flooded into the area from the backbone, however, area 2 must be able to send routes learned from the IGRP backbone into area 0.
27. Change the OSPF cost of using an Ethernet interface to 90 and the serial interface cost to 580 across the entire network. Do not use the OSPF cost command.
28. R4 should only see a single advertisement in its routing table from area 3.
29. Advertise all the loopback interfaces on R1 and R2, as well as loopback 1 on R3 via OSPF. Don't configure or run OSPF on any of these loopback interfaces, however, and don't propagate these networks outside of the area.
30. Configure the routers so that the source of each routing update is authenticated. Area 0 should use MD5 authentication, while area 3 and area 1 should use simple authentication. The key should be cisco, plus the area number. For example, the key for area 0 is cisco0.
31. Configure DDR between R1 and R6. R6 should call R1 when R6's serial connection is lost to R5. The backup should be invoked five seconds after the primary is lost. Once the primary is back, the router should wait 20 seconds before switching to the primary. Configure DDR so that R1 accepts R6's call, authenticates R6, and then securely calls R6 back.
32. Run OSPF on the ISDN circuit between R1 and R6. Make sure that all the routes are learned via R6 during backup.
33. Make sure that periodic OSPF Hello advertisements and LSAs are suppressed and that the link is only brought up by OSPF when a topology change occurs.
34. Because R6 is attached to the backbone, it should advertise a default route into the domain.
35. Configure R6 to run IGRP to the backbone. Only send IGRP messages out interface Ethernet 1/0.
36. R6 should only take in the following networks from the backbone:

```
152.12.0.0
152.13.0.0
152.14.0.0
152.15.0.0
```

37. Advertise the routes learned from the backbone in the OSPF domain. Also advertise as few of the networks as needed.
38. In addition to 152.1.11.0 and 152.1.12.0, which are already being advertised in the IGRP backbone, R6 should advertise the following networks.

```
152.1.1.0
152.1.2.0
```

Case Study 4: IS-IS

IOS Requirements

Most of the routing protocols discussed became available in IOS 10.0, but all labs were performed using IOS 11.2.

Equipment Needed

The following equipment is needed to perform this case study. Table 4-8 lists the interface requirements for each router.

- One Catalyst 5500 Ethernet switch.
- One terminal server with one Ethernet and one serial port.
- One Cisco router with one serial port and one Ethernet port.
- One Cisco router with an Ethernet port and two serial ports.
- One Cisco router with an Ethernet port, one serial port, and one ISDN BRI interface.
- One Cisco router with one serial port and one ISDN BRI.
- Cisco IOS 11.2.
- A PC running a terminal emulation program for connecting to the console port of the terminal server
- Six Ethernet cables.
- Three Cisco DTE/DCE crossover cables.
- One Cisco Rolled cable.

Table 4-8

Interface requirements for each router

R1	1 Serial, 1 Ethernet
R2	1 Serial, 1 Ethernet (terminal server)
R3	2 Serials, 1 Ethernet
R4	1 Serial, 1 Ethernet, 1 BRI
R5	1 Serial, 1 BRI

Physical Connectivity Diagram

Figure 4-23 shows the physical connectivity for the routers in this case study.

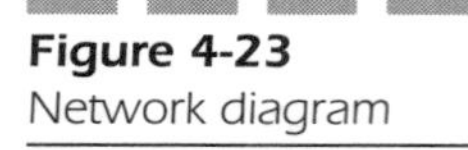

Figure 4-23
Network diagram

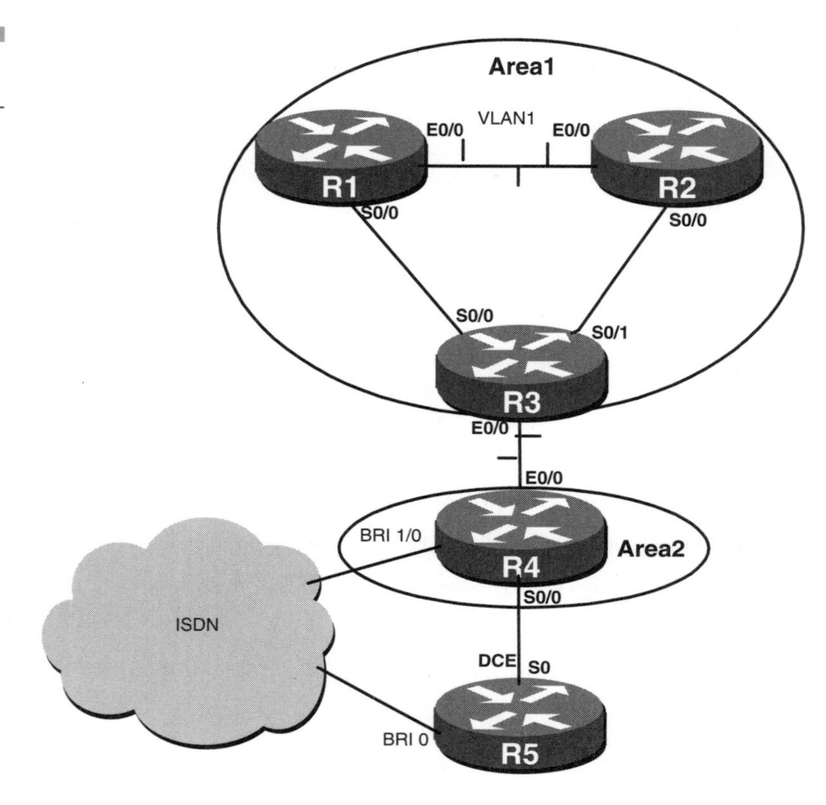

Notes

- All IP addresses will be assigned from the class B network 152.1.0.0.
- No static or default routes should be used in this case study unless explicitly called for.
- Read through the entire case study before beginning.
- Make sure you save your configurations regularly.
- All Level 1 passwords should be set to cisco; all Level 2 passwords should be set to ocsic.
- All area passwords should be set to cisco1 and all domain passwords should be set to ocsic1.
- The domain name on the Catalyst should be CCIE_Lab.

Questions

1. Cable the routers per Figure 4-23.
2. Erase any existing configuration from the routers and set each router's name. The name will be the corresponding router number; for example, router 1 will be R1.
3. Configure the terminal server (R2) so that all routers can be accessed by name through it. The port number of the terminal server attaches to the corresponding router. That is, port 1 of the terminal server goes to the console port of R1, while the Catalyst switch goes to port number 9. Use the Ethernet IP address as the IP address for reverse telneting.
4. Place Ethernet interface 0/0 of R1 and R2 in VLAN 1. The routers should be attached in ascending order, staring with port 1 on the Catalyst switch.
5. Assign IP addresses to Ethernet 0/0 of R1 and R2 from network 152.1.1.0. The subnet will contain 64 hosts. Assign the addresses in ascending order starting with R1.
6. Configure a loopback interface on R1, R2, and R3. The IP address should be 152.1.1.x, where x is the router name and has a 32-bit mask. For example, the loopback on R1 is 152.1.1.1.
7. Assign IP addresses to the interfaces connecting R1 and R3. Do not use more addresses than needed, and utilize the next available subnet. The addresses should be assigned in ascending order starting with R1.
8. Assign IP addresses to the interfaces connecting R2 and R3. Do not use more addresses than needed, and utilize the next available subnet. The addresses should be assigned in ascending order starting with R1.
9. Place Ethernet interface 0/0 of R3 and R4 in VLAN 2. The routers should be attached to the Catalyst switch in ascending order.
10. Assign an IP address to Ethernet 0/0 of R3 and R4. The subnet will contain 58 hosts. Assign the addresses in ascending order starting with R3.
11. Enable IS-IS on the Ethernet interfaces connecting R1 and R2. The interfaces will be in area 1, the Net will be 00.0001, and the system ID will be the MAC address of the Ethernet interface.
12. Enable IS-IS on the serial interfaces connecting R2 to R3 and R3 to R1. The interfaces will be in area 00.0001 and the system ID will be the MAC address of the Ethernet interface.
13. Enable IS-IS on the Ethernet interfaces connecting R3 and R4. The interfaces will be in area 2, the net will be 00.0002, and the system ID will be the MAC address of the Ethernet interface.
14. Make sure that R2 is elected the DR router for the broadcast network.
15. Configure R1 and R2 so that they only act as L1 routers.

16. R1 and R2 should be able to reach the Ethernet network connecting R3 to R4.
17. Advertise the loopback interfaces via IS-IS. The network should be advertised as internal by the router and no IS-IS packets should be sent over the loopback interfaces.
18. Configure R1 so that it authenticates its neighbor before forming an adjacency. The password used should be cisco.
19. Configure R3 so that it authenticates all L1 link state information within area 1. The password used should be cisco2.
20. Configure the IS-IS domain so that all L2 LSPs are authenticated.
21. Assign the following IP addresses to R4 and R5:

R4 Interface S0/0	152.1.2.1 /24
R4 Interface Bri0	152.1.3.1 /24
R5 Interface Bri0	152.1.3.2 /24
R5 Interface S0	152.1.2.2 /24
R5 Interface Loopback 0	152.1.4.1 /24

23. Configure RIP on R4 and R5. Do not send any RIP update packets out the interface connection R4 to R3.
24. Make sure that all routers can reach all the interfaces on all the other routers.
25. Configure DDR between R4 and R5. R4 should call R5 when the serial connection is lost. The backup should be invoked five seconds after the primary is lost, and once the primary is back, the router should switch back to the primary.
26. Make sure that RIP advertisements don't keep the link up if no other traffic is present.

Case Study 5: DLSW and Bridging

IOS Requirements

All routers in this case study were running IOS 11.2 or higher.

Equipment Needed

The following equipment is needed to perform this case study:

- One Cisco router with two Ethernet ports and one serial port (R1)
- One Cisco router with two serial ports and an Ethernet port (R2)
- One Cisco router with two serial ports and a Token Ring interface (R3)
- One Cisco router with two serial ports, one Ethernet port, and one Fast Ethernet port (R4)
- One Cisco router with two serial ports, two Ethernet ports, and one ISDN port (R5)
- One Cisco router with one serial port, one Ethernet port, and one ISDN port (R6)
- One Cisco terminal server
- A Cisco Catalyst switch
- A Cisco IOS image on each router capable of supporting DLSW, multicast, and bridging
- A PC running a terminal emulation program for connecting to the console port of the terminal server
- Two Ethernet crossover cables
- An Ethernet hub
- Six Ethernet cables
- Five Cisco DTE/DCE crossover cables
- Two ISDN BRI circuits and two ISDN BRI cables
- Four workstations, three with Ethernet connections and one with a Token Ring connection

Obtaining ISDN Circuits

This case study requires two ISDN BRI circuits. You can either purchase the two ISDN circuits from your local telecommunications carrier or obtain an ISDN test switch, which you can use to provision your own BRI circuits. An Adtran Atlas 800 desktop ISDN switch is the option that was used for this case study. Details on configuring this ISDN switch for this case study can be found at the end of this chapter.

Physical Connectivity Diagram

Figure 5-1 shows the physical connectivity for the routers in your case study. Although it is not shown in the figure, an additional router, R7, is used as a terminal server.

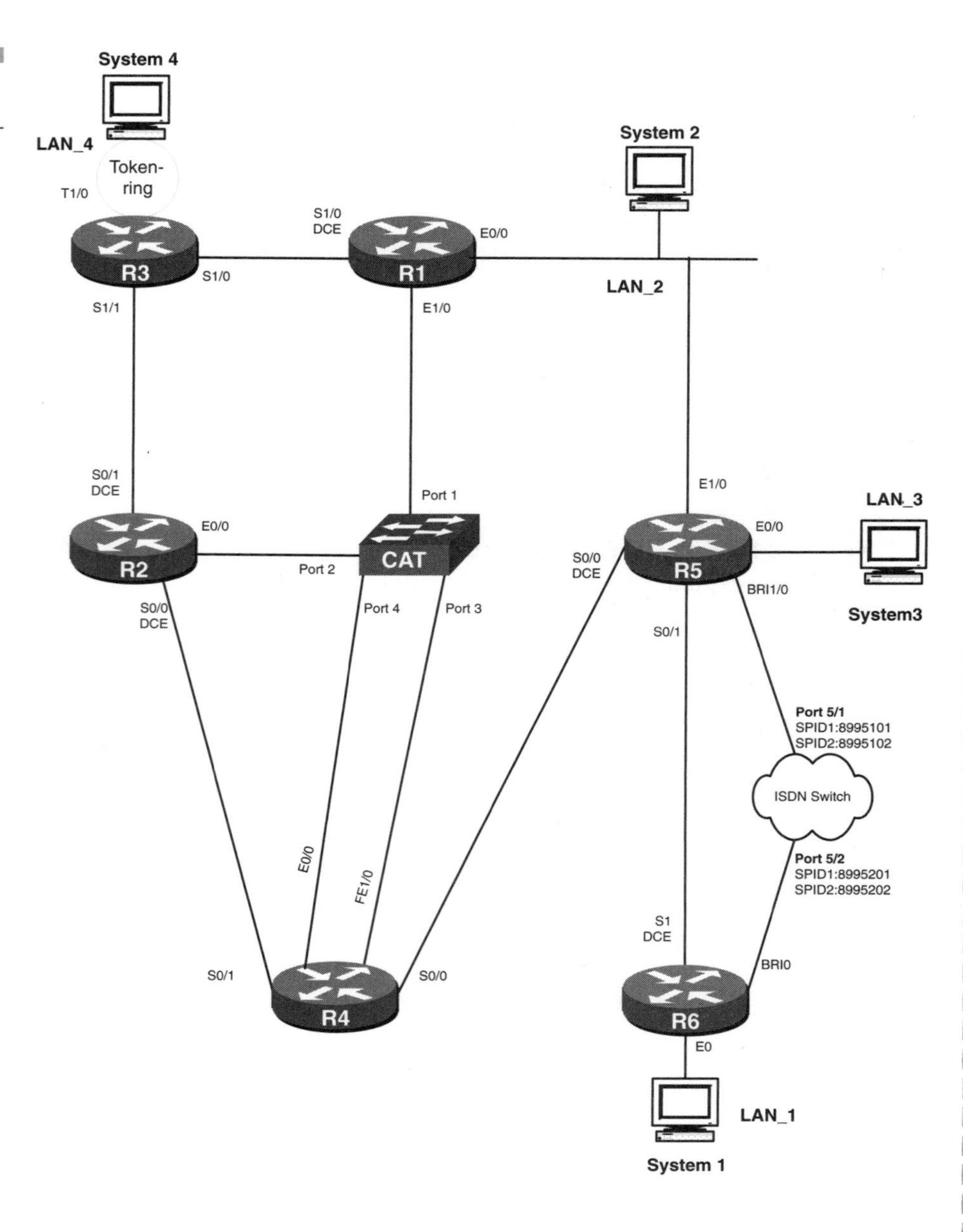

Figure 5-1 Physical connectivity

Questions

1. Cable the routers per Figure 5-1. Make sure that the DTE/DCE cables are connected per the figure.
2. Configure the terminal server so that all routers can be accessed by name through it. The port number of the terminal server attaches to the corresponding router; that is, port 1 of the terminal server goes to the console port of R1.
3. Set each router hostname per Figure 5-1.
4. Set the enable password on all routers to be **cisco**.
5. Configure the IP addressing on your network. All addresses should be assigned from the 152.3.0.0 network. Assign a loopback address to each router; use a 24-bit mask. The third octet should be the router number. For example, the loopback address on R1 should be 152.3.1.1 /24 .
6. LAN_3, LAN_2, and LAN_1 should be designed to handle 60 hosts each.
7. LAN_4 should be designed to handle 10 hosts.
8. The ISDN interfaces should be able to handle four hosts.
9. All serial interfaces, as well as the Ethernet core, should use as small a subnet as possible.
10. Configure R2's Ethernet interface to reside on VLAN 2 and configure R1's Ethernet interface to reside on VLAN 1.
11. Configure port 5 on the Catalyst switch to monitor all traffic from ports 1 and 2.
12. Verify and record the version of code running on the Catalyst switch.
13. Change the switching mode of the Catalyst to Store-and-Forward.
14. Change the baud rate of the console port of the Catalyst switch to 19,200 bits per second.
15. Configure the network to run OSPF. Use PPP encapsulation on all serial links.
16. Configure DLSW between LAN_4 (connected to R3) and LAN_2 (connected to R1 and R5). R1 should be the preferred peer to R3. If R1 fails, then R3 should automatically peer to R5.
17. Configure R3 so that its attached workstations will only be able to see two other workstations on the network. The two workstations that R3 can allow access to will have NetBIOS names: System3 and System9.
18. Configure transparent bridging between LAN_3 on R5 and LAN_1 on R6. Assume that stations on each of these LANs are running NetBEUI traffic.
19. Configure dial backup between R5 and R6. If the serial link fails between R5 and R6, the ISDN link should provide a backup path between the workstations on LAN_3 and LAN_1.
20. Configure R6 so that a workstation on LAN_4 with a source MAC address of 0000.8635.46e1 should have its packets rejected.

21. Enable IPX routing on R2 and R1.
22. Configure integrated routing and bridging between R2, R1, and R4. IP traffic should be routed and IPX traffic should be bridged.
23. Configure PIM Sparse mode on R1, R4, R5, and R6. Make the RP address 152.3.10.13 on R1.
24. Configure a static multicast group on R4. Make sure it is reachable and pingable from all routers.

Case Study 6: IPSEC, VoIP, and ATM LANE

IOS Requirements

This lab requires IOS 12.0 with support for IPSec, VoIP, and ATM LANE.

Equipment Needed

The following equipment is needed to perform this case study.

- One Cisco router with one Ethernet interface and two FXS voice ports
- One Cisco router with one Ethernet interface and two serial interfaces
- One Cisco router with four serial interfaces and one FXS voice port
- One Cisco router with two Ethernet interfaces and one serial interface
- One Cisco terminal server with one serial interface and one Ethernet interface
- Two Cisco routers with one Ethernet interface and one ATM OC-3 interface
- One Catalyst 5500 switch
- One LS1010 Lightstream ATM switch with two OC-3 ports
- A PC running a terminal emulation program for connecting to the console port of the terminal server
- Seven Ethernet cables
- Four Cisco DTE/DCE crossover cables
- Two single mode or multimode fiber cables
- Three analog phones

Table 6-3

Interface requirements for each router

R1	2 serial, 1 Ethernet
R2	1 Ethernet, 2 FXS
R3	4 serial, 1 FXS
R4	1 serial, 2 Ethernet
R5	1 serial, 1 Ethernet (terminal server)
R6	1 Ethernet, 1 ATM OC-3
R7	1 Ethernet, 1 ATM OC-3

Physical Connectivity Diagram

Figure 6-5 shows the physical connectivity for the routers in your case study.

Notes

- All IP addresses will be assigned from the class B network 135.25.0.0.
- No static or default routes should be used in this case study unless explicitly called for.
- Use Process ID 64 for all OSPF configurations.
- Read through the entire case study before beginning.
- Make sure you save your configurations regularly.
- The VTP domain name on the Catalyst should be CCIE_Lab.
- All required passwords should be set to cisco.

Questions

1. Cable the routers per Figure 6-5. Make sure that the correct ports are configured for DTE/DCE (the DCE clock should be set to 800K on all routers).
2. Erase any existing configuration from the routers and set each router's name. The name will be the corresponding router number; for example, router 1 will be R1.
3. Set the enable secret on each router to be cisco.
4. Configure the terminal server (R5) so that all routers can be accessed by name through it. The port number of the terminal server attaches to the corresponding router; that is, port

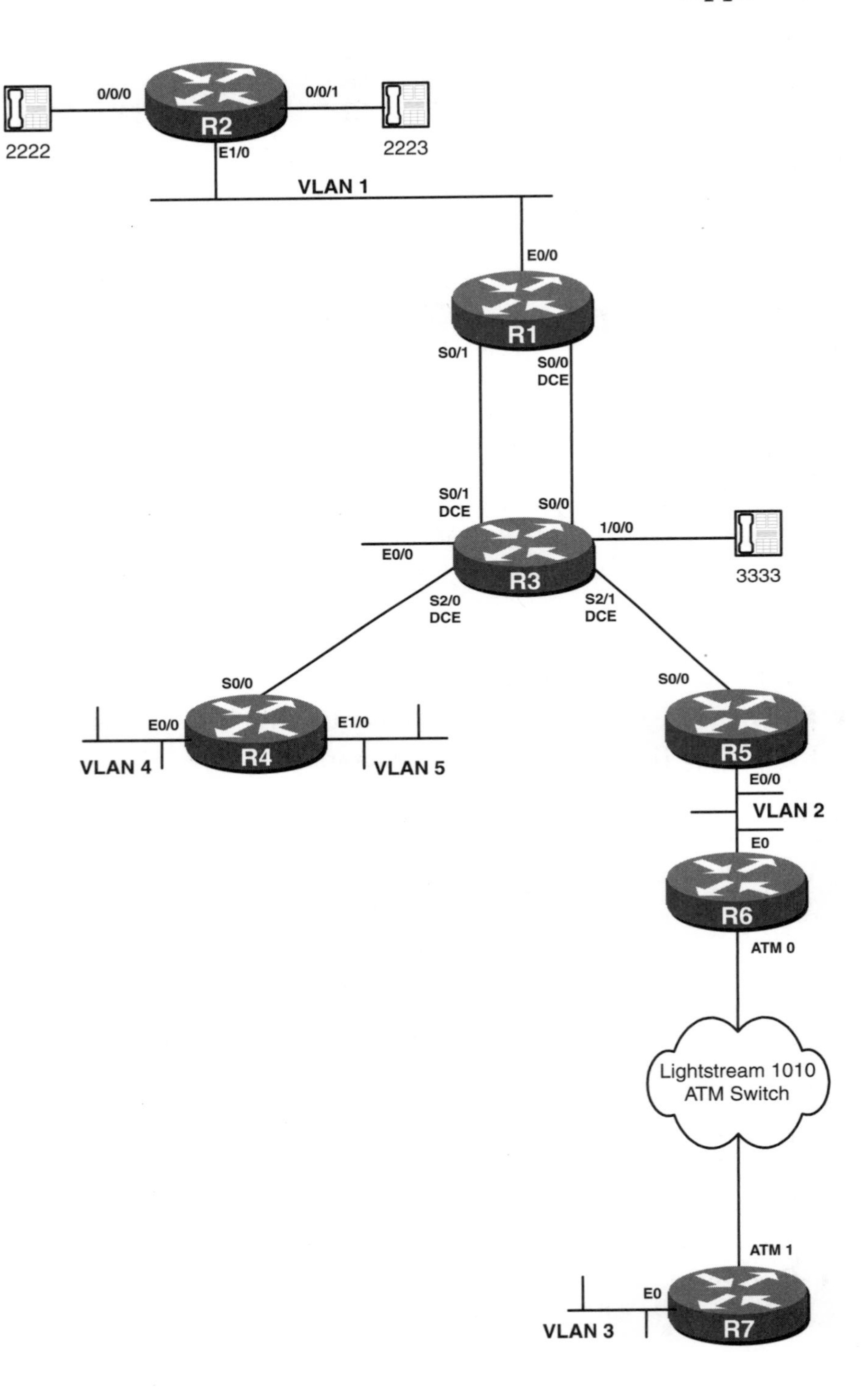

Figure 6-5 Physical connectivity diagram

1 of the terminal server goes to the console port of R1, and the Catalyst switch goes to port number 9. Use the loopback address as the IP address for reverse telneting.

5. Configure loopback 0 on all the routers. The IP address should be 135.25.x.1 where x is the router number. Use a 32-bit mask.
6. Assign IP addresses to the serial interface between R3 and R4, and R3 and R5. Use the subnet 135.25.11.0 and do not allocate more addresses than needed. Use the first available subnet and assign the addresses in ascending order, starting with the connection between R3 and R4.
7. Assign IP addresses to the serial interface between R1 and R3. Use the subnet 135.25.9.0. Do not allocate more addresses than needed. Use the first available subnet on the S0/0 interfaces and the second available subnet on the S0/1 interfaces.
8. Place the following interfaces in the assigned VLANs (ports should be attached to the Catalyst in ascending order, starting with R1):
   - E1/0 of R2 and E0/0 of R1 in VLAN 1
   - E0/0 of R5 and E0 of R6 in VLAN 2
   - E0 of R7 in VLAN 3
   - E0/0 of R4 in VLAN 4
   - E1/0 of R4 in VLAN 5
9. Assign private IP addresses of 10.1.1.1 /24 to E0/0 of R4.
10. Assign an IP address to the LAN between R2 and R1 that enables 30 useable host addresses. Assign the IP addresses in an ascending order starting with R2.
11. R4 contains a FTP server that needs to be accessed via networks off R3 and R5. The subnet contains only the FTP server, so assign the next available subnet from 135.25.11.0. Do not allocate more addresses than needed and use the first available subnet.
12. Place all networks, except for VLAN 4, VLAN 2, VLAN 3, R6, and R7, in OSPF Area 0.
13. Configure R4, R3, and R5 so that all telnet traffic originating from the router uses the source IP address of loopback 0.
14. Configure R4, R5, and R3 to accept telnet access. Only permit telnet traffic that is sourced from the loopback interface of any of the three routers.
15. Configure R3 and R4 so that when a telnet session is established from one router to the other, the session is encrypted. IKE should be used to create the SA using peer-share. The loopback interfaces also should be used as the peer addresses.
16. Configure R4 so that any packets originating from the 10.1.1.0 address have their source IP address changed to the IP address of Serial 0/0.
17. Assign private IP addresses of 135.25.12.1 /24 to E0/0 of R5 and place the interface in area 0.
18. Configure R3 and R4 so that traffic between E0/0 of R4 to E0/0 of R3 is encrypted and authenticated. IKE should be used to create the SA using peer-share. Triple Des should be

used for encryption and MD5 should be used for authentication. The IP address of the serial interface should be used as the peer and the key should be cisco1.

19. Configure R6 and R7 for LANE. R6 should be the LECS, LES, and the BUS, while R7 should be the LEC. Use mkt as the emulated ELAN name. Configure the LECS so that it participates in SSRP. The IP address of ATM0 R6 is 172.16.0.1 /16, and the IP address of R7 ATM0 R7 is 172.16.0.2 /16.
20. Assign IP address 2.2.2.2 /24 to Ethernet 0 of R7, 1.1.1.1 /24 to Ethernet 0 of R6, and 1.1.1.2 to E0/0 of R5. Make sure that 2.2.2.2 can reach 1.1.1.1. Do not enable routing on R6 or R7 and do not use static routes.
21. Configure R2 and R3 to handle voice traffic. Also configure the routers with the directory numbers shown.
22. Configure R2 so that when the handset at extension 2222 is picked up, it places a call to extension 3333.
23. Configure R3 so that extension 3333 can place a call to either extension 2222 or extension 2223.
24. Configure R3 so that you can only dial 611 to reach extension 2222. You should still have to dial 2223 to reach extension 2223.
25. Configure custom queuing on R2 so that voice is given 75 percent of the bandwidth and data is given 25 percent of the bandwidth.
26. Configure priority queuing on R3 to make voice high priority.
27. Configure RSVP between R1 and R3 S0/0 interfaces so that the maximum single flow allowed is equal to 400kb/sec.
28. Configure RSVP between R1 and R3 S0/1 interfaces so that the maximum single flow allowed is equal to 100kb/sec.
29. Configure IP precedence on R3. Calls placed to extension 2222 should have an IP Prec bit set to critical (5), while calls placed to extension 2223 should have an IP Prec bit set to priority (1).
30. You want to monitor the round-trip delay between R3 and R2. Configure the network so that a monitor packet is sent every 10 seconds for 30 minutes.
31. Configure R2 so that all other routers in the network can load R2's IOS image.
32. Configure the network so that traffic coming from R2 into R1 meets the following requirements:

- The first 496,000 bits/sec of traffic with a source IP address of 135.25.2.1 should be transmitted with an IP precedence of 6. Additional traffic should be sent with an IP precedence of 3.
- The first 96,000 bits/sec of Web traffic should be transmitted. Additional Web traffic should be dropped.
- Up to 200,000 bits/sec of other types of traffic should be transmitted with an IP precedence of 1. Additional traffic should be dropped.

INDEX

Symbols

A

B

C

D

M

N

O

P

T

U

V

W

X–Z

INTERNATIONAL CONTACT INFORMATION

AUSTRALIA
McGraw-Hill Book Company Australia Pty. Ltd.
TEL +61-2-9417-9899
FAX +61-2-9417-5687
http://www.mcgraw-hill.com.au
books-it_sydney@mcgraw-hill.com

CANADA
McGraw-Hill Ryerson Ltd.
TEL +905-430-5000
FAX +905-430-5020
http://www.mcgrawhill.ca

GREECE, MIDDLE EAST, NORTHERN AFRICA
McGraw-Hill Hellas
TEL +30-1-656-0990-3-4
FAX +30-1-654-5525

MEXICO (Also serving Latin America)
McGraw-Hill Interamericana Editores S.A. de C.V.
TEL +525-117-1583
FAX +525-117-1589
http://www.mcgraw-hill.com.mx
fernando_castellanos@mcgraw-hill.com

SINGAPORE (Serving Asia)
McGraw-Hill Book Company
TEL +65-863-1580
FAX +65-862-3354
http://www.mcgraw-hill.com.sg
mghasia@mcgraw-hill.com

SOUTH AFRICA
McGraw-Hill South Africa
TEL +27-11-622-7512
FAX +27-11-622-9045
robyn_swanepoel@mcgraw-hill.com

UNITED KINGDOM & EUROPE (Excluding Southern Europe)
McGraw-Hill Publishing Company
TEL +44-1-628-502500
FAX +44-1-628-770224
http://www.mcgraw-hill.co.uk
computing_neurope@mcgraw-hill.com

ALL OTHER INQUIRIES Contact:
Osborne/McGraw-Hill
TEL +1-510-549-6600
FAX +1-510-883-7600
http://www.osborne.com
omg_international@mcgraw-hill.com